Discrete Fractional Calculus

Christopher Goodrich • Allan C. Peterson

# Discrete Fractional Calculus

Springer

Christopher Goodrich
Department of Mathematics
Creighton Preparatory School
Omaha, NE, USA

Allan C. Peterson
Department of Mathematics
University of Nebraska–Lincoln
Lincoln, NE, USA

ISBN 978-3-319-25560-6 ISBN 978-3-319-25562-0 (eBook)
DOI 10.1007/978-3-319-25562-0

Library of Congress Control Number: 2015958659

Mathematics Subject Classification (2010): 26A33, 26E70, 34N05, 39A99, 39A10, 39A12, 34E10

Springer Cham Heidelberg New York Dordrecht London


Printed on acid-free paper

Springer International Publishing AG Switzerland is part of Springer Science+Business Media (www.springer.com)

*We dedicate this book to:*
*Barbara, Ben, Nate, Maddie, and Lily*
*and*
*Tina, Carla, David, and Carrie.*

# Preface

The continuous fractional calculus has a long history within the broad area of mathematical analysis. Indeed, it is nearly as old as the familiar integer-order calculus. Since its inception, it can be traced back to a question L'Hôpital had asked Leibniz in 1695 regarding the meaning of a one-half derivative; it was not until the 1800s that a firm theoretical foundation for the fractional calculus was provided. Nowadays the fractional calculus is studied both for its theoretical interest as well as its use in applications.

In spite of the existence of a substantial mathematical theory of the continuous fractional calculus, there was really no substantive parallel development of a discrete fractional calculus until very recently. Within the past five to seven years however, there has been a surge of interest in developing a discrete fractional calculus. This development has demonstrated that discrete fractional calculus has a number of unexpected difficulties and technical complications.

In this text we provide the first comprehensive treatment of the discrete fractional calculus with up-to-date references. We believe that students who are interested in learning about discrete fractional calculus will find this text to be a useful starting point. Moreover, experienced researchers, who wish to have an up-to-date reference for both discrete fractional calculus and on many related topics of current interest, will find this text instrumental.

Furthermore, we present this material in a particularly novel way since we simultaneously treat the fractional- and integer-order difference calculus (on a variety of time scales, including both the usual forward and backwards difference operators). Thus, the spirit of this text is quite modern so that the reader can not only acquire a solid foundation in the classical topics of the discrete calculus, but is also introduced to the exciting recent developments that bring them to the frontiers of the subject. This dual approach should be very useful for a variety of readers with a diverse set of backgrounds and interests.

There are several ways in which this book could be used as part of a formal course, and we have designed the text to be quite flexible and accommodating in its use. For example, if one prefers, it is possible to use this text for an introductory course in difference equations with the inclusion of discrete fractional calculus. In this case coverage of the first two chapters provide a basic introduction to the delta calculus, including the fractional calculus on the time scale of integers. We also recommend Sects. 7.2 and 7.3 if time permits.

On the other hand, if students have some basic knowledge of the difference calculus or if two semesters are available, then usage of this text in a number of other courses is possible. For example, for students with a background in elementary real analysis, one can cover Chaps. 1 and 2 more quickly and then skip to Chaps. 6 and 7 which present some basic results for fractional boundary value problems (FBVPs). If one is already familiar with the basics of the fractional calculus, then Chaps. 6 and 7 together with some of the current literature indicated in the references could easily form the basis for a seminar in the current theory of FBVPs. By contrast, if one has two semesters available, one can cover Chaps. 1–5 carefully, which will provide a very thorough introduction to both the discrete fractional calculus as well as the integer-order time scales calculus.

In short, there are a myriad of courses for which this text can serve either a primary or secondary role. And, in particular, the text has been designed so that, effectively, any chapter after Chaps. 1 and 2 can be freely omitted or included at one's discretion.

Regarding the specific content of the book, we note that in the first chapter of this book we develop the basic delta discrete calculus using the accepted standard notation. We define the forward difference operator $\Delta$ and develop the discrete calculus for this operator. When one applies this difference operator to the power functions, exponential function, trigonometric functions, and hyperbolic functions one often gets very complicated functions and these formulas are quite often not useful. In this book we define these functions in such a way that the formulas are very nice, and they actually resemble the formulas that we know from the continuous calculus. Many applications and interesting problems involving these functions are given.

In Chap. 2 we first introduce the discrete delta fractional calculus and then study the (delta) Laplace transform, which is a special case of the Laplace transform studied in the book by Bohner and Peterson [62]; we do not assume the reader has any knowledge of the material in that book. The delta Laplace transform is equivalent under a transformation to the well-known $Z$-transform, but we prefer the definition of the Laplace transform given here, which has the property that many of the Laplace transform formulas are analogous to the Laplace transform formulas in the continuous setting. We show how we can use the (delta) Laplace transform to enable us to solve certain initial value problems for difference equations and summation equations. We then develop several properties of this transform in the fractional calculus setting, giving a precise treatment of domains of convergence along the way. We then apply the Laplace transform method to solve fractional initial value problems and fractional summation equations.

In Chap. 3 we develop the calculus for the discrete nabla difference operator $\nabla$ (backwards difference operator). Once again, the appropriate power functions, exponential function, trigonometric functions, and hyperbolic functions are defined and their properties are derived. The nabla fractional calculus is also developed and a formula relating the nabla and delta fractional calculus is proved. In Chap. 4 the quantum calculus (or $q$-calculus) is given. The quantum calculus has important applications to quantum theory and to combinatorics, and this chapter provides a broad introduction to the basic theory of the $q$-calculus. Moreover, we also provide a brief introduction to the concept of fractional $q$-derivatives and integrals. Finally, in Chap. 5 we present the concept of a mixed time scale, which allows us to treat in a unified way a number of individual time scales and associated operators, e.g., the $q$-difference operator and the forward difference operator. This chapter will provide the reader with an introduction to the basic theory of the area such as the exponential and trigonometric functions on mixed time scales, the Laplace transform, and the application of these concepts to solving initial and boundary value problems.

The final two chapters of this text, Chaps. 6 and 7, focus on the theory of FBVPs; as such, these two chapters require more mathematical maturity than the first five. In general, and, furthermore, we assume that the reader has the relevant familiarity from the first half of the book. Thus, for example, in Chap. 7 we assume that the reader is familiar with Chap. 2 regarding the fractional delta calculus. In particular, in Chap. 6 the study of Green's functions and boundary value problems for fractional self-adjoint equations is given. Self-adjoint operators are an important classical area of differential equations and in that setting are well known to have a very pleasing mathematical theory. In Chap. 6 we present some of the known results in the discrete fractional setting, and this presentation will amply demonstrate the number of open questions that remain in this theory. Finally, in Chap. 7 the nonlocal structure of the fractional difference operator (in the delta case only) is explored in a variety of manifestations. For example, we discuss in what ways the sign of the fractional difference (for various orders) affects the behavior of the functions to which the difference is applied (e.g, monotonicity- and convexity-type results). As we show, there are some substantial and surprising differences in the case of the fractional delta operator. Furthermore, we examine how explicit nonlocal elements in discrete fractional boundary value problems may interact with the implicit nonlocal structure of the fractional difference operator, and we examine how to analyze such problems. All in all, in this final chapter of the book we aim to give the reader a sense of the tremendous complexity and mathematical richness that these nonlocal structures induce.

Finally, we should like to point out that we have included a great many exercises in this book, and the reader is encouraged to attempt as many of these as possible. To maximize the flexibility of this text as well as its potential use in independent study, we have included answers to many of the exercises.

We would like to thank Chris Ahrendt, Elvan Akin, Douglas Anderson, Ferhan Atici, Tanner Auch, Pushp Awasthi, Martin Bohner, Abigail Brackins, Paul Eloe, Lynn Erbe, Alex Estes, Scott Gensler, Julia St. Goar, Johnny Henderson, Wu Hongwu, Michael Holm, Wei Hu, Areeba Ikram, Baoguo Jia, Raziye Mert, Gordon Woodward, Rong Kun Zhuang, and the REU students Kevin Ahrendt, Lucas Castle, David Clark, Lydia DeWolf, James St. Dizier, Nicky Gaswick, Jeff Hein, Jonathan Lai, Liam Mazurowski, Sam McCarthy, Brent McKain, Kelsey Mitchell, Kaitlin Speer, Kathryn Yochman, Emily Obudzinski, Matt Olsen, Timothy Rolling, Richard

Ross, Sarah Stanley, Dominic Veconi, Cory Wright and Kathryn Yochman, for their influence on this book. Finally, we would like to thank Ann Kostant and our Springer Executive Editor, Elizabeth Loew, and her assistants for the accomplished handling of our manuscript.

Omaha, NE, USA Christopher Goodrich
Lincoln, NE, USA Allan C. Peterson

# Contents

# Chapter 1
# Basic Difference Calculus

## 1.1 Introduction

In this section we introduce the basic delta calculus that will be useful for our later results. Frequently, the functions we consider will be defined on a set of the form

$$\mathbb{N}_a := \{a, a+1, a+2, \dots\},$$

where $a \in \mathbb{R}$, or a set of the form

$$\mathbb{N}_a^b := \{a, a+1, a+2, \dots, b\},$$

where $a, b \in \mathbb{R}$ and $b - a$ is a positive integer.

**Definition 1.1.** Assume $f : \mathbb{N}_a^b \to \mathbb{R}$. If $b > a$, then we define the **forward difference operator** $\Delta$ by

$$\Delta f(t) := f(t+1) - f(t)$$

for $t \in \mathbb{N}_a^{b-1}$.

Note that in Definition 1.1 we make a slight abuse of notation by writing $\Delta f(t)$, as we shall do throughout this text. Technically, it would be more precise to write $(\Delta f)(t)$ to emphasize that $\Delta f$ is a function that is being evaluated at the point $t$. However, as long as one understands this true meaning of the notation, then we see no harm in using the simpler-to-read notation $\Delta f(t)$.

**Definition 1.2.** We define the **forward jump operator** $\sigma$ on $\mathbb{N}_a^{b-1}$ by

$$\sigma(t) = t + 1.$$



C. Goodrich, A.C. Peterson, *Discrete Fractional Calculus*,
DOI 10.1007/978-3-319-25562-0_1

It is often convenient to use the notation $f^\sigma$ to denote the function defined by the composition $f \circ \sigma$, that is

$$f^\sigma(t) = (f \circ \sigma)(t) = f(\sigma(t)) = f(t+1),$$

for $t \in \mathbb{N}_a^{b-1}$. Also, the operator $\Delta^n$, $n = 1, 2, 3, \ldots$ is defined recursively by $\Delta^n f(t) = \Delta(\Delta^{n-1} f(t))$ for $t \in \mathbb{N}_a^{b-n}$, where we assume the integer $b - a \geq n$. Finally, $\Delta^0$ denotes the identity operator, i.e., $\Delta^0 f(t) = f(t)$.

In the following theorem we give several important properties of the forward difference operator.

**Theorem 1.3.** *Assume $f, g : \mathbb{N}_a^b \to \mathbb{R}$ and $\alpha, \beta \in \mathbb{R}$, then for $t \in \mathbb{N}_a^{b-1}$*

(i) $\Delta\alpha = 0$;
(ii) $\Delta\alpha f(t) = \alpha\Delta f(t)$;
(iii) $\Delta[f+g](t) = \Delta f(t) + \Delta g(t)$;
(iv) $\Delta\alpha^{t+\beta} = (\alpha - 1)\alpha^{t+\beta}$;
(v) $\Delta[fg](t) = f(\sigma(t))\Delta g(t) + \Delta f(t) g(t)$;
(vi) $\Delta\left(\frac{f}{g}\right)(t) = \frac{g(t)\Delta f(t) - f(t)\Delta g(t)}{g(t)g(\sigma(t))}$,

*where in (vi) we assume $g(t) \neq 0$, $t \in \mathbb{N}_a^b$.*

*Proof.* We will just prove (iv) and the **quotient rule** (vi). Since

$$\Delta\alpha^{t+\beta} = \alpha^{t+1+\beta} - \alpha^{t+\beta} = (\alpha - 1)\alpha^{t+\beta}$$

we have that (iv) holds. To see that the quotient rule (vi) holds, note that

$$\begin{aligned}
\Delta\left(\frac{f}{g}\right)(t) &= \frac{f(t+1)}{g(t+1)} - \frac{f(t)}{g(t)} \\
&= \frac{f(t+1)g(t) - f(t)g(t+1)}{g(t)g(t+1)} \\
&= \frac{g(t)[f(t+1) - f(t)] - f(t)[g(t+1) - g(t)]}{g(t)g(\sigma(t))} \\
&= \frac{g(t)\Delta f(t) - f(t)\Delta g(t)}{g(t)g(\sigma(t))}.
\end{aligned}$$

The proof of the **product rule** (v) is Exercise 1.2. □

Due to the fact that (ii) and (iii) hold in Theorem 1.3 we say $\Delta$ is a **linear operator**.

Next, we define the falling function.

**Definition 1.4 (Falling Function).** For $n$ a positive integer we define the **falling function**, $t^{\underline{n}}$, read $t$ to the $n$ falling, by

$$t^{\underline{n}} := t(t-1)(t-2)\cdots(t-n+1).$$

Also we let $t^{\underline{0}} := 1$.

The falling function is defined so that the following power rule holds.

**Theorem 1.5 (Power Rule).** *The power rule*

$$\Delta t^{\underline{n}} = nt^{\underline{n-1}},$$

*holds for* $n = 1, 2, 3, \cdots$.

*Proof.* Assume $n$ is a positive integer and consider

$$\begin{aligned}\Delta t^{\underline{n}} &= (t+1)^{\underline{n}} - t^{\underline{n}} \\ &= (t+1)t(t-1)\cdots(t-n+2) - t(t-1)(t-2)\cdots(t-n+1) \\ &= t(t-1)(t-2)\cdots(t-n+2)[(t+1)-(t-n+1)] \\ &= nt^{\underline{n-1}}.\end{aligned}$$

This completes the proof. □

A very important function in mathematics is the **gamma function** which is defined as follows.

**Definition 1.6 (Gamma Function).** The gamma function is defined by

$$\Gamma(z) = \int_0^\infty e^{-t}t^{z-1}dt$$

for those complex numbers $z$ for which the real part of $z$ is positive (it can be shown that the above improper integral converges for all such $z$).

Integrating by parts we get that

$$\begin{aligned}\Gamma(z+1) &= \int_0^\infty e^{-t}t^z dt \\ &= [-e^{-t}t^z]_{t\to 0+}^{t\to\infty} - \int_0^\infty (-e^{-t})zt^{z-1}dt \\ &= z\Gamma(z)\end{aligned}$$

when the real part of $z$ is positive. We then use the very important formula

$$\Gamma(z+1) = z\Gamma(z) \tag{1.1}$$

to extend the domain of the gamma function to all complex numbers $z \neq 0, -1, -2, \cdots$. Also note that since it can be shown that $\lim_{z\to 0}|\Gamma(z)| = \infty$ it follows from (1.1) that

$$\lim_{z\to -n} |\Gamma(z)| = \infty, \quad n = 0, 1, 2, \ldots,$$

which is a fundamental property of the gamma function which we will use from time to time. Another well-known important consequence of (1.1) is that

$$\Gamma(n+1) = n!, \quad n = 0, 1, 2, \cdots.$$

Because of this, the gamma function is known as a generalization of the factorial function.

Note that for $n$ a positive integer

$$\begin{aligned} t^{\underline{n}} &= t(t-1)\cdots(t-n+1) \\ &= \frac{t(t-1)\cdots(t-n+1)\Gamma(t-n+1)}{\Gamma(t-n+1)} \\ &= \frac{\Gamma(t+1)}{\Gamma(t-n+1)}. \end{aligned}$$

Motivated by this above calculation, we extend the domain of the falling function in the following definition.

**Definition 1.7.** The (generalized) **falling function** is defined by

$$t^{\underline{r}} := \frac{\Gamma(t+1)}{\Gamma(t-r+1)}$$

for those values of $t$ and $r$ such that the right-hand side of this equation makes sense. We then extend this definition by making the common convention that $t^{\underline{r}} = 0$ when $t - r + 1$ is a nonpositive integer and $t + 1$ is not a nonpositive integer. We also use the convention given in Oldham and Spanier [152, equation (1.3.4)] that

$$\frac{\Gamma(-n)}{\Gamma(-N)} = (-n-1)^{\underline{N-n}} = (-1)^{N-n}\frac{N!}{n!},$$

where $n$ and $N$ are nonnegative integers.

The motivation for the first convention in Definition 1.7 is that whenever $t - r + 1$ is a nonpositive integer and $t + 1$ is not a nonpositive integer, then

$$\lim_{s\to t} s^{\underline{r}} = \lim_{s\to t} \frac{\Gamma(s+1)}{\Gamma(s-r+1)} = 0.$$

A similar remark motivates the second convention mentioned in Definition 1.7. Whenever these conventions are used one should always verify the conclusion by taking appropriate limits. This step will usually not be included in our calculations.

Next we state and prove the generalized power rules.

**Theorem 1.8 (Power Rules).** *The following (generalized) power rules*

$$\Delta(t+\alpha)^{\underline{r}} = r(t+\alpha)^{\underline{r-1}}, \tag{1.2}$$

*and*

$$\Delta(\alpha-t)^{\underline{r}} = -r(\alpha-\sigma(t))^{\underline{r-1}}, \tag{1.3}$$

*hold, whenever the expressions in these two formulas are well defined.*

*Proof.* Consider

$$\begin{aligned}
\Delta(t+\alpha)^{\underline{r}} &= (t+\alpha+1)^{\underline{r}} - (t+\alpha)^{\underline{r}} \\
&= \frac{\Gamma(t+\alpha+2)}{\Gamma(t+\alpha+2-r)} - \frac{\Gamma(t+\alpha+1)}{\Gamma(t+\alpha+1-r)} \\
&= \frac{[(t+\alpha+1)-(t+\alpha+1-r)]\Gamma(t+\alpha+1)}{\Gamma(t+\alpha+2-r)} \\
&= r\frac{\Gamma(t+\alpha+1)}{\Gamma(t+\alpha-r+2)} \\
&= r(t+\alpha)^{\underline{r-1}}.
\end{aligned}$$

Hence (1.2) holds.

To see that (1.3) holds, consider

$$\begin{aligned}
\Delta(\alpha-t)^{\underline{r}} &= \Delta\left(\frac{\Gamma(\alpha-t+1)}{\Gamma(\alpha-t+1-r)}\right) \\
&= \frac{\Gamma(\alpha-t)}{\Gamma(\alpha-t-r)} - \frac{\Gamma(\alpha-t+1)}{\Gamma(\alpha-t+1-r)} \\
&= [(\alpha-t-r)-(\alpha-t)]\frac{\Gamma(\alpha-t)}{\Gamma(\alpha-t+1-r)} \\
&= -r\frac{\Gamma(\alpha-t)}{\Gamma(\alpha-t+1-r)} \\
&= -r(\alpha-\sigma(t))^{\underline{r-1}}.
\end{aligned}$$

Hence the power rule (1.3) holds. □

Note that when $n \geq k \geq 0$ are integers, then the binomial coefficient satisfies

$$\binom{n}{k} := \frac{n!}{(n-k)!k!} = \frac{n(n-1)\cdots(n-k+1)}{k!} = \frac{n^{\underline{k}}}{\Gamma(k+1)}.$$

Motivated by this we next define the (generalized) binomial coefficient as follows.

**Definition 1.9.** The **(generalized) binomial coefficient** $\binom{t}{r}$ is defined by

$$\binom{t}{r} := \frac{t^{\underline{r}}}{\Gamma(r+1)}$$

for those values of $t$ and $r$ so that the right-hand side is well defined. Here we also use the convention that if the denominator is undefined, but the numerator is defined, then $\binom{n}{k} = 0$.

**Theorem 1.10.** *The following hold*

(i) $\Delta\binom{t}{r} = \binom{t}{r-1}$;
(ii) $\Delta\binom{r+t}{t} = \binom{r+t}{t+1}$;
(iii) $\Delta\Gamma(t) = (t-1)\Gamma(t)$,

*whenever these expressions make sense.*

The proof of this theorem is left as an exercise (Exercise 1.13).

## 1.2 Delta Exponential Function

In this section we want to study the delta exponential function that plays a similar role in the delta calculus on $\mathbb{N}_a$ that the exponential function $e^{pt}$, $p \in \mathbb{R}$, does in the continuous calculus. Keep in mind that when $p$ is a constant, $x(t) = e^{pt}$ is the unique solution of the initial value problem

$$x' = px, \quad x(0) = 1.$$

For the delta exponential function we would like to consider functions in the set of **regressive functions** defined by

$$\mathcal{R} = \{p : \mathbb{N}_a \to \mathbb{R} \text{ such that } 1 + p(t) \neq 0 \text{ for } t \in \mathbb{N}_a\}.$$

Some of the results that we give will be true if in the definition of regressive functions we consider complex-valued functions instead of real-valued functions. We leave it to the reader to note when this is true.

We then define the delta exponential function corresponding to a function $p \in \mathcal{R}$, based at $s \in \mathbb{N}_a$, to be the unique solution (why does $p \in \mathcal{R}$ guarantee uniqueness?), $e_p(t,s)$, of the initial value problem

$$\Delta x(t) = p(t)x(t), \tag{1.4}$$

$$x(s) = 1. \tag{1.5}$$

**Theorem 1.11.** *Assume $p \in \mathcal{R}$ and $s \in \mathbb{N}_a$. Then*

$$e_p(t,s) = \begin{cases} \prod_{\tau=s}^{t-1}[1+p(\tau)], & t \in \mathbb{N}_s \\ \prod_{\tau=t}^{s-1}[1+p(\tau)]^{-1}, & t \in \mathbb{N}_a^{s-1}. \end{cases} \tag{1.6}$$

*Here, by a standard convention on products, it is understood that for any function $h$ that*

$$\prod_{\tau=s}^{s-1} h(\tau) := 1.$$

*Proof.* We solve the IVP (1.4), (1.5) to get a formula for $e_p(t,s)$. Solving (1.4) for $x(t+1)$ we get

$$x(t+1) = [1+p(t)]x(t), \quad t \in \mathbb{N}_a. \tag{1.7}$$

Letting $t = s$ in (1.7) and using the initial condition (1.5) we get

$$x(s+1) = [1+p(s)]x(s) = [1+p(s)].$$

Next, letting $t = s+1$ in (1.7) we get

$$x(s+2) = [1+p(s+1)]x(s+1) = [1+p(s)][1+p(s+1)].$$

Proceeding in this fashion we get

$$e_p(t,s) = \prod_{\tau=s}^{t-1}[1+p(\tau)] \tag{1.8}$$

for $t \in \mathbb{N}_{s+1}$. In the product in (1.8), it is understood that the index $\tau$ takes on the values $s, s+1, s+2, \ldots, t-1$. By convention $e_p(s,s) = \prod_{\tau=s}^{s-1}[1+p(\tau)] = 1$. Next assume $t \in \mathbb{N}_a^{s-1}$. Solving (1.7) for $x(t)$ we get

$$x(t) = \frac{1}{1+p(t)}x(t+1). \tag{1.9}$$

Letting $t = s-1$ in (1.9), we get

$$x(s-1) = \frac{1}{1+p(s-1)}x(s) = \frac{1}{1+p(s-1)}.$$

Next, letting $t = s-2$ in (1.9), we get

$$x(s-2) = \frac{1}{1+p(s-2)}x(s-1) = \frac{1}{[1+p(s-2)]\,[1+p(s-1)]}.$$

Continuing in this manner we get

$$x(t) = \prod_{\tau=t}^{s-1}[1+p(\tau)]^{-1}, \quad t \in \mathbb{N}_a^{s-1}.$$

□

Theorem 1.11 gives us the following example.

*Example 1.12.* If $p(t) = p$ is a constant with $p \neq -1$ (note this constant function is in $\mathcal{R}$), then from (1.6)

$$e_p(t,s) := (1+p)^{t-s}, \quad t \in \mathbb{N}_a.$$

*Example 1.13.* Find $e_p(t,1)$ if $p(t) = t-1$, $t \in \mathbb{N}_1$. First note that $1+p(t) = t \neq 0$ for $t \in \mathbb{N}_1$, so $p \in \mathcal{R}$. From (1.6) we get

$$e_p(t,1) = \prod_{\tau=1}^{t-1}\tau = (t-1)!$$

for $t \in \mathbb{N}_1$.

It is easy to prove the following theorem.

**Theorem 1.14.** *If $p \in \mathcal{R}$, then a general solution of*

$$\Delta y(t) = p(t)y(t), \quad t \in \mathbb{N}_a$$

*is given by*

$$y(t) = ce_p(t,a), \quad t \in \mathbb{N}_a,$$

*where $c$ is an arbitrary constant.*

The following example is an interesting application using an exponential function.

*Example 1.15.* According to folklore, Peter Minuit in 1626 purchased Manhattan Island for goods worth \$24. If at the beginning of 1626 the \$24 could have been invested at an annual interest rate of 7% compounded quarterly, what would it have been worth at the end of the year 2014. Let $y(t)$ be the value of the investment after $t$ quarters of a year. Then $y(t)$ satisfies the equation

$$y(t+1) = y(t) + \frac{.07}{4}\, y(t)$$
$$= y(t) + .0175\, y(t).$$

Thus $y$ is a solution of the IVP

$$\Delta y(t) = .0175\, y(t), \quad y(0) = 24.$$

Using Theorem 1.14 and the initial condition we get that

$$y(t) = 24\, e_{.0175}(t, 0) = 24(1.0175)^t.$$

It follows that

$$y(1552) = 24(1.0175)^{1552} \approx 1.18 \times 10^{13}$$

(about 11.8 trillion dollars!).

We now develop some properties of the (delta) exponential function $e_p(t, a)$. To motivate our later results, consider, for $p, q \in \mathcal{R}$ the product

$$\begin{aligned} e_p(t,a)e_q(t,a) &= \prod_{\tau=a}^{t}(1 + p(\tau)) \prod_{\tau=a}^{t}(1 + q(\tau)) \\ &= \prod_{\tau=a}^{t}[1 + p(\tau)][1 + q(\tau)] \\ &= \prod_{\tau=a}^{t}[1 + (p(t) + q(t) + p(t)q(t))] \\ &= \prod_{\tau=a}^{t}[1 + (p \oplus q)(\tau)], \quad \text{if } (p \oplus q)(t) := p(t) + q(t) + p(t)q(t) \\ &= e_{p \oplus q}(t, a). \end{aligned}$$

Hence we get the law of exponents

$$e_p(t, a)e_q(t, a) = e_{p\oplus q}(t, a)$$

holds for $p, q \in \mathcal{R}$, provided

$$p \oplus q := p + q + pq.$$

**Theorem 1.16.** *If we define the* **circle plus addition**, $\oplus$, *on* $\mathcal{R}$ *by*

$$p \oplus q := p + q + pq,$$

*then* $\mathcal{R}$, $\oplus$ *is an Abelian group.*

*Proof.* First to see the closure property is satisfied, note that if $p, q \in \mathcal{R}$, then $1 + p(t) \neq 0$ and $1 + q(t) \neq 0$ for $t \in \mathbb{N}_a$. It follows that

$$1 + (p \oplus q)(t) = 1 + [p(t) + q(t) + p(t)q(t)] = [1 + p(t)][1 + q(t)] \neq 0$$

for $t \in \mathbb{N}_a$, and hence $p \oplus q \in \mathcal{R}$.

Next the zero function $0 \in \mathcal{R}$ as $1 + 0 = 1 \neq 0$. Also

$$0 \oplus p = 0 + p + 0 \cdot p = p, \quad \text{for all } p \in \mathcal{R},$$

so the zero function 0 is the additive identity element in $\mathcal{R}$.

To show that every element in $\mathcal{R}$ has an additive inverse let $p \in \mathcal{R}$. Then set $q = \frac{-p}{1+p}$ and note that since

$$1 + q(t) = 1 + \frac{-p(t)}{1 + p(t)} = \frac{1}{1 + p(t)} \neq 0$$

for $t \in \mathbb{N}_a$, so $q \in \mathcal{R}$ and we also have that

$$p \oplus q = p \oplus \frac{-p}{1+p} = p + \frac{-p}{1+p} + \frac{-p^2}{1+p} = p - p = 0$$

so $q$ is the **additive inverse** of $p$. For $p \in \mathcal{R}$, we use the following notation for the additive inverse of $p$:

$$\ominus p := \frac{-p}{1+p}. \tag{1.10}$$

The fact that the addition $\oplus$ is associative and commutative is Exercise 1.19. □

We can now define circle minus subtraction on $\mathcal{R}$ in the standard way that subtraction is defined in terms of addition.

**Definition 1.17.** We define circle minus subtraction on $\mathcal{R}$ by

$$p \ominus q := p \oplus [\ominus q].$$

It can be shown (Exercise 1.18) that if $p, q \in \mathcal{R}$ then

$$(p \ominus q)(t) = \frac{p(t) - q(t)}{1 + q(t)}, \quad t \in \mathbb{N}_a.$$

The next theorem gives us several properties of the exponential function $e_p(t, s)$, based at $s \in \mathbb{N}_a$.

**Theorem 1.18.** *Assume $p, q \in \mathcal{R}$ and $t, s, r \in \mathbb{N}_a$. Then*

(i) $e_0(t,s) = 1$ *and* $e_p(t,t) = 1$;
(ii) $e_p(t,s) \neq 0, \quad t \in \mathbb{N}_a$;
(iii) *if* $1 + p > 0$, *then* $e_p(t,s) > 0$;
(iv) $\Delta e_p(t,s) = p(t)\, e_p(t,s)$;
(v) $e_p^\sigma(t,s) = e_p(\sigma(t), s) = [1 + p(t)]e_p(t,s)$;
(vi) $e_p(t,s)e_p(s,r) = e_p(t,r)$;
(viii) $e_p(t,s)e_q(t,s) = e_{p\oplus q}(t,s)$;
(viii) $e_{\ominus p}(t,s) = \frac{1}{e_p(t,s)}$;
(ix) $\frac{e_p(t,s)}{e_q(t,s)} = e_{p\ominus q}(t,s)$;
(x) $e_p(t,s) = \frac{1}{e_p(s,t)}$.

*Proof.* We prove many of these properties when $s = a$ and leave it to the reader to show that the same results hold for any $s \in \mathbb{N}_a$. By the definition of the exponential we have that (i) and (iv) hold. To see that (ii) holds when $s = a$ note that since $p \in \mathcal{R}$, $1 + p(t) \neq 0$ for $t \in \mathbb{N}_a$ and hence we have that

$$e_p(t,a) = \prod_{\tau=a}^{t-1}[1 + p(\tau)] \neq 0,$$

for $t \in \mathbb{N}_a$. The proof of (iii) is similar to the proof of (ii).

Since

$$\begin{aligned} e_p(\sigma(t), a) &= \prod_{\tau=a}^{\sigma(t)-1}[1 + p(\tau)] \\ &= \prod_{\tau=a}^{t}[1 + p(\tau)] \\ &= [1 + p(t)]e_p(t,a), \end{aligned}$$

we have that (v) holds when $s = a$.

We only show (vi) holds when $t \geq s \geq r$ and leave the other cases to the reader. In particular, we merely observe that

$$\begin{aligned} e_p(t,s)e_p(s,r) &= \prod_{\tau=s}^{t-1}[1 + p(\tau)]\prod_{\tau=r}^{s-1}[1 + p(\tau)] \\ &= \prod_{\tau=r}^{t-1}[1 + p(\tau)] \\ &= e_p(t,r). \end{aligned}$$

We proved (vii) holds when with $s = a$, earlier to motivate the definition of the circle plus addition. To see that (viii) holds with $s = a$ note that

$$\begin{aligned} e_{\ominus p}(t,a) &= \prod_{\tau=a}^{t-1}[1 + (\ominus p)(\tau)] \\ &= \prod_{\tau=a}^{t-1} \frac{1}{1+p(\tau)} \\ &= \frac{1}{\prod_{\tau=a}^{t-1}[1+p(\tau)]} \\ &= \frac{1}{e_p(t,a)}. \end{aligned}$$

Since

$$\frac{e_p(t,a)}{e_q(t,a)} = e_p(t,a)e_{\ominus q}(t,a) = e_{p\oplus[\ominus q]}(t,a) = e_{p\ominus q}(t,a),$$

we have (ix) holds when $s = a$. Since

$$e_p(t,a) = \prod_{s=a}^{t-1}[1+p(s)] = \frac{1}{\prod_{s=a}^{t-1}[1+p(s)]^{-1}} = \frac{1}{e_p(a,t)},$$

we have that (x) holds. □

Before we derive some other properties of the exponential function we give another example where we use an exponential function.

*Example 1.19.* Assume initially that the number of bacteria in a culture is $P_0$ and after one hour the number of bacteria present is $\frac{3}{2}P_0$. Find the number of bacteria, $P(t)$, present after $t$ hours. How long does it take for the number of bacteria to triple? Experiments show that $P(t)$ satisfies the IVP (why is this plausible?)

$$\Delta P(t) = kP(t), \quad P(0) = P_0.$$

Solving this IVP we get from Theorem 1.14 that

$$P(t) = P_0 e_k(t,0) = P_0(1+k)^t.$$

Using the fact that $P(1) = \frac{3}{2}P_0$ we get $1 + k = \frac{3}{2}$. It follows that

$$P(t) = P_0\left(\frac{3}{2}\right)^t, \quad t \in \mathbb{N}_0.$$

Let $t_0$ be the amount of time it takes for the population of the bacteria to triple. Then

$$P(t_0) = P_0 \left(\frac{3}{2}\right)^{t_0} = 3P_0,$$

which implies that

$$t_0 = \frac{\ln(3)}{\ln(1.5)} \approx 2.71 \text{ hours}.$$

The set of positively regressive functions, $\mathcal{R}^+$, is defined by

$$\mathcal{R}^+ := \{p \in \mathcal{R} : 1 + p(t) > 0, \quad t \in \mathbb{N}_a\}.$$

Note that by Theorem 1.18, part (iii), we have that if $p \in \mathcal{R}^+$, then $e_p(t, a) > 0$ for $t \in \mathbb{N}_a$. It is easy to see (Exercise 1.20) that $(\mathcal{R}^+, \oplus)$ is a subgroup of $(\mathcal{R}, \oplus)$.

We next define the circle dot scalar multiplication $\odot$ on $\mathcal{R}^+$.

**Definition 1.20.** The circle dot scalar multiplication, $\odot$, is defined on $\mathcal{R}^+$ by

$$\alpha \odot p = (1 + p)^\alpha - 1.$$

**Theorem 1.21.** *If $\alpha \in \mathbb{R}$ and $p \in \mathcal{R}^+$, then*

$$e_p^\alpha(t, a) = e_{\alpha \odot p}(t, a)$$

*for $t \in \mathbb{N}_a$.*

*Proof.* Consider

$$\begin{aligned} e_p^\alpha(t, a) &= \left\{\prod_{\tau=a}^{t-1}[1 + p(\tau)]\right\}^\alpha \\ &= \prod_{\tau=a}^{t-1}[1 + p(\tau)]^\alpha \\ &= \prod_{\tau=a}^{t-1}\{1 + [(1 + p(\tau))^\alpha - 1]\} \\ &= \prod_{\tau=a}^{t-1}[1 + (\alpha \odot p)(\tau)] \\ &= e_{\alpha \odot p}(t, a). \end{aligned}$$

This completes the proof. □

The following lemma will be used in the proof of the next theorem.

**Lemma 1.22.** *If $p, q \in \mathcal{R}$ and*

$$e_p(t, a) = e_q(t, a), \quad t \in \mathbb{N}_a,$$

*then $p = q$.*

*Proof.* Assume $p, q \in \mathcal{R}$ and $e_p(t, a) = e_q(t, a)$ for $t \in \mathbb{N}_a$. It follows that

$$p(t)\, e_p(t, a) = q(t)\, e_q(t, a), \quad t \in \mathbb{N}_a.$$

Dividing by $e_p(t, a) = e_q(t, a)$ we get that $p = q$. □

**Theorem 1.23.** *The set of positively regressive functions $\mathcal{R}^+$, with the addition $\oplus$, and the scalar multiplication $\odot$ is a vector space.*

*Proof.* We just prove two of the properties of a vector space and leave the rest of the proof (see Exercise 1.25) to the reader. First we show that the distributive law

$$(\alpha + \beta) \odot p = (\alpha \odot p) \oplus (\beta \odot p)$$

holds for $\alpha, \beta \in \mathbb{R}$, $p \in \mathcal{R}^+$. This follows from

$$\begin{aligned} e_{(\alpha+\beta)\odot p}(t, a) &= e_p^{\alpha+\beta}(t, a) \\ &= e_p^{\alpha}(t, a) e_p^{\beta}(t, a) \\ &= e_{\alpha\odot p}(t, a) e_{\beta\odot p}(t, a) \\ &= e_{(\alpha\odot p)\oplus(\beta\odot p)}(t, a) \end{aligned}$$

and an application of Lemma 1.22. Next we show that $1 \odot p = p$ for all $p \in \mathcal{R}^+$. This follows from

$$e_{1\odot p}(t, a) = e_p^1(t, a) = e_p(t, a)$$

and an application of Lemma 1.22. □

## 1.3 Delta Trigonometric Functions

In this section we introduce the delta hyperbolic sine and cosine functions, the delta sine and delta cosine functions and give some of their properties. First, we define the delta hyperbolic sine and cosine functions.

**Definition 1.24.** Assume $\pm p \in \mathcal{R}$. Then the **delta hyperbolic sine** and the **delta hyperbolic cosine** functions are defined as follows:

$$\cosh_p(t,a) := \frac{e_p(t,a)+e_{-p}(t,a)}{2}, \quad \sinh_p(t,a) := \frac{e_p(t,a)-e_{-p}(t,a)}{2}$$

for $t \in \mathbb{N}_a$.

The following theorem gives various properties of the delta hyperbolic sine and cosine functions.

**Theorem 1.25.** *Assume* $\pm p \in \mathcal{R}$. *Then*

(i) $\cosh_p(a,a) = 1, \quad \sinh_p(a,a) = 0$;
(ii) $\cosh_p^2(t,a) - \sinh_p^2(t,a) = e_{-p^2}(t,a), \quad t \in \mathbb{N}_a$;
(iii) $\Delta \cosh_p(t,a) = p(t) \sinh_p(t,a), \quad t \in \mathbb{N}_a$;
(iv) $\Delta \sinh_p(t,a) = p(t) \cosh_p(t,a), \quad t \in \mathbb{N}_a$;
(v) $\cosh_{-p}(t,a) = \cosh_p(t,a), \quad t \in \mathbb{N}_a$;
(vi) $\sinh_{-p}(t,a) = -\sinh_p(t,a), \quad t \in \mathbb{N}_a$;
(vii) $e_p(t,a) = \cosh_p(t,a) + \sinh_p(t,a), \quad t \in \mathbb{N}_a$.

*Proof.* Clearly (i) holds. To see that (ii) holds note that

$$\begin{aligned}\cosh_p^2(t,a) - \sinh_p^2(t,a) &= \frac{(e_p(t,a)+e_{-p}(t,a))^2 - (e_p(t,a)-e_{-p}(t,a))^2}{4}\\ &= e_p(t,a)e_{-p}(t,a)\\ &= e_{p\oplus(-p)}(t,a)\\ &= e_{-p^2}(t,a).\end{aligned}$$

To see that (iii) holds, consider

$$\begin{aligned}\Delta \cosh_p(t,a) &= \frac{1}{2}\Delta e_p(t,a) + \frac{1}{2}\Delta e_{-p}(t,a)\\ &= \frac{1}{2}[p(t)e_p(t,a) - p(t)e_{-p}(t,a)]\\ &= p(t) \sinh_p(t,a).\end{aligned}$$

The proof of (iv) is similar. The proofs of (v) and (vi) are trivial. The formula in part (vii) we call the **(delta) hyperbolic Euler's formula** and its proof follows from the definitions of the hyperbolic sine and hyperbolic cosine functions. □

We next define the delta sine and cosine functions.

**Definition 1.26.** For $\pm ip \in \mathcal{R}$, we define the **delta sine function** and **delta cosine function** as follows:

$$\cos_p(t,a) = \frac{e_{ip}(t,a)+e_{-ip}(t,a)}{2}, \quad \sin_p(t,a) = \frac{e_{ip}(t,a)-e_{-ip}(t,a)}{2i}$$

for $t \in \mathbb{N}_a$.

The following theorem gives some relationships between the delta trigonometric functions and the delta hyperbolic trigonometric functions.

**Theorem 1.27.** *Assume* $\pm p \in \mathcal{R}$. *Then*

(i) $\sin_{ip}(t,a) = i \sinh_p(t,a)$;
(ii) $\cos_{ip}(t,a) = \cosh_p(t,a)$;
(iii) $\sinh_{ip}(t,a) = i \sin_p(t,a)$;
(iv) $\cosh_{ip}(t,a) = \cos_p(t,a)$,

*for* $t \in \mathbb{N}_a$.

*Proof.* To see that (i) holds note that

$$\begin{aligned}\sin_{ip}(t,a) &= \frac{1}{2i}[e_{i^2p}(t,a) - e_{-i^2p}(t,a)] \\ &= i\frac{e_p(t,a) - e_{-p}(t,a)}{2} \\ &= i \sinh_p(t,a).\end{aligned}$$

The proofs of (ii)–(iv) are similar. □

The following theorem gives various properties of the delta sine and cosine functions.

**Theorem 1.28.** *Assume* $\pm ip \in \mathcal{R}$. *Then*

(i) $\cos_p(a,a) = 1, \quad \sin_p(a,a) = 0$;
(ii) $\cos_p^2(t,a) + \sin_p^2(t,a) = e_{p^2}(t,a), \quad t \in \mathbb{N}_a$;
(iii) $\Delta \cos_p(t,a) = -p(t) \sin_p(t,a), \quad t \in \mathbb{N}_a$;
(iv) $\Delta \sin_p(t,a) = p(t) \cos_p(t,a), \quad t \in \mathbb{N}_a$;
(v) $\cos_{-p}(t,a) = \cos_p(t,a), \quad t \in \mathbb{N}_a$;
(vi) $\sin_{-p}(t,a) = -\sin_p(t,a), \quad t \in \mathbb{N}_a$;
(vii) $e_{ip}(t,a) = \cos_p(t,a) + i \sin_p(t,a), \quad t \in \mathbb{N}_a$.

*Proof.* The proof of this theorem follows from Theorems 1.25 and 1.27. □

We call the formula

$$e_{ip}(t,a) = \cos_p(t,a) + i \sin_p(t,a), \quad t \in \mathbb{N}_a \tag{1.11}$$

in part (vii) of Theorem 1.28 the (delta) Euler's formula.

## 1.4 Second Order Linear Equations with Constant Coefficients

The nonhomogeneous second order linear difference equation is given by

$$\Delta^2 y(t) + p(t)\Delta y(t) + q(t)y(t) = f(t), \quad t \in \mathbb{N}_a, \tag{1.12}$$

where we assume that $p(t) \neq q(t) + 1$, for $t \in \mathbb{N}_a$. In this section we will see that we can easily solve the corresponding second order linear homogeneous equation with constant coefficients

$$\Delta^2 y(t) + p\Delta y(t) + qy(t) = 0, \quad t \in \mathbb{N}_a, \tag{1.13}$$

where we assume the constants $p, q \in \mathbb{R}$ satisfy $p \neq 1 + q$.

First we prove an existence-uniqueness theorem for solutions of initial value problems (IVPs) for (1.12).

**Theorem 1.29.** *Assume that $p, q, f : \mathbb{N}_a \to \mathbb{R}$, $p(t) \neq 1 + q(t)$, $t \in \mathbb{N}_a$, $A, B \in \mathbb{R}$, and $t_0 \in \mathbb{N}_a$. Then the IVP*

$$\Delta^2 y(t) + p(t)\Delta y(t) + q(t)y(t) = f(t), \quad t \in \mathbb{N}_a, \quad y(t_0) = A, \quad y(t_0 + 1) = B, \tag{1.14}$$

*has a unique solution $y(t)$ on $\mathbb{N}_a$.*

*Proof.* Expanding equation (1.12) out we have first solving for $y(t + 2)$ and then solving for $y(t)$ that

$$y(t + 2) = [2 - p(t)]y(t + 1) - [1 - p(t) + q(t)]y(t) + f(t) \tag{1.15}$$

and, since $p(t) \neq 1 + q(t)$, $t \in \mathbb{N}_a$,

$$y(t) = \frac{2 - p(t)}{1 - p(t) + q(t)}y(t + 1) - \frac{1}{1 - p(t) + q(t)}y(t + 2) - \frac{f(t)}{1 - p(t) + q(t)}. \tag{1.16}$$

If we let $t = t_0$ in (1.15), then equation (1.12) holds at $t = t_0$ iff

$$y(t_0 + 2) = [2 - p(t_0)]B - [1 - p(t_0) + q(t_0)]A + f(t_0).$$

Hence, the solution of the IVP (1.14) is uniquely determined at $t_0 + 2$. But using the equation (1.15) evaluated at $t = t_0 + 1$, we have that the unique values of the solution at $y(t_0 + 1)$ and $y(t_0 + 2)$ uniquely determines the value of the solution at $t_0+3$. By induction we get that the solution of the IVP (1.14) is uniquely determined on $\mathbb{N}_{t_0}$. On the other hand, if $t_0 > a$, then using equation (1.16) with $t = t_0 - 1$, we have that

$$y(t_0 - 1) = \frac{1}{1 - p(t_0 - 1) + q(t_0 - 1)}\left[[2 - p(t_0 - 1)]A - B - f(t_0 - 1)\right].$$

Hence the solution of the IVP (1.14) is uniquely determined at $t_0 - 1$. Proceeding in this manner we have by mathematical induction that the solution of the IVP (1.14) is uniquely determined on $\mathbb{N}_a^{t_0}$. Hence the result follows. □

*Remark 1.30.* It follows from Theorem 1.29, that if $p(t) \neq 1 + q(t)$, $t \in \mathbb{N}_a$, then the general solution of the linear homogeneous equation

$$\Delta^2 y(t) + p(t)\Delta y(t) + q(t)y(t) = 0$$

is given by

$$y(t) = c_1 y_1(t) + c_2 y_2(t), \quad t \in \mathbb{N}_a,$$

where $y_1(t)$, $y_2(t)$ are any two linearly independent solutions of (1.13) on $\mathbb{N}_a$.

We now show we can solve the second order linear homogeneous equation (1.13) with constant coefficients.

**Theorem 1.31 (Distinct Roots).** *Assume $p \neq 1 + q$ and $\lambda_1 \neq \lambda_2$ (possibly complex) are solutions (called the characteristic values of* (1.13)) *of the characteristic equation*

$$\lambda^2 + p\lambda + q = 0.$$

*Then*

$$y(t) = c_1 e_{\lambda_1}(t, a) + c_2 e_{\lambda_2}(t, a)$$

*is a general solution of* (1.13).

*Proof.* Assume $\lambda_1$, $\lambda_2$ are characteristic values of (1.13). Then the characteristic equation of (1.13) is given by

$$\lambda^2 - (\lambda_1 + \lambda_2)\lambda + \lambda_1\lambda_2 = 0.$$

It follows that $p = -(\lambda_1 + \lambda_2)$ and $q = \lambda_1\lambda_2$. Hence $q + 1 - p = (\lambda_1 + 1)(\lambda_2 + 1) \neq 0$, since $p \neq q + 1$. Hence, we have that $\lambda_1, \lambda_2 \neq -1$ and so $e_{\lambda_i}(t, a)$, $i = 1, 2$, are well defined. Since

$$\Delta^2 e_{\lambda_i}(t, a) + p\,\Delta e_{\lambda_i}(t, a) + q\,e_{\lambda_i}(t, a) = [\lambda_i^2 + p\lambda_i + q]e_{\lambda_i}(t, a) = 0,$$

we have that $e_{\lambda_i}(t, a)$, $i = 1, 2$, are solutions of (1.13). Since these two solutions are linearly independent on $\mathbb{N}_a$, we have that

$$y(t) = c_1 e_{\lambda_1}(t, a) + c_2 e_{\lambda_2}(t, a)$$

is a general solution of (1.13) on $\mathbb{N}_a$. □

*Example 1.32 (Fibonacci Numbers).* The Fibonacci numbers $F(t)$, $t = 1, 2, 3, \cdots$ are defined recursively by

$$\begin{aligned} F(t+2) &= F(t) + F(t+1), \quad t \in \mathbb{N}_1, \\ F(1) &= 1 = F(2). \end{aligned}$$

The Fibonacci sequence is given by

$$1, \quad 1, \quad 2, \quad 3, \quad 5, \quad 8, \quad 13, \quad 21, \quad 34, \quad \cdots.$$

Fibonacci used this to model the population of pairs of rabbits under certain assumptions. To find $F(t)$, note that $F(t)$ is the solution of the IVP

$$\begin{aligned} \Delta^2 F(t) + \Delta F(t) - F(t) &= 0, \quad t \in \mathbb{N}_1, \\ F(1) &= 1 = F(2). \end{aligned}$$

To solve this IVP we first get that the characteristic equation is

$$\lambda^2 + \lambda - 1 = 0.$$

Hence, the characteristic values are

$$\lambda_1 = \frac{-1+\sqrt{5}}{2}, \quad \lambda_2 = \frac{-1-\sqrt{5}}{2}.$$

It follows that

$$\begin{aligned} F(t) &= c_1 e_{\lambda_1}(t,1) + c_2 e_{\lambda_2}(t,1) \\ &= c_1 \left(\frac{1+\sqrt{5}}{2}\right)^{t-1} + c_2 \left(\frac{1-\sqrt{5}}{2}\right)^{t-1}. \end{aligned} \tag{1.17}$$

Applying the initial conditions we get the system

$$\begin{aligned} F(1) &= 1 = c_1 + c_2 \\ F(2) &= 1 = c_1 \left(\frac{1+\sqrt{5}}{2}\right) + c_2 \left(\frac{1-\sqrt{5}}{2}\right). \end{aligned}$$

Solving this system for $c_1$ and $c_2$, using (1.17) and simplifying we get

$$F(t) = \frac{1}{\sqrt{5}} \left(\frac{1+\sqrt{5}}{2}\right)^t - \frac{1}{\sqrt{5}} \left(\frac{1-\sqrt{5}}{2}\right)^t,$$

for $t \in \mathbb{N}_1$. Note $F(t)$ is the integer nearest to $\frac{1}{\sqrt{5}}\left(\frac{1+\sqrt{5}}{2}\right)^t$. It follows that $F(20)$ is the integer nearest

$$\frac{1}{\sqrt{5}}\left(\frac{1+\sqrt{5}}{2}\right)^{20} \approx 6765.00003,$$

which gives us that $F(20) = 6765$.

Usually, we want to find all real-valued solutions of (1.13). When a characteristic root $\lambda_1$ is complex, $e_{\lambda_1}(t,a)$ is a complex-valued solution. In the next theorem we show how to use this complex-valued solution to find two linearly independent real-valued solutions on $\mathbb{N}_a$.

**Theorem 1.33 (Complex Roots).** *If the characteristic values are $\lambda = \alpha \pm i\beta$, $\beta > 0$ and $\alpha \neq -1$, then a general solution of* (1.13) *is given by*

$$y(t) = c_1 e_\alpha(t,a)\cos_\gamma(t,a) + c_2 e_\alpha(t,a)\sin_\gamma(t,a),$$

*where $\gamma := \frac{\beta}{1+\alpha}$.*

*Proof.* First we show that if $\alpha \pm i\beta$, $\beta > 0$ are complex characteristic values of (1.13), then the condition $p \neq q+1$ is satisfied. In this case the characteristic equation for (1.13) is given by

$$\lambda^2 - 2\alpha\lambda + \alpha^2 + \beta^2 = 0.$$

It follows that $p = -2\alpha$ and $q = \alpha^2+\beta^2$. Therefore, $1+q-p = (\alpha+1)^2+\beta^2 \neq 0$. This implies that $p \neq q+1$. By Theorem 1.31, we have that $y(t) = e_{\alpha+i\beta}(t,a)$ is a complex-valued solution of (1.13). Using

$$\alpha + i\beta = \alpha \oplus i\frac{\beta}{1+\alpha} = \alpha \oplus i\gamma,$$

we get that

$$y(t) = e_{\alpha+i\beta}(t,a) = e_{\alpha\oplus i\gamma}(t,a) = e_\alpha(t,a)e_{i\gamma}(t,a).$$

It follows from the (delta) Euler's formula (1.11) that

$$\begin{aligned} y(t) &= e_\alpha(t,a)e_{i\gamma}(t,a) \\ &= e_\alpha(t,a)[\cos_\gamma(t,a) + i\sin_\gamma(t,a)] \\ &= y_1(t) + iy_2(t) \end{aligned}$$

is a solution of (1.13). But since $p$ and $q$ are real, we have that the real part, $y_1(t) = e_\alpha(t,a)\cos_\gamma(t,a)$, and the imaginary part, $y_2(t) = e_\alpha(t,a)\sin_\gamma(t,a)$, of $y(t)$ are solutions of (1.13). Since $p \neq q+1$ and $y_1(t), y_2(t)$ are linearly independent on $\mathbb{N}_a$, we get from Remark 1.30 that

$$y(t) = c_1 e_\alpha(t,a)\cos_\gamma(t,a) + c_2 e_\alpha(t,a)\sin_\gamma(t,a)$$

is a general solution of (1.13) on $\mathbb{N}_a$. □

*Example 1.34.* Solve the difference equation

$$\Delta^2 y(t) - 2\Delta y(t) + 2y(t) = 0, \quad t \in \mathbb{N}_a. \tag{1.18}$$

The characteristic equation is

$$\lambda^2 - 2\lambda + 2 = 0$$

and so the characteristic values are $\lambda = 1 \pm i$. Hence using Theorem 1.33, we get

$$y(t) = c_1 e_1(t,a)\cos_{\frac{1}{2}}(t,a) + c_2 e_1(t,a)\sin_{\frac{1}{2}}(t,a)$$

is a general solution of (1.18) on $\mathbb{N}_a$.

The previous theorem (Theorem 1.33) excluded the case when the characteristic values of (1.13) are $-1 \pm i\beta$, where $\beta > 0$. The next theorem considers this case.

**Theorem 1.35.** *If the characteristic values of* (1.13) *are* $-1 \pm i\beta$, *where* $\beta > 0$, *then a general solution of* (1.13) *is given by*

$$y(t) = c_1\beta^{t-a}\cos\left[\frac{\pi}{2}(t-a)\right] + c_2\beta^{t-a}\sin\left[\frac{\pi}{2}(t-a)\right],$$

$t \in \mathbb{N}_a$.

*Proof.* First note that by the first part of the proof of Theorem 1.33 we have that $p \neq q+1$. Since $-1+i\beta$ is a characteristic root of (1.13), we have that $y(t) = e_{-1+i\beta}(t,a)$ is a complex-valued solution of (1.13). Now

$$\begin{aligned} y(t) &= e_{-1+i\beta}(t,a) \\ &= (i\beta)^{t-a} \\ &= \left(\beta e^{i\frac{\pi}{2}}\right)^{t-a} \\ &= \beta^{t-a}\cos\left[\frac{\pi}{2}(t-a)\right] + i\beta^{t-a}\sin\left[\frac{\pi}{2}(t-a)\right]. \end{aligned}$$

It follows that

$$y_1(t) = \beta^{t-a} \cos\left[\frac{\pi}{2}(t-a)\right], \quad y_2(t) = \beta^{t-a} \sin\left[\frac{\pi}{2}(t-a)\right]$$

are solutions of (1.13). Since these solutions are linearly independent on $\mathbb{N}_a$, we have that

$$y(t) = c_1 \beta^{t-a} \cos\left[\frac{\pi}{2}(t-a)\right] + c_2 \beta^{t-a} \sin\left[\frac{\pi}{2}(t-a)\right],$$

is a general solution of (1.13). □

*Example 1.36.* Solve the delta linear difference equation

$$\Delta^2 y(t) + 2\Delta y(t) + 5y(t) = 0, \quad t \in \mathbb{N}_0.$$

The characteristic equation is $\lambda^2 + 2\lambda + 5 = 0$, so the characteristic values are $\lambda = -1 \pm 2i$. It follows from Theorem 1.35 that

$$y(t) = c_1 2^t \cos\left(\frac{\pi}{2}t\right) + c_2 2^t \sin\left(\frac{\pi}{2}t\right),$$

is a general solution on $\mathbb{N}_0$.

**Theorem 1.37 (Double Root).** *Assume $p \neq 1 + q$, and $\lambda_1 = \lambda_2 = r$ is a double root of the characteristic equation. Then*

$$y(t) = c_1 e_r(t, a) + c_2 (t-a) e_r(t, a)$$

*is a general solution of* (1.13).

*Proof.* Since $\lambda_1 = r$ is a characteristic value, we have $y_1(t) = e_r(t, a)$ is a solution of (1.13). Since $\lambda_1 = \lambda_2 = r$, we have that the characteristic equation for (1.13) is

$$(\lambda - r)^2 = \lambda^2 - 2r\lambda + r^2 = 0.$$

Hence, in this case, (1.13) has the form

$$\Delta^2 y(t) - 2r\Delta y(t) + r^2 y(t) = 0.$$

From Exercise 1.32, we have that $y_2(t) = (t-a)e_r(t, a)$ is a second solution of (1.13) on $\mathbb{N}_a$. Since these two solutions are linearly independent on $\mathbb{N}_a$, we have that

$$y(t) = c_1 e_r(t, a) + c_2 (t-a) e_r(t, a)$$

is a general solution of (1.13). □

*Example 1.38.* Evaluate the $t$ by $t$ determinant of the following tridiagonal matrix

$$M(t) := \begin{bmatrix} 4\,4\,0\,0 \cdots 0\,0\,0 \\ 1\,4\,4\,0 \cdots 0\,0\,0 \\ 0\,1\,4\,4 \cdots 0\,0\,0 \\ 0\,0\,1\,4 \cdots 0\,0\,0 \\ \vdots\;\vdots\;\vdots\;\vdots\;\ddots\;\vdots\;\vdots\;\vdots \\ 0\,0\,0\,0 \cdots 4\,4\,0 \\ 0\,0\,0\,0 \cdots 1\,4\,4 \\ 0\,0\,0\,0 \cdots 0\,1\,4 \end{bmatrix}$$

for $t \in \mathbb{N}_1$. For example, we have that

$$M(1) := \begin{bmatrix} 4 \end{bmatrix}, \; M(2) := \begin{bmatrix} 4\,4 \\ 1\,4 \end{bmatrix}, \text{ and } M(3) := \begin{bmatrix} 4\,4\,0 \\ 1\,4\,4 \\ 0\,1\,4 \end{bmatrix}.$$

Let $D(t)$ be the value of the determinant of $M(t)$. Expanding the $t+2$ by $t+2$ determinant $D(t+2)$ along its first row we get

$$D(t+2) = 4D(t+1) - 4D(t), \quad t \in \mathbb{N}_1.$$

Note that $D(1) = 4$ and $D(2) = 12$. It follows that if we define $D(0) = 1$, then we have

$$D(t+2) - 4D(t+1) + 4D(t) = 0, \quad t \in \mathbb{N}_0.$$

It then follows that $D(t)$ is the solution of the IVP

$$\Delta^2 D(t) - 2\Delta D(t) + D(t) = 0, \quad t \in \mathbb{N}_0$$
$$D(0) = 1, \quad D(1) = 4.$$

The characteristic equation is $\lambda^2 - 2\lambda + 1 = 0$, and thus $\lambda_1 = \lambda_2 = 1$ are the characteristic values. Hence by Theorem 1.37,

$$\begin{aligned} D(t) &= c_1 e_1(t,0) + c_2(t-0)e_1(t,0) \\ &= c_1 2^t + c_2 t 2^t. \end{aligned}$$

Using the initial conditions we get the system of equations

$$D(0) = 1 = c_1$$
$$D(1) = 4 = 2c_1 + 2c_2.$$

Solving this system we get that $c_1 = c_2 = 1$ and hence

$$D(t) = 2^t + t2^t.$$

The reader should check this answer for a few values of $t$.

*Example 1.39.* Find the determinant $D(t)$, $t \in \mathbb{N}_1$ of the $t \times t$ matrix that has zeros down the diagonal, 2's down the superdiagonal, 8's down the subdiagonal. This leads to solving the IVP

$$D(t+2) + 16D(t) = 0, \quad t \in \mathbb{N}_0$$
$$D(0) = 1, \quad D(1) = 0.$$

If we tried to use Theorem 1.33 to solve the difference equation $D(t+2)+16D(t) = 0$ we would write the equation $D(t+2) + 16D(t) = 0$ in the form

$$\Delta^2 D(t) + 2\Delta D(t) + 17D(t) = 0.$$

The characteristic values for this equation are $-1 \pm 4i$. But Theorem 1.33 does not apply since the real part of $-1 \pm 4i$ is $-1$. Applying Theorem 1.44 we get that a general solution of $D(t+2) + 16D(t) = 0$ is given by

$$D(t) = c_1 4^t \cos\left(\frac{\pi}{2}t\right) + c_2 4^t \cos\left(\frac{\pi}{2}t\right), \quad t \in \mathbb{N}_0.$$

Applying the initial conditions we get

$$D(t) = 4^t \cos\left(\frac{\pi}{2}t\right),$$

for $t \in \mathbb{N}_1$.

Sometimes it is convenient to know how to solve the second order linear homogeneous difference equation when it is of the form

$$y(t+2) + cy(t+1) + dy(t) = 0, \tag{1.19}$$

where $c, d \in \mathbb{R}$ and $d \neq 0$ without first writing (1.19) (as we did in Examples 1.32 and 1.38) in the form (1.13). Exercise 1.36 shows that the difference equation 1.13 with $p \neq 1 + q$ is equivalent to the difference equation 1.19 with $d \neq 0$. Similar to the proof of Theorem 1.31 we can prove (Exercise 1.29) the theorem.

**Theorem 1.40 (Distinct Roots).** *Assume $d \neq 0$ and $r_1$, $r_2$ are distinct roots of the equation*

$$r^2 + cr + d = 0.$$

*Then*

$$y(t) = c_1 r_1^t + c_2 r_2^t$$

*is a general solution of* (1.19).

*Example 1.41.* Solve the difference equation

$$u(t+2) - 5u(t+1) + 6u(t) = 0. \tag{1.20}$$

Solving $r^2 - 5r + 6 = (r-2)(r-3) = 0$, we get $r_1 = 2$, $r_2 = 3$. It follows from Theorem 1.40 that

$$u(t) = c_1 2^t + c_2 3^t$$

is a general solution of (1.20).

Similar to the proof of Theorem 1.37 one could prove (see Exercise 1.30) the following theorem.

**Theorem 1.42 (Double Root).** *Assume $d \neq 0$ and $r$ is a double root of $r^2 + cr + d = 0$. Then*

$$y(t) = c_1 r^t + c_2 t r^t$$

*is a general solution of* (1.19).

*Example 1.43.* Solve the difference equation

$$u(t+2) + 4u(t+1) + 4u(t) = 0. \tag{1.21}$$

Solving the equation $r^2 + 4r + 4 = (r+2)^2 = 0$, we get $r_1 = r_2 = -2$. It follows from Theorem 1.42 that

$$u(t) = c_1(-2)^t + c_2 t(-2)^t$$

is a general solution of (1.21).

**Theorem 1.44 (Complex Roots).** *Assume $d \neq 0$ and $\alpha \pm i\beta$, $\beta > 0$ are complex roots of $r^2 + cr + d = 0$. Then*

$$y(t) = c_1 r^t \cos(\theta t) + c_2 r^t \sin(\theta t),$$

*where $r = \sqrt{\alpha^2 + \beta^2}$ and $\theta = \mathrm{Tan}^{-1}\frac{\beta}{\alpha}$ if $\alpha \neq 0$ and $\theta = \frac{\pi}{2}$ if $\alpha = 0$ is a general solution of* (1.19).

*Proof.* Since $r = \alpha + i\beta$, $\beta > 0$ is a solution of $r^2 + cr + d = 0$, we have by Theorem 1.40 that $y(t) = (\alpha + i\beta)^t$ is a (complex-valued) solution of (1.19). Let $\theta := Tan^{-1}\frac{\beta}{\alpha}$ if $\alpha \neq 0$ and let $\theta = \frac{\pi}{2}$ if $\alpha = 0$. Then if $r := \sqrt{\alpha^2 + \beta^2}$ we have that

$$\begin{aligned} y(t) &= (\alpha + i\beta)^t \\ &= \left(re^{i\theta}\right)^t \\ &= r^t e^{i\theta t} \\ &= r^t \cos(\theta t) + ir^t \sin(\theta t). \end{aligned}$$

Since the real and imaginary parts of $y(t)$ are linearly independent solutions of (1.19), we have that

$$y(t) = c_1 r^t \cos(\theta t) + c_2 r^t \sin(\theta t)$$

is a general solution of (1.19). □

Next we briefly discuss the method of annihilators for solving certain nonhomogeneous difference equations. For an arbitrary function $f : \mathbb{N}_a \to \mathbb{R}$ we define the operator $E$ by

$$Ef(t) = f(t+1), \quad t \in \mathbb{N}_a.$$

Then $E^n := E \cdot E^{n-1}$ for $n \in \mathbb{N}_1$. We say the polynomial in $E$,

$$p(E) := E^n + a_1 E^{n-1} + \cdots + a_n I,$$

annihilates $f : \mathbb{N}_a \to \mathbb{R}$ provided $p(E)f(t) = 0$ for $t \in \mathbb{N}_a$. Similarly we say the polynomial in the operator $\Delta$,

$$p(\Delta) := \Delta^n + a_1 \Delta^{n-1} + \cdots + a_n I,$$

annihilates $f : \mathbb{N}_a \to \mathbb{R}$ provided $p(\Delta)f(t) = 0$

*Example 1.45.* Here are some simple annihilators for various functions:

(i) $(E - rI)r^t = 0$;
(ii) $(\Delta - rI)e_r(t, a) = 0$;
(iii) $(E - rI)^2 tr^t = 0$;
(iv) $(\Delta^2 - p^2)\cosh_p(t, a) = 0$;
(v) $(\Delta^2 - p^2)\sinh_p(t, a) = 0$;
(vi) $(\Delta^2 + p^2)\cosh_p(t, a) = 0$;
(vii) $(\Delta^2 + p^2)\sinh_p(t, a) = 0$;
(viii) $\Delta^n (t-a)^k = 0$, for integers $n \geq k \geq 0$.

We now give some simple examples where we use the method of annihilators to solve various difference equations.

*Example 1.46.* Solve the first order linear equation

$$y(t+1) - 3y(t) = 5^t, \quad t \in \mathbb{Z}. \tag{1.22}$$

First we write this equation in the form

$$(E - 3I)y(t) = 5^t.$$

Since $E - 5I$ annihilates the right-hand side, we multiply each side by the operator $E - 5I$ to get

$$(E - 5I)(E - 3I)y(t) = 0, \quad t \in \mathbb{Z}.$$

It follows that a solution of (1.22) must be of the form

$$y(t) = c_1 3^t + c_2 5^t, \quad t \in \mathbb{Z}.$$

Substituting this into equation (1.22) we get

$$c_1 3^{t+1} + c_2 5^{t+1} - c_1 3^{t+1} - 3c_2 5^t = 5^t, \quad t \in \mathbb{Z}.$$

Hence we see that we must have $c_2 = \frac{1}{2}$. It follows that

$$y(t) = c_1 3^t + \frac{1}{2} \cdot 5^t, \quad t \in \mathbb{Z}$$

is a general solution of (1.22).

*Example 1.47.* Solve the second order linear nonhomogeneous difference equation

$$y(t+2) - 6y(t+1) + 8y(t) = 16 \cdot 4^t, \quad t \in \mathbb{N}_0. \tag{1.23}$$

by the annihilator method. The difference equation (1.23) can be written in the form

$$(E - 2I)(E - 4I)y(t) = 16 \cdot 4^t, \quad t \in \mathbb{N}_0.$$

Multiplying both sides by the operator $E - 4I$ we get that

$$(E - 2I)(E - 4I)^2 y(t) = 0.$$

Hence $y(t)$ must have the form

$$y(t) = c_1 2^t + c_2 4^t + c_3 t 4^t, \quad t \in \mathbb{N}_0.$$

Substituting this into the difference equation (1.23) we get after simplification that $8c_3 = 16$. Hence $c_3 = 2$ and we have that

$$y(t) = c_1 2^t + c_2 4^t + 2t4^t, \quad t \in \mathbb{N}_0$$

is a general solution of (1.23).

Also we can use the method of annihilators to solve certain nonhomogeneous equations of the form

$$\Delta^2 y(t) + p\Delta y(t) + qy(t) = f(t),$$

where $p, q$ are real constants with $p \neq 1 + q$, as is shown in the following example.

*Example 1.48.* Use the method of annihilators to solve the nonhomogeneous equation

$$\Delta^2 y(t) - 3\Delta y(t) + 2y(t) = 4^{t-a}, \quad t \in \mathbb{N}_a. \tag{1.24}$$

The equation (1.24) can be written in the form

$$(\Delta - I)(\Delta - 2I)y(t) = e_3(t, a).$$

Multiplying both sides by the operator $(\Delta - 3I)$ we get that solutions of (1.24) are solutions of

$$(\Delta - I)(\Delta - 2I)(\Delta - 3I)y(t) = 0.$$

The values of the characteristic equation $(\lambda - 1)(\lambda - 2)(\lambda - 3) = 0$ are $\lambda_1 = 1$, $\lambda_2 = 2$, $\lambda_3 = 3$. Hence all solutions of (1.24) are of the form

$$y(t) = c_1 e_1(t, a) + c_2 e_2(t, a) + c_3 e_3(t, a).$$

Substituting this into the equation we get that we must have $2c_3 e_3(t, a) = e_3(t, a)$ which gives us that $c_3 = 2$. Hence the general solution of (1.24) is given by

$$y(t) = c_1 e_1(t, a) + c_2 e_2(t, a) + 2e_3(t, a), \quad t \in \mathbb{N}_a.$$

## 1.5 The Delta Integral

First we define the delta definite integral.

**Definition 1.49.** Assume $f : \mathbb{N}_a \to \mathbb{R}$ and $c \leq d$ are in $\mathbb{N}_a$, then

$$\int_c^d f(t)\Delta t := \sum_{t=c}^{d} f(t), \tag{1.25}$$

where by convention $\sum_{t=c}^{c-k} f(t) := 0$, whenever $k \in \mathbb{N}_1$. We will define this to be the case even when $f(t)$ is not defined for some (or all) values $t \in \mathbb{N}_{c-k}$. In the sum in (1.25) it is understood that the index $t$ takes on the values $c, c+1, c+2, \ldots, d-1$ when $d > c$.

Note that the value of the integral $\int_c^d f(t)\Delta t$ does not depend on the value $f(d)$. The following theorem gives some properties of this delta integral.

**Theorem 1.50.** *Assume $f, g : \mathbb{N}_a \to \mathbb{R}$, $b, c, d \in \mathbb{N}_a$, $b \le c \le d$, and $\alpha \in \mathbb{R}$. Then*

(i) $\int_b^c \alpha f(t)\Delta t = \alpha \int_b^c f(t)\Delta t$;
(ii) $\int_b^c (f(t) + g(t))\Delta t = \int_b^c f(t)\Delta t + \int_b^c g(t)\Delta t$;
(iii) $\int_b^b f(t)\Delta t = 0$;
(iv) $\int_b^d f(t)\Delta t = \int_b^c f(t)\Delta t + \int_c^d f(t)\Delta t$;
(v) $|\int_b^c f(t)\Delta t| \le \int_b^c |f(t)|\Delta t$;
(vi) *if* $F(t) := \int_b^t f(s)\Delta s$, *for* $t \in \mathbb{N}_b^c$, *then* $\Delta F(t) = f(t)$, $t \in \mathbb{N}_b^{c-1}$;
(vii) *if* $f(t) \ge g(t)$, *for* $t \in \mathbb{N}_b^{c-1}$, *then* $\int_b^c f(t)\Delta t \ge \int_b^c g(t)\Delta t$.

*Proof.* Most of these properties of the integral hold since the corresponding properties for sums hold. We leave the proof of this theorem to the reader. □

**Definition 1.51.** Assume $f : \mathbb{N}_a^b \to \mathbb{R}$. We say $F$ is an **antidifference** of $f$ on $\mathbb{N}_a^b$ provided

$$\Delta F(t) = f(t), \quad t \in \mathbb{N}_a^{b-1}.$$

Since $\Delta\left(\frac{1}{2}3^t\right) = 3^t$, $t \in \mathbb{N}_a$, we have that $F(t) = \frac{1}{2}3^t$ is an antidifference of $f(t) = 3^t$ on $\mathbb{N}_a$.

**Theorem 1.52.** *If $f : \mathbb{N}_a^b \to \mathbb{R}$ and $G(t)$ is an antidifference of $f(t)$ on $\mathbb{N}_a^b$, then $F(t) = G(t) + C$, where $C$ is an arbitrary constant, is a general antidifference of $f(t)$ on $\mathbb{N}_a^b$.*

*Proof.* Assume $G(t)$ is an antidifference of $f(t)$ on $\mathbb{N}_a^b$. Let $F(t) := G(t) + C$, $t \in \mathbb{N}_a^b$, where $C$ is a constant. Then

$$\Delta F(t) = \Delta G(t) = f(t), \quad t \in \mathbb{N}_a^b,$$

and so $F(t)$ is an antidifference of $f(t)$ on $\mathbb{N}_a^b$. Next assume $F(t)$ is an antidifference of $f(t)$ on $\mathbb{N}_a^b$. Then

$$\Delta[F(t) - G(t)] = \Delta F(t) - \Delta G(t) = f(t) - f(t) = 0$$

for $t \in \mathbb{N}_a^{b-1}$. This implies (Exercise 1.1) $F(t) - G(t) = C$, for $t \in \mathbb{N}_a^b$, where $C$ is a constant. Hence $F(t) := G(t) + C$, for $t \in \mathbb{N}_a^b$. □

*Example 1.53.* Find the number of regions, $R(n)$, the plane is divided into by $n$ lines, where no two lines are parallel and no three lines intersect at the same point. Note that

$$R(1) = 2, \quad R(2) = 4, \quad R(3) = 7, \quad R(4) = 11.$$

To find $R(n)$ for all $n \in \mathbb{N}_1$, first convince yourself that for any $n \in \mathbb{N}_1$,

$$R(n+1) = R(n) + n + 1.$$

It follows that

$$\Delta R(n) = n + 1 = (n+1)^{\underline{1}}.$$

Since $\frac{1}{2}(n+1)^{\underline{2}}$ is an antidifference of $(n+1)^{\underline{1}}$, we have by Theorem 1.52 that

$$R(n) = \frac{1}{2}(n+1)^{\underline{2}} + C.$$

Using $R(1) = 2$ we get that $C = 1$ and hence

$$R(n) = \frac{1}{2}(n+1)^{\underline{2}} + 1, \quad t \in \mathbb{N}_1.$$

**Definition 1.54.** If $f : \mathbb{N}_a \to \mathbb{R}$, then the delta indefinite integral of $f$ is defined by

$$\int f(t)\Delta t := F(t) + C,$$

where $F$ is an antidifference of $f$ and $C$ is an arbitrary constant.

It is easy to verify that

$$\int \alpha f(t)\Delta t = \alpha \int f(t)\Delta t$$

and

$$\int (f(t) + g(t))\Delta t = \int f(t)\Delta t + \int g(t)\Delta t.$$

Any formula for a delta derivative gives us a formula for an indefinite delta integral, so we have the following theorem.

**Theorem 1.55.** *Assume $p$, $r$, $\alpha$ are constants. Then the following hold:*

(i) $\int (t-\alpha)^{\underline{r}}\Delta t = \frac{1}{r+1}(t-\alpha)^{\underline{r+1}} + C, \quad r \neq -1;$
(ii) $\int e_p(t,a)\Delta t = \frac{1}{p}e_p(t,a) + C, \quad p \neq 0, -1;$

(iii) $\int \cosh_p(t,a)\Delta t = \frac{1}{p}\sinh_p(t,a) + C, \quad p \neq 0, \pm 1;$
(iv) $\int \sinh_p(t,a)\Delta t = \frac{1}{p}\cosh_p(t,a) + C, \quad p \neq 0, \pm 1;$
(v) $\int \cos_p(t,a)\Delta t = \frac{1}{p}\sin_p(t,a) + C, \quad p \neq 0, \pm i;$
(vi) $\int \sin_p(t,a)\Delta t = -\frac{1}{p}\cos_p(t,a) + C, \quad p \neq 0, \pm i;$
(vii) $\int (\alpha - \sigma(t))^{\underline{r}}\Delta t = \frac{-1}{r+1}(\alpha - t)^{\underline{r+1}} + C, \quad r \neq -1;$
(viii) $\int \binom{t}{r}\Delta t = \binom{t}{r+1} + C;$
(ix) $\int \binom{r+t}{t}\Delta t = \binom{r+t}{t-1} + C;$
(x) $\int (t-1)\Gamma(t)\Delta t = \Gamma(t) + C;$
(xi) $\int \alpha^t \Delta t = \frac{1}{\alpha - 1}\alpha^t + C, \quad \alpha \neq 1,$

*where C is an arbitrary constant.*

**Theorem 1.56 (Fundamental Theorem for the Difference Calculus).** *Assume $f : \mathbb{N}_a^b \to \mathbb{R}$ and $F(t)$ is any antidifference of $f(t)$ on $\mathbb{N}_a^b$. Then*

$$\int_a^b f(t)\Delta t = \int_a^b \Delta F(t)\Delta t = F(t)|_a^b.$$

*(Here we use the common notation $F(t)|_a^b := F(b) - F(a)$.)*

*Proof.* Assume $F(t)$ is any antidifference of $f(t)$ on $\mathbb{N}_a^b$. Let

$$G(t) := \int_a^t f(s)\Delta s, \quad t \in \mathbb{N}_a^b.$$

Then by Theorem 1.50 (vi), $G(t)$ is an antidifference of $f(t)$ on $\mathbb{N}_a^b$. Hence by Theorem 1.52, $F(t) = G(t) + C$, where $C$ is a constant. Then

$$\begin{aligned} F(t)|_a^b &= F(b) - F(a) \\ &= [(G(b) + C) - (G(a) + C)] \\ &= G(b) - G(a) \\ &= \int_a^b f(t)\Delta t. \end{aligned}$$

This completes the proof. □

*Example 1.57.* Use a delta integral to find the sum of the squares of the first $n$ positive integers. Using $k^2 = k^{\underline{2}} + k^{\underline{1}}$ we get

$$\begin{aligned} \sum_{k=1}^{n} k^2 &= \int_1^{n+1} k^2 \Delta k \\ &= \int_1^{n+1} (k^{\underline{2}} + k^{\underline{1}})\Delta k \end{aligned}$$

$$= \left[\frac{1}{3}k^{\underline{3}} + \frac{1}{2}k^{\underline{2}}\right]_1^{n+1}$$
$$= \frac{1}{3}(n+1)^{\underline{3}} + \frac{1}{2}(n+1)^{\underline{2}}$$
$$= \frac{(n+1)n(n-1)}{3} + \frac{(n+1)n}{2}$$
$$= \frac{n(n+1)(2n+1)}{6}.$$

Using the product rules in Exercise 1.2 we can prove the following integration by parts theorem.

**Theorem 1.58 (Integration by Parts).** *Given two functions $u, v : \mathbb{N}_a \to \mathbb{R}$ and $b, c \in \mathbb{N}_a$, $b < c$, we have the integration by parts formulas*

$$\int_b^c u(t)\Delta v(t)\Delta t = u(t)v(t)\Big|_b^c - \int_b^c v(\sigma(t))\Delta u(t)\Delta t. \tag{1.26}$$

$$\int_b^c u(\sigma(t))\Delta v(t)\Delta t = u(t)v(t)\Big|_b^c - \int_b^c v(t)\Delta u(t)\Delta t. \tag{1.27}$$

*Example 1.59.* Evaluate

$$\int t4^t \Delta t,$$

where we consider $t4^t$ for $t \in \mathbb{N}_0$. Note that

$$\int t4^t \Delta t = \int te_3(t,0)\Delta t.$$

To set up integration by parts let

$$u(t) = t, \quad \Delta v(t) = e_3(t,0).$$

We then use

$$\Delta u(t) = 1, \quad v(t) = \frac{1}{3}e_3(t,0), \quad v(\sigma(t)) = \frac{4}{3}e_3(t,0)$$

and the integration by parts formula (1.26) to get

$$\begin{aligned}\int t4^t \Delta t &= \int te_3(t,0)\Delta t \\ &= \frac{1}{3}te_3(t,0) - \frac{4}{3}\int e_3(t,0)\Delta t \\ &= \frac{1}{3}te_3(t,0) - \frac{4}{9}e_3(t,0) + C \\ &= \frac{1}{3}t4^t - \frac{4}{9}4^t + C.\end{aligned}$$

## 1.6 Discrete Taylor's Theorem

In this section we want to prove the discrete version of Taylor's Theorem. First we study the discrete (delta) Taylor monomials and give some of their properties. We will see that these discrete Taylor monomials will appear in the discrete Taylor's Theorem. These (delta) Taylor monomials take the place of the Taylor monomials $\frac{(t-s)^n}{n!}$ in the continuous calculus.

**Definition 1.60.** We define the discrete **Taylor monomials** (based at $s \in \mathbb{N}_a$), $h_n(t,s)$, $n \in \mathbb{N}_0$ by

$$h_n(t,s) = \frac{(t-s)^{\underline{n}}}{n!}, \quad t \in \mathbb{N}_a.$$

In particular if $s = a$, then

$$h_n(t,a) = \frac{(t-a)^{\underline{n}}}{n!}, \quad t \in \mathbb{N}_a.$$

**Theorem 1.61.** *The Taylor monomials satisfy the following:*

(i) $h_0(t,a) = 1, \quad t \in \mathbb{N}_a;$
(ii) $h_n(t,t) = 0, \quad t \in \mathbb{N}_a, \quad n \in \mathbb{N}_1;$
(iii) $\Delta h_{n+1}(t,a) = h_n(t,a), \quad t \in \mathbb{N}_a, \quad n \in \mathbb{N}_0;$
(iv) $\int h_n(t,a)\Delta t = h_{n+1}(t,a) + C, t \in \mathbb{N}_a, \quad n \in \mathbb{N}_0;$
(v) $\Delta_s h_{n+1}(t,s) = -h_n(t,\sigma(s)), \quad t \in \mathbb{N}_a, \quad n \in \mathbb{N}_0;$
(vi) $\int h_n(t,\sigma(s))\Delta s = -h_{n+1}(t,s) + C, \quad t \in \mathbb{N}_a, \quad n \in \mathbb{N}_0,$

*where $C$ is a constant.*

*Proof.* We will only prove part (v). Since

$$\begin{aligned}
\Delta_s h_{n+1}(t,s) &= \Delta_s \frac{(t-s)^{\underline{n+1}}}{(n+1)!} \\
&= \frac{1}{(n+1)!}\Delta_s (t-s)^{\underline{n+1}} \\
&= -\frac{1}{n!}(t-\sigma(s))^{\underline{n}}, \quad \text{by} \quad (1.3) \\
&= -\frac{(t-\sigma(s))^{\underline{n}}}{n!} \\
&= -h_n(t,\sigma(s)),
\end{aligned}$$

we have that (v) holds. □

Now we state and prove the discrete Taylor's Theorem.

**Theorem 1.62 (Taylor's Formula).** *Assume $f : \mathbb{N}_a \to \mathbb{R}$ and $n \in \mathbb{N}_0$. Then*

$$f(t) = p_n(t) + R_n(t), \quad t \in \mathbb{N}_a,$$

*where the n-th degree Taylor polynomial, $p_n(t)$, is given by*

$$p_n(t) := \sum_{k=0}^{n} \Delta^k f(a)\frac{(t-a)^{\underline{k}}}{k!} = \sum_{k=0}^{n} \Delta^k f(a) h_k(t,a)$$

*and the Taylor remainder, $R_n(t)$, is given by*

$$R_n(t) = \int_a^t \frac{(t-\sigma(s))^{\underline{n}}}{n!}\Delta^{n+1} f(s)\Delta s = \int_a^t h_n(t,\sigma(s))\Delta^{n+1} f(s)\Delta s,$$

*for $t \in \mathbb{N}_a$.*

*Proof.* If $n = 0$, then

$$\begin{aligned}
p_0(t) + R_0(t) &= f(a)h_0(t,a) + \int_a^t h_0(t,\sigma(s))\Delta f(s)\Delta s \\
&= f(a) + \int_a^t \Delta f(s)\Delta s \\
&= f(a) + f(t) - f(a) = f(t).
\end{aligned}$$

Hence Taylor's Theorem holds for $n = 0$. Now assume that $n \geq 1$. We will apply the second integration by parts formula in Theorem 1.58, namely (1.27), to

$$R_n(t) = \int_a^t h_n(t,\sigma(s))\Delta^{n+1} f(s)\Delta s.$$

To do this we set

$$u(\sigma(s)) = h_n(t, \sigma(s)), \quad \Delta v(s) = \Delta^{n+1} f(s),$$

then it follows that

$$u(s) = h_n(t, s), \quad v(s) = \Delta^n f(s).$$

Using Theorem 1.61, (v), we get

$$\Delta_s u(s) = -h_{n-1}(t, \sigma(s)).$$

Hence we get from the integration by parts formula (1.27), that

$$\begin{aligned}
R_n(t) &= \int_a^t h_n(t, \sigma(s)) \Delta^{n+1} f(s) \Delta s \\
&= h_n(t, s) \Delta^n f(s) \Big|_{s=a}^{s=t} + \int_a^t h_{n-1}(t, \sigma(s)) \Delta^n f(s) \Delta s \\
&= -\Delta^n f(a) h_n(t, a) + \int_a^t h_{n-1}(t, \sigma(s)) \Delta^n f(s) \Delta s.
\end{aligned}$$

If $n \geq 2$, then again we apply the integration by parts formula (1.27), to get

$$\begin{aligned}
R_n(t) &= -\Delta^n f(a) h_n(t, a) + h_{n-1}(t, s) \Delta^{n-1} f(s) \Big|_{s=a}^{s=t} \\
&\quad + \int_a^t h_{n-2}(t, \sigma(s)) \Delta^{n-1} f(s) \Delta s \\
&= -\Delta^n f(a) h_n(t, a) - \Delta^{n-1} f(a) h_{n-1}(t, a) \\
&\quad + \int_a^t h_{n-2}(t, \sigma(s)) \Delta^{n-1} f(s) \Delta s.
\end{aligned}$$

By induction on $n$, we get

$$\begin{aligned}
R_n(t) &= -\sum_{k=1}^{n} \Delta^k f(a) h_k(t, a) + \int_a^t h_0(t, \sigma(s)) \Delta f(s) \Delta s \\
&= -\sum_{k=1}^{n} \Delta^k f(a) h_k(t, a) + f(t) - f(a) \\
&= -\sum_{k=0}^{n} \Delta^k f(a) h_k(t, a) + f(t).
\end{aligned}$$

Solving for $f(t)$ we get the desired result. □

**Definition 1.63.** If $f : \mathbb{N}_a \to \mathbb{R}$, then we call

$$\sum_{k=0}^{\infty} \Delta^k f(a) \frac{(t-a)^{\underline{k}}}{k!} = \sum_{k=0}^{\infty} \Delta^k f(a) h_k(t,a)$$

the (formal) **Taylor series** of $f$ based $t = a$.

The following theorem gives us some Taylor series for various functions.

**Theorem 1.64.** *Assume $p$ is a constant. Then the following hold:*

(i) *If $p \neq -1$, then $e_p(t,a) = \sum_{n=0}^{\infty} p^n h_n(t,a)$;*
(ii) *If $p \neq \pm 1$, then $\cosh_p(t,a) = \sum_{n=0}^{\infty} p^{2n} h_{2n}(t,a)$;*
(iii) *If $p \neq \pm 1$, then $\sinh_p(t,a) = \sum_{n=0}^{\infty} p^{2n+1} h_{2n+1}(t,a)$;*
(iv) *If $p \neq \pm i$, then $\sin_p(t,a) = \sum_{n=0}^{\infty} (-1)^n p^{2n+1} h_{2n+1}(t,a)$;*
(v) *If $p \neq \pm i$, then $\cos_p(t,a) = \sum_{n=0}^{\infty} (-1)^n p^{2n} h_{2n}(t,a)$;*

*for all $t \in \mathbb{N}_a$.*

*Proof.* We first prove part (i). Since $\Delta^k e_p(t,a) = p^k e_p(t,a)$ for each $k \in \mathbb{N}$, we have that the Taylor series for $e_p(t,a)$ is given by

$$\sum_{n=0}^{\infty} \Delta^n e_p(a,a) h_n(t,a) = \sum_{n=0}^{\infty} p^n h_n(t,a).$$

To show that the above Taylor series converges to $e_p(t,a)$, for each $t \in \mathbb{N}_a$, it suffices to show that the remainder term, $R_n(t)$, in Taylor's formula, satisfies

$$\lim_{n\to\infty} R_n(t) = 0$$

for each fixed $t \in \mathbb{N}_a$.

So fix $t \in \mathbb{N}_a$ and consider

$$\begin{aligned} |R_n(t)| &= \left| \int_a^t h_n(t,\sigma(s)) \Delta^{n+1} e_p(s,a) \Delta s \right| \\ &= \left| \int_a^t h_n(t,\sigma(s)) p^{n+1} e_p(s,a) \Delta s \right| \\ &\leq \int_a^t h_n(t,\sigma(s)) |p|^{n+1} |e_p(s,a)| \Delta s \end{aligned}$$

Since $t$ is fixed, there is a constant $C$ such that

$$|e_p(s,a)| \leq C, \quad s \in \mathbb{N}_a^{t-1}.$$

Hence

$$\begin{aligned}|R_n(t)| &\le C\int_a^t h_n(t,\sigma(s))|p|^{n+1}\Delta s\\ &= C|p|^{n+1}\int_a^t h_n(t,\sigma(s))\Delta s\\ &= -C|p|^{n+1}h_{n+1}(t,s)\Big|_{s=a}^{s=t}\\ &= C|p|^{n+1}h_{n+1}(t,a)\\ &= C|p|^{n+1}\frac{(t-a)^{\underline{n+1}}}{(n+1)!}.\end{aligned}$$

Since $(t-a)^{\underline{n+1}} = 0$, for $n \ge t-a$, we have that $R_n(t) = 0$ for $n \ge t-a$, so for each fixed $t$, $\lim_{n\to\infty} R_n(t) = 0$. Hence,

$$e_p(t,a) = \sum_{n=0}^{\infty} p^n h_n(t,a), \quad t \in \mathbb{N}_a.$$

To see that (ii) holds for $t \in \mathbb{N}_a$, note that for $p \ne \pm 1$

$$\begin{aligned}\cosh_p(t,a) &= \frac{e_p(t,a)+e_{-p}(t,a)}{2}\\ &= \sum_{k=0}^{\infty}\frac{p^k+(-p)^k}{2}h_k(t,a)\\ &= \sum_{n=0}^{\infty}p^{2n}h_{2n}(t,a).\end{aligned}$$

Similarly, since $\sinh_p(t,a)$, $\sin_p(t,a)$, and $\cos_p(t,a)$, are defined in terms of exponential functions, parts (iii)–(v) follow easily from part (i). □

We next show that Taylor's Theorem gives us a variation of constants formula.

**Theorem 1.65 (Variation of Constants Formula).** *Assume $f : \mathbb{N}_a \to \mathbb{R}$. Then the unique solution of the IVP*

$$\begin{aligned}\Delta^n y(t) &= f(t), \quad t \in \mathbb{N}_a\\ \Delta^k y(a) &= c_k, \quad 0 \le k \le n-1,\end{aligned}$$

*where $c_k$, $0 \le k \le n-1$, are given constants is given by*

$$y(t) = \sum_{k=0}^{n-1} c_k h_k(t,a) + \int_a^t h_{n-1}(t,\sigma(s))f(s)\Delta s,$$

*for* $t \in \mathbb{N}_a$.

*Proof.* The proof of uniqueness is similar to the proof of Theorem 1.29. Using Taylor's Theorem 1.62, we get the solution, $y(t)$, of the given IVP is given by

$$\begin{aligned} y(t) &= p_{n-1}(t) + R_{n-1}(t) \\ &= \sum_{k=0}^{n-1} \Delta^k y(a) h_k(t,a) + \int_a^t h_{n-1}(t,\sigma(s))\Delta^n y(s)\Delta s \\ &= \sum_{k=0}^{n-1} c_k h_k(t,a) + \int_a^t h_{n-1}(t,\sigma(s))f(s)\Delta s \end{aligned}$$

for $t \in \mathbb{N}_a$. □

We now give a very elementary example to illustrate the variation of constants formula.

*Example 1.66.* Use the integer variation of constants formula to solve the IVP

$$\Delta^2 y(t) = 3^t, \quad t \in \mathbb{N}_0$$
$$y(0) = \Delta y(0) = 0.$$

Using the variation of constants formula in Theorem 1.65, the solution of the given IVP is given by

$$y(t) = \int_0^t h_1(t,\sigma(s))3^s\Delta s = \int_0^t h_1(t,\sigma(s))e_2(s,0)\Delta s.$$

Integrating by parts we get

$$\begin{aligned} y(t) &= \frac{1}{2}(t-s)e_2(s,0)|_{s=0}^{t} + \frac{1}{2}\int_0^t e_2(s,0)\Delta s \\ &= -\frac{1}{2}t + \frac{1}{4}[e_2(s,0)]_{s=0}^{t} \\ &= -\frac{1}{2}t + \frac{1}{4}[e_2(t,0) - 1] \\ &= -\frac{1}{2}t - \frac{1}{4} + \frac{1}{4}3^t, \end{aligned}$$

for $t \in \mathbb{N}_0$. Of course one could easily solve the IVP in this example by twice integrating both sides of $\Delta^2 y(t) = 3^t$ from 0 to $t$ (Exercise 1.52).

## 1.7 First Order Linear Difference Equations

In this section we show how to solve the first order linear equation

$$\Delta y(t) = p(t)y(t) + q(t), \quad t \in \mathbb{N}_a, \tag{1.28}$$

where we assume $p, q : \mathbb{N}_a \to \mathbb{R}$ and $p(t) \neq -1$ for $t \in \mathbb{N}_a$.

We will use the following Leibniz formula to find a variation of constants formula for (1.28).

**Theorem 1.67 (Leibniz Formula).** *Assume $f : \mathbb{N}_a \times \mathbb{N}_a \to \mathbb{R}$. Then*

$$\Delta\left(\int_a^t f(t,s)\Delta s\right) = \int_a^t \Delta_t f(t,s)\Delta s + f(\sigma(t), t), \quad t \in \mathbb{N}_a. \tag{1.29}$$

*Proof.* We have that

$$\begin{aligned}
\Delta\left(\int_a^t f(t,s)\Delta s\right) &= \int_a^{t+1} f(t+1,s)\Delta s - \int_a^t f(t,s)\Delta s \\
&= \int_a^t [f(t+1,s) - f(t,s)]\Delta s + \int_t^{t+1} f(t+1,s)\Delta s \\
&= \int_a^t \Delta_t f(t,s)\Delta s + \int_t^{t+1} f(t+1,s)\Delta s \\
&= \int_a^t \Delta_t f(t,s)\Delta s + f(\sigma(t), t),
\end{aligned}$$

which completes the proof. □

**Theorem 1.68 (Variation of Constants Formula).** *Assume $p \in \mathcal{R}$ and $q : \mathbb{N}_a \to \mathbb{R}$. Then the unique solution of the IVP*

$$\begin{aligned}
\Delta y(t) &= p(t)y(t) + q(t), \quad t \in \mathbb{N}_a \\
y(a) &= A
\end{aligned}$$

*is given by*

$$y(t) = Ae_p(t,a) + \int_a^t e_p(t, \sigma(s))q(s)\Delta s,$$

*for $t \in \mathbb{N}_a$.*

*Proof.* The proof (see Exercise 1.56) of the uniqueness of solutions of IVPs for this case is similar to the proof of Theorem 1.29. Let

$$y(t) = Ae_p(t,a) + \int_a^t e_p(t,\sigma(s))q(s)\Delta s, \quad t \in \mathbb{N}_a.$$

Using the Leibniz formula (1.29), we get

$$\begin{aligned}\Delta y(t) &= Ap(t)e_p(t,a) + \int_a^t p(t)e_p(t,\sigma(s))q(s)\Delta s + e_p(\sigma(t),\sigma(t))q(t)\\ &= p(t)\left[Ae_p(t,a) + \int_a^t e_p(t,\sigma(s))q(s)\Delta s\right] + q(t)\\ &= p(t)y(t) + q(t).\end{aligned}$$

Also $y(a) = A$. □

Of course, it is always possible to compute solutions of difference equations by direct step by step computation from the difference equation. We next give an interesting example due to Gautschi [87] (and appearing in Kelley and Peterson [134, 135]) that illustrates that round off error can be a serious problem.

*Example 1.69 (Gautschi [87]).* First we solve the IVP

$$\begin{aligned}\Delta y(t) &= (t-1)y(t) + 1, \quad t \in \mathbb{N}_1\\ y(1) &= 1 - e.\end{aligned}$$

Note that $p(t) := t-1$ is a regressive function on $\mathbb{N}_1$. Using the variation of constants formula in Theorem 1.68, we get that the solution of our given IVP is given by

$$\begin{aligned}y(t) &= (1-e)e_{t-1}(t,1) + \int_1^t e_{t-1}(t,\sigma(s)) \cdot 1\Delta s\\ &= e_{t-1}(t,1)\left[1 - e + \int_1^t e_{t-1}(1,\sigma(s))\Delta s\right]\\ &= e_{t-1}(t,1)\left[1 - e + \int_1^t \frac{1}{e_{t-1}(\sigma(s),1)}\Delta s\right].\end{aligned}$$

From Example 1.13, we have that $e_{t-1}(t,1) = (t-1)!$. Hence

$$y(t) = (t-1)!\left[1 - e + \int_1^t \frac{1}{(\sigma(s)-1)!}\Delta s\right]$$

$$= (t-1)!\left[1 - e + \sum_{s=1}^{t-1} \frac{1}{s!}\right]$$
$$= -(t-1)! \sum_{k=t}^{\infty} \frac{1}{k!}.$$

Note that this solution is negative on $\mathbb{N}_1$. Now if one was to approximate the initial value $1 - e$ in this IVP by a finite decimal expansion, it can be shown that the solution $z(t)$ of this new IVP satisfies $\lim_{t\to\infty} z(t) = \infty$ and hence $z(t)$ is not a good approximation for the actual solution. For example, if $z(t)$ solves the IVP

$$\Delta z(t) = (1-t)z(t) + 1, \quad t \in \mathbb{N}_1$$
$$z(1) = -1.718,$$

then $z(2) = -.718$, $z(3) = -.436$, $z(4) = -.308$, $z(5) = -.232$, $z(6) = -.16$, $z(7) = .04$ and after that $z(t)$ increases rapidly with $\lim_{t\to\infty} z(t) = \infty$. Hence $z(t)$ is not a good approximation to the actual solution $y(t)$ of our original IVP.

A general solution of the linear equation (1.28) is given by adding a general solution of the corresponding homogeneous equation $\Delta y(t) = p(t)y(t)$ to a particular solution to the nonhomogeneous difference equation (1.28). Hence by Theorem 1.14 and Theorem 1.68

$$y(t) = ce_p(t,a) + \int_a^t e_p(t,\sigma(s))q(s)\Delta s$$

is a general solution of (1.28). We use this fact in the following example.

*Example 1.70.* Find a general solution of the linear difference equation

$$\Delta y(t) = (\ominus 2)y(t) + t, \quad t \in \mathbb{N}_0. \tag{1.30}$$

Note that the constant function $\ominus 2$ is a regressive function on $\mathbb{N}_0$. The general solution of (1.30) is given by

$$y(t) = ce_p(t,a) + \int_a^t e_p(t,\sigma(s))q(s)\Delta s$$
$$= ce_{\ominus 2}(t,0) + \int_0^t s\, e_{\ominus 2}(t,\sigma(s))\Delta s$$
$$= ce_{\ominus 2}(t,0) + \int_0^t s\, e_2(\sigma(s),t)\Delta s$$
$$= ce_{\ominus 2}(t,0) + 3\int_0^t s\, e_2(s,t)\Delta s.$$

Integrating by parts we get

$$
\begin{aligned}
y(t) &= ce_{\ominus 2}(t,0) + \frac{3}{2} s e_2(s,t)\Big|_{s=0}^{t} - \frac{3}{2}\int_0^t e_2(\sigma(s),t)\Delta s \\
&= ce_{\ominus 2}(t,0) + \frac{3}{2}t - \frac{9}{2}\int_0^t e_2(s,t)\Delta s \\
&= ce_{\ominus 2}(t,0) + \frac{3}{2}t - \frac{9}{4}e_2(s,t)\Big|_0^t \\
&= ce_{\ominus 2}(t,0) + \frac{3}{2}t - \frac{9}{4} + \frac{9}{4}e_2(0,t) \\
&= \alpha e_{\ominus 2}(t,0) + \frac{3}{2}t - \frac{9}{4} \\
&= \alpha\left(\frac{1}{3}\right)^t + \frac{3}{2}t - \frac{9}{4}.
\end{aligned}
$$

## 1.8 Second Order Linear Equations (Variable Coefficients)

In this section we will show how we can solve some second order linear equations (with possibly variable coefficients) by the method of factoring. As a special case of our factoring method we will solve Euler–Cauchy difference equations. We start with the following example to illustrate the method of factoring.

*Example 1.71.* Solve the difference equation

$$\Delta^2 y(t) - (t+2)\Delta y(t) + 2ty(t) = 0, \quad t \in \mathbb{N}_0. \tag{1.31}$$

The method we use to solve this equation is called the method of factoring. We first write our difference equation in the form

$$\Delta[\Delta y(t) - 2y(t)] - t[\Delta y(t) - 2y(t)] = 0$$

which also can be written in the factored form (using $I$ as the identity operator)

$$(\Delta - tI)(\Delta - 2I)y(t) = 0.$$

Note that from Exercise 1.60 the operators $\Delta - tI$ and $\Delta - 2I$ do not commute, so one has to be careful when one uses this method of factoring. Letting $y(t)$ be a solution of our given difference equation and setting $v(t) = (\Delta - 2I)y(t)$, we get that

$$(\Delta - tI)v(t) = 0$$

which gives us the equation

$$\Delta v(t) = tv(t).$$

By Theorem 1.14,

$$v(t) = c_2 e_t(t, 0) = c_2 \prod_{s=0}^{t-1}(1 + s) = c_2 t!.$$

Since $v(t) = (\Delta - 2I)y(t)$ we have that

$$\Delta y(t) = 2y(t) + c_2 t!.$$

Using the variation of constants formula in Theorem 1.68, we get that

$$\begin{aligned} y(t) &= c_1 e_2(t, 0) + \int_0^t e_2(t, \sigma(s)) c_2 s! \Delta s \\ &= c_1 3^t + c_2 \int_0^t 3^{t-s-1} s! \Delta s \\ &= c_1 3^t + c_2 3^{t-1} \int_0^t \left(\frac{s!}{3^s}\right) \Delta s \\ &= c_1 3^t + c_2 3^{t-1} \sum_{s=0}^{t-1} \frac{s!}{3^s}. \end{aligned}$$

Hence

$$y(t) = \alpha 3^t + \beta 3^t \sum_{k=0}^{t-1} \frac{k!}{3^k}$$

is a general solution of the difference equation (1.31).

We next show how we can use this method of factoring to solve the **Euler–Cauchy difference equation**

$$t\sigma(t)\Delta^2 y(t) + ct\Delta y(t) + dy(t) = 0, \quad t \in \mathbb{N}_a, \tag{1.32}$$

where $c$ and $d$ are real constants. We assume that

$$1 + \frac{d}{t\sigma(t)} - \frac{c}{\sigma(t)} \neq 0, \quad t \in \mathbb{N}_a. \tag{1.33}$$

Since (1.33) holds, we have from Theorem 1.29 that solutions of IVPs for (1.32) are unique and solutions exist on $\mathbb{N}_a$. Also by Remark 1.30 we have that if $y_1(t)$ and $y_2(t)$ are linearly independent solutions of (1.32) on $\mathbb{N}_a$, then

$$y(t) = c_1 y_1(t) + c_2 y_2(t), \quad t \in \mathbb{N}_a$$

is a general solution of (1.32). We call the equation

$$r(r-1) + cr + d = 0 \tag{1.34}$$

or equivalently

$$r^2 + (c-1)r + d = 0 \tag{1.35}$$

the characteristic equation of the Euler–Cauchy difference equation (1.32) and the solutions of (1.34) and (1.35) the characteristic values (roots). Let $\alpha, \beta$ be the characteristic values, then the characteristic equation is given by

$$(r-\alpha)(r-\beta) = r^2 - (\alpha+\beta)r + \alpha\beta = 0.$$

Hence

$$c - 1 = -(\alpha+\beta), \quad d = \alpha\beta.$$

Note that

$$\begin{aligned} 1 + \frac{d}{t\sigma(t)} - \frac{c}{\sigma(t)} &= 1 + \frac{\alpha\beta}{t\sigma(t)} - \frac{1-\alpha-\beta}{\sigma(t)} \\ &= \frac{t\sigma(t) - (1-\alpha-\beta)t + \alpha\beta}{t\sigma(t)} \\ &= \frac{t^2 + (\alpha+\beta)t + \alpha\beta}{t\sigma(t)} \\ &= \frac{(t+\alpha)(t+\beta)}{t\sigma(t)} \\ &= \frac{t\left(1+\frac{\alpha}{t}\right)\left(1+\frac{\beta}{t}\right)}{\sigma(t)}. \end{aligned}$$

Hence we see that (1.33) holds if and only if

$$\frac{\alpha}{t} \in \mathcal{R}, \quad \frac{\beta}{t} \in \mathcal{R}.$$

We now show how to factor the left-hand side of equation (1.32) assuming that (1.33) holds. Let $y : \mathbb{N}_a \to \mathbb{C}$. Then, for $t \in \mathbb{N}_a$,

$$\begin{aligned}
&t\sigma(t)\Delta^2 y(t) + ct\Delta y(t) + dy(t) \\
&= t\sigma(t)\Delta^2 y(t) + (1 - \alpha - \beta)t\Delta y(t) + \alpha\beta\, y(t) \\
&= \{[t\sigma(t)\Delta^2 y(t) + t\Delta y(t)] - \beta t\Delta y(t)\} - \alpha[t\Delta y(t) - \beta y(t)] \\
&= t\Delta[t\Delta y(t) - \beta y(t)] - \alpha[t\Delta y(t) - \beta y(t)] \\
&= (t\Delta - \alpha I)\,(t\Delta - \beta I)\, y(t),
\end{aligned}$$

where $I$ is the identity operator on the space of functions defined on $\mathbb{N}_a$. We call the difference equation

$$(t\Delta - \alpha I)\,(t\Delta - \beta I)\, y(t) = 0 \tag{1.36}$$

the factored form of the Euler–Cauchy equation.

We will show that to solve the Euler–Cauchy equation we just need to find the characteristic values of (1.32). Next we solve the Euler–Cauchy equation by solving the factored form of the Euler–Cauchy equation (1.36).

**Theorem 1.72 (Distinct Roots).** *Assume $\alpha \neq \beta$ are the characteristic values of the Euler–Cauchy difference equation* (1.32) *and $\frac{\alpha}{t}, \frac{\beta}{t} \in \mathcal{R}$. Then a general solution of the Euler–Cauchy equation* (1.32) *is given by*

$$y(t) = c_1 e_{\frac{\alpha}{t}}(t, a) + c_2 e_{\frac{\beta}{t}}(t, a),$$

*for $t \in \mathbb{N}_a$.*

*Proof.* Note from the factored form of the Euler–Cauchy equation, if

$$(t\Delta - \beta I)\, y(t) = 0,$$

then $y(t)$ is a solution of the Euler–Cauchy equation (1.32). Since $y_2(t) = e_{\frac{\beta}{t}}(t, a)$ is a solution of

$$\Delta y(t) = \frac{\beta}{t} y(t)$$

we have that $y_2(t) = e_{\frac{\beta}{t}}(t, a)$ is also a solution of (1.32). Now by Exercise 1.61, we can write the factored equation in the form

$$(t\Delta - \beta I)(t\Delta - \alpha I)\, y(t) = 0.$$

Hence, by the above argument $y_1(t) = e_{\frac{\alpha}{t}}(t, a)$ is a solution of (1.32). Since $\alpha \neq \beta$ the solutions $y_1(t)$, $y_2(t)$ are linearly independent solutions on $\mathbb{N}_a$. Hence a general solution of the Euler–Cauchy equation (1.32) is given by

$$y(t) = c_1 e_{\frac{\alpha}{t}}(t, a) + c_2 e_{\frac{\beta}{t}}(t, a),$$

for $t \in \mathbb{N}_a$. □

*Example 1.73 (Distinct Real Roots).* Solve the Euler–Cauchy difference equation

$$t\sigma(t)\Delta^2 y(t) - 5t\Delta y(t) + 8y(t) = 0, \quad t \in \mathbb{N}_1. \tag{1.37}$$

The characteristic equation is

$$r^2 - 6r + 8 = (r-2)(r-4) = 0$$

and so the characteristic values are $r_1 = 2$, $r_2 = 4$. It follows from Theorem 1.72 that a general solution of (1.37) is given by

$$\begin{aligned} y(t) &= c_1 e_{\frac{2}{t}}(t, 1) + c_2 e_{\frac{4}{t}}(t, 1) \\ &= c_1 \prod_{s=1}^{t-1}\left(1 + \frac{2}{s}\right) + c_2 \prod_{s=1}^{t-1}\left(1 + \frac{4}{s}\right) \\ &= c_1 \prod_{s=1}^{t-1}\left(\frac{s+2}{s}\right) + c_2 \prod_{s=1}^{t-1}\left(\frac{s+4}{s}\right) \\ &= c_1 \frac{(t+1)^{\underline{2}}}{2!} + c_2 \frac{(t+3)^{\underline{4}}}{4!} \\ &= a_1 (t+1)^{\underline{2}} + a_2 (t+3)^{\underline{4}} \end{aligned}$$

for $t \in \mathbb{N}_1$.

**Theorem 1.74 (Double Root).** *Assume $\alpha$ is a double root of the characteristic equation for the Euler–Cauchy difference equation* (1.32) *and $\frac{\alpha}{t} \in \mathcal{R}$. Then a general solution of the Euler–Cauchy difference equation* (1.32) *is given by*

$$y(t) = c_1 e_{\frac{\alpha}{t}}(t, a) + c_2 e_{\frac{\alpha}{t}}(t, a) \sum_{s=a}^{t-1} \frac{1}{s+\alpha},$$

*for $t \in \mathbb{N}_a$.*

*Proof.* In this case the factored Euler–Cauchy equation is given by

$$(t\Delta - \alpha I)(t\Delta - \alpha I)y(t) = 0.$$

It follows that $y_1(t) = e_{\frac{\alpha}{t}}(t,a)$ is a solution of this equation. To get a second solution note that by considering the factored equation we get that if $y(t)$ satisfies the difference equation

$$(t\Delta - \alpha I)y(t) = e_{\frac{\alpha}{t}}(t,a),$$

then $y$ is a solution. In particular, let $y_2$ be the solution of the IVP

$$\Delta y(t) = \frac{\alpha}{t}y(t) + \frac{1}{t}e_{\frac{\alpha}{t}}(t,a), \quad y(a) = 0.$$

By the variation of constants formula in Theorem 1.68,

$$\begin{aligned}
y_2(t) &= \int_a^t e_{\frac{\alpha}{t}}(t,\sigma(s))\frac{1}{s}e_{\frac{\alpha}{t}}(s,a)\Delta s \\
&= e_{\frac{\alpha}{t}}(t,a)\int_a^t \frac{1}{s}e_{\frac{\alpha}{t}}(a,\sigma(s))e_{\frac{\alpha}{t}}(s,a)\Delta s \\
&= e_{\frac{\alpha}{t}}(t,a)\int_a^t \frac{1}{s}e_{\ominus\frac{\alpha}{t}}(\sigma(s),a)e_{\frac{\alpha}{t}}(s,a)\Delta s \\
&= e_{\frac{\alpha}{t}}(t,a)\int_a^t \left(\frac{1}{s+\alpha}\right)e_{\ominus\frac{\alpha}{t}}(s,a)e_{\frac{\alpha}{t}}(s,a)\Delta s \\
&= e_{\frac{\alpha}{t}}(t,a)\int_a^t \left(\frac{1}{s+\alpha}\right)\Delta s \\
&= e_{\frac{\alpha}{t}}(t,a)\sum_{s=a}^{t-1}\left(\frac{1}{s+\alpha}\right).
\end{aligned}$$

Since $y_1(t)$, $y_2(t)$ are linearly independent solutions of the given Euler–Cauchy equation we have that a general solution is given by

$$y(t) = c_1 e_{\frac{\alpha}{t}}(t,a) + c_2 e_{\frac{\alpha}{t}}(t,a)\sum_{s=a}^{t-1}\frac{1}{s+\alpha},$$

for $t \in \mathbb{N}_a$. □

*Example 1.75 (Double Root).* Solve the Euler–Cauchy difference equation

$$t\sigma(t)\Delta^2 y(t) - 3t\Delta y(t) + 4y(t) = 0, \quad t \in \mathbb{N}_a, \tag{1.38}$$

where $a > 0$. The characteristic equation is given by

$$r^2 - 4r + 4 = (r-2)^2 = 0,$$

so 2 is a double root of the characteristic equation. By Theorem 1.74 we have that a general solution of (1.38) is given by

$$y(t) = c_1 e_{\frac{2}{t}}(t,a) + c_2 e_{\frac{2}{t}}(t,a) \sum_{s=a}^{t-1} \frac{1}{s+2},$$

for $t \in \mathbb{N}_a$.

**Theorem 1.76 (Complex Roots).** *Assume $r_1 = \alpha + i\beta$, $r_2 = \alpha - i\beta$ are the characteristic values of the Euler–Cauchy difference equation* (1.32) *and $\frac{\alpha}{t} \in \mathcal{R}$. Then a general solution of* (1.32) *is given by*

$$y(t) = c_1 e_{\frac{\alpha}{t}}(t,a) \cos_{\frac{\beta}{t+\alpha}}(t,a) + c_2 e_{\frac{\alpha}{t}}(t,a) \sin_{\frac{\beta}{t+\alpha}}(t,a),$$

*for $t \in \mathbb{N}_a$.*

*Proof.* By the proof of Theorem 1.72, we have that $y(t) = e_{\frac{\alpha+i\beta}{t}}(t,a)$ is a (complex-valued) solution of the Euler–Cauchy difference equation (1.32). Since $\frac{\alpha}{t} \in \mathcal{R}$ it follows that $-\alpha \notin \mathbb{N}_a$ and hence $\frac{\beta}{t+\alpha}$ is well defined on $\mathbb{N}_a$. Now consider

$$\begin{aligned}\frac{\alpha}{t} \oplus \frac{i\beta}{t+\alpha} &= \frac{\alpha}{t} + \frac{i\beta}{t+\alpha} + \frac{\alpha}{t}\frac{i\beta}{t+\alpha} \\ &= \frac{\alpha}{t} + i\frac{\beta}{t}.\end{aligned}$$

Using Euler's formula (see Theorem 1.28, (vii) or Exercise 1.27), we have that

$$\begin{aligned}y(t) &= e_{\frac{\alpha+i\beta}{t}}(t,a) \\ &= e_{\frac{\alpha}{t}+i\frac{\beta}{t}}(t,a) \\ &= e_{\frac{\alpha}{t}\oplus\frac{i\beta}{t+\alpha}}(t,a) \\ &= e_{\frac{\alpha}{t}}(t,a) e_{\frac{i\beta}{t+\alpha}}(t,a) \\ &= e_{\frac{\alpha}{t}}(t,a) \cos_{\frac{\beta}{t+\alpha}}(t,a) + i e_{\frac{\alpha}{t}}(t,a) \sin_{\frac{\beta}{t+\alpha}}(t,a).\end{aligned}$$

Since we are assuming the constants $c, d$ in the Euler–Cauchy difference equation (1.32) are real, we have that

$$e_{\frac{\alpha}{t}} \cos_{\frac{\beta}{t+\alpha}}(t,a), \quad \sin_{\frac{\beta}{t+\alpha}}(t,a)$$

are real-valued solutions of the Euler–Cauchy difference equation (1.32). Since they are linearly independent on $\mathbb{N}_a$,

$$y(t) = c_1 e_{\frac{\alpha}{t}}(t,a) \cos_{\frac{\beta}{t+\alpha}}(t,a) + c_2 e_{\frac{\alpha}{t}}(t,a) \sin_{\frac{\beta}{t+\alpha}}(t,a)$$

is a general solution of the Euler–Cauchy difference equation (1.32). □

*Example 1.77 (Complex Roots).* Solve the Euler–Cauchy difference equation

$$t\sigma(t)\Delta^2 y(t) - 5t\Delta y(t) + 13y(t) = 0, \quad t \in \mathbb{N}_a.$$

The characteristic equation is

$$r^2 - 6r + 13 = 0,$$

and hence the characteristic values are $r = 3 \pm 2i$. From Theorem 1.76, we get that

$$y(t) = c_1 e_{\frac{3}{t}}(t,a) \cos_{\frac{2}{t+3}}(t,a) + c_2 e_{\frac{3}{t}}(t,a) \sin_{\frac{2}{t+3}}(t,a)$$

is a general solution.

For $n \geq 3$ M. Bohner and E. Akin (see [63, Chapter 2]) defined the Euler–Cauchy difference equation of arbitrary order by

$$(t\Delta - \alpha_1)(t\Delta - \alpha_2)(t\Delta - \alpha_n) = 0, \tag{1.39}$$

where we assume $\frac{\alpha_i}{t} \in \mathcal{R}$, $1 \leq i \leq n$. For the reader interested in Euler–Cauchy difference equations of order $n \geq 3$ see Exercise 1.67.

## 1.9 Vector Difference Equations

In this section, we will examine the properties of the linear vector difference equation with variable coefficients

$$\Delta y(t) = A(t)y(t) + f(t), \quad t \in \mathbb{N}_a, \tag{1.40}$$

and the corresponding homogeneous system

$$\Delta u(t) = A(t)u(t), \quad t \in \mathbb{N}_a, \tag{1.41}$$

where $f : \mathbb{N}_a \to \mathbb{R}_n$ and the real $n \times 1$ matrix function $A(t)$ will be assumed to be a regressive matrix function on $\mathbb{N}_a$ (that is $I + A(t)$ is nonsingular for all $t \in \mathbb{N}_a$). With these assumptions, it is easy to show that for any $t_0 \in \mathbb{N}_a$ the initial value problem

$$\Delta x(t) = A(t)x(t) + f(t)$$

$$x(t_0) = x_0,$$

where $x_0 \in \mathbb{R}^n$ is a given $n \times 1$ constant vector, has a unique solution on $\mathbb{N}_a$. To solve the nonhomogeneous difference equation (1.40) we will see that we first want to be able to solve the corresponding homogeneous difference equation (1.41). The matrix equation analogue of the homogeneous vector difference equation (1.41) is

$$\Delta U(t) = A(t)U(t), \quad t \in \mathbb{N}_a \tag{1.42}$$

where $U(t)$ is an $n \times n$ matrix function. Note that $U(t)$ is a solution of (1.42) if and only if each of its column vectors is a solution of (1.41). From the uniqueness of solutions to IVPs for the vector equation (1.40) we have that the matrix IVP

$$\Delta U(t) = A(t)U(t), \quad U(t_0) = U_0,$$

where $t_0 \in \mathbb{N}_a$ and $U_0$ is a given $n \times n$ constant matrix has a unique solution on $\mathbb{N}_a$.

**Theorem 1.78.** *Assume $A(t)$ is a regressive matrix function on $\mathbb{N}_a$. If $\Phi(t)$ is a solution of* (1.42)*, then either det $\Phi(t) \neq 0$ for all $t \in \mathbb{N}_a$ or det $\Phi(t) = 0$ for all $t \in \mathbb{N}_a$.*

*Proof.* Since $\Phi(t)$ is a solution of (1.42) on $\mathbb{N}_a$,

$$\Phi(t+1) = [I + A(t)]\Phi(t), \quad t \in \mathbb{N}_a.$$

Therefore,

$$\det \Phi(t+1) = \det\,[I + A(t)] \det \Phi(t), \tag{1.43}$$

for all $t \in \mathbb{N}_a$. Now either det $\Phi(a) \neq 0$ or det $\Phi(a) = 0$. Since det $[I + A(t)] \neq 0$ for all $t \in \mathbb{N}_a$, we have by (1.43) that if det $\Phi(a) \neq 0$, then det $\Phi(t) \neq 0$ for all $t \in \mathbb{N}_a$, while if det $\Phi(a) = 0$, then det $\Phi(t) = 0$ for all $t \in \mathbb{N}_a$. □

**Definition 1.79.** We say that $\Phi(t)$ is a **fundamental matrix** of the vector difference equation (1.41) provided $\Phi(t)$ is a solution of the matrix equation (1.42) and det $\Phi(t) \neq 0$ for $t \in \mathbb{N}_a$.

**Definition 1.80.** If $A$ is a regressive matrix function on $\mathbb{N}_a$, then we define the matrix exponential function, $e_A(t, t_0)$, based at $t_0 \in \mathbb{N}_a$ to be the unique solution of the matrix IVP

$$\Delta U(t) = A(t)U(t), \quad U(t_0) = I.$$

From Exercise 1.71 we have that $\Phi(t)$ is a fundamental matrix of $\Delta u(t) = A(t)u(t)$ if and only if its columns are $n$ linearly independent solutions of the vector equation $\Delta u(t) = A(t)u(t)$ on $\mathbb{N}_a$. To find a formula for the matrix exponential function, $e_A(t, t_0)$, we want to solve the IVP

$$U(t+1) = [I + A(t)]U(t), \quad t \in \mathbb{N}_a, \quad U(t_0) = I.$$

Iterating this equation we get

$$e_A(t, t_0) = \begin{cases} {}^*\prod_{s=t_0}^{t-1} [I + A(s)], & t \in \mathbb{N}_{t_0} \\ \prod_{s=t}^{t_0-1} [I + A(s)]^{-1}, & t \in \mathbb{N}_a^{t_0-1}, \end{cases}$$

where it is understood that $\prod_{s=t_0}^{t_0-1} [I + A(s)] = e_A(t_0, t_0) = I$ and for $t \in \mathbb{N}_{t_0+1}$

$${}^*\prod_{s=t_0}^{t-1} [I + A(s)] := [I + A(t-1)][I + A(t-2)] \cdots [I + A(t_0)].$$

*Example 1.81.* If $A$ is an $n \times n$ constant matrix and $I + A$ is invertible, then

$$e_A(t, t_0) = (I + A)^{t-t_0}, \quad t \in \mathbb{N}_a.$$

Similar to the proof of Theorem 1.16 one can prove (Exercise 1.73) the following theorem.

**Theorem 1.82.** *The set of all $n \times n$ regressive matrix functions on $\mathbb{N}_a$ with the addition, $\oplus$ defined by*

$$(A \oplus B)(t) := A(t) + B(t) + A(t)B(t), \quad t \in \mathbb{N}_a$$

*is a group. Furthermore, the additive inverse of a regressive matrix function $A$ defined on $\mathbb{N}_a$ is given by*

$$(\ominus A)(t) := -[I + A(t)]^{-1}A(t), \quad t \in \mathbb{N}_a.$$

In the next theorem we give several properties of the matrix exponential. To prove part (vii) of the this theorem we will use the following lemma.

**Lemma 1.83.** *Assume $Y(t)$ and $Y(\sigma(t))$ are invertible matrices. Then*

$$\Delta Y^{-1}(t) = -Y^{-1}(\sigma(t))\Delta Y(t)Y^{-1}(t) = -Y^{-1}(t)\Delta Y(t)Y^{-1}(\sigma(t)).$$

*Proof.* Taking the difference of both sides of $Y(t)Y^{-1}(t) = I$ we get that

$$Y(\sigma(t))\Delta Y^{-1}(t) + \Delta Y(t)Y^{-1}(t) = 0.$$

Solving this last equation for $\Delta Y^{-1}(t)$ we get that

$$\Delta Y^{-1}(t) = -Y^{-1}(\sigma(t))\Delta Y(t)Y^{-1}(t).$$

Similarly, one can use $Y^{-1}(t)Y(t) = I$ to get that

$$\Delta Y^{-1}(t) = -Y^{-1}(t)\Delta Y(t)Y^{-1}(\sigma(t)).$$

□

**Theorem 1.84.** *Assume $A$ and $B$ are regressive matrix functions on $\mathbb{N}_a$ and $s, r \in \mathbb{N}_a$. Then the following hold:*

(i) $\Delta e_A(t,s) = A(t)e_A(t,s)$;
(ii) $e_A(s,s) = I$;
(iii) $\det e_A(t,s) \neq 0$ *for* $t \in \mathbb{N}_a$;
(iv) $e_A(t,s)$ *is a fundamental matrix of* (1.41)*;*
(v) $e_A(\sigma(t),s) = [I + A(t)]e_A(t,s)$;
(vi) *(semigroup property)* $e_A(t,r)e_A(r,s) = e_A(t,s)$ *holds for* $t, r, s \in \mathbb{N}_a$;
(vii) $e_A^{-1}(t,s) = e^*_{\ominus A^*}(t,s)$;
(viii) $e_A(t,s) = e_A^{-1}(s,t) = e^*_{\ominus A^*}(s,t)$, *where $A^*$ denotes the conjugate transpose of the matrix $A$*;
(ix) $B(t)e_A(t,t_0) = e_A(t,t_0)B(t)$, *if $A(t)$ and $B(\tau)$ commute for all $t, \tau \in \mathbb{N}_a$*;
(x) $e_A(t,s)e_B(t,s) = e_{A\oplus B}(t,s)$, *if $A(t)$ and $B(\tau)$ commute for all $t, \tau \in \mathbb{N}_a$.*

*Proof.* Note that (i) and (ii) follow from the definition of the matrix exponential. Part (iii) follows from Theorem 1.78 and part (ii). Parts (i) and (iii) imply part (iv) holds. Since $\Phi(\sigma(t)) = \Phi(t) + \Delta\Phi(t)$, we have that

$$e_A(\sigma(t),s) = e_A(t,s) + \Delta e_A(t,s) = [I + A(t)]e_A(t,s)$$

and hence (v) holds. To see that the semigroup property (vi) holds, fix $r, s \in \mathbb{N}_a$ and set $\Phi(t) = e_A(t,r)e_A(r,s)$. Then

$$\begin{aligned}\Delta\Phi(t) &= \Delta e_A(t,r)e_A(r,s) \\ &= A(t)e_A(t,r)e_A(r,s) \\ &= A(t)\Phi(t).\end{aligned}$$

Next we show that $\Phi(s) = e_A(s,r)e_A(r,s)) = I$. First note that if $r = s$, then $\Phi(s) = \Phi(s,s)\Phi(s,s) = I$. Hence we can assume that $s \neq r$. For the case $r > s \geq a$, we have that

$$\begin{aligned}\Phi(s) = e_A(s,r)e_A(r,s) &= \left(\prod_{\tau=s}^{r-1}[I + A(\tau)]^{-1}\right)\left({}^*\prod_{\tau=s}^{r-1}[I + A(\tau)]\right) \\ &= [I + A(s)]^{-1}\cdots[I + A(r-1)]^{-1}[I + A(r-1)]\cdots[I + A(s)] = I.\end{aligned}$$

Similarly, for the case $s > r \geq a$ one can show that $\Phi(s) = I$. Hence, by the uniqueness of solutions for IVPs we get that $e_A(t,r)e_A(r,s) = e_A(t,s)$. To see that (vii) holds, fix $s \in \mathbb{N}_a$ and let

$$Y(t) := \left[e_A^{-1}(t,s)\right]^*, \quad t \in \mathbb{N}_a.$$

Then

$$\begin{aligned}
\Delta Y(t) &= \left[\Delta e_A^{-1}(t,s)\right]^* \\
&= -\left[e_A^{-1}(\sigma(t),s)\Delta e_A(t,s)e_A^{-1}(t,s)\right]^* \\
&= -\left[e_A^{-1}(\sigma(t),s)A(t)\right]^* \\
&= -\left[([I+A(t)]e_A(t,s))^{-1}A(t)\right]^* \\
&= -\left[e_A^{-1}(t,s)[I+A(t)]^{-1}A(t)\right]^* \\
&= -A^*(t)[I+A^*(t)]^{-1}\left[e_A^{-1}(t,s)\right]^* \\
&= (\ominus A^*)(t)Y(t).
\end{aligned}$$

Since $\left[e_A^{-1}(t,s)\right]^*$ and $e_{\ominus A^*}(t,s)$ satisfy the same matrix IVP we get

$$\left(e_A^{-1}(t,s)\right)^* = e_{\ominus A^*}(t,s).$$

Taking the conjugate transpose of both sides of this last equation we get that part (vii) holds. The proof of (viii) is Exercise 1.78 and the proof of (ix) is Exercise 1.79.

To see that (x) holds, let $\Phi(t) = e_A(t,s)e_B(t,s)$, $t \in \mathbb{N}_a$. Then by the product rule

$$\begin{aligned}
\Delta\Phi(t) &= \Delta\left[e_A(t,s)e_B(t,s)\right] \\
&= e_A(\sigma(t),s)\Delta e_B(t,s) + \Delta e_A(t,s)e_B(t,s) \\
&= [1+A(t)]e_A(t,s)B(t)e_B(t,s) + A(t)e_A(t,s)e_B(t,s) \\
&= [A(t)+B(t)+A(t)B(t)]e_A(t,s)e_B(t,s) \\
&= [(A\oplus B)(t)]\Phi(t)
\end{aligned}$$

for $t \in \mathbb{N}_a$. Since $\Phi(s) = I$, we have that $\Phi(t)$ and $e_{A\oplus B}(t,s)$ satisfy the same matrix IVP. Hence, by the uniqueness theorem for solutions of matrix IVPs we get the desired result $e_A(t,s)e_B(t,s) = e_{A\oplus B}(t,s)$. □

Now for any nonsingular matrix $U_0$, the solution $U(t)$ of (1.42) with $U(t_0) = U_0$ is a fundamental matrix of (1.41), so there are always infinitely many fundamental matrices of (1.41). In particular, if $A$ is a regressive matrix function on $\mathbb{N}_a$, then $\Phi(t) = e_A(t,t_0)$ is a fundamental matrix of the vector equation $\Delta u(t) = A(t)u(t)$.

The following theorem characterizes fundamental matrices for (1.41).

**Theorem 1.85.** *If* $\Phi(t)$ *is a fundamental matrix for* (1.41)*, then* $\Psi(t)$ *is another fundamental matrix if and only if there is a nonsingular constant matrix* $C$ *such that*

$$\Psi(t) = \Phi(t)C,$$

*for* $t \in \mathbb{N}_a$.

*Proof.* Let $\Psi(t) = \Phi(t)C$, where $\Phi(t)$ is a fundamental matrix of (1.41) and $C$ is nonsingular constant matrix. Then $\Psi(t)$ is nonsingular for all $t \in \mathbb{N}_a$, and

$$\begin{aligned}\Delta\Psi(t) &= \Delta\Phi(t)C \\ &= A(t)\Phi(t)C \\ &= A(t)\Psi(t).\end{aligned}$$

Therefore $\Psi(t)$ is a fundamental matrix of (1.41).

Conversely, assume $\Phi(t)$ and $\Psi(t)$ are fundamental matrices of (1.41). For some $t_0 \in \mathbb{N}_a$, let

$$C := \Phi^{-1}(t_0)\Psi(t_0).$$

Then $\Psi(t)$ and $\Phi(t)C$ are both solutions of (1.42) satisfying the same initial condition at $t_0$. By uniqueness,

$$\Psi(t) = \Phi(t)C,$$

for all $t \in \mathbb{N}_a$. □

The proof of the following theorem is similar to that of Theorem 1.85 and is left as an exercise (Exercise 1.68).

**Theorem 1.86.** *If* $\Phi(t)$ *is a fundamental matrix of* (1.41)*, then the general solution of* (1.41) *is given by*

$$u(t) = \Phi(t)c,$$

*where* $c$ *is an arbitrary constant column vector.*

Hence we see to solve the vector equation (1.41) we just need to find the fundamental matrix $\Phi(t) = e_A(t, t_0)$. We will set off to prove the Putzer algorithm (Theorem 1.88) which will give us a nice formula $e_A(t, 0)$, for $t \in \mathbb{N}_0$ when $A$ is a constant $n \times n$ matrix. In the proof of this theorem we will use the Cayley–Hamilton Theorem which states that every square constant matrix satisfies its own characteristic equation. We now give an example to illustrate this important theorem.

*Example 1.87.* Show directly that the matrix

$$A = \begin{bmatrix} 2 & -1 \\ 3 & -4 \end{bmatrix}$$

satisfies its own characteristic equation. The characteristic equation of $A$ is

$$\lambda^2 + 2\lambda - 5 = 0.$$

Then

$$\begin{aligned} A^2 + 2A - 5I &= \begin{bmatrix} 1 & 2 \\ -6 & 13 \end{bmatrix} + \begin{bmatrix} 4 & -2 \\ 6 & -8 \end{bmatrix} - \begin{bmatrix} 5 & 0 \\ 0 & 5 \end{bmatrix} \\ &= \begin{bmatrix} 0 & 0 \\ 0 & 0 \end{bmatrix} \end{aligned}$$

and so $A$ does satisfy its own characteristic equation.

The proof of the next theorem is motivated by the fact that by the Cayley–Hamilton Theorem an $n$ by $n$ constant matrix $A$ can be written as a linear combination of the matrices $I$, $A$, $A^2$, …, $A^{n-1}$ and therefore every nonnegative integer power $A^t$ of $A$ can also be written as a linear combination of $I$, $A$, $A^2$, …, $A^{n-1}$.

**Theorem 1.88 (Putzer's Algorithm).** *Let $\lambda_1, \lambda_2, \dots, \lambda_n$ be the (not necessarily distinct) eigenvalues of the constant n by n matrix A, with each eigenvalue repeated as many times as its multiplicity. Define the matrices $M_k$, $0 \le k \le n$, recursively by*

$$\begin{aligned} M_0 &= I \\ M_k &= (A - \lambda_k I)M_{k-1}, \quad 1 \le k \le n. \end{aligned}$$

*Then*

$$A^t = \sum_{k=0}^{n-1} p_{k+1}(t) M_k, \quad t \in \mathbb{N}_0,$$

*where the $p_k(t)$, $1 \le k \le n$ are chosen so that*

$$\begin{bmatrix} p_1(t+1) \\ p_2(t+1) \\ \vdots \\ p_n(t+1) \end{bmatrix} = \begin{bmatrix} \lambda_1 & 0 & 0 & \cdots & 0 \\ 1 & \lambda_2 & 0 & \cdots & 0 \\ 0 & 1 & \lambda_3 & \cdots & 0 \\ \vdots & \ddots & \ddots & \ddots & \vdots \\ 0 & \cdots & 0 & 1 & \lambda_n \end{bmatrix} \begin{bmatrix} p_1(t) \\ p_2(t) \\ \vdots \\ p_n(t) \end{bmatrix} \tag{1.44}$$

*and*

$$\begin{bmatrix} p_1(0) \\ p_2(0) \\ \vdots \\ p_n(0) \end{bmatrix} = \begin{bmatrix} 1 \\ 0 \\ \vdots \\ 0 \end{bmatrix}. \tag{1.45}$$

*Proof.* Let the matrices $M_k$, $0 \leq k \leq n$ be defined as in the statement of this theorem. Since for each fixed $t \geq 0$, $A^t$ is a linear combination of $I, A, A^2, \ldots, A^{n-1}$, we also have that for each fixed $t$, $A^t$ is a linear combination of $M_0, M_1, M_2, \ldots, M_{n-1}$, that is

$$A^t = \sum_{i=0}^{n-1} p_{k+1}(t) M_k$$

for $t \geq 0$. It remains to show that the $p_k$'s are as in the statement of this theorem. Since $A^{t+1} = A \cdot A^t$, we have that

$$\begin{aligned}
\sum_{k=0}^{n-1} p_{k+1}(t+1) M_k &= A \sum_{k=0}^{n-1} p_{k+1}(t) M_k \\
&= \sum_{k=0}^{n-1} p_{k+1}(t) \left[ A M_k \right] \\
&= \sum_{k=0}^{n-1} p_{k+1}(t) \left[ M_{k+1} + \lambda_{k+1} M_k \right] \\
&= \sum_{k=0}^{n-1} p_{k+1}(t) M_{k+1} + \sum_{k=0}^{n-1} \lambda_{k+1} p_{k+1}(t) M_k \\
&= \sum_{k=1}^{n-1} p_k(t) M_k + \sum_{k=0}^{n-1} p_{k+1}(t) \lambda_{k+1} M_k, \\
&= \lambda_1 p_1(t) M_0 + \sum_{k=1}^{n-1} \left[ p_k(t) + \lambda_{k+1} p_{k+1}(t) \right] M_k, \qquad (1.46)
\end{aligned}$$

where in the second to the last step we have replaced $k$ by $k-1$ in the first sum and used the fact that (by the Cayley–Hamilton Theorem) $M_n = 0$. Note that equation (1.46) is satisfied if $p_k(t)$, $k = 1, 2, \ldots, n$, are chosen to satisfy the system (1.44). Since $A^0 = I = p_1(0)I + \cdots + p_n(0)M_{n-1}$, we must have (1.45) is satisfied. □

The following example shows how we can use the Putzer algorithm to find the exponential function $e_A(t,0)$ when $A$ is a constant matrix. This method is called finding the matrix exponential $e_A(t,0)$ using the Putzer algorithm.

*Example 1.89.* Use the Putzer algorithm (Theorem 1.88) to find $e_A(t,0)$, $t \in \mathbb{N}_0$, where

$$A := \begin{bmatrix} 1 & 2 \\ 1 & 2 \end{bmatrix}.$$

Note $e_A(t,0) = \prod_{\tau=0}^{t-1} [I + A] = (I + A)^t$. So to find $e_A(t,0)$ we just need to find $B^t$ where

$$B := I + A = \begin{bmatrix} 2 & 2 \\ 1 & 3 \end{bmatrix}.$$

We now apply Putzer's algorithm (Theorem 1.88) to find $B^t$. The characteristic equation for $B$ is given by $\lambda^2 - 5\lambda + 4 = (\lambda - 1)(\lambda - 4) = 0$. Hence the eigenvalues of $B$ are given by $\lambda_1 = 1$, $\lambda_2 = 4$. It follows that $M_0 = I$ and

$$M_1 = B - \lambda_1 I = \begin{bmatrix} 1 & 2 \\ 1 & 2 \end{bmatrix}.$$

To find $p_1(t)$ we now solve the IVP

$$p_1(t+1) = \lambda_1 p(t) = p_1(t), \quad p_1(0) = 1.$$

It follows that $p_1(t) = 1$. Next to find $p_2(t)$ we solve the IVP

$$p_2(t+1) = p_1(t) + \lambda_2 p_2(t) = 1 + 4p_2(t), \quad p_2(0) = 0.$$

This gives us the IVP

$$\Delta p_2(t) = 3p_2(t) + 1, \quad p_2(0) = 0.$$

Using the variation of constants formula in Theorem 1.68 we get

$$\begin{aligned} p_2(t) &= \int_0^t e_3(t, \sigma(s))\Delta s \\ &= \int_0^t e_{\ominus 3}(\sigma(s), t)\Delta s \\ &= \int_0^t [1 + \ominus 3] e_{\ominus 3}(s, t)\Delta s \end{aligned}$$

$$
\begin{aligned}
&= \frac{1}{4}\int_0^t e_{\ominus 3}(s,t)\Delta s \\
&= -\frac{1}{3}e_{\ominus 3}(s,t)|_{s=0}^{s=t} \\
&= -\frac{1}{3}e_{\ominus 3}(t,t) + \frac{1}{3}e_{\ominus 3}(0,t) \\
&= -\frac{1}{3} + \frac{1}{3}e_3(t,0) \\
&= -\frac{1}{3} + \frac{1}{3}4^t.
\end{aligned}
$$

It follows that

$$
\begin{aligned}
e_A(t,0) &= p_1(t)M_0 + p_2(t)M_1 \\
&= \begin{bmatrix} 1 & 0 \\ 0 & 1 \end{bmatrix} + \left(-\frac{1}{3} + \frac{1}{3}4^t\right)\begin{bmatrix} 1 & 2 \\ 1 & 2 \end{bmatrix} \\
&= \frac{1}{3}\begin{bmatrix} 2+4^t & -2+2\cdot 4^t \\ -1+4^t & 1+2\cdot 4^t \end{bmatrix}.
\end{aligned}
$$

It follows from this that

$$
y(t) = c_1\begin{bmatrix} 2+4^t \\ -1+4^t \end{bmatrix} + c_2\begin{bmatrix} -2+2\cdot 4^t \\ 1+2\cdot 4^t \end{bmatrix}
$$

is a general solution of

$$
\Delta y(t) = \begin{bmatrix} 1 & 2 \\ 1 & 2 \end{bmatrix} y(t), \quad t \in \mathbb{N}_0.
$$

*Example 1.90.* Use Putzer's algorithm for finding the matrix exponential $e_A(t,0)$ to solve the vector equation

$$
\Delta u(t) = Au(t), \quad t \in \mathbb{N}_0,
$$

where $A$ is the regressive matrix given by

$$
A = \begin{bmatrix} 1 & 1 \\ -1 & 3 \end{bmatrix}.
$$

Let $B := I + A$, then $e_A(t,0) = [I+A]^t = B^t$, where

$$
B = \begin{bmatrix} 2 & 1 \\ -1 & 4 \end{bmatrix}.
$$

The characteristic equation of the constant matrix $B$ is given by

$$\lambda^2 - 6\lambda + 9 = 0$$

and so the characteristic values are $\lambda_1 = \lambda_2 = 3$. It follows that

$$M_0 = I = \begin{bmatrix} 1 & 0 \\ 0 & 1 \end{bmatrix} \quad \text{and} \quad M_1 = (B - 3I)\, M_0 = \begin{bmatrix} -1 & 1 \\ -1 & 1 \end{bmatrix}.$$

Next we solve the IVP

$$p(t+1) = \begin{bmatrix} 3 & 0 \\ 1 & 3 \end{bmatrix} p(t), \quad p(0) = \begin{bmatrix} 1 \\ 0 \end{bmatrix}.$$

Hence $p_1(t)$ solves the IVP

$$p_1(t+1) = 3p_1(t), \quad p_1(0) = 1.$$

Since $\Delta p_1(t) = 2p_1(t)$, $p_1(0) = 1$, we have that $p_1(t) = e_2(t, 0) = 3^t$. Also $p_2(t)$ solves the IVP

$$p_2(t+1) = p_1(t) + 3p_2(t), \quad p_2(0) = 0.$$

It follows that $p_2(t)$ solves the IVP

$$\Delta p_2(t) = 2p_2(t) + e_2(t, 0), \quad p_2(0) = 0.$$

Using the variation of constants formula in Theorem 1.68 we get that

$$\begin{aligned}
p_2(t) &= \int_0^t e_2(t, \sigma(\tau)) e_2(\tau, 0) \Delta\tau \\
&= \int_0^t e_{\ominus 2}(\sigma(\tau), t) e_2(\tau, 0) \Delta\tau \\
&= \frac{1}{3} \int_0^t e_{\ominus 2}(\tau, t) e_2(\tau, 0) \Delta\tau \\
&= \frac{1}{3} \int_0^t e_{\ominus 2}(\tau, t) e_2(\tau, 0) \Delta\tau \\
&= \frac{1}{3} \int_0^t e_2(t, \tau) e_2(\tau, 0) \Delta\tau \\
&= \frac{1}{3} e_2(t, 0) \int_0^t 1 \Delta\tau
\end{aligned}$$

$$= \frac{1}{3} te_2(t,0)$$

$$= \frac{1}{3} t3^t$$

Hence by Putzer's algorithm

$$\begin{aligned}
e_A(t,0) &= B^t \\
&= p_1(t)M_0 + p_2(t)M_1 \\
&= 3^t \begin{bmatrix} 1 & 0 \\ 0 & 1 \end{bmatrix} + \frac{1}{3} t3^t \begin{bmatrix} -1 & 1 \\ -1 & 1 \end{bmatrix} \\
&= 3^t \begin{bmatrix} 1 - \frac{1}{3}t & \frac{1}{3}t \\ -\frac{1}{3}t & 1 + \frac{1}{3}t \end{bmatrix}.
\end{aligned}$$

Since $e_A(t,0)$ is a fundamental matrix of $\Delta u(t) = Au(t)$, we have by Theorem 1.86 that

$$\begin{aligned}
u(t) &= 3^t \begin{bmatrix} 1 - \frac{1}{3}t & \frac{1}{3}t \\ -\frac{1}{3}t & 1 + \frac{1}{3}t \end{bmatrix} c \\
&= c_1 3^t \begin{bmatrix} 1 - \frac{1}{3}t \\ -\frac{1}{3}t \end{bmatrix} + c_2 3^t \begin{bmatrix} \frac{1}{3}t \\ 1 + \frac{1}{3}t \end{bmatrix}
\end{aligned}$$

is a general solution.

Fundamental matrices can be used to solve the nonhomogeneous equation (1.40).

**Theorem 1.91 (Variation of Parameters (Constants)).** *Assume* $\Phi(t)$ *is a fundamental matrix of* (1.41). *Then the unique solution of* (1.40) *that satisfies the initial condition* $y(a) = y_0$ *is given by the variation of parameters formula*

$$y(t) = \Phi(t)\Phi^{-1}(a)y_0 + \Phi(t) \int_a^t \Phi^{-1}(s+1)f(s)\Delta s, \tag{1.47}$$

*for* $t \in \mathbb{N}_a$.

*Proof.* Let $y(t)$ be given by (1.47) for $t \in \mathbb{N}_a$. Using the vector version of the Leibniz formula (1.29), we have

$$\begin{aligned}
\Delta y(t) = \; & \Delta\Phi(t)\Phi^{-1}(a)y_0 + \Delta\Phi(t) \int_a^t \Phi^{-1}(s+1)f(s)\Delta s \\
& + \Phi(t+1)\Phi^{-1}(t+1)f(t)
\end{aligned}$$

$$= A(t)\Phi(t)\Phi^{-1}(a)y_0 + A(t)\Phi(t)\int_a^t \Phi^{-1}(s+1)f(s)\Delta s + f(t)$$

$$= A(t)\left[\Phi(t)\Phi^{-1}(a)y_0 + \Phi(t)\int_a^t \Phi^{-1}(s+1)f(s)\Delta s\right] + f(t)$$

$$= A(t)y(t) + f(t).$$

Consequently, $y(t)$ defined by (1.47) is a solution of the nonhomogeneous equation, and also we have that $y(a) = y_0$. □

A special case of the above theorem is the following result.

**Theorem 1.92.** *Assume $A(t)$ is a regressive matrix function on $\mathbb{N}_a$ and assume $f : \mathbb{N}_a \to \mathbb{R}^n$. Then the unique solution of the IVP*

$$\Delta y(t) = A(t)y(t) + f(t), \quad t \in \mathbb{N}_a$$

$$y(a) = y_0$$

*is given by the variation of constants formula*

$$y(t) = e_A(t,a)y_0 + \int_a^t e_A(t,\sigma(s))f(s)\Delta s,$$

*for $t \in \mathbb{N}_a$.*

*Proof.* Since $\Phi(t) = e_A(t,a)$ is a fundamental matrix of $\Delta y(t) = A(t)y(t)$, we have by Theorem 1.91, that the solution of our IVP in the statement of this theorem is given by

$$y(t) = e_A(t,a)e_A^{-1}(a,a)y_0 + e_A(t,a)\int_a^t e_A^{-1}(\sigma(s),a)f(s)\Delta s$$

$$= e_A(t,a)y_0 + \int_a^t e_A(t,a)e_A(a,\sigma(s))f(s)\Delta s$$

$$= e_A(t,a)y_0 + \int_a^t e_A(t,\sigma(s))f(s)\Delta s,$$

where in the last two steps we used properties (viii) and (vi) in Theorem 1.84. □

*Example 1.93.* Solve the system

$$u(t+1) = \begin{bmatrix} 0 & 1 \\ -2 & -3 \end{bmatrix} u(t) + \left(\frac{2}{3}\right)^t \begin{bmatrix} 1 \\ -2 \end{bmatrix}, \quad t \in \mathbb{N}_0,$$

$$u(0) = \begin{bmatrix} 1 \\ 1 \end{bmatrix}.$$

From Exercise 1.72, we can choose

$$\Phi(t) = \begin{bmatrix} (-2)^t & (-1)^t \\ (-2)^{t+1} & (-1)^{t+1} \end{bmatrix}$$
$$= (-1)^t \begin{bmatrix} 2^t & 1 \\ -2^{t+1} & -1 \end{bmatrix}.$$

Then

$$\Phi^{-1}(t) = \frac{(-1)^t}{2^t} \begin{bmatrix} -1 & -1 \\ 2^{t+1} & 2^t \end{bmatrix}.$$

From (1.47), we have for $t \geq 0$,

$$u(t) = (-1)^t \begin{bmatrix} 2^t & 1 \\ -2^{t+1} & -1 \end{bmatrix} \left( \begin{bmatrix} -2 \\ 3 \end{bmatrix} + \sum_{s=0}^{t-1} \begin{bmatrix} -.5(-3)^{-s} \\ 0 \end{bmatrix} \right)$$
$$= (-1)^t \begin{bmatrix} 2^t & 1 \\ -2^{t+1} & -1 \end{bmatrix} \left( \begin{bmatrix} -2 \\ 3 \end{bmatrix} + \begin{bmatrix} .375((-3)^{-t} - 1) \\ 0 \end{bmatrix} \right)$$
$$= (-1)^t \begin{bmatrix} 2^t & 1 \\ -2^{t+1} & -1 \end{bmatrix} \begin{bmatrix} -.125((-3)^{1-t} + 19) \\ 3 \end{bmatrix}.$$

## 1.10 Stability of Linear Systems

At the outset of this section we will be concerned with the stability of the trivial solution of the vector difference equation

$$y(t+1) = Ay(t), \quad t \in \mathbb{N}_a, \tag{1.48}$$

where $A$ is an $n \times n$ constant matrix. By the trivial solution of (1.48) we mean the solution $y(t) \equiv 0$, $t \in \mathbb{N}_a$ (here by context we know 0 denotes the zero vector). First we define what we mean by the stability of the trivial solution on $\mathbb{N}_a$. We will adopt the notation that $y(t, z)$ denotes the unique solution of the IVP

$$y(t+1) = Ay(t), \quad y(a) = z, \quad z \in \mathbb{R}^n.$$

**Definition 1.94.** Let $\|\cdot\|$ be a norm on $\mathbb{R}^n$. We say the trivial solution of (1.48) is **stable** on $\mathbb{N}_a$ provided given any $\epsilon > 0$, there is a $\delta > 0$ such that $\|y(t, z)\| < \epsilon$ on $\mathbb{N}_a$ if $\|z\| < \delta$. If this is not the case we say the trivial solution of (1.48) is **unstable** on $\mathbb{N}_a$. If the trivial solution is stable on $\mathbb{N}_a$ and $\lim_{t \to \infty} y(t) = 0$ for every solution $y$ of (1.48), then we say the trivial solution of (1.48) is **globally asymptotically stable** on $\mathbb{N}_a$.

We will use the following remark in the proof of the next theorem.

*Remark 1.95.* An important result [137, Theorem 2.54] in analysis gives that for any $n \times n$ constant matrix $M$, there is a constant $D > 0$, depending on $M$ and the norm $\|\cdot\|$ on $\mathbb{R}^n$, so that

$$\|Mz\| \le D\|z\|$$

for all $z \in \mathbb{R}^n$.

**Theorem 1.96.** *If the eigenvalues of A satisfy* $|\lambda_k| < 1$, $1 \le k \le n$, *then the trivial solution of* (1.48) *is globally asymptotically stable on* $\mathbb{N}_a$.

*Proof.* We will just prove this theorem for the case $a = 0$. Let $r := \max\{|\lambda_k| : 1 \le k \le n\}$ and fix $\delta$ so that $0 \le r < \delta < 1$. From the Putzer algorithm (Theorem 1.88), the solution $y(t, z)$ of (1.48) satisfying $y(0, z) = z$ is given by

$$y(t, z) = A^t z = \sum_{k=0}^{n-1} p_{k+1}(t) M_k z, \quad t \in \mathbb{N}_0. \tag{1.49}$$

We now show that for each $1 \le k \le n$ there is a constant $B_k > 0$ such that

$$|p_k(t)| \le B_k \delta^t, \quad t \in \mathbb{N}_0. \tag{1.50}$$

By (1.44),

$$|p_1(t+1)| \le r|p_1(t)|, \quad t \in \mathbb{N}_0.$$

Iterating this inequality and using $p_1(0) = 1$, we have

$$|p_1(t)| \le r^t, \quad t \in \mathbb{N}_0.$$

Hence if we let $B_1 = 1$ and use the fact that $r < \delta$ we have that

$$|p_1(t)| \le B_1 \delta^t, \quad t \in \mathbb{N}_0.$$

Hence (1.50) holds for $k = 1$. We next show that there is a constant $B_2 > 0$ such that

$$|p_2(t)| \le B_2 \delta^t, \quad t \in \mathbb{N}_0.$$

From (1.44) we get

$$\begin{aligned} |p_2(t+1)| &\le r|p_2(t)| + |p_1(t)| \\ &\le r|p_2(t)| + r^t. \end{aligned}$$

It follows from iteration and $p_2(0) = 0$ that

$$\begin{aligned} |p_2(t)| &\le t \cdot r^{t-1} \\ &\le \frac{t}{\delta}\left(\frac{r}{\delta}\right)^{t-1} \delta^t \end{aligned}$$

for $t \in \mathbb{N}_0$. L'Hôpital's rule implies that

$$\lim_{t\to\infty} \frac{t}{\delta}\left(\frac{r}{\delta}\right)^{t-1} = 0,$$

so there is a constant $B_2 > 0$ so that

$$|p_2(t)| \le B_2\delta^t, \quad t \in \mathbb{N}_0.$$

Hence (1.50) holds for $k = 2$. Similarly, we can show that for $t \in \mathbb{N}_0$

$$|p_3(t)| \le \frac{t(t-1)}{2} r^{t-2},$$

from which it follows that there is a $B_3$ so that

$$|p_3(t)| \le B_3\delta^t, \quad t \in \mathbb{N}_0.$$

Continuing in this manner, we obtain constants $B_k > 0$, $1 \le k \le n$ so that

$$|p_k(t)| \le B_k\delta^t, \quad t \in N_0,$$

for $k = 1, 2, \cdots, n$. Using Remark 1.95 we have there are constants $D_k$ such that

$$\|M_k z\| \le D_k\|z\|, \quad 1 \le k \le n$$

for all $z \in \mathbb{R}^n$. Using this and (1.49) we have that for $t \in \mathbb{N}_0$,

$$\begin{aligned} \|y(t,z)\| &\le \sum_{k=0}^{n-1} |p_{k+1}(t)| \, \|M_k z\| \\ &\le \left(\sum_{k=0}^{n-1} B_{k+1}D_k\right) \|z\|\delta^t \\ &\le C\delta^t\|z\| \end{aligned} \tag{1.51}$$

where $C := \sum_{k=0}^{n-1} B_{k+1}D_k$. It follows from (1.51) that the trivial solution is stable on $\mathbb{N}_0$. Since $0 < \delta < 1$, it also follows from (1.51) that $\lim_{t\to\infty} y(t,z) = 0$. Hence the trivial solution of (1.48) is globally asymptotically stable on $\mathbb{N}_0$. □

*Example 1.97.* Consider the vector difference equation

$$u(t+1) = \begin{bmatrix} 1 & -5 \\ .25 & -1 \end{bmatrix} u(t). \tag{1.52}$$

The characteristic equation for $A = \begin{bmatrix} 1 & -5 \\ .25 & -1 \end{bmatrix}$ is $\lambda^2 + \frac{1}{4} = 0$ and hence the eigenvalues of $A$ are $\lambda_1 = \frac{i}{2}$ and $\lambda_2 = -\frac{i}{2}$. Since

$$|\lambda_1| = |\lambda_2| = \frac{1}{2} < 1,$$

we have by Theorem 1.96 the trivial solution of (1.52) is globally asymptotically stable on $\mathbb{N}_0$.

In the next theorem we give conditions under which the trivial solution of (1.48) is unstable on $\mathbb{N}_a$.

**Theorem 1.98.** *If there is an eigenvalue, $\lambda_0$, of $A$ satisfying $|\lambda_0| > 1$, then the trivial solution of* (1.48) *is unstable on $\mathbb{N}_a$.*

*Proof.* Assume $\lambda_0$ is an eigenvalue of $A$ so that $|\lambda_0| > 1$. Let $v_0$ be a corresponding eigenvector. Then $y_0(t) = \lambda_0^{t-a} v_0$ is a solution of equation (1.48) on $\mathbb{N}_a$, and

$$\lim_{t\to\infty} \|y_0(t)\| = \lim_{t\to\infty} |\lambda|^{t-a} \|v_0\| = \infty.$$

This implies that the trivial solution of (1.48) is unstable on $\mathbb{N}_a$. □

*Example 1.99.* Consider the vector difference equation

$$y(t+1) = \begin{bmatrix} -.5 & 3 \\ .5 & -1 \end{bmatrix} y(t). \tag{1.53}$$

The characteristic equation for $A = \begin{bmatrix} -.5 & 3 \\ .5 & -1 \end{bmatrix}$ is $\lambda^2 + \frac{3}{2}\lambda - 1 = 0$ and so the eigenvalues are $\lambda_1 = .5$, $\lambda_2 = -2$, Since $|\lambda_2| = 2 > 1$ we have by Theorem 1.98 that the trivial solution of (1.53) is unstable on $\mathbb{N}_0$.

In the next theorem we give conditions on the matrix $A$ which implies the trivial solution is stable on $\mathbb{N}_0$.

**Theorem 1.100.** *Let $\lambda_1, \lambda_2, \ldots, \lambda_n$ be the eigenvalues of A. Assume $|\lambda_k| \le 1$ and whenever $|\lambda_k| = 1$, then $\lambda_k$ is a simple eigenvalue of A. Then the trivial solution of* (1.48) *is stable on $\mathbb{N}_a$.*

*Proof.* We prove this theorem for the case $a = 0$. If all the eigenvalues of $A$ satisfy $|\lambda_i| < 1$, then by Theorem 1.96 we have that the trivial solution of (1.48) is stable on $\mathbb{N}_0$. Now assume there is at least one eigenvalue of $A$ with modulus one. Without loss of generality we can order the eigenvalues of $A$ so that $|\lambda_i| = 1$ for $i = 1, \ldots, k-1$, where $2 \le k \le n$ and $|\lambda_i| < 1$ for $i = k, \ldots, n$. From equations (1.44) and (1.45),

$$p_1(t) = \lambda_1^t.$$

Next, $p_2$ satisfies

$$p_2(t+1) = \lambda_2 p_2(t) + \lambda_1^t,$$

so (as in the annihilator method)

$$(E - \lambda_1 I)(E - \lambda_2 I)p_2(t) = 0.$$

Since $\lambda_1 \neq \lambda_2$,

$$p_2(t) = B_{12}\lambda_1^t + B_{22}\lambda_2^t,$$

for some constants $B_{12}$, $B_{22}$. Continuing in this way, we have

$$p_i(t) = B_{1i}\lambda_1^t + \cdots + B_{ii}\lambda_i^t$$

for $i = 1, \ldots, k-1$. Consequently, there is a constant $D > 0$ so that

$$|p_i(t)| \le D$$

for $i = 1, \ldots, k-1$ and $t \in \mathbb{N}_0$.

From (1.44), $p_k(t+1) = \lambda_k p_k(t) + p_{k-1}(t)$ and hence

$$|p_k(t+1)| \le |\lambda_k||p_k(t)| + D, \quad t \in \mathbb{N}_0.$$

Choose $\delta = \max\{|\lambda_k|, |\lambda_{k+1}|, \ldots, |\lambda_n|\} < 1$. Then

$$|p_k(t+1)| \le \delta|p_k(t)| + D.$$

By iteration and the initial condition $p_k(0) = 0$,

$$\begin{aligned} |p_k(t)| &\le D\sum_{j=0}^{t-1}\delta^j \\ &\le D\sum_{j=0}^{\infty}\delta^j \\ &= \frac{D}{1-\delta} \end{aligned}$$

for $t \in \mathbb{N}_0$. In a similar manner, we find that there is a constant $D^*$ so that

$$|p_i(t)| \leq D^*$$

for $i = 1, 2, \ldots, n$ and $t \in \mathbb{N}_0$.

From Theorem 1.88, the solution of equation (1.48) satisfying $u(0) = u_0$, is given by

$$u(t) = \sum_{i=0}^{n-1} p_{i+1}(t) M_i u_0$$

and it follows that

$$\begin{aligned}\|u(t)\| &\leq D^* \sum_{i=0}^{n-1} \|M_i u_0\| \\ &\leq C\|u_0\|\end{aligned}$$

for $t \in \mathbb{N}_0$ and some $C > 0$. □

*Example 1.101.* Consider the system

$$u(t+1) = \begin{bmatrix} \cos\theta & \sin\theta \\ -\sin\theta & \cos\theta \end{bmatrix} u(t), \quad t \in \mathbb{N}_a, \tag{1.54}$$

where $\theta$ is a real number. For each $\theta$ the eigenvalues of the coefficient matrix in (1.54) are $\lambda_{1,2} = e^{\pm i\theta}$. Since $|\lambda_1| = |\lambda_2| = 1$ and both eigenvalues are simple, we have by Theorem 1.100 that the trivial solution of (1.54) is stable on $\mathbb{N}_a$. From linear algebra the coefficient matrix in (1.54) is called a rotation matrix. When a vector $u$ is multiplied by this coefficient matrix, the resulting vector has the same length as $u$, but its direction is $\theta$ radians clockwise from $u$. Consequently, every solution $u$ of the system has all of its values on a circle centered at the origin of radius $|u(a)|$. This also tells us that the trivial solution of (1.54) is stable on $\mathbb{N}_a$, but not globally asymptotically stable on $\mathbb{N}_a$.

## 1.11 Floquet Systems

In this section we consider the so-called **Floquet system**

$$u(t+1) = A(t)u(t), \quad t \in \mathbb{Z}_\alpha, \tag{1.55}$$

where $\alpha \in \mathbb{R}$ and

$$\mathbb{Z}_\alpha := \{\ldots, \alpha-2, \alpha-1, \alpha, \alpha+1, \alpha+2, \ldots\},$$

and we assume that $A(t)$ is an $n \times n$ matrix function which has minimum positive period $p$ ($p$ is an integer). We say that $p$ is the **prime period** of $A(t)$. Here are two simple scalar examples that are indicative of the behavior of general Floquet systems.

*Example 1.102.* Since $a(t) := 2 + (-1)^t$, $t \in \mathbb{Z}_0$ is periodic with prime period $p = 2$, the equation

$$u(t+1) = [2 + (-1)^t]u(t), \quad t \in \mathbb{Z}_0 \tag{1.56}$$

is a scalar Floquet system with prime period $p = 2$. Equation (1.56) can be written in the form

$$\Delta u(t) = [1 + (-1)^t]u(t).$$

By Theorem 1.14 the general solution of (1.56) is given by

$$u(t) \;= ce_h(t, 0),$$

where $h(t) := 1 + (-1)^t$, $t \in \mathbb{Z}_0$. It follows from (1.8) that for $t \in \mathbb{N}_0$

$$\begin{aligned} u(t) &= c \prod_{\tau=0}^{t-1} [1 + h(\tau)] \\ &= c \prod_{\tau=0}^{t-1} [2 + (-1)^\tau] \\ &= \begin{cases} (\sqrt{3})^t, & \text{if } t \text{ is even} \\ (\sqrt{3})^{t+1}, & \text{if } t \text{ is odd.} \end{cases} \end{aligned}$$

It is easy to check that the last expression above is also true for negative integers. Define the $p = 2$ periodic function $r$ by

$$r(t) := \begin{cases} c, & \text{if } t \text{ is even} \\ c\sqrt{3}, & \text{if } t \text{ is odd,} \end{cases}$$

for $t \in \mathbb{Z}_0$ and put $b := \sqrt{3}$. Then we have that all solutions of (1.56) are of the form

$$u(t) = r(t)b^t, \quad t \in \mathbb{Z}_0.$$

Compare this with formula (1.59) in Floquet's Theorem 1.105.

In some cases we need $b$ to be a complex number to write the general solution in the form $u(t) = r(t)b^t$ as the next example shows.

*Example 1.103.* Since $a(t) := (-1)^t$ is periodic with prime period $p = 2$, the equation

$$u(t+1) = (-1)^t u(t), \quad t \in \mathbb{Z}_0$$

is a scalar Floquet system. The general solution is given by

$$u(t) = \alpha(-1)^{\frac{t(t-1)}{2}}, \quad t \in \mathbb{Z}_0.$$

We can write this solution in the form

$$u(t) = r(t)b^t, \quad t \in \mathbb{Z}_0,$$

where

$$r(t) = \alpha(-1)^{\frac{t^2}{2}}, \quad t \in \mathbb{Z}_0$$

is periodic with period 2 and $b = -i$. Compare this with the formula (1.59) in Floquet's Theorem 1.105.

In preparation for the proof of Floquet's Theorem, we need the following result concerning logarithms of nonsingular matrices.

**Lemma 1.104.** *Assume $C$ is a nonsingular matrix and $p$ is a positive integer. Then there is a nonsingular matrix $B$ such that*

$$B^p = C.$$

*Proof.* We will prove this theorem only for $2 \times 2$ matrices. First consider the case where $C$ has two linearly independent eigenvectors. In this case, by the Jordan canonical form theorem, there is a nonsingular matrix $Q$ so that

$$C = Q^{-1}JQ,$$

where

$$J = \begin{bmatrix} \lambda_1 & 0 \\ 0 & \lambda_2 \end{bmatrix},$$

and $\lambda_1, \lambda_2$ ($\lambda_1 = \lambda_2$ is possible) are the eigenvalues of $C$. Now we want to find a matrix $B$ such that

$$B^p = C = Q^{-1}JQ.$$

Equivalently, we want to pick $B$ so that

$$QB^pQ^{-1} = J,$$

or

$$(QBQ^{-1})^p = J.$$

Then we need to choose $B$ so that

$$QBQ^{-1} = \begin{bmatrix} (\lambda_1)^{\frac{1}{p}} & 0 \\ 0 & (\lambda_2)^{\frac{1}{p}} \end{bmatrix},$$

so

$$B = Q^{-1} \begin{bmatrix} (\lambda_1)^{\frac{1}{p}} & 0 \\ 0 & (\lambda_2)^{\frac{1}{p}} \end{bmatrix} Q.$$

Finally, we consider the case where $C$ has only one linearly independent eigenvector. In this case, by the Jordan canonical form theorem, there is a nonsingular matrix $Q$ so that

$$C = Q^{-1}JQ,$$

where

$$J = \begin{bmatrix} \lambda_1 & 1 \\ 0 & \lambda_1 \end{bmatrix},$$

and $\lambda_1$ is the eigenvalue of $C$.

Let's try to find a matrix $B$ of the form

$$B = Q^{-1} \begin{bmatrix} a & b \\ 0 & a \end{bmatrix} Q \tag{1.57}$$

so that $B^p = C$.

Then

$$\begin{aligned} B^p &= \left(Q^{-1} \begin{bmatrix} a & b \\ 0 & a \end{bmatrix} Q\right)^p \\ &= Q^{-1} \begin{bmatrix} a & b \\ 0 & a \end{bmatrix}^p Q \\ &= Q^{-1} \left\{ aI + \begin{bmatrix} 0 & b \\ 0 & 0 \end{bmatrix} \right\}^p Q, \end{aligned}$$

where $I$ is the $2 \times 2$ identity matrix. Using the binomial theorem, we get

$$
\begin{aligned}
B^p &= Q^{-1}\left\{a^p I + pa^{p-1}\begin{bmatrix} 0 & b \\ 0 & 0 \end{bmatrix}\right\} Q \\
&= Q^{-1}\begin{bmatrix} a^p & pa^{p-1}b \\ 0 & a^p \end{bmatrix} Q \\
&= C = Q^{-1}\begin{bmatrix} \lambda_1 & 1 \\ 0 & \lambda_1 \end{bmatrix} Q,
\end{aligned}
$$

if $a$ and $b$ are picked to satisfy

$$a^p = \lambda_1 \quad \text{and} \quad pa^{p-1}b = 1.$$

Solving for $a$ and $b$ we get from (1.57) that

$$B = Q^{-1}\begin{bmatrix} \lambda_1^{\frac{1}{p}} & \frac{1}{p}\lambda_1^{\frac{1}{p}-1} \\ 0 & \lambda_1^{\frac{1}{p}} \end{bmatrix} Q$$

is the desired expression for $B$. □

**Theorem 1.105 (Discrete Floquet's Theorem).** *If $\Phi(t)$ is a fundamental matrix for the Floquet system (1.55), then $\Phi(t+p)$ is also a fundamental matrix and $\Phi(t+p) = \Phi(t)C$, $t \in \mathbb{Z}_\alpha$, where*

$$C = \Phi^{-1}(\alpha)\Phi(\alpha + p). \tag{1.58}$$

*Furthermore, there is a nonsingular matrix function $P(t)$ and a nonsingular constant matrix $B$ such that*

$$\Phi(t) = P(t)B^{t-\alpha}, \quad t \in \mathbb{Z}_\alpha, \tag{1.59}$$

*where $P(t)$ is periodic on $\mathbb{Z}_\alpha$ with period $p$.*

*Proof.* Assume $\Phi(t)$ is a fundamental matrix for the Floquet system (1.55). If $\Psi(t) := \Phi(t+p)$, then $\Psi(t)$ is nonsingular for all $t \in \mathbb{Z}_\alpha$, and

$$
\begin{aligned}
\Psi(t+1) &= \Phi(t+p+1) \\
&= A(t+p)\Phi(t+p) \\
&= A(t)\Psi(t),
\end{aligned}
$$

for $t \in \mathbb{Z}_\alpha$. Hence $\Psi(t) = \Phi(t+p)$ is a fundamental matrix for the vector difference equation (1.55). Since $\Psi(t)$ and $\Phi(t)$ are both fundamental matrices of (1.55), we have by Theorem 1.85, there is a nonsingular constant matrix $C$ such that

$$\Psi(t) = \Phi(t+p) = \Phi(t)C, \quad t \in \mathbb{Z}_\alpha.$$

Letting $t = \alpha$ and solving for $C$, we get that equation (1.58) holds. By Lemma 1.104, there is a nonsingular matrix $B$ so that $B^p = C$. Let

$$P(t) := \Phi(t)B^{-(t-\alpha)}, \quad t \in \mathbb{Z}_\alpha. \tag{1.60}$$

Note that $P(t)$ is nonsingular for all $t \in \mathbb{Z}_\alpha$ and since

$$\begin{aligned} P(t+p) &= \Phi(t+p)B^{-(t-\alpha+p)} \\ &= \Phi(t)CB^{-p}B^{-(t-\alpha)} \\ &= \Phi(t)B^{-(t-\alpha)} \\ &= P(t), \end{aligned}$$

$P(t)$ is periodic with period $p$. Solving equation (1.60) for $\Phi(t)$, we get equation (1.59). □

**Definition 1.106.** Let $\Phi(t)$ and $C$ be as in Floquet's theorem (Theorem 1.105). Then the eigenvalues $\mu$ of the matrix $C$ are called the **Floquet multipliers** of the Floquet system (1.55).

Since fundamental matrices of a linear system are not unique (see Theorem 1.85), we must show that the Floquet multipliers are well defined. Let $\Phi(t)$ and $\Psi(t)$ be fundamental matrices for the Floquet system (1.55) and let

$$C_1 = \Phi^{-1}(\alpha)\Phi(\alpha+p) \quad \text{and} \quad C_2 = \Psi^{-1}(\alpha)\Psi(\alpha+p).$$

It remains to show that $C_1$ and $C_2$ have the same eigenvalues. By Theorem 1.85 there is a nonsingular constant matrix $F$ so that

$$\Psi(t) = \Phi(t)F, \quad t \in \mathbb{Z}_\alpha.$$

Hence,

$$\begin{aligned} C_2 &= \Psi^{-1}(\alpha)\Psi(\alpha+p) \\ &= [\Phi(\alpha)F]^{-1}[\Phi(\alpha+p)F] \\ &= F^{-1}\Phi^{-1}(\alpha)\Phi(\alpha+p)F \\ &= F^{-1}C_1F. \end{aligned}$$

Since

$$\begin{aligned}\det(C_2 - \lambda I) &= \det(F^{-1}C_1F - \lambda I)\\ &= \det F^{-1}(C_1 - \lambda I)F\\ &= \det(C_1 - \lambda I),\end{aligned}$$

$C_1$ and $C_2$ have the same characteristic polynomial and therefore have the same eigenvalues. Hence Floquet multipliers are well defined.

**Theorem 1.107.** *The Floquet multipliers of the Floquet system* (1.55) *are the eigenvalues of the matrix*

$$D := A(\alpha + p - 1)A(\alpha + p - 2)\cdots A(\alpha).$$

*Proof.* To see this let $\Phi(t)$ be the fundamental matrix of the Floquet system (1.55) satisfying $\Phi(\alpha) = I$. Then the Floquet multipliers are the eigenvalues of

$$D = \Phi^{-1}(\alpha)\Phi(\alpha + p) = \Phi(\alpha + p).$$

Iterating the equation

$$\Phi(t + 1) = A(t)\Phi(t),$$

we get that

$$\begin{aligned}D = \Phi(\alpha + p) &= [A(\alpha + p - 1)A(\alpha + p - 2)\cdots A(\alpha)]\Phi(\alpha)\\ &= A(\alpha + p - 1)A(\alpha + p - 2)\cdots A(\alpha),\end{aligned}$$

which is the desired result. □

Here are some simple examples of Floquet multipliers. Note that in the scalar case we use $d$ instead of $D$.

*Example 1.108.* For the scalar equation

$$u(t + 1) = (-1)^t u(t), \quad t \in \mathbb{Z}_0,$$

the coefficient function $a(t) = (-1)^t$ has prime period $p = 2$, and $d = a(1)\,a(0) = -1$, so $\mu = -1$ is the Floquet multiplier.

*Example 1.109.* Find the Floquet multipliers for the Floquet system

$$u(t + 1) = \begin{bmatrix} 0 & 1 \\ (-1)^t & 0 \end{bmatrix} u(t), \quad t \in \mathbb{Z}_0.$$

The coefficient matrix $A(t)$ is periodic with prime period $p = 2$, so

$$D = A(1)A(0)$$
$$= \begin{bmatrix} 0 & 1 \\ -1 & 0 \end{bmatrix} \begin{bmatrix} 0 & 1 \\ 1 & 0 \end{bmatrix}$$
$$= \begin{bmatrix} 1 & 0 \\ 0 & -1 \end{bmatrix}.$$

Consequently, $\mu_1 = 1$, $\mu_2 = -1$ are the Floquet multipliers.

The following theorem demonstrates why the term multiplier is appropriate.

**Theorem 1.110.** *The number $\mu$ is a Floquet multiplier for the Floquet system* (1.55), *if and only if there is a nontrivial solution $u(t)$ of* (1.55) *such that*

$$u(t + p) = \mu u(t), \quad t \in \mathbb{Z}_\alpha.$$

*Furthermore, if $\mu_1, \mu_2, \cdots, \mu_k$ are distinct Floquet multipliers, then there are $k$ linearly independent solutions, $u_i(t), 1 \leq i \leq k$ of the Floquet system* (1.55) *on $\mathbb{Z}_\alpha$ satisfying*

$$u_i(t + p) = \mu_i u_i(t), \quad t \in \mathbb{Z}_\alpha, \quad 1 \leq i \leq k.$$

*Proof.* Assume $\mu_0$ is a Floquet multiplier of (1.55). Then $\mu_0$ is an eigenvalue of the matrix $C$ given by equation (1.58). Let $u_0$ be an eigenvector of $C$ corresponding to $\mu_0$, and $\Phi(t)$ be a fundamental matrix for (1.55). Define

$$u(t) := \Phi(t)u_0 \quad t \in \mathbb{Z}_\alpha.$$

Then $u(t)$ is a nontrivial solution of (1.55), and from Floquet's theorem $\Phi(t + p) = \Phi(t)C$. Hence, we have

$$\begin{aligned} u(t + p) &= \Phi(t + p)u_0 \\ &= \Phi(t)Cu_0 \\ &= \Phi(t)\mu_0 u_0 \\ &= \mu_0 u(t), \end{aligned}$$

for $t \in \mathbb{Z}_\alpha$. The proof of the converse is essentially reversing the above steps. The proof of the last statement in this theorem is Exercise 1.89. □

In Example 1.109 we saw that 1 and $-1$ were Floquet multipliers. Theorem 1.110 implies that there are linearly independent solutions that are periodic with periods 2 and 4. The next theorem shows how a Floquet system can be transformed into an autonomous system.

**Theorem 1.111.** *Let $\Phi(t) = P(t)B^{t-\alpha}$, $t \in \mathbb{Z}_\alpha$, be as in Floquet's theorem. Then $y(t)$ is a solution of the Floquet system (1.55) if and only if*

$$z(t) = P^{-1}(t)y(t), \quad t \in \mathbb{Z}_\alpha$$

*is a solution of the autonomous system*

$$z(t+1) = Bz(t), \quad t \in \mathbb{Z}_\alpha.$$

*Proof.* Assume $y(t)$ is a solution of the Floquet system (1.55). Then there is a column vector $w$ so that

$$y(t) = \Phi(t)w = P(t)B^{t-\alpha}w, \quad t \in \mathbb{Z}_\alpha.$$

Let $z(t) := P^{-1}(t)y(t)$, $t \in \mathbb{Z}_\alpha$. Then $z(t) = P^{-1}(t)y(t) = B^{t-\alpha}w$. It follows that $z(t)$ is a solution of $z(t+1) = Bz(t)$, $t \in \mathbb{Z}_\alpha$. The converse can be proved by reversing the above steps. □

*Example 1.112.* In this example we determine the asymptotic behavior of two linearly independent solutions of the Floquet system

$$u(t+1) = \begin{bmatrix} 0 & \frac{2+(-1)^t}{2} \\ \frac{2-(-1)^t}{2} & 0 \end{bmatrix} u(t), \quad t \in \mathbb{Z}_0.$$

First we find the Floquet multipliers for this Floquet system. For this system, $p = 2$ and thus

$$\begin{aligned} D &= A(1)A(0) \\ &= \begin{bmatrix} 0 & \frac{1}{2} \\ \frac{3}{2} & 0 \end{bmatrix} \begin{bmatrix} 0 & \frac{3}{2} \\ \frac{1}{2} & 0 \end{bmatrix} \\ &= \begin{bmatrix} \frac{1}{4} & 0 \\ 0 & \frac{9}{4} \end{bmatrix}. \end{aligned}$$

Hence the Floquet multipliers are $\mu_1 = \frac{1}{4}$ and $\mu_2 = \frac{9}{4}$. Since $|\mu_1| = \frac{1}{4} < 1$ and $|\mu_2| = \frac{9}{4} > 1$, we get there are two linearly independent solutions $u_1(t)$, $u_2(t)$ on $\mathbb{Z}_0$ satisfying

$$\lim_{t\to\infty} \|u_1(t)\| = 0, \quad \lim_{t\to\infty} \|u_2(t)\| = \infty.$$

Using Theorem 1.111 one can prove (see Exercise 1.92) the following stability theorem for Floquet systems.

**Theorem 1.113.** *Let $\mu_1, \mu_2, \ldots, \mu_n$ be the Floquet multipliers of the Floquet system* (1.55). *Then the trivial solution is*

(i) *globally asymptotically stable iff* $\|\mu_i\| < 1$, $1 \le i \le n$;
(ii) *stable provided* $\|\mu_i\| \le 1$, $1 \le i \le n$ *and whenever* $\|\mu_i\| = 1$, *then* $\mu_i$ *is a simple eigenvalue;*
(iii) *unstable provided there is an* $i_0$, $1 \le i_0 \le n$, *such that* $\|\mu_i\| > 1$.

*Example 1.114.* By Theorem 1.113 the trivial solution of the Floquet system in Example 1.112 is globally asymptotically stable.

We conclude this section by mentioning that although in this chapter we have explored a substantial introduction to the classical difference calculus, there are naturally many stones we have left unturned. So, for the reader who is interested in more advanced and specialized techniques from the classical theory of difference equations, we encourage him or her to consult the book by Kelley and Peterson [135] for a multitude of related results.

## 1.12 Exercises

**1.1.** Show that if $f : \mathbb{N}_a^b \to \mathbb{R}$ satisfies $\Delta f(t) = 0$ for $t \in \mathbb{N}_a^{b-1}$, then $f(t) = C$ for $t \in \mathbb{N}_a^b$, where $C$ is a constant.

**1.2.** Prove the product rules

(i) $\Delta\left(f(t)g(t)\right) = f(\sigma(t))\Delta g(t) + \Delta f(t)g(t)$;
(ii) $\Delta\left(f(t)g(t)\right) = f(t)\Delta g(t) + \Delta f(t)g(\sigma(t))$.

Why does (i) imply (ii)?

**1.3.** Show that $\Gamma(1) = 1$ and that for any positive integer that $\Gamma(n+1) = n!$.

**1.4.** Show that for $x$ a real variable that $\lim_{x\to 0+} \Gamma(x) = +\infty$.

**1.5.** Show that $\Gamma(1/2) = \sqrt{\pi}$. **Hint** first show that

$$\left(\Gamma(1/2)\right)^2 = 4\int_0^\infty \int_0^\infty e^{-x^2-y^2}\,dx\,dy.$$

**1.6.** By Exercise 1.5, $\Gamma(1/2) = \sqrt{\pi}$. Find $\Gamma(5/2)$ and $\Gamma(-3/2)$.

**1.7.** Use the definition of the (generalized) falling function (Definition 1.7) to show that for $n \in \mathbb{N}_1$,

$$t^{\underline{-n}} = \frac{1}{(t+1)(t+2)\cdots(t+n)}.$$

Then use this expression directly (do not use the gamma function) to prove

$$\Delta(t^{\underline{-n}}) = -n(t^{\underline{-(n+1)}}),$$

$t \neq -1, -2, -3, \ldots, -n-1$.

**1.8.** Show that $\nu^{\underline{\nu}} = \Gamma(\nu+1)$ for $\nu \neq 0, -1, -2, -3, \ldots$, and that $(\nu-k)^{\underline{\nu}} = 0$, $\nu - k \neq -1, -2, -3, \cdots, k = 1, 2, 3, \cdots$.

**1.9.** Show that

$$(t-\mu)t^{\underline{\mu}} = t^{\underline{\mu+1}},$$

whenever both sides of this equation are well defined.

**1.10.** For integers $m$ and $n$ satisfying $m > n \geq 0$, evaluate the binomial coefficient $\binom{n}{m}$.

**1.11.** Evaluate each of the following binomial coefficients:

(i) $\binom{t}{t}$ for $t \neq -1, -2, -3, \ldots$;
(ii) $\binom{1}{\frac{3}{2}}$;
(iii) $\binom{\frac{1}{2}}{\frac{3}{2}}$;
(iv) $\binom{\sqrt{2}+2}{\sqrt{2}}$.

**1.12.** Prove each of the following:

(i) $\binom{t}{r} = \binom{t}{t-r}$;
(ii) $\binom{t}{r} = \frac{t}{r}\binom{t-1}{r-1}$;
(iii) $\binom{t}{r} = \binom{t-1}{r} + \binom{t-1}{r-1}$;
(iv) $\binom{-\nu}{k} = (-1)^k\binom{k+\nu-1}{\nu-1}$;
(v) $\binom{\mu+\nu}{\nu} = \binom{\mu+\nu}{\mu}$.

where in (iv) $\nu > 0$ and $k \in \mathbb{N}_0$.

**1.13.** Prove Theorem 1.10.

**1.14.** For each of the following, find $e_p(t, a)$ given that

(i) $p(t) = 7, \quad t \in \mathbb{N}_a, \quad a = 5$;
(ii) $p(t) = \frac{2}{t+1}, \quad t \in \mathbb{N}_a, \quad a = 0$;
(iii) $p(t) = \frac{4}{t}, \quad t \in \mathbb{N}_1, \quad a = 1$;
(iv) $p(t) = \frac{3-t}{(t+1)(t+7)}, \quad t \in \mathbb{N}_0, \quad a = 0$.

**1.15.** Let $P(t)$ be the population of a bacteria in a culture after $t$ hours. Assuming that $P$ satisfies the IVP

$$\Delta P(t) = 9P(t), \quad P(0) = 5{,}000$$

use Theorem 1.14 to find a formula for $P(t)$.

**1.16 (Compound Interest).** A bank pays interest with an annual interest rate of 8% and interest is compounded 4 times a year. If \$100 is invested, how much money do you have after 20 years?

**1.17 (Radioactive Decay).** Let $R(t)$ be the amount of the radioactive isotope Pb-209 present at time $t$. Assume initially that $R_0$ is the amount of Pb-209 present and that the change in the amount of Pb-209 each hour is proportional to the amount present at the beginning of that hour and the half life of Pb-209 is 3.3 hours. Find a formula for $R(t)$. How long does it take for 70% of the Pb-209 to decay?

**1.18.** Show that if $p, q \in \mathcal{R}$, then

$$(p \ominus q)(t) = \frac{p(t) - q(t)}{1 + q(t)}, \quad t \in \mathbb{N}_a.$$

**1.19.** Complete the proof of Theorem 1.16 by showing that the addition $\oplus$ on $\mathcal{R}$ is associative and commutative.

**1.20.** Show that the set of positively regressive functions $\mathcal{R}^+$ with the addition $\oplus$ is a subgroup of the set of regressive functions $\mathcal{R}$.

**1.21.** Prove Theorem 1.21 if $a$ is replaced by $s$, where $s \in \mathbb{N}_a$.

**1.22.** Prove parts (iv) and (v) of Theorem 1.25.

**1.23.** Prove part (ii) of Theorem 1.27.

**1.24.** Prove that if $p \in \mathcal{R}$ and $n \in \mathbb{N}_1$, then

$$n \odot p = p \oplus p \oplus \cdots \oplus p,$$

where the right-hand side has $n$ terms.

**1.25.** Prove that if $p, q \in \mathcal{R}^+$ and $\alpha, \beta \in \mathbb{R}$, then the following hold:

(i) $\alpha \odot (\beta \odot p) = (\alpha\beta) \odot p$;
(ii) $\alpha \odot (p \oplus q) = (\alpha \odot p) \oplus (\alpha \odot q)$.

**1.26.** Prove Theorem 1.28 directly (do not use Theorem 1.27) from the definitions of $\cos_p(t, a)$ and $\sin_p(t, a)$.

**1.27.** Derive Euler's formula (1.11)

$$e_{ip}(t, a) = \cos_p(t, a) + i \sin_p(t, a), \quad t \in \mathbb{N}_a.$$

Also derive the hyperbolic analogue of Euler's formula

$$e_p(t, a) = \cosh_p(t, a) + \sinh_p(t, a), \quad t \in \mathbb{N}_a.$$

**1.28.** Verify each of the following formulas for $p \neq \pm i$:

(i) $\cos_p(\sigma(t), a) = \cos_p(t, a) - p \sin_p(t, a), \quad t \in \mathbb{N}_a;$
(ii) $\sin_p(\sigma(t), a) = \sin_p(t, a) + p \cos_p(t, a), \quad t \in \mathbb{N}_a;$
(iii) $\cos_p(t-1, a) = \frac{1}{1+p^2}[\cos_p(t, a) + p \sin_p(t, a)], \quad t \in \mathbb{N}_{a+1};$
(iv) $\sin_p(t-1, a) = \frac{1}{1+p^2}[\sin_p(t, a) - p \cos_p(t, a)], \quad t \in \mathbb{N}_{a+1}.$

**1.29.** Prove Theorem 1.40.

**1.30.** Prove Theorem 1.42.

**1.31.** Using a delta integral (see Example 1.57) find

(i) the sum of the first $n$ positive integers;
(ii) the sum of the cubes of the first $n$ positive integers.

**1.32.** Show by direct substitution that $y(t) = (t-a)e_r(t, a)$, $r \neq -1$, is a solution of the second order linear equation $\Delta^2 y(t) - 2r\Delta y(t) + r^2 y(t) = 0$.

**1.33.** Solve each of the following difference equations:

(i) $\Delta^2 y(t) - 5\Delta y(t) + 6y(t) = 0, \quad t \in \mathbb{N}_0;$
(ii) $\Delta^2 y(t) + 2\Delta y(t) - 8y(t) = 0, \quad t \in \mathbb{N}_0;$
(iii) $\Delta^2 y(t) - 2\Delta y(t) + 5y(t) = 0, \quad t \in \mathbb{N}_a;$
(iv) $\Delta^2 y(t) + y(t) = 0, \quad t \in \mathbb{N}_0;$
(v) $\Delta^2 y(t) + 6\Delta y(t) + 9y(t) = 0, \quad t \in \mathbb{N}_0;$
(vi) $\Delta^2 y(t) + 8\Delta y(t) + 16y(t) = 0, \quad t \in \mathbb{N}_0.$

**1.34.** Solve each of the following difference equations:

(i) $u(t+2) + 4u(t+1) - 5u(t) = 0, \quad t \in \mathbb{Z};$
(ii) $u(t+2) - 4u(t+1) + 8u(t) = 0, \quad t \in \mathbb{Z};$
(iii) $u(t+3) - u(t+2) - 8u(t+1) + 12u(t) = 0, \quad t \in \mathbb{N}_0.$

**1.35.** Solve each of the following linear difference equations:

(i) $\Delta^2 y(t) + 2\Delta y(t) + 2y(t) = 0, \quad t \in \mathbb{N}_a;$
(ii) $\Delta^2 y(t) + 2\Delta y(t) + 10y(t) = 0, \quad t \in \mathbb{N}_a.$

**1.36.** Show that a second order linear homogeneous equation of the form $\Delta^2 y(t) + p(t)\Delta y(t) + q(t)y(t) = 0$ with $p(t) \neq 1 + q(t)$ is equivalent to an equation of the form $y(t+2) + c(t)y(t+1) + d(t)y(t) = 0$ with $d(t) \neq 0$.

**1.37 (Real Roots).** Find the value of the determinant of the $t$ by $t$ matrix with all 4's on the diagonal, 1's on the superdiagonal, 3's on the subdiagonal, and 0's elsewhere.

**1.38 (Complex Roots).** Find the value of the determinant of the $t$ by $t$ matrix with all $-2$'s on the diagonal, 4's on the superdiagonal, 1's on the subdiagonal, and 0's elsewhere.

**1.39 (Complex Roots).** Find the value of the determinant of the $t$ by $t$ matrix with all 2's on the diagonal, 4's on the superdiagonal, 1's on the subdiagonal, and 0's elsewhere.

**1.40.** What would you guess are general solutions of each of the following:

(i) $\Delta^3 y(t) - 6\Delta^2 y(t) + 11\Delta y(t) - 6y(t) = 0, \quad t \in \mathbb{N}_0;$
(ii) $\Delta^3 y(t) - \Delta^2 y(t) - 8\Delta y(t) + 12y(t) = 0, \quad t \in \mathbb{N}_0;$
(iii) $\Delta^3 y(t) - 7\Delta^2 y(t) + 16\Delta y(t) - 10y(t) = 0, \quad t \in \mathbb{N}_0?$

**1.41.** Show that if $F(t)$ is the $t$-th term in the Fibonacci sequence (see Example 1.32), then

$$\lim_{t\to\infty} \frac{F(t+1)}{F(t)} = \frac{1+\sqrt{5}}{2}.$$

The ratio $\frac{1+\sqrt{5}}{2}$ is known as the "golden section" and was considered by the ancient Greeks to be the most aesthetically pleasing ratio of the length of a rectangle to its width.

**1.42.** In how many ways can you tile a $1 \times n$, hallway, $n \geq 2$, if you have green $1 \times 1$ tiles and red and yellow $1 \times 2$ tiles?

**1.43.** Solve the following difference equations:

(i) $u(t+2) + 2u(t+1) - 8u(t) = 0, \quad t \in \mathbb{N}_0;$
(ii) $u(t+2) - 6u(t+1) + 9u(t) = 0 \quad t \in \mathbb{N}_0;$
(iii) $u(t+2) + 2u(t+1) + 4u(t) = 0, \quad t \in \mathbb{N}_0.$

**1.44.** Solve the following difference equations:

(i) $u(t+2) + 8u(t+1) - 9u(t) = 0, \quad t \in \mathbb{N}_0;$
(ii) $u(t+2) + 9u(t) = 0 \quad t \in \mathbb{N}_0;$
(iii) $u(t+3) + u(t+2) - 8u(t+1) - 12u(t) = 0, \quad t \in \mathbb{N}_0.$

**1.45.** Solve the following difference equations:

(i) $u(t+2) - 3u(t+1) - 10u(t) = 0, \quad t \in \mathbb{N}_0;$
(ii) $u(t+2) - 8u(t+1) + 16u(t) = 0 \quad t \in \mathbb{N}_0;$
(iii) $u(t+2) - 4u(t+1) + 16u(t) = 0 \quad t \in \mathbb{N}_0;$
(iv) $u(t+3) - 3u(t+2) - 9u(t+1) + 27u(t) = 0, \quad t \in \mathbb{N}_0.$

**1.46.** Use the method of annihilators to solve the following difference equations:

(i) $y(t+2) - 5y(t+1) + 4y(t) = 4^t, \quad t \in \mathbb{N}_0;$
(ii) $y(t+2) - y(t+1) - 6y(t) = 5^t, \quad t \in \mathbb{N}_0;$
(iii) $y(t+2) - 2y(t+1) - 8y(t) = 2(4)^t, \quad t \in \mathbb{N}_0.$

**1.47.** Use the method of annihilators to solve the following difference equations:

(i) $y(t+2) - 3y(t+1) + 2y(t) = 3^t; \quad t \in \mathbb{N}_0;$
(ii) $y(t+2) - 3y(t+1) - 4y(t) = 4^t, \quad t \in \mathbb{N}_0;$
(iii) $y(t+2) - 6y(t+1) + 8y(t) = 3^t, \quad t \in \mathbb{N}_0.$

**1.48.** Use integration by parts to evaluate each of the following:

(i) $\int t^2 3^t \Delta t, \quad t \in \mathbb{N}_0$;
(ii) $\sum_{s=0}^{t-1} s2^s, \quad t \in \mathbb{N}_0$;
(iii) $\int \binom{t}{2}^2 \Delta t, \quad t \in \mathbb{N}_0$.

**1.49.** Evaluate $\sum_{k=0}^{n} k5^k$ for $n \in \mathbb{N}_0$.

**1.50.** Use integration by parts to evaluate each of the following:

(i) $\int \binom{t}{2}\binom{t}{5}\Delta t, \quad t \in \mathbb{N}_0$;
(ii) $\int \frac{t}{(t+1)(t+2)(t+3)} \Delta t, \quad t \in \mathbb{N}_0$;
(iii) $\sum_{k=1}^{n-1} \frac{k}{2^k}, \quad n \in \mathbb{N}_0$.

**1.51.** Show directly (do not use Theorem 1.62) that if $f : \mathbb{N}_a \to \mathbb{R}$ is a polynomial of degree $n$, then $f(t) = f(a) + \Delta f(a)h_1(t,a) + \cdots + \Delta^n f(a)h_n(t,a)$, here $h_k(t,a)$, $k \in \mathbb{N}_0$ are the Taylor monomials.

**1.52.** Solve the IVP in Example 1.66 by twice integrating both sides of $\Delta^2 y(t) = 3^t$ from 0 to $t$.

**1.53 (Tower of Hanoi Problem).** Assume you have three vertical pegs with $n$ rings of different sizes on the first peg with larger rings below smaller ones. Find the minimum number of moves, $y(n)$, that it takes in moving the $n$ rings on the first peg to the third peg. A move consists of transferring a single ring from one peg to another peg with the restriction that a larger ring cannot be placed on a smaller ring. (**Hint:** Find a first order linear equation that $y(n)$ satisfies and use Theorem 1.68 to find $y(n)$.)

**1.54.** Suppose that at the beginning of each year we deposit \$2,000 dollars in an IRA account that pays an annual interest rate of 4%. Find an IVP that the amount of money, $y(t)$, that we have in the account after $t$ years satisfies and use Theorem 1.68 to find $y(t)$. How much money do we have in the account after 25 years?

**1.55.** Suppose at the beginning of each year that we deposit \$3,000 dollars in an IRA account that pays an annual interest rate of 5%. How much money, $y(t)$, will we have in our IRA account at the end of the $t$-th year?

**1.56.** Prove that the IVP in Theorem 1.68 has a unique solution on $\mathbb{N}_a$.

**1.57 (Newton's Law of Cooling).** A small object of temperature 70 degrees F is placed at time $t = 0$ in a large body of water with constant temperature 40 degrees F. After 10 minutes the temperature of the object is 60 degrees F. Experiments indicate that during each minute the change in the temperature of the object is proportional to the difference of the temperature of the object and the water at the beginning of that minute. What is the temperature of the object after 5 minutes? When will the temperature of the object be 50 degrees F?

**1.58.** Solve each of the following first order linear difference equations

(i) $\Delta y(t) = 2y(t) + 3^t, \quad t \in \mathbb{N}_0;$
(ii) $y(t+1) - 4y(t) = 4^t\binom{t}{5}, \quad t \in \mathbb{N}_0;$
(iii) $\Delta y(t) = p(t)y(t) + e_p(t,a), \quad t \in \mathbb{N}_a;$
(iv) $\Delta y(t) - \frac{1}{t}y(t) = t^2 - 1, \quad t \in \mathbb{N}_1,$

where in (iii) we assume $p \in \mathcal{R}$.

**1.59.** Solve each of the following first order linear difference equations

(i) $y(t+1) - 3y(t) = 4^t, \quad t \in \mathbb{N}_0;$
(ii) $\Delta y(t) = 2y(t) + 3^t, \quad t \in \mathbb{N}_0.$

**1.60.** Show that the operators $\Delta - tI$ and $\Delta - 2I$, where $I$ is the identity operator, as operators on the set of functions mapping $\mathbb{N}_a$ to $\mathbb{R}$ do not commute.

**1.61.** Show that the operators $t\Delta - \alpha I$ and $t\Delta - \beta I$ commute, where $I$ is the identity operator and $\alpha$ and $\beta$ are constants, as operators on the set of functions mapping $\mathbb{N}_a$ to $\mathbb{C}$.

**1.62.** Use the method of factoring to solve each of the following difference equations:

(i) $\Delta^2 y(t) - 5\Delta y(t) + 6y(t) = 0, \quad t \in \mathbb{N}_a;$
(ii) $\Delta^2 y(t) + 6\Delta y(t) + 9y(t) = 0, \quad t \in \mathbb{N}_a;$
(iii) $\Delta^2 y(t) - \frac{1+3t}{t}\Delta y(t) + \frac{3}{t}y(t) = 0, \quad t \in \mathbb{N}_1;$
(iv) $\Delta^2 y(t) - (t+4)\Delta y(t) + (3t-1)y(t) = 0, \quad t \in \mathbb{N}_0.$

**1.63.** Use the method of factoring to solve each of the following difference equations:

(i) $y(t+2) - (t+4)y(t+1) + (2t+2)y(t) = 0, \quad t \in \mathbb{N}_0;$
(ii) $u(t+2) - (t+3)u(t+1) + 2tu(t) = 0, \quad t \in \mathbb{N}_1.$

**1.64.** Solve the following Euler–Cauchy difference equations:

(i) $t\sigma(t)\Delta^2 y(t) - 5t\Delta y(t) + 9y(t) = 0, \quad t \in \mathbb{N}_1;$
(ii) $t\sigma(t)\Delta^2 y(t) - 9t\Delta y(t) + 25y(t) = 0, \quad t \in \mathbb{N}_1.$

**1.65.** Solve the following Euler–Cauchy difference equations:

(i) $t(t+1)\Delta^2 y(t) + 4t\Delta y(t) + 2y(t) = 0, \quad t \in \mathbb{N}_2;$
(ii) $t\sigma(t)\Delta^2 y(t) + 9t\Delta y(t) + 16y(t) = 0, \quad t \in \mathbb{N}_5;$
(iii) $t\sigma(t)\Delta^2 y(t) - 6t\Delta y(t) + 12y(t) = 0, \quad t \in \mathbb{N}_1.$

**1.66.** Solve the following Euler–Cauchy difference equations:

(i) $t\sigma(t)\Delta^2 y(t) - 5t\Delta y(t) + 3y(t) = 0, \quad t \in \mathbb{N}_1;$
(ii) $(t\Delta - 5I)(t\Delta - 2I)\, y(t) = 0, \quad t \in \mathbb{N}_1;$
(iii) $t\sigma(t)\Delta^2 y(t) + t\Delta y(t) + 4y(t) = 0, \quad t \in \mathbb{N}_1.$

**1.67.** Solve the following Euler–Cauchy difference equations:

(i) $(t\Delta - 2I)\,(t\Delta - 3I)\,(t\Delta - 4I)\,y(t) = 0, \quad t \in \mathbb{N}_1;$
(ii) $(t\Delta - 4I)\,(t\Delta - 2I)\,(t\Delta - 2I)\,y(t) = 0, \quad t \in \mathbb{N}_1.$

**1.68.** Prove Theorem 1.86.

**1.69.** Use the variation of constants formula as in Example 1.66 to solve the IVP

$$\Delta^2 y(t) = t^{\underline{3}}, \quad t \in \mathbb{N}_0,$$
$$y(0) = y(1) = 0.$$

**1.70.** Use the variation of constants formula as in Example 1.66 to solve the IVP

$$\Delta^2 y(t) = \cos_p(t, 0), \quad t \in \mathbb{N}_0$$
$$y(0) = y(1) = 0,$$

where $p \neq 0, \pm i$ is a constant.

**1.71.** Assume $A(t)$ is a regressive matrix function on $\mathbb{N}_a$. Show that $\Phi(t)$ is a fundamental matrix of $\Delta u(t) = A(t)u(t)$ if and only if its columns are $n$ linearly independent solutions of the vector equation $\Delta u(t) = A(t)u(t)$ on $\mathbb{N}_a$.

**1.72.** Show that

$$\Phi(t) := \begin{bmatrix} (-2)^t & (-1)^t \\ (-2)^{t+1} & (-1)^{t+1} \end{bmatrix}$$
$$= (-1)^t \begin{bmatrix} 2^t & 1 \\ -2^{t+1} & -1 \end{bmatrix}$$

is a fundamental matrix of the system

$$u(t+1) = \begin{bmatrix} 0 & 1 \\ -2 & -3 \end{bmatrix} u(t).$$

**1.73.** Prove Theorem 1.82.

**1.74.** Prove that if $A(t)$ is a regressive matrix function on $\mathbb{N}_a$, then

$$(\ominus A)(t) = -A(t)[I + A(t)]^{-1}$$

for $t \in \mathbb{N}_a$.

**1.75.** Show that if $Y(t)$ is invertible for $t \in \mathbb{N}_a$, then

$$\Delta\left[Y^{-1}(t)\right] = -Y^{-1}(t)\Delta Y(t)\,(Y^{\sigma}(t))^{-1},$$

for $t \in \mathbb{N}_a$.

**1.76.** For each of the following show directly that the given matrix satisfies its own characteristic equation.

(i) $A = \begin{bmatrix} 2 & 1 & 3 \\ -1 & 2 & 0 \\ 1 & -2 & 3 \end{bmatrix}$;

(ii) $A = \begin{bmatrix} a & b \\ c & d \end{bmatrix}$.

**1.77.** Find 2 by 2 matrices $A$ and $B$ such that $A^tB^t \neq (AB)^t$ for some $t \geq 1$. Show that if two $n$ by $n$ matrices $C$ and $D$ commute, then $(CD)^t = C^tD^t$ for $t = 0, 1, 2, \cdots$.

**1.78.** Prove part (viii) of Theorem 1.84, that is

$$e_A(t, s) = e_A^{-1}(s, t) = e^*_{\ominus A^*}(s, t),$$

where $A^*$ denotes the conjugate transpose of the matrix $A$.

**1.79.** Prove part (ix) of Theorem 1.84, that is $B(t)e_A(t, t_0) = e_A(t, t_0)B(t)$, if $A(t)$ and $B(\tau)$ commute for all $t, \tau \in \mathbb{N}_a$.

**1.80.** Solve each of the following systems:

(i) $u(t+1) = \begin{bmatrix} 1 & -1 \\ 1 & 1 \end{bmatrix} u(t)$;

(ii) $u(t+1) = \begin{bmatrix} 1 & -4 \\ 2 & -3 \end{bmatrix} u(t)$;

(iii) $u(t+1) = \begin{bmatrix} 8 & -4 & 0 \\ 9 & -4 & 0 \\ 2 & -1 & 3 \end{bmatrix} u(t)$;

(iv) $u(t+1) = \begin{bmatrix} 1 & 0 & 0 \\ 1 & 0 & 1 \\ 0 & 1 & 0 \end{bmatrix} u(t)$.

**1.81.** Solve each of the following IVPs:

(i) $u(t+1) = \begin{bmatrix} 1 & 1 \\ -1 & 3 \end{bmatrix} u(t), \quad u(0) = \begin{bmatrix} 2 \\ -1 \end{bmatrix}$;

(ii) $u(t+1) = \begin{bmatrix} 2 & 0 \\ 0 & 1 \end{bmatrix} u(t) + \begin{bmatrix} 2^t \\ 3^t \end{bmatrix}, \quad u(0) = \begin{bmatrix} 1 \\ 2 \end{bmatrix}$;

(iii) $u(t+1) = \begin{bmatrix} 2 & 1 \\ -1 & 4 \end{bmatrix} u(t) + \begin{bmatrix} 3^{-t} \\ 0 \end{bmatrix}, \quad u(0) = \begin{bmatrix} 1 \\ -1 \end{bmatrix}$.

**1.82.** Solve each of the following IVPs:

(i) $u(t+1) = \begin{bmatrix} 4 & 1 \\ -1 & 2 \end{bmatrix} u(t) + \begin{bmatrix} 1 \\ 2 \end{bmatrix}, \quad u(0) = \begin{bmatrix} 1 \\ -1 \end{bmatrix}$;

(ii) $u(t+1) = \begin{bmatrix} 2 & 2 \\ 2 & -1 \end{bmatrix} u(t) + \begin{bmatrix} 1 \\ 0 \end{bmatrix}, \quad u(0) = \begin{bmatrix} 1 \\ 1 \end{bmatrix};$

(iii) $u(t+1) = \begin{bmatrix} -1 & 4 \\ -3 & 6 \end{bmatrix} u(t) + \begin{bmatrix} 0 \\ 3^t \end{bmatrix}, \quad u(0) = \begin{bmatrix} 1 \\ 1 \end{bmatrix}.$

**1.83.** Use Putzer's algorithm (see Example 1.89) to find $e_A(t, 0)$ for each of the following:

(i) $A = \begin{bmatrix} 3 & 1 \\ -1 & 1 \end{bmatrix};$

(ii) $A = \begin{bmatrix} 0 & -1 \\ 1 & 2 \end{bmatrix};$

(iii) $A = \begin{bmatrix} 2 & 1 \\ -1 & 2 \end{bmatrix}.$

**1.84.** Let

$$A = \begin{bmatrix} 0 & 1 \\ -c & -d \end{bmatrix}.$$

Show that if the matrix $A$ has a multiple characteristic root with modulus one, then the vector equation $y(t+1) = Ay(t)$, $t \in \mathbb{N}_0$, has an unbounded solution and hence the trivial solution of $y(t+1) = Ay(t)$ is unstable on $\mathbb{N}_0$. Relate this example to Theorem 1.100.

**1.85.** Prove the following result. Assume $A$ is an $n \times n$ constant matrix and $f : \mathbb{N}_0 \to \mathbb{R}^n$. Then the solution of the IVP

$$u(t+1) = Au(t) + f(t), \quad u(0) = u_0 \tag{1.61}$$

where $u_0$ is a given $n \times 1$ constant vector has a unique solution given by

$$u(t) = A^t u_0 + \int_0^t A^{t-s-1} f(s) \Delta s, \quad t \in \mathbb{N}_0.$$

**1.86.** Without solving $y(t+1) = Ay(t)$, $t \in \mathbb{N}_a$, determine the stability of the trivial solution of $y(t+1) = Ay(t)$ on $\mathbb{N}_a$ for each of the following cases:

(i) $A = \begin{bmatrix} 0 & 1 \\ -\frac{1}{2} & 1 \end{bmatrix};$

(ii) $A = \begin{bmatrix} 0 & 1 \\ \frac{1}{6} & \frac{1}{6} \end{bmatrix};$

(iii) $A = \begin{bmatrix} \frac{1}{2} & 0 & 0 \\ 0 & \frac{1}{2} & \frac{2}{3} \\ 0 & -\frac{2}{3} & \frac{1}{2} \end{bmatrix}.$

**1.87.** Find the Floquet multiplier of each of the following difference equations:

(i) $y(t+1) = \cos(\frac{2\pi t}{3})y(t)$;
(ii) $y(t+1) = \frac{2+(-1)^t}{3}y(t)$.

**1.88.** Find the Floquet multipliers of the Floquet system

$$u(t+1) = \begin{bmatrix} 0 & 1 + \frac{3+(-1)^t}{2} \\ \frac{3-(-1)^t}{2} & 0 \end{bmatrix} u(t).$$

**1.89.** Show that if $\mu_1, \mu_2, \ldots, \mu_k$ are distinct Floquet multipliers, then there are $k$ linearly independent solutions, $y_i(t)$, $1 \leq i \leq k$, of the Floquet system (1.55) on $\mathbb{Z}$ satisfying

$$y_i(t+p) = \mu_i y_i(t), \quad t \in \mathbb{Z}, \ 1 \leq i \leq k.$$

**1.90.** Find the Floquet multipliers of the Floquet system

$$u(t+1) = \begin{bmatrix} 1 & \sin\left(\frac{\pi}{2}t\right) \\ \cos\left(\frac{\pi}{2}t\right) & 1 \end{bmatrix} u(t), \quad t \in \mathbb{Z}_0.$$

**1.91.** Find the Floquet multipliers of the Floquet system

$$u(t+1) = \begin{bmatrix} 1 & \sin\left(\frac{2\pi}{3}t\right) \\ \cos\left(\frac{2\pi}{3}t\right) & 1 \end{bmatrix} u(t), \quad t \in \mathbb{Z}_0.$$

**1.92.** Prove Theorem 1.113.

# Chapter 2
# Discrete Delta Fractional Calculus and Laplace Transforms

## 2.1 Introduction

At the outset of this chapter we will be concerned with the (delta) Laplace transform, which is a special case of the Laplace transform studied in the book by Bohner and Peterson [62]. We will not assume the reader has any knowledge of the material in that book. The delta Laplace transform is equivalent under a transformation to the $Z$-transform, but we prefer the definition of the Laplace transform given here, which has the property that many of the Laplace transform formulas will be analogous to the Laplace transform formulas in the continuous setting. We will show how we can use the (delta) Laplace transform to solve initial value problems for difference equations and to solve summation equations. We then develop the discrete delta fractional calculus. Finally, we apply the Laplace transform method to solve fractional initial value problems and fractional summation equations.

The continuous fractional calculus has been well developed (see the books by Miller and Ross [147], Oldham and Spanier [152], and Podlubny [153]). But only recently has there been a great deal of interest in the discrete fractional calculus (see the papers by Atici and Eloe [32–36], Goodrich [88–96], Miller and Ross [146], and M. Holm [123–125]). More specifically, the discrete delta fractional calculus has been recently studied by a variety of authors such as Atici and Eloe [31, 32, 34, 35], Goodrich [88, 89, 91, 92, 94, 95], Miller and Ross [147], and M. Holm [123–125]. As we shall see in this chapter, one of the peculiarities of the delta fractional difference is its domain shifting properties. This property makes, in certain ways, the study of the delta fractional difference more complicated than its nabla counterpart, as a comparison of the present chapter to Chap. 3 will demonstrate.



C. Goodrich, A.C. Peterson, *Discrete Fractional Calculus*,
DOI 10.1007/978-3-319-25562-0_2

## 2.2 The Delta Laplace Transform

In this section we develop properties of the (delta) Laplace transform. First we give an abstract definition of this transform.

**Definition 2.1 (Bohner–Peterson [62]).** Assume $f : \mathbb{N}_a \to \mathbb{R}$. Then we define the (delta) **Laplace transform** of $f$ based at $a$ by

$$\mathcal{L}_a\{f\}(s) = \int_a^\infty e_{\ominus s}(\sigma(t), a)f(t)\Delta t$$

for all complex numbers $s \neq -1$ such that this improper integral converges.

The following theorem gives two useful expressions for the Laplace transform of $f$.

**Theorem 2.2.** *Assume $f : \mathbb{N}_a \to \mathbb{R}$. Then*

$$\mathcal{L}_a\{f\}(s) = F_a(s) := \int_0^\infty \frac{f(a+k)}{(s+1)^{k+1}}\Delta k \tag{2.1}$$

$$= \sum_{k=0}^{\infty} \frac{f(a+k)}{(s+1)^{k+1}}, \tag{2.2}$$

*for all complex numbers $s \neq -1$ such that this improper integral (infinite series) converges.*

*Proof.* To see that (2.1) holds note that

$$\begin{aligned}\mathcal{L}_a\{f\}(s) &= \int_a^\infty e_{\ominus s}(\sigma(t), a)f(t)\Delta t \\ &= \sum_{t=a}^{\infty} e_{\ominus s}(\sigma(t), a)f(t) \\ &= \sum_{t=a}^{\infty} [1 + \ominus s]^{\sigma(t)-a} f(t) \\ &= \sum_{t=a}^{\infty} \frac{f(t)}{(1+s)^{t-a+1}} \\ &= \sum_{k=0}^{\infty} \frac{f(a+k)}{(1+s)^{k+1}}.\end{aligned}$$

This also gives us that

$$\mathcal{L}_a\{f\}(s) = \int_0^\infty \frac{f(a+k)}{(1+s)^{k+1}}\Delta k.$$

□

To find functions such that the Laplace transform exists on a nonempty set we make the following definition.

**Definition 2.3.** We say that a function $f : \mathbb{N}_a \to \mathbb{R}$ is of **exponential order** $r > 0$ (at $\infty$) if there exists a constant $A > 0$ such that

$$|f(t)| \leq Ar^t, \quad \text{for } t \in \mathbb{N}_a, \quad \text{sufficiently large.}$$

Now we can prove the following existence theorem.

**Theorem 2.4 (Existence Theorem).** *Suppose $f : \mathbb{N}_a \to \mathbb{R}$ is of exponential order $r > 0$. Then $\mathcal{L}_a\{f\}(s)$ converges absolutely for $|s+1| > r$.*

*Proof.* Assume $f : \mathbb{N}_a \to \mathbb{R}$ is of exponential order $r > 0$. Then there is a constant $A > 0$ and an $m \in \mathbb{N}_0$ such that for each $t \in \mathbb{N}_{a+m}$, $|f(t)| \leq Ar^t$. Hence for $|s+1| > r$,

$$\begin{aligned}
\sum_{k=m}^{\infty}\left|\frac{f(k+a)}{(s+1)^{k+1}}\right| &= \sum_{k=m}^{\infty}\frac{|f(k+a)|}{|s+1|^{k+1}} \\
&\leq \sum_{k=m}^{\infty}\frac{Ar^{k+a}}{|s+1|^{k+1}} \\
&= \frac{Ar^a}{|s+1|}\sum_{k=m}^{\infty}\left(\frac{r}{|s+1|}\right)^k \\
&= \frac{Ar^a}{|s+1|}\frac{\left(\frac{r}{|s+1|}\right)^m}{1-\left(\frac{r}{|s+1|}\right)} \\
&= \frac{A}{|s+1|^m}\frac{r^{a+m}}{|s+1|-r} \\
&< \infty.
\end{aligned}$$

Hence, the Laplace transform of $f$ converges absolutely for $|s+1| > r$. □

We will see later (see Remark 2.57) that the converse of Theorem 2.4 does not hold in general.

In this chapter, we will usually consider functions $f$ of some exponential order $r > 0$, ensuring that the Laplace transform of $f$ does in fact converge somewhere in the complex plane—specifically, it converges for all complex numbers outside the closed ball of radius $r$ centered at negative one, that is, for $|s+1| > r$. We will abuse the notation by sometimes writing $\mathcal{L}_a\{f(t)\}(s)$ instead of the preferred notation $\mathcal{L}_a\{f\}(s)$.

*Example 2.5.* Clearly, $e_p(t,a)$, $p \neq -1$, a constant, is of exponential order $r = |1+p| > 0$. Therefore, we have for $|s+1| > r = |1+p|$,

$$\begin{aligned}\mathcal{L}_a\left\{e_p(t,a)\right\}(s) &= \mathcal{L}_a\left\{(1+p)^{t-a}\right\}(s)\\ &= \sum_{k=0}^{\infty}\frac{(1+p)^k}{(s+1)^{k+1}}\\ &= \frac{1}{s+1}\sum_{k=0}^{\infty}\left(\frac{p+1}{s+1}\right)^k\\ &= \frac{1}{s+1}\left(\frac{1}{1-\frac{p+1}{s+1}}\right)\\ &= \frac{1}{s-p}.\end{aligned}$$

Hence

$$\mathcal{L}_a\{e_p(t,a)\}(s) = \frac{1}{s-p}, \quad |s+1| > |1+p|.$$

An important special case ($p=0$) of the above formula is

$$\mathcal{L}_a\{1\}(s) = \frac{1}{s}, \quad \text{for} \quad |s+1| > 1.$$

In the next theorem we see that the Laplace transform operator $\mathcal{L}_a$ is a linear operator.

**Theorem 2.6 (Linearity).** *Suppose $f, g : \mathbb{N}_a \to \mathbb{R}$ and the Laplace transforms of $f$ and $g$ converge for $|s+1| > r$, where $r > 0$, and let $c_1, c_2 \in \mathbb{C}$. Then the Laplace transform of $c_1 f + c_2 g$ converges for $|s+1| > r$ and*

$$\mathcal{L}_a\{c_1 f + c_2 g\}(s) = c_1\mathcal{L}_a\{f\}(s) + c_2\mathcal{L}_a\{g\}(s), \tag{2.3}$$

*for $|s+1| > r$.*

*Proof.* Since $f, g : \mathbb{N}_a \to \mathbb{R}$ and the Laplace transforms of $f$ and $g$ converge for $|s+1| > r$, where $r > 0$, we have that for $|s+1| > r$

$$\begin{aligned} c_1\mathcal{L}_a\{f\}(s) &+ c_2\mathcal{L}_a\{g\}(s) \\ &= c_1\sum_{k=0}^{\infty}\frac{f(a+k)}{(s+1)^{k+1}} + c_2\sum_{k=0}^{\infty}\frac{g(a+k)}{(s+1)^{k+1}} \\ &= \sum_{k=0}^{\infty}\frac{(c_1f+c_2g)(a+k)}{(s+1)^{k+1}} \\ &= \mathcal{L}_a\{c_1f+c_2g\}(s). \end{aligned}$$

This completes the proof. □

The following uniqueness theorem is very useful.

**Theorem 2.7 (Uniqueness).** *Assume $f, g : \mathbb{N}_a \to \mathbb{R}$ and there is an $r > 0$ such that*

$$\mathcal{L}_a\{f\}(s) = \mathcal{L}_a\{g\}(s)$$

*for $|s+1| > r$. Then*

$$f(t) = g(t), \quad \textit{for all} \quad t \in \mathbb{N}_a.$$

*Proof.* By hypothesis we have that

$$\mathcal{L}_a\{f\}(s) = \mathcal{L}_a\{g\}(s)$$

for $|s+1| > r$. This implies that

$$\sum_{k=0}^{\infty}\frac{f(a+k)}{(s+1)^{k+1}} = \sum_{k=0}^{\infty}\frac{g(a+k)}{(s+1)^{k+1}}$$

for $|s+1| > r$. It follows from this that

$$f(a+k) = g(a+k), \quad k \in \mathbb{N}_0,$$

and this completes the proof. □

Next we give the Laplace transforms of the (delta) hyperbolic sine and cosine functions.

**Theorem 2.8.** *Assume $p \neq \pm 1$ is a constant. Then*

(i) $\mathcal{L}_a\{\cosh_p(t,a)\}(s) = \frac{s}{s^2-p^2}$;
(ii) $\mathcal{L}_a\{\sinh_p(t,a)\}(s) = \frac{p}{s^2-p^2}$,

*for $|s+1| > \max\{|1+p|, |1-p|\}$.*

*Proof.* To see that (ii) holds, consider

$$\begin{aligned}\mathcal{L}_a\{\sinh_p(t,a)\}(s) &= \frac{1}{2}\left[\mathcal{L}_a\{e_p(t,a)\}(s) - \mathcal{L}\{e_{-p}(t,a)\}(s)\right]\\ &= \frac{1}{2}\frac{1}{s-p} - \frac{1}{2}\frac{1}{s+p}\\ &= \frac{p}{s^2-p^2}\end{aligned}$$

for $|s+1| > \max\{|1+p|, |1-p|\}$. The proof of (i) is similar (see Exercise 2.5). □

Next, we give the Laplace transforms of the (discrete) sine and cosine functions.

**Theorem 2.9.** *Assume $p \neq \pm i$. Then*

(i) $\mathcal{L}_a\{\cos_p(t,a)\}(s) = \frac{s}{s^2+p^2}$;
(ii) $\mathcal{L}_a\{\sin_p(t,a)\}(s) = \frac{p}{s^2+p^2}$,

*for $|s+1| > \max\{|1+ip|, |1-ip|\}$.*

*Proof.* To see that (i) holds, note that

$$\begin{aligned}\mathcal{L}_a\{\cos_p(t,a)\}(s) &= \mathcal{L}_a\{\cosh_{ip}(t,a)\}(s)\\ &= \frac{1}{2}\left[\mathcal{L}_a\{e_{ip}(t,a)\}(s) + \mathcal{L}\{e_{-ip}(t,a)\}(s)\right]\\ &= \frac{1}{2}\frac{1}{s-ip} + \frac{1}{2}\frac{1}{s+ip}\\ &= \frac{s}{s^2+p^2},\end{aligned}$$

for $|s+1| > \max\{|1+ip|, |1-ip|\}$. For the proof of part (ii) see Exercise 2.6. □

**Theorem 2.10.** *Assume $\alpha \neq -1$ and $\frac{\beta}{1+\alpha} \neq \pm 1$. Then*

(i) $\mathcal{L}_a\{e_\alpha(t,a)\cosh_{\frac{\beta}{1+\alpha}}(t,a)\}(s) = \frac{s-\alpha}{(s-\alpha)^2-\beta^2}$;
(ii) $\mathcal{L}_a\{e_\alpha(t,a)\sinh_{\frac{\beta}{1+\alpha}}(t,a)\}(s) = \frac{\beta}{(s-\alpha)^2-\beta^2}$,

*for $|s+1| > \max\{|1+\alpha+\beta|, |1+\alpha-\beta|\}$.*

*Proof.* To see that (i) holds, for $|s+1| > \max\{|1+\alpha+\beta|, |1+\alpha-\beta|\}$, consider

$$\begin{aligned}&\mathcal{L}_a\{e_\alpha(t,a)\cosh_{\frac{\beta}{1+\alpha}}(t,a)\}(s)\\ &\quad= \frac{1}{2}\mathcal{L}_a\{e_\alpha(t,a)e_{\frac{\beta}{1+\alpha}}(t,a)\}(s) + \frac{1}{2}\mathcal{L}_a\{e_\alpha(t,a)e_{\frac{-\beta}{1+\alpha}}(t,a)\}(s)\\ &\quad= \frac{1}{2}\mathcal{L}_a\{e_{\alpha\oplus\frac{\beta}{1+\alpha}}(t,a)\}(s) + \frac{1}{2}\mathcal{L}_a\{e_{\alpha\oplus\frac{-\beta}{1+\alpha}}(t,a)\}(s)\end{aligned}$$

$$= \frac{1}{2}\mathcal{L}_a\{e_{\alpha+\beta}(t,a)\}(s) + \frac{1}{2}\mathcal{L}_a\{e_{\alpha-\beta}(t,a)\}(s)$$
$$= \frac{1}{2}\frac{1}{s-\alpha-\beta} + \frac{1}{2}\frac{1}{s-\alpha+\beta}$$
$$= \frac{s-\alpha}{(s-\alpha)^2-\beta^2}.$$

The proof of (ii) is Exercise 2.7. □

Similar to the proof of Theorem 2.10 one can prove the following theorem.

**Theorem 2.11.** *Assume* $\alpha \neq -1$ *and* $\frac{\beta}{1+\alpha} \neq \pm i$. *Then*

(i) $\mathcal{L}_a\{e_\alpha(t,a)\cos_{\frac{\beta}{1+\alpha}}(t,a)\}(s) = \frac{s-\alpha}{(s-\alpha)^2+\beta^2}$;

(ii) $\mathcal{L}_a\{e_\alpha(t,a)\sin_{\frac{\beta}{1+\alpha}}(t,a)\}(s) = \frac{\beta}{(s-\alpha)^2+\beta^2}$,

*for* $|s+1| > \max\{|1+\alpha+i\beta|, |1+\alpha-i\beta|\}$.

When solving certain difference equations one frequently uses the following theorem.

**Theorem 2.12.** *Assume that* $f$ *is of exponential order* $r > 0$. *Then for any positive integer* $N$

$$\mathcal{L}_a\left\{\Delta^N f\right\}(s) = s^N F_a(s) - \sum_{j=0}^{N-1} s^j \Delta^{N-1-j} f(a), \tag{2.4}$$

*for* $|s+1| > r$.

*Proof.* By Exercise 2.2 we have for each positive integer $N$, the function $\Delta^N f$ is of exponential order $r$. Hence, by Theorem 2.4 the Laplace transform of $\Delta^N f$ for each $N \geq 1$ exists for $|s+1| > r$. Now integrating by parts we get

$$\mathcal{L}_a\{\Delta f\}(s) = \int_a^\infty e_{\ominus s}(\sigma(t),a)\Delta f(t)\Delta t$$
$$= e_{\ominus s}(t,a)f(t)|_a^{b\to\infty} - \int_a^\infty \ominus s e_{\ominus s}(t,a)f(t)\Delta t$$
$$= -f(a) + s\int_a^\infty e_{\ominus s}(\sigma(t),a)f(t)\Delta t$$
$$= sF_a(s) - f(a)$$

for $|s+1| > r$. Hence (2.4) holds for $N = 1$. Now assume $N \geq 1$ and (2.4) holds. Then

$$\mathcal{L}_a\{\Delta^{N+1}f\}(s) = \mathcal{L}_a\{\Delta\left(\Delta^N f\right)\}(s)$$
$$= s\mathcal{L}_a\{\Delta^N f\}(s) - \Delta^N f(a)$$

$$= s\left[s^N F_a(s) - \sum_{j=0}^{N-1} s^j \Delta^{N-1-j} f(a)\right] - \Delta^N f(a)$$

$$= s^{N+1} F_a(s) - \sum_{j=0}^{(N+1)-1} s^j \Delta^{(N+1)-1-j} f(a).$$

Hence (2.4) holds for each positive integer by mathematical induction. □

The following example is an application of formula (2.4).

*Example 2.13.* Use Laplace transforms to solve the IVP

$$\Delta^2 y(t) - 3\Delta y(t) + 2y(t) = 2\cdot 4^t, \quad t \in \mathbb{N}_0$$
$$y(0) = 2, \quad \Delta y(0) = 4.$$

Assume $y(t)$ is the solution of the above IVP. We have, by taking the Laplace transform of both sides of the difference equation in this example,

$$[s^2 Y_0(s) - sy(0) - \Delta y(0)] - 3[sY_0(s) - y(0)] + 2Y_0(s) = \frac{2}{s-3}.$$

Applying the initial conditions and simplifying we get

$$(s^2 - 3s + 2)Y_0(s) = 2s - 2 + \frac{2}{s-3}.$$

Further simplification leads to

$$(s-1)(s-2)Y_0(s) = \frac{2(s-2)^2}{s-3}.$$

Hence

$$Y_0(s) = \frac{2(s-2)}{(s-1)(s-3)}$$
$$= \frac{1}{s-1} + \frac{1}{s-3}.$$

It follows that the solution of our IVP is given by

$$y(t) = e_1(t,0) + e_3(t,0)$$
$$= 2^t + 4^t, \quad t \in \mathbb{N}_0.$$

Now that we see that our solution is of exponential order we see that the steps we did above are valid.

The following corollary gives us a useful formula for solving certain summation (delta integral) equations.

**Corollary 2.14.** *Assume $f : \mathbb{N}_a \to \mathbb{R}$ is of exponential order $r > 1$. Then*

$$\mathcal{L}_a\left\{\int_a^t f(\tau)\Delta\tau\right\}(s) = \frac{1}{s}\mathcal{L}_a\{f\}(s) = \frac{F_a(s)}{s}$$

*for $|s+1| > r$.*

*Proof.* Since $f : \mathbb{N}_a \to \mathbb{R}$ is of exponential order $r > 1$, we have by Exercise 2.3 that the function $h$ defined by

$$h(t) := \int_a^t f(\tau)\Delta\tau, \quad t \in \mathbb{N}_a$$

is also of exponential order $r > 1$. Hence the Laplace transform of $h$ exists for $|s+1| > r$. Then

$$\begin{aligned}\mathcal{L}_a\{f\}(s) &= \mathcal{L}_a\{\Delta h\}(s)\\ &= s\mathcal{L}_a\{h\}(s) - h(a)\\ &= s\mathcal{L}_a\left\{\int_a^t f(\tau)\Delta\tau\right\}(s).\end{aligned}$$

It follows that

$$\mathcal{L}_a\left\{\int_a^t f(\tau)\Delta\tau\right\}(s) = \frac{1}{s}\mathcal{L}_a\{f\}(s) = \frac{F_a(s)}{s}$$

for $|s+1| > r$. □

*Example 2.15.* Solve the summation equation

$$y(t) = 2\cdot 4^t + 2\sum_{k=0}^{t-1} y(k), \quad t \in \mathbb{N}_0. \tag{2.5}$$

Equation (2.5) can be written in the equivalent form

$$y(t) = 2\cdot e_3(t,0) + 2\int_0^t y(k)\Delta k, \quad t \in \mathbb{N}_0. \tag{2.6}$$

Taking the Laplace transform of both sides of (2.6) we get, using Corollary 2.14,

$$Y_0(s) = \frac{2}{s-3} + \frac{2}{s}\,Y_0(s).$$

Solving for $Y_0(s)$ we get

$$Y_0(s) = \frac{2s}{(s-2)(s-3)} = \frac{6}{s-3} - \frac{4}{s-2}.$$

It follows that

$$y(t) = 6e_3(t,0) - 4e_2(t,0) = 6 \cdot 4^t - 4 \cdot 3^t, \quad t \in \mathbb{N}_0.$$

is the solution of (2.5).

Next we introduce the Dirac delta function and find its Laplace transform.

**Definition 2.16.** Let $c \in \mathbb{N}_a$. We define the Dirac delta function at $c$ on $\mathbb{N}_a$ by

$$\delta_c(t) = \begin{cases} 1, & t = c \\ 0, & t \neq c. \end{cases}$$

**Theorem 2.17.** *Assume $c \in \mathbb{N}_a$. Then*

$$\mathcal{L}_a\{\delta_c\}(s) = \frac{1}{(s+1)^{c-a+1}} \quad \textit{for} \quad |s+1| > 0.$$

*Proof.* For $|s+1| > 0$,

$$\mathcal{L}_a\{\delta_c\}(s) = \sum_{k=0}^{\infty} \frac{\delta_c(a+k)}{(s+1)^{k+1}} = \frac{1}{(s+1)^{c-a+1}}.$$

This completes the proof. □

Next we define the unit step function and later find its Laplace transform.

**Definition 2.18.** Let $c \in \mathbb{N}_a$. We define the **unit step function** on $\mathbb{N}_a$ by

$$u_c(t) = \begin{cases} 0, & t \in \mathbb{N}_a^{c-1} \\ 1, & t \in \mathbb{N}_c. \end{cases}$$

We now prove the following shifting theorem.

**Theorem 2.19 (Shifting Theorem).** *Let $c \in \mathbb{N}_a$ and assume the Laplace transform of $f : \mathbb{N}_a \to \mathbb{R}$ exists for $|s+1| > r$. Then the following hold:*

(i) $\mathcal{L}_a\{f(t-(c-a))u_c(t)\}(s) = \frac{1}{(s+1)^{c-a}}\mathcal{L}_a\{f\}(s)$;

(ii) $\mathcal{L}_a\{f(t+(c-a))\}(s) = (s+1)^{c-a}\left[\mathcal{L}_a\{f\}(s) - \sum_{k=0}^{c-a-1}\frac{f(a+k)}{(s+1)^{k+1}}\right]$,

*for $|s+1| > r$. (In (i) we have the convention that $f(t-(c-a))u_c(t) = 0$ for $t \in \mathbb{N}_a^{c-1}$ if $c \geq a+1$.)*

*Proof.* To see that (i) holds, consider

$$\begin{aligned}
\mathcal{L}_a\{f(t+a-c)u_c(t)\}(s) &= \sum_{k=0}^{\infty}\frac{f(2a+k-c)u_c(a+k)}{(s+1)^{k+1}} \\
&= \sum_{k=c-a}^{\infty}\frac{f(2a+k-c)}{(s+1)^{k+1}} \\
&= \sum_{k=0}^{\infty}\frac{f(2a+k+c-a-c)}{(s+1)^{k+c-a+1}} \\
&= \sum_{k=0}^{\infty}\frac{f(a+k)}{(s+1)^{k+c-a+1}} \\
&= \frac{1}{(s+1)^{c-a}}\sum_{k=0}^{\infty}\frac{f(a+k)}{(s+1)^{k+1}} \\
&= \frac{1}{(s+1)^{c-a}}\mathcal{L}_a\{f\}(s)
\end{aligned}$$

for $|s+1| > r$.

Part (ii) holds since

$$\begin{aligned}
\mathcal{L}_a\{f(t+(c-a))\}(s) &= \sum_{k=0}^{\infty}\frac{f(a+k+c-a)}{(s+1)^{k+1}} \\
&= \sum_{k=0}^{\infty}\frac{f(k+c)}{(s+1)^{k+1}} \\
&= \sum_{k=c-a}^{\infty}\frac{f(a+k)}{(s+1)^{k-c+a+1}} \\
&= (s+1)^{c-a}\sum_{k=c-a}^{\infty}\frac{f(a+k)}{(s+1)^{k+1}}
\end{aligned}$$

$$= (s+1)^{c-a}\left[\sum_{k=0}^{\infty}\frac{f(a+k)}{(s+1)^{k+1}} - \sum_{k=0}^{c-a-1}\frac{f(a+k)}{(s+1)^{k+1}}\right]$$

$$= (s+1)^{c-a}\left[\mathcal{L}_a\{f\}(s) - \sum_{k=0}^{c-a-1}\frac{f(a+k)}{(s+1)^{k+1}}\right]$$

for $|s+1| > r$. □

In the following example we will use part (i) of Theorem 2.19 to solve an IVP.

*Example 2.20.* Solve the IVP

$$\Delta y(t) - 3y(t) = 2\delta_{50}(t), \quad t \in \mathbb{N}_0$$
$$y(0) = 5.$$

Taking the Laplace transform of both sides, we get

$$sY_0(s) - y(0) - 3Y_0(s) = \frac{2}{(s+1)^{51}}.$$

Using the initial condition and solving for $Y_0(s)$ we have that

$$Y_0(s) = \frac{5}{s-3} + \frac{2}{s-3}\frac{1}{(s+1)^{51}}.$$

Taking the inverse transform of both sides we get the desired solution

$$y(t) = 5e_3(t,0) + 2e_3(t-51,0)u_{51}(t)$$
$$= 5(4^t) + 2(4)^{t-51}u_{51}(t), \quad t \in \mathbb{N}_0.$$

In the following example we will use part (ii) of Theorem 2.19 to solve an IVP.

*Example 2.21.* Use Laplace transforms to solve the IVP

$$y(t+2) + y(t+1) - 6y(t) = 0, \quad t \in \mathbb{N}_0$$
$$y(0) = 5, \quad y(1) = 2.$$

Assume $y(t)$ is the solution of this IVP and take the Laplace transform of both sides of the given difference equation to get (using part (ii) of Theorem 2.19) that

$$(s+1)^2\left[Y_0(s) - \frac{5}{s+1} - \frac{2}{(s+1)^2}\right] + (s+1)\left[Y_0(s) - \frac{5}{s+1}\right] - 6Y_0(s) = 0.$$

Solving for $Y_0(s)$ we get

$$Y_0(s) = \frac{5s+12}{(s-1)(s+4)} = \frac{17}{5}\frac{1}{s-1} + \frac{8}{5}\frac{1}{s+4}.$$

Taking the inverse transform of both sides we get

$$y(t) = \frac{17}{5}e_1(t,0) + \frac{8}{5}e_{-4}(t,0) = \frac{17}{5}2^t + \frac{8}{5}(-3)^t, \quad t \in \mathbb{N}_0.$$

**Theorem 2.22.** *The following hold for $n \geq 0$:*

(i) $\mathcal{L}_a\{h_n(t,a)\}(s) = \frac{1}{s^{n+1}}$ *for* $|s+1| > 1$;
(ii) $\mathcal{L}_a\{(t-a)^{\underline{n}}\}(s) = \frac{n!}{s^{n+1}}$ *for* $|s+1| > 1$.

*Proof.* The proof of this theorem follows from Corollary 2.14 and the fact that $\mathcal{L}\{1\}(s) = \frac{1}{s}$ for $|s+1| > 1$. □

## 2.3 Fractional Sums and Differences

The following theorem will motivate the definition of the $n$-th integer sum, which will in turn motivate the definition of the $\nu$-th fractional sum. We will then define the $\nu$-th fractional difference in terms of the $\nu$-th fractional sum.

**Theorem 2.23 (Repeated Summation Rule).** *Let $f : \mathbb{N}_a \to \mathbb{R}$ be given, then*

$$\int_a^t \int_a^{\tau_1} \cdots \int_a^{\tau_{n-1}} f(\tau_n)\Delta\tau_n \cdots \Delta\tau_2\Delta\tau_1 = \int_a^t h_{n-1}(t,\sigma(s))f(s)\Delta s. \tag{2.7}$$

*Proof.* We will prove this by induction on $n$ for $n \geq 1$. The case $n = 1$ is trivially true. Assume (2.7) holds for some $n \geq 1$. It remains to show that (2.7) then holds when $n$ is replaced by $n+1$. To this end, let

$$y(t) := \int_a^t \int_a^{\tau_1} \cdots \int_a^{\tau_{n-1}} \int_a^{\tau_n} f(\tau_{n+1})\Delta\tau_{n+1}\Delta\tau_n \cdots \Delta\tau_2\Delta\tau_1.$$

Let $g(\tau_n) = \int_a^{\tau_n} f(\tau_{n+1})\Delta\tau_{n+1}$, then it follows from the induction assumption that

$$y(t) = \int_a^t h_{n-1}(t, \sigma(s))g(s)\Delta s$$
$$= \int_a^t u(s)\Delta v(s)\Delta s,$$

where

$$u(s) := g(s), \quad \Delta v(s) = h_{n-1}(t, \sigma(s)).$$

It follows (using Theorem 1.61, (v)) that

$$\Delta u(s) = f(s) \quad v(s) = -h_n(t, s), \quad v(\sigma(s)) = -h_n(t, \sigma(s)).$$

Hence, integrating by parts, it follows that

$$y(t) = -h_n(t, s) \int_a^s f(\tau_{n+1})\Delta \tau_{n+1}\Big|_a^t$$
$$+ \int_a^t h_n(t, \sigma(s))f(s)\Delta s$$
$$= \int_a^t h_n(t, \sigma(s))f(s)\Delta s.$$

This completes the proof. □

Motivated by (2.7), we define the $n$-th integer sum $\Delta_a^{-n} f(t)$ for positive integers $n$, by

$$\Delta_a^{-n} f(t) = \int_a^t h_{n-1}(t, \sigma(s))f(s)\Delta s.$$

But, since

$$h_{n-1}(t, \sigma(s)) = 0, \quad s = t-1, t-2, \cdots, t-n+1,$$

we obtain

$$\Delta_a^{-n} f(t) = \int_a^{t-n+1} h_{n-1}(t, \sigma(s))f(s)\Delta s, \tag{2.8}$$

which we consider the correct form of the $n$-th integer sum of $f(t)$. Before we use the definition (2.8) of the $n$-th integer sum to motivate the definition of the $\nu$-th fractional sum, we define the $\nu$-th fractional Taylor monomial as follows.

**Definition 2.24.** The $\nu$-th fractional Taylor monomial based at $s$ is defined by

$$h_\nu(t,s) = \frac{(t-s)^{\underline{\nu}}}{\Gamma(\nu+1)},$$

whenever the right-hand side is well defined.

We can now define the $\nu$-th fractional sum.

**Definition 2.25.** Assume $f : \mathbb{N}_a \to \mathbb{R}$ and $\nu > 0$. Then the $\nu$-th fractional sum of $f$ (based at $a$) is defined by

$$\begin{aligned}\Delta_a^{-\nu} f(t) &:= \int_a^{t-\nu+1} h_{\nu-1}(t,\sigma(\tau))f(\tau)\Delta\tau \\ &= \sum_{\tau=a}^{t-\nu} h_{\nu-1}(t,\sigma(\tau))f(\tau),\end{aligned}$$

for $t \in \mathbb{N}_{a+\nu}$. Note that by our convention on delta integrals (sums) we can extend the domain of $\Delta_a^{-\nu} f$ to $\mathbb{N}_{a+\nu-N}$, where $N$ is the unique positive integer satisfying $N-1 < \nu \leq N$, by noting that

$$\Delta_a^{-\nu} f(t) = 0, \quad t \in \mathbb{N}_{a+\nu-N}^{a+\nu-1}.$$

The expression "fractional sum" is actually is misnomer as we define the $\nu$-th fractional sum of a function for any $\nu > 0$. Expressions like $\Delta_a^{\sqrt{3}} y(t)$ and $\Delta_a^{\pi} y(t)$ are well defined.

*Remark 2.26.* Note that the value of the $\nu$-th fractional sum of $f$ based at $a$ is a linear combination of $f(a), f(a+1), \cdots, f(t-\nu)$, where the coefficient of $f(t-\nu)$ is one. In particular one can check that $\Delta_a^{-\nu} f(t)$ has the form

$$\Delta_a^{-\nu} f(t) = h_{\nu-1}(t,\sigma(a))f(a) + \cdots + \nu f(t-\nu-1) + f(t-\nu). \tag{2.9}$$

The following formulas concerning the fractional Taylor monomials generalize the integer version of this theorem (Theorem 1.61).

**Theorem 2.27.** *Let* $t, s \in \mathbb{N}_a$. *Then*

(i) $h_\nu(t,t) = 0$
(ii) $\Delta h_\nu(t,a) = h_{\nu-1}(t,a)$;
(iii) $\Delta_s h_\nu(t,s) = -h_{\nu-1}(t,\sigma(s))$;
(iv) $\int h_\nu(t,a)\Delta t = h_{\nu+1}(t,a) + C$;
(v) $\int h_\nu(t,\sigma(s))\Delta s = -h_{\nu+1}(t,s) + C$,

*whenever these expressions make sense.*

*Proof.* To see that (iii) holds, note that

$$\begin{aligned}
\Delta_s h_\nu(t,s) &= h_\nu(t,s+1) - h_\nu(t,s) \\
&= \frac{(t-s-1)^{\underline{\nu}}}{\Gamma(\nu+1)} - \frac{(t-s)^{\underline{\nu}}}{\Gamma(\nu+1)} \\
&= \frac{\Gamma(t-s)}{\Gamma(t-s-\nu)\Gamma(\nu+1)} - \frac{\Gamma(t-s+1)}{\Gamma(t-s+1-\nu)\Gamma(\nu+1)} \\
&= \Big[(t-s-\nu) - (t-s)\Big]\frac{\Gamma(t-s)}{\Gamma(\nu+1)\Gamma(t-s-\nu+1)} \\
&= -\frac{(\nu+1)\Gamma(t-s)}{\Gamma(\nu)\Gamma(t-s-\nu+1)} \\
&= -\frac{\Gamma(t-s)}{\Gamma(\nu)\Gamma(t-s-\nu+1)} \\
&= -\frac{(t-\sigma(s))^{\underline{\nu-1}}}{\Gamma(\nu)} \\
&= -h_{\nu-1}(t,\sigma(s)).
\end{aligned}$$

The rest of the proof of this theorem is Exercise 2.16. □

*Example 2.28.* Using the definition of the fractional sum (Definition 2.25), find $\Delta_0^{-\frac{1}{2}}1$.

Using Theorem 2.27, part (v), we get

$$\begin{aligned}
\Delta_0^{-\frac{1}{2}}1 &= \int_0^{t+\frac{1}{2}} h_{-\frac{1}{2}}(t,\sigma(s)) \cdot 1\,\Delta s \\
&= -h_{\frac{1}{2}}(t,s)\Big|_{s=0}^{s=t+\frac{1}{2}} \\
&= -h_{\frac{1}{2}}(t,t+\frac{1}{2}) + h_{\frac{1}{2}}(t,0) \\
&= -\frac{(-\frac{1}{2})^{\underline{\frac{1}{2}}}}{\Gamma(\frac{3}{2})} + \frac{t^{\underline{\frac{1}{2}}}}{\Gamma(\frac{3}{2})} \\
&= \frac{2}{\sqrt{\pi}}\, t^{\underline{\frac{1}{2}}}.
\end{aligned}$$

Later we will give a formula (2.16) that also gives us this result.

Next we define the fractional difference in terms of the fractional sum.

**Definition 2.29.** Assume $f : \mathbb{N}_a \to \mathbb{R}$ and $\nu > 0$. Choose a positive integer $N$ such that $N-1 < \nu \leq N$. Then we define the $\nu$-th fractional difference by

$$\Delta_a^\nu f(t) := \Delta^N \Delta_a^{-(N-\nu)} f(t), \quad t \in \mathbb{N}_{a+N-\nu}.$$

Note that our fractional difference agrees with our prior understanding of whole-order differences—that is, for any $\nu = N \in \mathbb{N}_0$

$$\Delta_a^\nu f(t) := \Delta^N \Delta_a^{-(N-\nu)} f(t) = \Delta^N \Delta_a^{-0} f(t) = \Delta^N f(t), \tag{2.10}$$

for $t \in \mathbb{N}_a$. This is called the **Riemann–Liouville** definition of the $\nu$-th delta fractional difference.

*Remark 2.30.* We will see in the proof of Theorem 2.35 below that the value of the fractional difference $\Delta_a^\nu f(t)$ depends on the values of $f$ on $\mathbb{N}_{a+\nu-N}^{t+\nu}$. This full history nature of the value of the $\nu$-th fractional difference of $f$ is one of the important features of this fractional difference. In contrast if one is studying an $n$-th order difference equation, the term $\Delta^n f(t)$ only depends on the values of $f$ at the $n+1$ points $t, t+1, t+2, \cdots, t+n$.

*Example 2.31.* Use Definition 2.29 to find $\Delta_0^{\frac{1}{2}} 1$. Using Example 2.28, we have that

$$\begin{aligned} \Delta_0^{\frac{1}{2}} 1 &= \Delta \Delta_0^{-\frac{1}{2}} 1 \\ &= \Delta \frac{2}{\sqrt{\pi}} t^{\underline{\frac{1}{2}}} \\ &= \frac{1}{\sqrt{\pi}} t^{\underline{-\frac{1}{2}}}. \end{aligned}$$

Later we will give a formula (see (2.22)) that also gives us this result.

The following **Leibniz formulas** will be very useful.

**Lemma 2.32 (Leibniz Formulas).** *Assume* $f : \mathbb{N}_{a+\mu} \times \mathbb{N}_a \to \mathbb{R}$. *Then*

$$\Delta \left[ \int_a^{t-\mu+1} f(t,\tau) \Delta\tau \right] = \int_a^{t-\mu+1} \Delta_t f(t,\tau)\Delta\tau + f(t+1, t-\mu+1) \tag{2.11}$$

*and*

$$\Delta \left[ \int_a^{t-\mu+1} f(t,\tau) \Delta\tau \right] = \int_a^{t-\mu+2} \Delta_t f(t,\tau)\Delta\tau + f(t, t-\mu+1) \tag{2.12}$$

*for* $t \in \mathbb{N}_{a+\mu}$, *where the* $\Delta_t f(t,s)$ *inside the integral means the difference of* $f(t,\tau)$ *with respect to* $t$.

*Proof.* To see that (2.11) holds, note that, for $t \in \mathbb{N}_{a+\mu}$,

$$\Delta\left[\int_a^{t-\mu+1} f(t,\tau)\Delta\tau\right] = \int_a^{t-\mu+2} f(t+1,\tau)\Delta\tau - \int_a^{t-\mu+1} f(t,\tau)\Delta\tau$$
$$= \int_a^{t-\mu+1} \Delta_t f(t,\tau)\Delta\tau + f(t+1, t+1-\mu).$$

The proof of (2.12) is Exercise 2.19. □

In the next theorem we give a very useful formula for $\Delta_a^\nu f(t)$. We call this formula the alternate definition of $\Delta_a^\nu f(t)$ (see Holm [123, 124]).

**Theorem 2.33.** *Let $f : \mathbb{N}_a \to \mathbb{R}$ and $\nu > 0$ be given, with $N-1 < \nu \leq N$. Then*

$$\Delta_a^\nu f(t) := \begin{cases} \int_a^{t+\nu+1} h_{-\nu-1}(t,\sigma(\tau))f(\tau)\Delta\tau, & N-1<\nu<N \\ \Delta^N f(t), & \nu = N \end{cases} \tag{2.13}$$

*for $t \in \mathbb{N}_{a+N-\nu}$.*

*Proof.* First note that if $\nu = N \in \mathbb{N}_0$, then using (2.10), we have that

$$\Delta_a^\nu f(t) = \Delta^N \Delta_a^{-(N-\nu)} f(t) = \Delta^N \Delta_a^{-0} f(t) = \Delta^N f(t).$$

Now assume $N-1 < \nu < N$. Our proof of (2.13) will follow from $N$ applications of the Leibniz formula (2.12). To see this we have for $t \in \mathbb{N}_{a+N-\nu}$,

$$\Delta_a^\nu f(t) = \Delta^N \Delta_a^{-(N-\nu)} f(t)$$
$$= \Delta^N \left[\int_a^{t-(N-\nu)+1} h_{N-\nu-1}(t,\sigma(\tau))f(\tau)\Delta\tau\right]$$
$$= \Delta^{N-1}\cdot\Delta\left[\int_a^{t-(N-\nu)+1} h_{N-\nu-1}(t,\sigma(\tau))f(\tau)\Delta\tau\right].$$

Using the Leibniz rule (2.12), we get

$$\Delta_a^\nu f(t) = \Delta^{N-1}\left[\int_a^{t-(N-\nu-1)+1} h_{N-\nu-2}(t,\sigma(\tau))f(\tau)\Delta\tau \right.$$
$$\left. + h_{N-\nu-1}(t, t-(N-\nu-2))f(t-(N-\nu-1))\right]$$
$$= \Delta^{N-1}\left[\int_a^{t-(N-\nu-1)+1} h_{N-\nu-2}(t,\sigma(\tau))f(\tau)\Delta\tau\right].$$

Applying the Leibniz formula (2.12) again we get

$$\Delta_a^\nu f(t) = \Delta^{N-2}\left[\int_a^{t-(N-\nu-2)+1} h_{N-\nu-3}(t,\sigma(\tau))f(\tau)\Delta\tau \right.$$
$$\left. + h_{N-\nu-2}(t, t-(N-\nu-3))f(t-(N-\nu-2))\right]$$
$$= \Delta^{N-2}\left[\int_a^{t-(N-\nu-2)+1} h_{N-\nu-3}(t,\sigma(\tau))f(\tau)\Delta\tau\right].$$

Repeating these steps $N-2$ more times, we find that

$$\Delta_a^\nu f(t) = \Delta^{N-N}\left[\int_a^{t-(N-\nu-N)+1} h_{N-\nu-N-1}(t,\sigma(\tau))f(\tau)\Delta\tau \right.$$
$$\left. + h_{N-\nu-N}(t, t-(N-\nu-(N+1)))f(t-(N-\nu-N))\right]$$
$$= \int_a^{t+\nu+1} h_{-\nu-1}(t,\sigma(\tau))f(\tau)\Delta\tau + h_{-\nu}(t,t+\nu+1)f(t+\nu)$$
$$= \int_a^{t+\nu+1} h_{-\nu-1}(t,\sigma(\tau))f(\tau)\Delta\tau.$$

This completes the proof. □

*Remark 2.34.* By Theorem 2.33 we get for all $\nu > 0$, $\nu \notin \mathbb{N}_1$ that the formula for $\Delta_a^\nu f(t)$ can be obtained from the formula for $\Delta_a^{-\nu} f(t)$ in Definition 2.25 by replacing $\nu$ by $-\nu$ and vice-versa, but the domains are different.

**Theorem 2.35 (Existence-Uniqueness Theorem).** *Assume $q, f : \mathbb{N}_0 \to \mathbb{R}$, $\nu > 0$ and $N$ is a positive integer such that $N-1 < \nu \le N$. Then the initial value problem*

$$\Delta_{\nu-N}^\nu y(t) + q(t)y(t+\nu-N) = f(t), \quad t \in \mathbb{N}_0 \tag{2.14}$$
$$y(\nu-N+i) = A_i, \quad 0 \le i \le N-1, \tag{2.15}$$

*where $A_i$, $0 \le i \le N-1$, are given constants, has a unique solution on $\mathbb{N}_{\nu-N}$.*

*Proof.* Note that by Remark 2.26, for each fixed $t$, $\Delta_{\nu-N}^{-(N-\nu)} y(t)$ is a linear combination of $y(\nu-N), y(\nu-N+1), \cdots, y(t-N+\nu)$ with the coefficient of $y(t-N+\nu)$ being one. Since

$$\Delta^{\nu}_{\nu-N}y(t) = \Delta^N \Delta^{-(N-\nu)}_{\nu-N}y(t),$$

we have for each fixed $t$, $\Delta^{\nu}_{\nu-N}y(t)$ is a linear combination of $y(\nu-N), y(\nu-N+1), \cdots, y(t+\nu)$, where the coefficient of $y(t+\nu)$ is one. Now define $y(t)$ on $\mathbb{N}^{\nu-1}_{\nu-N}$ by the initial conditions (2.15). Then note that $y(t)$ satisfies the fractional difference equation (2.14) at $t = 0$ iff

$$\Delta^{\nu}_{\nu-N}y(0) + q(0)y(\nu - N) = f(0).$$

But this holds iff

$$(\cdots)y(\nu - N) + (\cdots)y(\nu - N + 1) + \cdots + y(\nu) + q(0)y(\nu - N) = f(0),$$

which is equivalent to the equation

$$(\cdots)A_0 + (\cdots)A_1 + \cdots + (\cdots)A_{n-1} + y(\nu) + q(0)A_0 = f(0).$$

Hence if we define $y(\nu)$ to be the solution of this last equation, then $y(t)$ satisfies the fractional difference equation at $t = 0$. Summarizing, we have shown that knowing $y(t)$ at the points $\nu - N + i$, $0 \le i \le N - 1$ uniquely determines what the value of the solution is at the next point $\nu$. Next one uses the fact that the values of $y(t)$ on $\mathbb{N}^{\nu}_{\nu-N}$ uniquely determine the value of the solution at $\nu + 1$. An induction argument shows that the solution is uniquely determined on $\mathbb{N}_{\nu-N}$. □

*Remark 2.36.* We could easily extend Theorem 2.35 to the case when $f, q : \mathbb{N}_a \to \mathbb{R}$ instead of the special case $a = 0$ that we considered in Theorem 2.35. Also, the term $q(t)y(t+\nu-N)$ in equation (2.14) could be replaced by $q(t)y(t+\nu-N+i)$ for any $0 \le i \le N-1$. Note that we picked the nice set $\mathbb{N}_0$ so that the fractional difference equation needs to be satisfied for all $t \in \mathbb{N}_0$, but then solutions are defined on the shifted set $\mathbb{N}_{\nu-N}$. By shifting the set on which the fractional difference equation is defined, we can evidently obtain solutions that are defined on the nicer set $\mathbb{N}_0$. In this book our convention when considering fractional difference equations is to assume the fractional difference equation is satisfied for $t \in \mathbb{N}_a$ and the solutions are defined on $\mathbb{N}_{a+\nu-N}$.

In a standard manner one gets the following result that follows from Theorem 2.35.

**Theorem 2.37.** *Assume $q : \mathbb{N}_0 \to \mathbb{R}$. Then the homogeneous fractional difference equation*

$$\Delta^{\nu}_{\nu-N}u(t) + q(t)u(t + \nu - N) = 0, \quad t \in \mathbb{N}_{\nu-N}$$

*has $N$ linearly independent solutions $u_i(t)$, $1 \le i \le N$, on $\mathbb{N}_0$ and*

$$u(t) = c_1u_1(t) + c_2u_2(t) + \cdots + c_Nu_N(t),$$

*where $c_1, c_2, \cdots, c_N$ are arbitrary constants, is a general solution of this homogeneous fractional difference equation on $\mathbb{N}_0$. Furthermore, if in addition, $y_p(t)$ is a particular solution of the nonhomogeneous fractional difference equation* (2.14) *on $\mathbb{N}_0$, then*

$$y(t) = c_1 u_1(t) + c_2 u_2(t) + \cdots + c_N u_N(t) + y_p(t),$$

*where $c_1, c_2, \cdots, c_N$ are arbitrary constants, is a general solution of the nonhomogeneous fractional difference equation* (2.14).

## 2.4 Fractional Power Rules

Using the Leibniz formula we will prove the following fractional sum power rule. Later in this chapter (see Theorem 2.71) we will use discrete Laplace transforms to give an easier proof of this theorem. Later we will see that the fractional difference power rule (Theorem 2.40) will follow from this fractional sum power rule.

**Theorem 2.38 (Fractional Sum Power Rule).** *Assume $\mu \geq 0$ and $\nu > 0$. Then*

$$\Delta_{a+\mu}^{-\nu}(t-a)^{\underline{\mu}} = \frac{\Gamma(\mu+1)}{\Gamma(\mu+\nu+1)}(t-a)^{\underline{\mu+\nu}} \tag{2.16}$$

*for $t \in \mathbb{N}_{a+\mu+\nu}$.*

*Proof.* Let

$$g_1(t) := \frac{\Gamma(\mu+1)}{\Gamma(\mu+\nu+1)}(t-a)^{\underline{\mu+\nu}},$$

and

$$g_2(t) := \Delta_{a+\mu}^{-\nu}(t-a)^{\underline{\mu}} = \sum_{s=a+\mu}^{t-\nu} h_{\nu-1}(t, \sigma(s))(s-a)^{\underline{\mu}}, \tag{2.17}$$

for $t \in \mathbb{N}_{a+\mu+\nu}$. To complete the proof we will show that both of these functions satisfy the initial value problem

$$(t-a-(\mu+\nu)+1)\Delta g(t) = (\mu+\nu)g(t) \tag{2.18}$$

$$g(a+\mu+\nu) = \Gamma(\mu+1). \tag{2.19}$$

Since

$$\begin{aligned} g_1(a+\mu+\nu) &= \frac{\Gamma(\mu+1)}{\Gamma(\mu+\nu+1)}(\mu+\nu)^{\underline{\mu+\nu}} \\ &= \Gamma(\mu+1) \end{aligned}$$

and

$$g_2(a+\mu+\nu) = \frac{1}{\Gamma(\nu)} \sum_{s=a+\mu}^{a+\mu} (a+\mu+\nu-\sigma(s))^{\underline{\nu-1}}(s-a)^{\underline{\mu}}$$

$$= \frac{1}{\Gamma(\nu)}(\nu-1)^{\underline{\nu-1}}\mu^{\underline{\mu}}$$

$$= \Gamma(\mu+1)$$

we have that $g_i(t)$, $i = 1, 2$ both satisfy the initial condition (2.19).

We next show that $g_1(t)$ satisfies the difference equation (2.18). Note that

$$\Delta g_1(t) = (\mu+\nu)\frac{\Gamma(\mu+1)}{\Gamma(\mu+\nu+1)}(t-a)^{\underline{\mu+\nu-1}}.$$

Multiplying both sides by $t-a-(\mu+\nu)+1$ we obtain

$$\begin{aligned}
&(t-a-(\mu+\nu)+1)\Delta g_1(t)\\
&= (\mu+\nu)\frac{\Gamma(\mu+1)}{\Gamma(\mu+\nu+1)}[t-a-(\mu+\nu-1)](t-a)^{\underline{\mu+\nu-1}}\\
&= (\mu+\nu)\frac{\Gamma(\mu+1)}{\Gamma(\mu+\nu+1)}(t-a)^{\underline{\mu+\nu}} \qquad \text{by Exercise (1.9)}\\
&= (\mu+\nu)g_1(t)
\end{aligned}$$

for $t \in \mathbb{N}_{a+\mu+\nu}$. That is, $g_1(t)$ is a solution of (2.18).

It remains to show that $g_2(t)$ satisfies (2.18). Using (2.17) we have that

$$\begin{aligned}
&g_2(t)\\
&= \frac{1}{\Gamma(\nu)} \sum_{s=a+\mu}^{t-\nu} [(t-\sigma(s))-(\nu-2)](t-\sigma(s))^{\underline{\nu-2}}(s-a)^{\underline{\mu}}\\
&= \frac{1}{\Gamma(\nu)} \sum_{s=a+\mu}^{t-\nu} [(t-a-(\mu+\nu)+1)-(s-a-\mu)](t-\sigma(s))^{\underline{\nu-2}}(s-a)^{\underline{\mu}}\\
&= \frac{t-a-(\mu+\nu)+1}{\Gamma(\nu)} \sum_{s=a+\mu}^{t-\nu} (t-\sigma(s))^{\underline{\nu-2}}(s-a)^{\underline{\mu}}\\
&- \frac{1}{\Gamma(\nu)} \sum_{s=a+\mu}^{t-\nu} (t-\sigma(s))^{\underline{\nu-2}}(s-a-\mu)(s-a)^{\underline{\mu}}\\
&= h(t)-k(t),
\end{aligned}$$

where

$$h(t) := \frac{t-a-(\mu+\nu)+1}{\Gamma(\nu)} \sum_{s=a+\mu}^{t-\nu} (t-\sigma(s))^{\underline{\nu-2}}(s-a)^{\underline{\mu}}$$

and

$$\begin{aligned} k(t) &:= \frac{1}{\Gamma(\nu)} \sum_{s=a+\mu}^{t-\nu} (t-\sigma(s))^{\underline{\nu-2}}(s-a-\mu)(s-a)^{\underline{\mu}} \\ &= \frac{1}{\Gamma(\nu)} \sum_{s=a+\mu}^{t-\nu} (t-\sigma(s))^{\underline{\nu-2}}(s-a)^{\underline{\mu+1}}. \end{aligned}$$

Using (2.17) and (2.11) we get

$$\begin{aligned} &\Delta g_2(t) \\ &= \frac{\nu-1}{\Gamma(\nu)} \sum_{s=a+\mu}^{t-\nu} (t-\sigma(s))^{\underline{\nu-2}}(s-a)^{\underline{\mu}} + \frac{1}{\Gamma(\nu)}(\nu-1)^{\underline{\nu-1}}(t+1-\nu-a)^{\underline{\mu}} \\ &= \frac{\nu-1}{\Gamma(\nu)} \sum_{s=a+\mu}^{t-\nu} (t-\sigma(s))^{\underline{\nu-2}}(s-a)^{\underline{\mu}} + (t+1-\nu-a)^{\underline{\mu}}. \end{aligned}$$

It follows that

$$(t-a+(\mu+\nu)+1)\Delta g_2(t) = (\nu-1)h(t) + (t+1-\nu-a)^{\underline{\mu+1}}. \tag{2.20}$$

Also, integrating by parts we get (here we also use Lemma 2.32)

$$\begin{aligned} k(t) &= \frac{1}{\Gamma(\nu)} \sum_{s=a+\mu}^{t-\nu} (t-\sigma(s))^{\underline{\nu-2}}(s-a)^{\underline{\mu+1}} \\ &= \frac{1}{\Gamma(\nu)} \left[ -\frac{(s-a)^{\underline{\mu+1}}(t-s)^{\underline{\nu-1}}}{\nu-1} \right]_{s=a+\mu}^{s=t+1-\nu} \\ &\quad + \frac{\mu+1}{(\nu-1)\Gamma(\nu)} \sum_{s=a+\mu}^{t-\nu} (t-\sigma(s))^{\underline{\nu-1}}(s-a)^{\underline{\mu}} \\ &= -\frac{(t+1-\nu-a)^{\underline{\mu+1}}}{\nu-1} + \frac{\mu+1}{(\nu-1)\Gamma(\nu)} \sum_{s=a+\mu}^{t-\nu} (t-\sigma(s))^{\underline{\nu-1}}(s-a)^{\underline{\mu}}. \end{aligned}$$

It follows that

$$(t+1-\nu-a)^{\underline{\mu+1}} = -(\nu-1)k(t) + (\mu+1)g_2(t). \tag{2.21}$$

Finally, from (2.21) and (2.20), we get

$$\begin{aligned}(t-a+(\mu+\nu)+1)\Delta g_2(t) &= (\nu-1)h(t) + (t+1-\nu-a)^{\underline{\mu+1}} \\ &= (\nu-1)h(t) - (\nu-1)k(t) + (\mu+1)g_2(t) \\ &= (\mu+\nu)g_2(t).\end{aligned}$$

This completes the proof. □

*Example 2.39.* Find

$$\Delta_{\frac{5}{2}}^{-\frac{3}{2}}(t-2)^{\underline{\frac{1}{2}}}, \quad t \in \mathbb{N}_2.$$

Consider

$$\begin{aligned}\Delta_{\frac{5}{2}}^{-\frac{3}{2}}(t-2)^{\underline{\frac{1}{2}}} &= \Delta_{2+\frac{1}{2}}^{-\frac{3}{2}}(t-2)^{\underline{\frac{1}{2}}} \\ &= \frac{\Gamma(\frac{3}{2})}{\Gamma(3)}(t-2)^{\underline{2}} \\ &= \frac{\sqrt{\pi}}{4}(t-2)^{\underline{2}} \\ &= \frac{\sqrt{\pi}}{4}(t^2-5t+6),\end{aligned}$$

for $t \in \mathbb{N}_2$.

**Theorem 2.40 (Fractional Difference Power Rule).** *Assume* $\mu > 0$ *and* $\nu \geq 0$, $N-1 < \nu < N$. *Then*

$$\Delta_{a+\mu}^{\nu}(t-a)^{\underline{\mu}} = \frac{\Gamma(\mu+1)}{\Gamma(\mu-\nu+1)}(t-a)^{\underline{\mu-\nu}} \tag{2.22}$$

*for* $t \in \mathbb{N}_{a+\mu+N-\nu}$.

*Proof.* To see that (2.22) holds, note that

$$\begin{aligned}\Delta_{a+\mu}^{\nu}(t-a)^{\underline{\mu}} &= \Delta^N \Delta_{a+\mu}^{-(N-\nu)}(t-a)^{\underline{\mu}} \\ &= \Delta^N\left(\frac{\Gamma(\mu+1)}{\Gamma(\mu+1+N-\nu)}(t-a)^{\underline{\mu+N-\nu}}\right)\end{aligned}$$

$$= \frac{\Gamma(\mu+1)}{\Gamma(\mu+1+N-\nu)} \Delta^N (t-a)^{\underline{\mu+N-\nu}}$$
$$= \frac{\Gamma(\mu+1)(\mu+N-\nu)^{\underline{N}}}{\Gamma(\mu+1+N-\nu)}(t-a)^{\underline{\mu-\nu}}$$
$$= \frac{\Gamma(\mu+1)}{\Gamma(\mu+1-\nu)}(t-a)^{\underline{\mu-\nu}}.$$

This completes the proof. □

*Example 2.41.* Find

$$\Delta_{\frac{5}{2}}^{\frac{1}{2}}(t-1)^{\underline{\frac{3}{2}}}, \quad t \in \mathbb{N}_1.$$

Consider

$$\Delta_{\frac{5}{2}}^{\frac{1}{2}}(t-1)^{\underline{\frac{3}{2}}} = \Delta_{1+\frac{3}{2}}^{\frac{1}{2}}(t-1)^{\underline{\frac{3}{2}}}$$
$$= \frac{\Gamma(\frac{5}{2})}{\Gamma(2)}(t-1)^{\underline{1}}$$
$$= \frac{3\sqrt{\pi}}{4}(t-1),$$

for $t \in \mathbb{N}_1$.

The fractional power rules in terms of Taylor monomials take a nice form as we see in the following theorem.

**Theorem 2.42.** *Assume $\mu > 0$, $\nu > 0$, then the following hold:*

(i) $\Delta_{a+\mu}^{-\nu} h_\mu(t,a) = h_{\mu+\nu}(t,a), \quad t \in \mathbb{N}_{a+\mu+\nu};$
(ii) $\Delta_{a+\mu}^{\nu} h_\mu(t,a) = h_{\mu-\nu}(t,a), \quad t \in \mathbb{N}_{a+\mu-\nu}.$

*Proof.* To see that (i) follows from Theorem 2.38 note that for $t \in \mathbb{N}_{a+\mu+\nu}$

$$\Delta_{a+\mu}^{-\nu} h_\mu(t,a) = \Delta_{a+\mu}^{-\nu} \frac{(t-a)^{\underline{\mu}}}{\Gamma(\mu+1)}$$
$$= \frac{1}{\Gamma(\mu+1)} \frac{\Gamma(\mu+1)}{\Gamma(\mu+\nu+1)}(t-a)^{\underline{\mu+\nu}}$$
$$= \frac{(t-a)^{\underline{\mu+\nu}}}{\Gamma(\mu+\nu+1)}$$
$$= h_{\mu+\nu}(t,a).$$

Similarly, part (ii) follows from Theorem 2.40 (see Exercise 2.22). □

**Theorem 2.43.** *Assume $\mu > 0$ and $N$ is a positive integer such that $N-1 < \mu \le N$. Then for any constant $a$*

$$x(t) = c_1(t-a)^{\underline{\mu-1}} + c_2(t-a)^{\underline{\mu-2}} + \cdots + c_N(t-a)^{\underline{\mu-N}}$$

*for all constants $c_1, c_2, \cdots, c_N$, is a solution of the fractional difference equation $\Delta^{\mu}_{a+\mu-N}y(t) = 0$ on $\mathbb{N}_{a+\mu-N}$.*

*Proof.* Let $\mu$ and $N$ be as in the statement of this theorem. If $\mu = N$, then for $1 \le k \le N$, we have that

$$\Delta^{\mu}_{a+\mu-N}(t-a)^{\underline{\mu-k}} = \Delta^{N}(t-a)^{\underline{N-k}} = 0.$$

Now assume that $N-1 < \mu < N$. Then we want to consider the expression

$$\Delta^{\mu}_{a+\mu-N}(t-a)^{\underline{\mu-k}}.$$

Note that since the subscript and the exponent do not match up in the correct way we cannot immediately apply formula (2.22) to the above expression. To compensate for this we do the following.

$$\begin{aligned}\Delta^{\mu}_{a+\mu-N}\,(t-a)^{\underline{\mu-k}} &= \sum_{s=a+\mu-N}^{t+\mu} h_{-\mu-1}(t,\sigma(s))(s-a)^{\underline{\mu-k}}\\ &= \sum_{s=a+\mu-k}^{t+\mu} h_{-\mu-1}(t,\sigma(s))(s-a)^{\underline{\mu-k}},\end{aligned}$$

since

$$(s-a)^{\underline{\mu-k}} = 0, \quad \text{for } s = a+\mu-N, a+\mu-N+1, \cdots, a+\mu-k-1.$$

Therefore, we have that

$$\begin{aligned}\Delta^{\mu}_{a+\mu-N}\,(t-a)^{\underline{\mu-k}} &= \Delta^{\mu}_{a+\mu-k}(t-a)^{\underline{\mu-k}}\\ &= \frac{\Gamma(\mu-k+1)}{\Gamma(1-k)}(t-a)^{\underline{-k}}\\ &= 0.\end{aligned}$$

The conclusion of the theorem then follows from the fact that $\Delta^{\mu}_{a}$ is a linear operator. □

It follows from Theorem 2.43 that

$$x(t) = a_1h_{\mu-1}(t,a) + a_2h_{\mu-2}(t,a) + \cdots + a_Nh_{\mu-N}(t,a)$$

is a general solution of $\Delta_{a+\mu-N}y(t) = 0$.

**Theorem 2.44 (Continuity of Fractional Differences).** *Let $f : \mathbb{N}_a \to \mathbb{R}$ be given. Then the fractional difference $\Delta_a^\nu f$ is continuous with respect to $\nu$ for $\nu > 0$. By this we mean for each fixed $m \in \mathbb{N}_0$,*

$$\Delta_a^\nu f(a + \lceil \nu \rceil - \nu + m),$$

*where $\lceil \nu \rceil$ denotes the ceiling of $\nu$, is continuous for $\nu > 0$.*

*Proof.* To prove this theorem it suffices to prove the following:

(i) $\Delta_a^\nu f(a + N - \nu + m)$ is continuous with respect to $\nu$ on $(N-1, N)$;
(ii) $\lim_{\nu \to N^-} \Delta_a^\nu f(a + N - \nu + m) = \Delta^N f(a + m)$;
(iii) $\lim_{\nu \to (N-1)^+} \Delta_a^\nu f(a + N - \nu + m) = \Delta^{N-1} f(a + m + 1)$.

First we show that (i) holds. For any fixed $\nu > 0$ with $N - 1 < \nu < N$, we have

$$\Delta_a^\nu f(a + N - \nu + m) = \sum_{s=a}^{t+\nu} h_{-\nu-1}(t, \sigma(s)) f(s) \Big|_{t=a+N-\nu+m}$$

$$= \sum_{s=a}^{a+N+m} h_{-\nu-1}(a + N - \nu + m, \sigma(s)) f(s)$$

$$= \sum_{s=a}^{a+N+m-1} h_{-\nu-1}(a + N - \nu + m, \sigma(s)) f(s) + f(a + N + m)$$

$$= \sum_{s=a}^{a+N+m-1} \frac{(a + N - \nu + m - \sigma(s))^{\underline{-\nu-1}}}{\Gamma(-\nu)} f(s) + f(a + N + m)$$

$$= \sum_{s=a}^{a+N+m-1} \frac{\Gamma(a + N - \nu + m - s)}{\Gamma(a + N + m - s + 1)\Gamma(-\nu)} f(s) + f(a + N + m)$$

$$= \sum_{s=a}^{a+N+m-1} \left( \frac{(a + N - \nu + m - s - 1) \cdots (-\nu)}{(a + N + m - s)!} f(s) \right)$$
$$+ f(a + N + m)$$

$$= \sum_{i=1}^{N+m} \left( \frac{(i - 1 - \nu) \cdots (-\nu + 1)(-\nu)}{i!} f(a + N + m - i) \right)$$
$$+ f(a + N + m).$$

It follows from this last expression that $\Delta_a^\nu f(a+N-\nu+m)$ is a continuous function of $\nu$, for $N-1<\nu<N$.

$$\begin{aligned}
&\lim_{\nu\to N^-} \Delta_a^\nu f(a+N-\nu+m)\\
&= \lim_{\nu\to N^-}\Big[\sum_{i=1}^{N+m}\Big(\frac{(i-1-\nu)\cdots(-\nu)}{i!}f(a+N+m-i)\Big)\\
&\qquad +f(a+N+m)\Big]\\
&= \sum_{i=1}^{N+m}\Big(\frac{(i-1-N)\cdots(-N)}{i!}f(a+N+m-i)\Big)+f(a+N+m)\\
&= \sum_{i=1}^{N}\Big(\frac{(i-1-N)\cdots(-N)}{i!}f(a+N+m-i)\Big)+f(a+N+m),\\
&= \sum_{i=1}^{N}\Big((-1)^i\frac{(N)\cdots(N-i+1)}{i!}f(a+N+m-i)\Big)+f(a+N+m)\\
&= \sum_{i=1}^{N}\Big((-1)^i\binom{N}{i}f(a+N+m-i)\Big)\\
&\qquad +f(a+N+m)\\
&= \sum_{i=0}^{N}(-1)^i\binom{N}{i}f(a+N+m-i)\\
&= \sum_{i=0}^{N}(-1)^i\binom{N}{i}f((a+m)+N-i)\\
&= \Delta^N f(a+m).
\end{aligned}$$

Hence, (ii) holds.

Finally, we show (iii) holds. To see this consider

$$\begin{aligned}
&\lim_{\nu\to(N-1)^+} \Delta_a^\nu f(a+N-\nu+m)\\
&= \lim_{\nu\to(N-1)^+}\Bigg[\sum_{i=1}^{N+m}\Big(\frac{(i-1-\nu)\cdots(-\nu)}{i!}f(a+N+m-i)\Big)\\
&\qquad\qquad +f(a+N+m)\Bigg]
\end{aligned}$$

$$= \sum_{i=1}^{N+m} \left( \frac{(i-N)\cdots(-N+1)}{i!} f(a+N+m-i) \right) + f(a+N+m)$$

$$= \sum_{i=1}^{N-1} \left( \frac{(i-N)\cdots(-N+1)}{i!} f(a+N+m-i) \right) + f(a+N+m)$$

$$= \sum_{i=1}^{N-1} \left( (-1)^i \frac{(N-1)\cdots(N-i)}{i!} f(a+N+m-i) \right)$$

$$+ f(a+N+m)$$

$$= \sum_{i=1}^{N-1} \left( (-1)^i \binom{N-1}{i} f(a+N+m-i) \right) + f(a+N+m)$$

$$= \sum_{i=0}^{N-1} \left( (-1)^i \binom{N-1}{i} f(a+m+1+(N-1)-i) \right)$$

$$= \Delta^{N-1} f(a+m+1).$$

Hence, (iii) holds. □

The binomial expression for $\Delta^N f(t)$ is given by

$$\Delta^N f(t) = \sum_{i=0}^{N} (-1)^i \binom{N}{i} f(t+N-i).$$

In the following theorem we give the binomial expressions for fractional differences and fractional sums.

**Theorem 2.45 (Fractional Binomial Formulas).** *Assume $N-1 < \nu \leq N$ and $f : \mathbb{N}_a \to \mathbb{R}$. Then*

$$\Delta_a^\nu f(t) = \sum_{k=0}^{t+\nu-a} (-1)^k \binom{\nu}{k} f(t+\nu-k), \qquad t \in \mathbb{N}_{a+N-\nu} \tag{2.23}$$

*and*

$$\Delta_a^{-\nu} f(t) = \sum_{k=0}^{t-a-\nu} (-1)^k \binom{-\nu}{k} f(t-\nu-k) \tag{2.24}$$

$$= \sum_{k=0}^{t-a-\nu} \binom{\nu+k-1}{k} f(t-\nu-k), \qquad t \in \mathbb{N}_{a+\nu}. \tag{2.25}$$

*Proof.* Assume $f : \mathbb{N}_a \to \mathbb{R}$ and $0 \leq \nu \leq N$. Fix $t \in \mathbb{N}_{a+N-\nu}$. Then $t = a + N - \nu + m$, for some $m \in \mathbb{N}_0$. Then

$$\begin{aligned}
\Delta_a^\nu f(t) &= \int_a^{t+\nu+1} h_{-\nu-1}(t, \sigma(\tau)) f(\tau) \Delta\tau \\
&= \sum_{\tau=a}^{t+\nu} \frac{(t-\sigma(\tau))^{\underline{-\nu-1}}}{\Gamma(-\nu)} f(\tau) \\
&= \sum_{\tau=a}^{t+\nu} \frac{\Gamma(t-\tau)}{\Gamma(t-\tau+\nu+1)\Gamma(-\nu)} f(\tau) \\
&= \sum_{\tau=a}^{a+N+m} \frac{\Gamma(a+N-\nu+m-\tau)}{\Gamma(a+N+m-\tau+1)\Gamma(-\nu)} f(\tau) \\
&= \sum_{\tau=0}^{N+m} \frac{\Gamma(N+m-\tau-\nu)}{\Gamma(N+m-\tau+1)\Gamma(-\nu)} f(a+\tau) \\
&= f(a+N+m) + \sum_{\tau=0}^{N+m-1} \frac{(N+m-1-\tau-\nu)\cdots(-\nu)}{\Gamma(N+m-\tau+1)} f(a+\tau) \\
&= f(a+N+m) \\
&\quad + \sum_{\tau=0}^{N+m-1} (-1)^{N+m-\tau} \frac{(\nu)\cdots(\nu-(N+m-\tau)+1)}{\Gamma(N+m-\tau+1)} f(a+\tau) \\
&= \sum_{\tau=0}^{N+m} (-1)^{N+m-\tau} \binom{\nu}{N+m-\tau} f(a+\tau) \\
&= \sum_{k=0}^{N+m} (-1)^k \binom{\nu}{k} f(a+N+m-k) \\
&= \sum_{k=0}^{N+m} (-1)^k \binom{\nu}{k} f((a+N-\nu+m)+\nu-k) \\
&= \sum_{k=0}^{t-a+\nu} (-1)^k \binom{\nu}{k} f(t+\nu-k).
\end{aligned}$$

Hence (2.23) holds. Since we can obtain the formula for $\Delta_a^{-\nu} f(t)$ from the formula for $\Delta_a^\nu f(t)$ by replacing $\nu$ by $-\nu$ we get that (2.24) holds with the appropriate change in domains. Finally, since

$$\binom{-\nu}{k} = (-1)^k \binom{\nu+k-1}{k},$$

(2.25) follows immediately from (2.24). □

Note that if we let $\nu = N$ in (2.23), we get the following integer binomial expression for $\Delta^N f(t)$, that is

$$\Delta^N f(t) = \sum_{k=0}^{N} (-1)^k \binom{N}{k} f(t+N-k), \qquad t \in \mathbb{N}_a.$$

## 2.5 Composition Rules

**Theorem 2.46 (Composition of Fractional Sums).** *Assume $f$ is defined on $\mathbb{N}_a$ and $\mu$, $\nu$ are positive numbers. Then*

$$\left[\Delta_{a+\nu}^{-\mu}\left(\Delta_a^{-\nu} f\right)\right](t) = \left(\Delta_a^{-(\mu+\nu)} f\right)(t) = \left[\Delta_{a+\mu}^{-\nu}\left(\Delta_a^{-\mu} f\right)\right](t)$$

*for $t \in \mathbb{N}_{a+\mu+\nu}$.*

*Proof.* For $t \in \mathbb{N}_{a+\mu+\nu}$, consider

$$\begin{aligned}
\left[\Delta_{a+\nu}^{-\mu}\left(\Delta_a^{-\nu} f\right)\right](t) &= \sum_{s=a+\nu}^{t-\mu} h_{\mu-1}(t,\sigma(s))\left(\Delta_a^{-\nu} f\right)(s) \\
&= \sum_{s=a+\nu}^{t-\mu} h_{\mu-1}(t,\sigma(s)) \sum_{r=a}^{s-\nu} h_{\nu-1}(s,\sigma(r)) f(r) \\
&= \frac{1}{\Gamma(\mu)\Gamma(\nu)} \sum_{s=a+\nu}^{t-\mu} \sum_{r=a}^{s-\nu} (t-\sigma(s))^{\underline{\mu-1}} (s-\sigma(r))^{\underline{\nu-1}} f(r) \\
&= \frac{1}{\Gamma(\mu)\Gamma(\nu)} \sum_{r=a}^{t-(\mu+\nu)} \sum_{s=r+\nu}^{t-\mu} (t-\sigma(s))^{\underline{\mu-1}} (s-\sigma(r))^{\underline{\nu-1}} f(r),
\end{aligned}$$

where in the last step we interchanged the order of summation. Letting $x = s - \sigma(r)$ we obtain

$$\begin{aligned}
&\left[\Delta_{a+\nu}^{-\mu}\left(\Delta_a^{-\nu} f\right)\right](t) \\
&= \frac{1}{\Gamma(\mu)\Gamma(\nu)} \sum_{r=a}^{t-(\mu+\nu)} \left[\sum_{x=\nu-1}^{t-\mu-r-1} (t-x-r-2)^{\underline{\mu-1}} x^{\underline{\nu-1}}\right] f(r)
\end{aligned}$$

$$= \frac{1}{\Gamma(\nu)} \sum_{r=a}^{t-(\mu+\nu)} \left[ \frac{1}{\Gamma(\mu)} \sum_{x=\nu-1}^{(t-r-1)-\mu} (t-r-1-\sigma(x))^{\underline{\mu-1}} x^{\underline{\nu-1}} \right] f(r)$$

$$= \frac{1}{\Gamma(\nu)} \sum_{r=a}^{t-(\mu+\nu)} \left[ \Delta_{\nu-1}^{-\mu} t^{\underline{\nu-1}} \right]_{t \to t-r-1} f(r).$$

But by Theorem 2.38

$$\Delta_{\nu-1}^{-\mu} t^{\underline{\nu-1}} = \frac{\Gamma(\nu)}{\Gamma(\nu+\mu)} t^{\underline{\mu+\nu-1}}$$

and therefore

$$\left[\Delta_{a+\nu}^{-\mu} \left(\Delta_a^{-\nu} f\right)\right](t) = \frac{1}{\Gamma(\nu)} \sum_{r=a}^{t-(\mu+\nu)} \frac{\Gamma(\nu)}{\Gamma(\mu+\nu)} (t-r-1)^{\underline{\mu+\nu-1}} f(r)$$

$$= \frac{1}{\Gamma(\mu+\nu)} \sum_{r=a}^{t-(\mu+\nu)} (t-\sigma(r))^{\underline{\mu+\nu-1}} f(r)$$

$$= \left(\Delta_a^{-(\mu+\nu)} f\right)(t),$$

$t \in \mathbb{N}_{a+\nu+\mu}$, which is one of the desired conclusions. Interchanging $\mu$ and $\nu$ in the above formula we also get the result

$$\left[\Delta_{a+\mu}^{-\nu} \left(\Delta_a^{-\mu} f\right)\right](t) = \left(\Delta_a^{-(\mu+\nu)} f\right)(t)$$

for $t \in \mathbb{N}_{a+\mu+\nu}$. □

In the next lemma we give composition rules for an integer difference with a fractional sum and with a fractional difference. Atici and Eloe proved (2.26) with the additional assumption that $0 < k < \nu$ and Holm [123, 125] proved (2.26) in this more general setting.

**Lemma 2.47.** *Assume* $f : \mathbb{N}_a \to \mathbb{R}$, $\nu > 0$, $N - 1 < \nu \leq N$. *Then*

$$\left[\Delta^k \left(\Delta_a^{-\nu} f\right)\right](t) = \left(\Delta_a^{k-\nu} f\right)(t), \qquad t \in \mathbb{N}_{a+\nu}. \tag{2.26}$$

*and*

$$\left[\Delta^k \left(\Delta_a^{\nu} f\right)\right](t) = \left(\Delta_a^{k+\nu} f\right)(t), \qquad t \in \mathbb{N}_{a+N-\nu}. \tag{2.27}$$

*Proof.* First we prove that

$$\left[\Delta^k \left(\Delta_a^{-k} f\right)\right](t) = f(t), \qquad t \in \mathbb{N}_{a+k} \tag{2.28}$$

by induction for $k \in \mathbb{N}_1$. For the base case we have

$$\Delta\Delta_a^{-1}f(t) = \Delta\left[\int_a^t f(\tau)\Delta\tau\right] = f(t)$$

for $t \in \mathbb{N}_{a+1}$. Now assume $k \geq 1$ and (2.28) holds. Then

$$\begin{aligned}
\Delta^{k+1}\Delta_a^{k+1}f(t) &= \Delta^{k+1}\Delta_{a+k}^{-1}\Delta_a^{-k}f(t) \qquad \text{using Theorem 2.46}\\
&= \Delta^k[\Delta\Delta_{a+k}^{-1}]\Delta_a^{-k}f(t)\\
&= \Delta^k\Delta_a^{-k}f(t) \qquad \text{by the base case with base } a+k\\
&= f(t) \qquad \text{by the induction assumption (2.28)}
\end{aligned}$$

for $t \in \mathbb{N}_{a+k+1}$. Therefore, for $k \geq N$

$$\Delta^k\Delta_a^{-N}f(t) = \Delta^{k-N}[\Delta^N\Delta_a^{-N}]f(t) = \Delta^{k-N}f(t)$$

and for $k < N$

$$\Delta^k\Delta_a^{-N}f(t) = \Delta^k\Delta_{a+N-k}^{-k}[\Delta_a^{-(N-k)}]f(t) = \Delta_a^{-(N-k)}f(t) = \Delta_a^{k-N}f(t)$$

for $t \in \mathbb{N}_{a+N}$. Hence for all $k \in \mathbb{N}_1$ we have that (2.26) holds for the case $\nu = N$. It is also true that (2.27) holds when $\nu = N$. Assume for the rest of this proof that $N - 1 < \nu < N$. We will now show by induction that (2.27) holds for $k \in \mathbb{N}_1$. For the base case $k = 1$ we have using the Leibniz rule (2.11)

$$\begin{aligned}
&\Delta\Delta_a^{\nu}f(t)\\
&= \Delta\left[\int_a^{t+\nu+1} h_{-\nu-1}(t,\sigma(\tau))f(\tau)\Delta\tau\right]\\
&= \int_a^{t+\nu+1} h_{-\nu-2}(t,\sigma(\tau))f(\tau)\Delta\tau + h_{\nu-1}(\sigma(t), t+\nu+1)f(t+\nu+1)\\
&= \int_a^{t+\nu+1} h_{-\nu-2}(t,\sigma(\tau))f(\tau)\Delta\tau + f(t+\nu+1)\\
&= \int_a^{t+\nu+2} h_{-\nu-2}(t,\sigma(\tau))f(\tau)\Delta\tau\\
&= \Delta_a^{-(-\nu-1)}f(t)\\
&= \Delta_a^{1+\nu}f(t).
\end{aligned}$$

Hence the base case

$$\Delta\Delta_a^{\nu}f(t) = \Delta_a^{1+\nu}f(t)$$

holds. Now assume $k \geq 1$ and

$$\Delta^k \Delta_a^\nu f(t) = \Delta_a^{k+\nu} f(t) \tag{2.29}$$

holds. It follows from the induction hypothesis (2.29) and the base case that

$$\begin{aligned} \Delta^{k+1} \Delta_a^\nu f(t) &= \Delta \Delta^k \Delta_a^{1+\nu} f(t) \\ &= \Delta \Delta_a^{k+\nu} f(t) \\ &= \Delta_a^{k+1+\nu} f(t). \end{aligned}$$

Hence (2.27) holds for all $k \in \mathbb{N}_1$. The proof of (2.26) is very similar and is left as an exercise (Exercise 2.23). □

We now prove a composition rule that appears in Holm [125] for a fractional difference with a fractional sum.

**Theorem 2.48.** *Assume $f : \mathbb{N}_a \to \mathbb{R}$, $\nu, \mu > 0$ and $N - 1 < \nu \leq N$, $N \in \mathbb{N}_1$. Then*

$$\Delta_{a+\mu}^\nu \Delta_a^{-\mu} f(t) = \Delta_a^{\nu-\mu} f(t), \qquad t \in \mathbb{N}_{a+\mu+N-\nu}. \tag{2.30}$$

*Proof.* Note that for $t \in \mathbb{N}_{a+\mu+N-\nu}$,

$$\begin{aligned} \Delta_{a+\mu}^\nu \Delta_a^{-\mu} f(t) &= \Delta^N \Delta_{a+\mu}^{-(N-\nu)} \Delta_a^{-\mu} f(t) \\ &= \Delta^N \Delta_a^{-(N-\nu+\mu)} f(t) \qquad \text{by Theorem 2.46} \\ &= \Delta_a^{N-(N-\nu+\mu)} f(t) \qquad \text{by (2.26)} \\ &= \Delta_a^{\nu-\mu} f(t). \end{aligned}$$

Hence (2.30) holds. □

*Remark 2.49.* From Theorem 2.46 we saw that we can take fractional sums of fractional sums by adding exponents and by Theorem 2.48 we can take fractional differences of fractional sums by adding exponents. The fundamental theorem of calculus gives us that

$$\Delta_a^{-1} \Delta f(\tau) = \int_a^t \Delta f(\tau) = f(t) - f(a) = \Delta_a^0 f(t) - f(a).$$

Hence we should not expect the fractional sum of a fractional difference can be obtained by adding exponents.

In the next theorem we give a formula for a fractional sum of an integer difference. The first formula in the following Theorem 2.50 is given in Atici et al. [34] and the second formula appears in Holm [125].

**Theorem 2.50.** *Assume $f : \mathbb{N}_a \to \mathbb{R}$, $k \in \mathbb{N}_0$ and $\nu, \mu > 0$ with $N - 1 < \mu \leq N$. Then*

$$\Delta_a^{-\nu} \Delta^k f(t) = \Delta_a^{k-\nu} f(t) - \sum_{j=0}^{k-1} h_{\nu-k+j}(t, a) \Delta^j f(a), \tag{2.31}$$

*for $t \in \mathbb{N}_{a+\nu}$, and*

$$\Delta_{a+N-\mu}^{-\nu} \Delta_a^\mu f(t) = \Delta_a^{\mu-\nu} f(t) - \sum_{j=0}^{N-1} h_{\nu-N+j}(t - N + \nu, a) \Delta_a^{j-(N-\mu)} f(a + N - \mu), \tag{2.32}$$

*for $t \in \mathbb{N}_{a+N-\mu+\nu}$.*

*Proof.* We first prove that (2.31) holds by induction for $k \in \mathbb{N}_1$. For the base case $k = 1$ we have using integration by parts and

$$h_{\nu-1}(t, t - \nu + 1) = 1 = h_{\nu-2}(t, t - \nu + 2)$$

that for $t \in \mathbb{N}_{a+\nu}$

$$\begin{aligned}
\Delta_a^{-\nu} \Delta f(t) &= \int_a^{t-\nu+1} h_{\nu-1}(t, \sigma(\tau)) \Delta f(\tau) \Delta\tau \\
&= h_{\nu-1}(t, \tau) f(t) \Big|_{\tau=a}^{t-\nu+1} + \int_a^{t-\nu+1} h_{\nu-2}(t, \sigma(\tau)) f(\tau) \Delta\tau \\
&= h_{\nu-1}(t, t - \nu + 1) f(t - \nu + 1) - h_{\nu-1}(t, a) f(a) \\
&\quad + \int_a^{t-\nu+1} h_{\nu-2}(t, \sigma(\tau)) \Delta\tau \\
&= f(t - \nu + 1) - h_{\nu-1}(t, a) f(a) + \int_a^{t-\nu+1} h_{\nu-2}(t, \sigma(\tau)) f(\tau) \Delta\tau \\
&= \int_a^{t-\nu+2} h_{\nu-2}(t, \sigma(\tau)) f(\tau) \Delta\tau - h_{\nu-1}(t, a) f(a) \\
&= \Delta_a^{1-\nu} f(t) - h_{\nu-1}(t, a) f(a)
\end{aligned}$$

which proves (2.31) for the base case $k = 1$. Now assume $k \geq 1$ and (2.31) holds for that $k$. Then we have that

$$\begin{aligned}
\Delta_a^{-\nu}\Delta^{k+1}f(t) &= \Delta_a^{-\nu}\Delta^k\Delta f(t)\\
&= \Delta_a^{k-\nu}\Delta f(t) - \sum_{j=0}^{k-1} h_{\nu-k-j}(t,a)\Delta^{j+1}f(a) \quad \text{(by (2.31))}\\
&= \Delta_a^{k-\nu}f(t) - \sum_{j=0}^{k-1} h_{\nu-k+j}(t,a)\Delta^{j+1}f(a) - h_{\nu-k-1}(t,a)f(a)\\
&= \Delta_a^{-\nu}\Delta_a^{k+1-\nu}f(t) - \sum_{j=0}^{k} h_{\nu-k-1+j}(t,a)\Delta^j f(a),
\end{aligned}$$

for $t \in \mathbb{N}_{a+N-\nu}$. Hence (2.31) holds. Next we show that (2.32) holds. To see this suppose now that $\nu > 0$ and $\mu > 0$ with $N-1 < \mu \le N$. Letting $g(t) = \Delta_a^{-(N-\mu)}f(t)$ and $b = a + N - \mu$ (the first point in the domain of $g$), we have for $t \in \mathbb{N}_{a+N-\mu+\nu}$,

$$\begin{aligned}
&\Delta_{a+N-\mu}^{-\nu}\Delta_a^{\mu}f(t)\\
&= \Delta_{a+N-\mu}^{-\nu}\Delta^N\left(\Delta_a^{-(N-\mu)}f(t)\right)\\
&= \Delta_{a+N-\mu}^{-\nu}\Delta^N g(t)\\
&= \Delta_{a+N-\mu}^{N-\nu}g(t) - \sum_{j=0}^{N-1} h_{\nu-N+j}(t,b)\Delta^j g(b) \qquad \text{by (2.32)}\\
&= \Delta_{a+N-\mu}^{N-\nu}\Delta_a^{-(N-\mu)}f(t) - \sum_{j=0}^{N-1} h_{\nu-N+j}(t,b)\Delta^j\Delta_a^{-(N-\mu)}f(b)\\
&= \Delta_a^{\mu-\nu}f(t) - \sum_{j=0}^{N-1} h_{\nu-N+j}(t-N+\nu,a)\Delta_a^{j-N+\mu}f(a+N-\mu),
\end{aligned}$$

where in this last step, we applied 2.31. □

Finally, we give a composition formula for composing two fractional differences. Note that the rule for this composition is nearly identical to the rule (2.32) for the composition $\Delta_{a+M-\mu}^{-\nu}\Delta_a^{\mu}$. Theorem 2.51 is given for the specific case $\mu \in \mathbb{N}_0$ by Atici and Eloe in [34].

**Theorem 2.51.** *Let $f : \mathbb{N}_a \to \mathbb{R}$ be given and suppose $\nu, \mu > 0$, with $N-1 < \nu \le N$ and $M-1 < \mu \le M$. Then for $t \in \mathbb{N}_{a+M-\mu+N-\nu}$,*

$$\begin{aligned}
\Delta_{a+M-\mu}^{\nu}\Delta_a^{\mu}f(t) &= \Delta_a^{\nu+\mu}f(t) -\\
&\sum_{j=0}^{M-1} h_{-\nu-M+j}(t-M+\mu,a)\Delta_a^{j-M+\mu}f(a+M-\mu) \qquad (2.33)
\end{aligned}$$

*for* $N-1<\nu<N$. *If* $\nu=N$, *then* (2.33) *simplifies to*

$$\Delta^{\nu}_{a+M-\mu}\Delta^{\mu}_{a}f(t)=\Delta^{\nu+\mu}_{a}f(t),\quad t\in\mathbb{N}_{a+M-\mu}.$$

*Proof.* Let $f,\nu$, and $\mu$ be given as in the statement of the theorem. Lemma 2.47 has already proven the case when $\nu=N$.

If $N-1<\nu<N$, then for $t\in\mathbb{N}_{a+M-\mu+N-\nu}$, we have

$$\begin{aligned}
&\Delta^{\nu}_{a+M-\mu}\Delta^{\mu}_{a}f(t)\\
&=\Delta^{N}\left[\Delta^{-(N-\nu)}_{a+M-\mu}\Delta^{\mu}_{a}f(t)\right], \text{ and now using (2.50),}\\
&=\Delta^{N}\left[\Delta_a^{-N+\nu+\mu}f(t)\right.\\
&\quad\left.-\sum_{j=0}^{M-1}\Delta_a^{j-M+\mu}f(a+M-\mu)h_{N-\nu-M+j}(t-M+\mu,a)\right]\\
&=\Delta_a^{\nu+\mu}\Delta^N h_{N-\nu-M+j}(t-M+\mu)f(t)-\\
&\sum_{j=0}^{M-1}\Delta_a^{j-M+\mu}f(a+M-\mu)\Delta^N h_{N-\nu-M+j}(t-M+\mu,a)\ \text{(Lemma 2.47)}\\
&=\Delta_a^{\nu+\mu}f(t)-\\
&\sum_{j=0}^{M-1}\Delta_a^{j-M+\mu}f(a+M-\mu)h_{-\nu-M+j}(t-M+\mu,a)\\
&=\Delta_a^{\nu+\mu}f(t)\\
&\quad-\sum_{j=0}^{M-1}\Delta_a^{j-M+\mu}h_{-\nu-M+\mu}(t-M+\mu)f(a+M-\mu).
\end{aligned}$$

□

**Theorem 2.52 (Variation of Constants Formula).** *Assume* $N\geq 1$ *is an integer and* $N-1<\nu\leq N$. *If* $f:\mathbb{N}_0\to\mathbb{R}$, *then the solution of the IVP*

$$\Delta^{\nu}_{\nu-N}y(t)=f(t),\quad t\in\mathbb{N}_0 \tag{2.34}$$

$$y(\nu-N+i)=0,\quad 0\leq i\leq N-1 \tag{2.35}$$

*is given by*

$$y(t)=\Delta_0^{-\nu}f(t)=\sum_{s=0}^{t-\nu}h_{\nu-1}(t,\sigma(s))f(s),\quad t\in\mathbb{N}_{\nu-N}.$$

*Proof.* Let

$$y(t) = \Delta_0^{-\nu} f(t) = \sum_{s=0}^{t-\nu} h_{\nu-1}(t, \sigma(s))\, f(s).$$

Then by our convention on sums

$$y(\nu - N + i) = \sum_{s=0}^{-N+i} h_{\nu-1}(\nu - N + i, \sigma(s))\, f(s) = 0$$

for $0 \leq i \leq N-1$, and hence the initial conditions (2.35) are satisfied.

Also, for $t \in \mathbb{N}_0$,

$$\begin{aligned}
\Delta_{\nu-N}^{\nu} y(t) &= \Delta^N \Delta_{\nu-N}^{-(N-\nu)} y(t) \\
&= \Delta^N \sum_{s=\nu-N}^{t-(N-\nu)} h_{N-\nu-1}(t, \sigma(s))\, y(s) \\
&= \Delta^N \sum_{s=\nu}^{t-(N-\nu)} h_{N-\nu-1}(t, \sigma(s))\, y(s),
\end{aligned}$$

where in the last step we used the initial conditions (2.35). Hence,

$$\begin{aligned}
\Delta_{\nu-N}^{\nu} y(t) &= \Delta^N \Delta_{\nu}^{-(N-\nu)} y(t) \\
&= \Delta^N \Delta_{0+\nu}^{-(N-\nu)} \Delta_0^{-\nu} f(t) \\
&= \Delta^N \Delta_0^{-N} f(t) \\
&= f(t).
\end{aligned}$$

Therefore $y$ is a solution of the fractional difference equation (2.34) on $\mathbb{N}_0$. □

Next we use the fractional variation of constants formula to solve a simple fractional IVP.

*Example 2.53.* Use the variation of constants formula in Theorem 2.52 to solve the fractional IVP

$$\Delta_{-\frac{1}{2}}^{\frac{1}{2}} y(t) = 5, \quad t \in \mathbb{N}_0$$

$$y\left(-\frac{1}{2}\right) = 3\sqrt{\pi}.$$

The solution of this IVP is defined on $\mathbb{N}_{-\frac{1}{2}}$. Note that the corresponding homogeneous fractional difference equation

$$\Delta^{\frac{1}{2}}_{-\frac{1}{2}}y(t) = 0, \quad t \in \mathbb{N}_0$$

has the general fractional equation form

$$\Delta^{\nu}_{a+\nu-N}y(t) = 0, \quad t \in \mathbb{N}_a$$

in Theorem 2.43, where

$$a = 0, \quad \nu = \frac{1}{2} \quad N = 1, \quad a + \nu - N = -\frac{1}{2}.$$

Hence $t^{\underline{-\frac{1}{2}}}$ is a solution of the homogeneous equation $\Delta^{\frac{1}{2}}_{-\frac{1}{2}}y(t) = 0$ and hence (using Theorem 2.52) a general solution of $\Delta^{\frac{1}{2}}_{-\frac{1}{2}}y(t) = 5$ is given by

$$\begin{aligned} y(t) &= ct^{\underline{-\frac{1}{2}}} + \Delta_0^{-\frac{1}{2}}5 \\ &= ct^{\underline{-\frac{1}{2}}} + 5\Delta_0^{-\frac{1}{2}}1 \end{aligned} \tag{2.36}$$

By formula (2.16) we have that

$$\Delta_0^{-\frac{1}{2}}1 = \Delta_0^{-\frac{1}{2}}t^{\underline{0}} = \frac{\Gamma(1)}{\Gamma(\frac{3}{2})}t^{\underline{\frac{1}{2}}} = \frac{2}{\sqrt{\pi}}t^{\underline{\frac{1}{2}}},$$

which is the expression that we got for $\Delta_0^{-\frac{1}{2}}1$ in Example 2.28. It follows from (2.36) that

$$y(t) = ct^{\underline{-\frac{1}{2}}} + \frac{10}{\sqrt{\pi}}t^{\underline{\frac{1}{2}}}.$$

Using the initial condition $y\left(-\frac{1}{2}\right) = 3\sqrt{\pi}$ we get that $c = 3$. Therefore, the solution of the given IVP is

$$y(t) = 3t^{\underline{-\frac{1}{2}}} + \frac{10}{\sqrt{\pi}}t^{\underline{\frac{1}{2}}},$$

for $t \in \mathbb{N}_{-\frac{1}{2}}$.

Also, it is often necessary to know how a shifted Laplace transform with respect to its base relates to the original Laplace transform with base $a$, as is described in the following theorem.

**Theorem 2.54.** *Let $m \in \mathbb{N}_0$ be given and suppose $f : \mathbb{N}_{a-m} \to \mathbb{R}$ and $g : \mathbb{N}_a \to \mathbb{R}$ are of exponential order $r > 0$. Then for $|s+1| > r$,*

$$\mathcal{L}_{a-m}\{f\}(s) = \frac{1}{(s+1)^m}\mathcal{L}_a\{f\}(s) + \sum_{k=0}^{m-1}\frac{f(a+k-m)}{(s+1)^{k+1}} \tag{2.37}$$

*and*

$$\mathcal{L}_{a+m}\{g\}(s) = (s+1)^m\,\mathcal{L}_a\{g\}(s) - \sum_{k=0}^{m-1}(s+1)^{m-1-k}\,g(a+k). \tag{2.38}$$

*Proof.* Let $f, g, r$, and $m$ be given as in the statement of this theorem. Then for $|s+1| > r$,

$$\begin{aligned}
\mathcal{L}_{a-m}\{f\}(s) &= \sum_{k=0}^{\infty}\frac{f(a-m+k)}{(s+1)^{k+1}}\\
&= \sum_{k=m}^{\infty}\frac{f(a-m+k)}{(s+1)^{k+1}} + \sum_{k=0}^{m-1}\frac{f(a-m+k)}{(s+1)^{k+1}}\\
&= \sum_{k=0}^{\infty}\frac{f(a+k)}{(s+1)^{k+m+1}} + \sum_{k=0}^{m-1}\frac{f(a+k-m)}{(s+1)^{k+1}}\\
&= \frac{1}{(s+1)^m}\mathcal{L}_a\{f\}(s) + \sum_{k=0}^{m-1}\frac{f(a+k-m)}{(s+1)^{k+1}},
\end{aligned}$$

and hence (2.37) holds.

Next, consider

$$\begin{aligned}
\mathcal{L}_{a+m}\{g\}(s) &= \sum_{k=0}^{\infty}\frac{g(a+m+k)}{(s+1)^{k+1}}\\
&= \sum_{k=m}^{\infty}\frac{g(a+k)}{(s+1)^{k-m+1}}\\
&= \sum_{k=0}^{\infty}\frac{g(a+k)}{(s+1)^{k-m+1}} - \sum_{k=0}^{m-1}\frac{g(a+k)}{(s+1)^{k-m+1}}\\
&= (s+1)^m\,\mathcal{L}_a\{g\}(s) - \sum_{k=0}^{m-1}(s+1)^{m-1-k}\,g(a+k),
\end{aligned}$$

and thus (2.38) holds. □

We leave it as an exercise to verify that applying formulas (2.37) and (2.38) yields

$$\mathcal{L}_{(a+m)-m}\{f\}(s) = \mathcal{L}_{(a-m)+m}\{f\}(s) = \mathcal{L}_a\{f\}(s),$$

for $|s+1| > r$.

Recall the definition of the fractional Taylor monomials (Definition 2.24).

**Definition 2.55.** For each $\mu \in \mathbb{R}\setminus(-\mathbb{N}_1)$, define the $\mu$-th order **Taylor monomial**, $h_\mu(t,a)$, by

$$h_\mu(t,a) := \frac{(t-a)^{\underline{\mu}}}{\Gamma(\mu+1)}, \quad \text{for} \quad t \in \mathbb{N}_a.$$

**Theorem 2.56.** *If $\mu \le 0$ and $\mu \notin (-\mathbb{N}_1)$, then $h_\mu(t,a)$ is bounded (and hence is of exponential order $r = 1$). If $\mu > 0$, then for every $r > 1$, $h_\mu(t,a)$ is of exponential order $r$.*

*Proof.* First consider the case that $\mu \le 0$ with $\mu \notin (-\mathbb{N}_0)$. Then for all large $t \in \mathbb{N}_a$,

$$h_\mu(t,a) = \frac{\Gamma(t-a+1)}{\Gamma(\mu+1)\Gamma(t-a+1-\mu)} \le \frac{1}{\Gamma(\mu+1)},$$

implying that $h_\mu$ is of exponential order one (i.e., bounded).

Next assume that $\mu > 0$, with $N \in \mathbb{N}_0$ chosen so that $N-1 < \mu \le N$. Then for any fixed $r > 1$,

$$\begin{aligned} h_\mu(t,a) = \frac{(t-a)^{\underline{\mu}}}{\Gamma(\mu+1)} &= \frac{\Gamma(t-a+1)}{\Gamma(\mu+1)\Gamma(t-a+1-\mu)} \\ &\le \frac{\Gamma(t-a+1)}{\Gamma(\mu+1)\Gamma(t-a+1-N)} \\ &= \frac{(t-a)\cdots(t-a-N+1)}{\Gamma(\mu+1)} \\ &\le \frac{(t-a)^N}{\Gamma(\mu+1)} \\ &\le \frac{r^t}{\Gamma(\mu+1)}, \end{aligned}$$

for sufficiently large $t \in \mathbb{N}_a$.

Therefore, $h_\mu(t,a)$ is of exponential order $r$ for each $\mu \in \mathbb{R}\setminus(-\mathbb{N}_1)$ and $r > 1$. It follows from Theorem 2.4 that $\mathcal{L}_a\left\{h_\mu(t,a)\right\}(s)$ exists for $|s+1| > 1$. □

*Remark 2.57.* Note that the fractional Taylor monomials, $h_\mu(t,a)$ for $\mu > 0$ are examples of functions that are of order $r$ for all $r > 1$, but are not of order 1 (see Exercise 2.4).

**Theorem 2.58.** *Let $\mu \in \mathbb{R}\setminus(-\mathbb{N}_1)$. Then*

$$\mathcal{L}_{a+\mu}\{h_\mu(t,a)\}(s) = \frac{(s+1)^\mu}{s^{\mu+1}} \tag{2.39}$$

*for $|s+1| > 1$.*

*Proof.* For $|s+1| > 1$, consider

$$\frac{(s+1)^\mu}{s^{\mu+1}} = \frac{1}{s+1}\left(\frac{s+1}{s}\right)^{\mu+1} = \frac{1}{s+1}\left(1-\frac{1}{s+1}\right)^{-\mu-1}.$$

Since $|\frac{1}{s+1}| < 1$, we have by the binomial theorem that

$$\begin{aligned}\frac{(s+1)^\mu}{s^{\mu+1}} &= \frac{1}{s+1}\sum_{k=0}^{\infty}(-1)^k\binom{-\mu-1}{k}\left(\frac{1}{s+1}\right)^k \\ &= \sum_{k=0}^{\infty}(-1)^k\binom{-\mu-1}{k}\frac{1}{(s+1)^{k+1}}.\end{aligned} \tag{2.40}$$

But

$$\begin{aligned}(-1)^k\binom{-\mu-1}{k} &= (-1)^k\frac{(-\mu-1)^{\underline{k}}}{k!} \\ &= (-1)^k\frac{(-\mu-1)(-\mu-2)\cdots(-\mu-k)}{k!} \\ &= \frac{(\mu+k)(\mu+k-1)\cdots(\mu+1)}{k!} \\ &= \frac{(\mu+k)^{\underline{k}}}{k!} \\ &= \binom{\mu+k}{k} = \binom{\mu+k}{\mu} \quad \text{by Exercise 1.12, (v)} \\ &= \frac{(\mu+k)^{\underline{\mu}}}{\Gamma(\mu+1)} \\ &= \frac{[(a+\mu+k)-a]^{\underline{\mu}}}{\Gamma(\mu+1)} \\ &= h_\mu(a+\mu+k,a).\end{aligned} \tag{2.41}$$

Using (2.40) and (2.41), we have that

$$\begin{aligned}\frac{(s+1)^{\mu}}{s^{\mu+1}} &= \sum_{k=0}^{\infty} \frac{h_{\mu}(a+\mu+k,a)}{(s+1)^{k+1}} \\ &= \mathcal{L}_{a+\mu}\left\{h_{\mu}(t,a)\right\}(s),\end{aligned}$$

for $|s+1|>1$. □

## 2.6 The Convolution Product

The following definition of the convolution product agrees with the convolution product defined for general time scales in [62], but it differs from the convolution product defined by Atici and Eloe in [32] (in the upper limit). We demonstrate several advantages of using Definition 2.59 in the following results.

**Definition 2.59.** Let $f,g:\mathbb{N}_a\to\mathbb{R}$ be given. Define the **convolution product** of $f$ and $g$ to be

$$(f*g)(t) := \sum_{r=a}^{t-1} f(r)g(t-\sigma(r)+a), \quad \text{for } t\in\mathbb{N}_a \tag{2.42}$$

(note that $(f*g)(a)=0$ by our convention on sums).

*Example 2.60.* For $p\neq 0,-1$, find the convolution product $e_p(t,a)*1$, and use your answer to find $\mathcal{L}\{e_p(t,a)*1\}(s)$. By the definition of the convolution product

$$\begin{aligned}(e_p(t,a)*1)(t) &= \sum_{r=a}^{t-1} e_p(r,a) \\ &= \int_a^t e_p(r,a)\Delta r \\ &= \frac{1}{p}e_p(r,a)\Big|_a^t \\ &= \frac{1}{p}e_p(t,a)-\frac{1}{p}.\end{aligned}$$

It follows that

$$\mathcal{L}_a\{e_p(t,a)*1\}(s) = \frac{1}{p}\frac{1}{s-p}-\frac{1}{p}\frac{1}{s} = \frac{1}{(s-p)s}.$$

Note from Example 2.60 we get that

$$\mathcal{L}_a\{e_p(t,a) * 1\}(s) = \frac{1}{(s-p)s} = \frac{1}{s-p}\frac{1}{s} = \mathcal{L}_a\{e_p(t,a)\}(s)\mathcal{L}_a\{1\}(s),$$

which is a special case of the following theorem which gives a formula for the Laplace transform of the convolution product of two functions. Later we will show that this formula is useful in solving fractional initial value problems. In this theorem we use the notation $F_a(s) := \mathcal{L}_a\{f\}(s)$, which was introduced earlier.

**Theorem 2.61 (Convolution Theorem).** *Let* $f, g : \mathbb{N}_a \to \mathbb{R}$ *be of exponential order* $r_0 > 0$. *Then*

$$\mathcal{L}_a\{f * g\}(s) = F_a(s)G_a(s), \quad \textit{for } |s+1| > r_0. \tag{2.43}$$

*Proof.* We have

$$\begin{aligned}
\mathcal{L}_a\{f * g\}(s) &= \sum_{k=0}^{\infty} \frac{(f*g)(a+k)}{(s+1)^{k+1}} = \sum_{k=1}^{\infty} \frac{(f*g)(a+k)}{(s+1)^{k+1}} \\
&= \sum_{k=1}^{\infty} \frac{1}{(s+1)^{k+1}} \sum_{r=a}^{a+k-1} f(r)g(a+k-\sigma(r)+a) \\
&= \sum_{k=1}^{\infty}\sum_{r=0}^{k-1} \frac{f(a+r)g(a+k-r-1)}{(s+1)^{k+1}} \\
&= \sum_{r=0}^{\infty}\sum_{k=0}^{\infty} \frac{f(a+r)g(a+k-r-1)}{(s+1)^{k+1}}.
\end{aligned}$$

Making the change of variables $\tau = k - r - 1$ gives us that

$$\begin{aligned}
\mathcal{L}_a\{f * g\}(s) &= \sum_{\tau=0}^{\infty}\sum_{r=0}^{\infty} \frac{f(a+r)g(a+\tau)}{(s+1)^{\tau+r+2}} \\
&= \sum_{r=0}^{\infty} \frac{f(a+r)}{(s+1)^{r+1}} \sum_{\tau=0}^{\infty} \frac{g(a+\tau)}{(s+1)^{\tau+1}} \\
&= F_a(s)G_a(s),
\end{aligned}$$

for $|s+1| > r_0$. □

*Example 2.62.* Solve the (Volterra) summation equation

$$y(t) = 3 + 12\sum_{r=0}^{t-1} \left[2^{t-r-1} - 1\right] y(r), \quad t \in \mathbb{N}_0 \tag{2.44}$$

using Laplace transforms. We can write equation (2.44) in the equivalent form

$$\begin{aligned} y(t) &= 3 + 12\sum_{r=0}^{t-1}[e_1(t-r-1,0)-1]y(r) \\ &= 3 + 12\,[(e_1(t,0)-1) * y(t)], \quad t \in \mathbb{N}_0. \end{aligned} \tag{2.45}$$

Taking the Laplace transform (based at 0) of both sides of (2.45), we obtain

$$\begin{aligned} Y_0(s) &= \frac{3}{s} + 12\left[\frac{1}{s-1} - \frac{1}{s}\right] Y_0(s) \\ &= \frac{3}{s} + \frac{12}{s(s-1)} Y_0(s). \end{aligned}$$

Solving for $Y_0(s)$, we get

$$\begin{aligned} Y_0(s) &= \frac{3(s-1)}{(s+3)(s-4)} \\ &= \frac{12/7}{s+3} + \frac{9/7}{s-4}. \end{aligned}$$

Taking the inverse Laplace transform of both sides, we get

$$\begin{aligned} y(t) &= \frac{12}{7}e_{-3}(t,0) + \frac{9}{7}e_4(t,0) \\ &= \frac{12}{7}(-2)^t + \frac{9}{7}\,5^t. \end{aligned}$$

## 2.7 Using Laplace Transforms to Solve Fractional Equations

When solving certain summation equations one uses the formula

$$\mathcal{L}_a\left\{\Delta_a^{-N} f\right\}(s) = \frac{F_a(s)}{s^N}, \tag{2.46}$$

where $N$ is a positive integer. Since the summation equation (2.5) can be written in the form

$$y(t) = 2 \cdot 4^t + 2\int_0^t y(s)\,\Delta s, \quad t \in \mathbb{N}_0,$$

this is an example of a summation equation for which we want to use the formula (2.46) with $N = 1$.

We will now set out to generalize formulas (2.4) and (2.46) to the fractional case so that we can solve fractional difference and summation equations using Laplace transforms.

We will show (see Theorem 2.65) that if $f : \mathbb{N}_a \to \mathbb{R}$ is of exponential order, then $\Delta_a^{-\nu}f$ and $\Delta_a^{\nu}f$ are of a certain exponential order and hence their Laplace transforms will exist. We will use the following lemma, which gives an estimate for $t^{\underline{\nu}}$ in the proof of Theorem 2.65.

**Lemma 2.63.** *Assume $\nu > -1$ and $N - 1 < \nu \leq N$. Then*

$$t^{\underline{\nu}} \leq t^N, \quad \text{for } t \text{ sufficiently large.} \tag{2.47}$$

*Proof.* In this proof we use the fact that $\Gamma(x) > 0$ for $x > 0$ and $\Gamma(x)$ is strictly increasing for $x \geq 2$. First consider the case $-1 < \nu \leq 0$. Then, since $t + 1 - \nu \geq t + 1$, we have for large $t$

$$\begin{aligned} t^{\underline{\nu}} &= \frac{\Gamma(t+1)}{\Gamma(t+1-\nu)} \\ &\leq 1 = t^0 = t^N. \end{aligned}$$

Next, consider the case $\nu > 0$. Then for large $t$ we have

$$t^{\underline{\nu}} = \frac{\Gamma(t+1)}{\Gamma(t+1-\nu)} \leq \frac{\Gamma(t+1)}{\Gamma(t+1-N)} = t\,(t-1)\cdots(t-(N-1)) \leq t^N.$$

This completes the proof. □

*Remark 2.64.* Thus far whenever we have considered a function $f : \mathbb{N}_a \to \mathbb{R}$, we have always taken the domain of $\Delta_a^{-\nu}f$ to be the set $\mathbb{N}_{a+\nu}$. However, it is sometimes convenient to take the domain of $\Delta_a^{-\nu}f$ to be the set $\mathbb{N}_{a+\nu-N}$, where $\nu > 0$, and $N - 1 < \nu \leq N$. By our convention on sums we see that

$$\Delta_a^{-\nu}f(a + \nu - N + k) = 0, \quad \text{for} \quad 0 \leq k \leq N - 1.$$

Later (see, for example, Theorem 2.67) we will consider both of the

$$\mathcal{L}_{a+\nu}\{\Delta_a^{-\nu}f\}(s) \quad \text{and} \quad \mathcal{L}_{a+\nu-N}\{\Delta_a^{-\nu}f\}(s).$$

Note that $\Delta_a^{-\nu}f : \mathbb{N}_{a+\nu} \to \mathbb{R}$ and $\Delta_a^{-\nu}f : \mathbb{N}_{a+\nu-N} \to \mathbb{R}$ are of the same exponential order. Theorem 2.67 will give a relationship between these two Laplace transforms.

**Theorem 2.65.** *Suppose that $f : \mathbb{N}_a \to \mathbb{R}$ is of exponential order $r \geq 1$, and let $\nu > 0$, $N - 1 < \nu \leq N$, be given. Then for each fixed $\epsilon > 0$, $\Delta_a^{-\nu}f : \mathbb{N}_{a+\nu} \to \mathbb{R}$, $\Delta_a^{-\nu}f : \mathbb{N}_{a+\nu-N} \to \mathbb{R}$, and $\Delta_a^{\nu}f : \mathbb{N}_{a+N-\nu} \to \mathbb{R}$ are of exponential order $r + \epsilon$.*

*Proof.* First we show if $f : \mathbb{N}_a \to \mathbb{R}$ is of exponential order $r = 1$, then $\Delta_a^{-\nu} f : \mathbb{N}_{a+\nu} \to \mathbb{R}$ is of exponential order $r = 1 + \epsilon$, for each $\epsilon > 0$. By Exercise 2.1 it suffices to show that $f$ is bounded on $\mathbb{N}_a$ implies $\Delta_a^{-\nu} f : \mathbb{N}_{a+\nu} \to \mathbb{R}$ is of exponential order $r = 1 + \epsilon$, for each $\epsilon > 0$. To this end assume

$$|f(t)| \leq N, \quad t \in \mathbb{N}_a.$$

Then, for $t \in \mathbb{N}_{a+\nu}$,

$$\begin{aligned}
|\Delta_a^{-\nu} f(t)| &= \left| \int_a^{t-\nu+1} h_{\nu-1}(t, \sigma(s)) f(s) \Delta s \right| \\
&\leq \int_a^{t-\nu+1} h_{\nu-1}(t, \sigma(s)) |f(s)| \Delta s \\
&\leq N \int_a^{t-\nu+1} h_{\nu-1}(t, \sigma(s)) \Delta s \\
&= -N h_\nu(t, s)\Big|_{s=a}^{s=t-\nu+1}, \qquad \text{by Theorem 2.27, part (v)} \\
&= -N h_\nu(t, t-\nu+1) + N h_\nu(t, a) \\
&= N h_\nu(t, a).
\end{aligned}$$

Since, by Theorem 2.56, $h_\nu(t, a)$ is of exponential order $1 + \epsilon$ for each $\epsilon > 0$, it follows that $\Delta_a^{-\nu} f : \mathbb{N}_{a+\nu} \to \mathbb{R}$ is of exponential order $1 + \epsilon$, for each $\epsilon > 0$.

Next assume $f$ is of exponential order $r > 1$, there exist an $A > 0$ and a $T \in \mathbb{N}_a$ such that

$$|f(t)| \leq A r^t, \quad \text{for all} \quad t \in \mathbb{N}_T. \tag{2.48}$$

For $t \in \mathbb{N}_{T+\nu}$, sufficiently large, consider

$$\begin{aligned}
\left|\Delta_a^{-\nu} f(t)\right| &= \left| \sum_{s=a}^{t-\nu} h_{\nu-1}(t, \sigma(s)) f(s) \right| \\
&\leq \sum_{s=a}^{t-\nu} h_{\nu-1}(t, \sigma(s)) |f(s)| \\
&= \sum_{s=a}^{T-1} h_{\nu-1}(t, \sigma(s)) |f(s)| + \sum_{s=T}^{t-\nu} h_{\nu-1}(t, \sigma(s)) |f(s)| \\
&\leq \left( \sum_{s=a}^{T-1} \frac{|f(s)|}{\Gamma(\nu)} \right) (t-a)^{N-1} + \frac{A(t-a)^{N-1}}{\Gamma(\nu)} \int_T^{t-\nu+1} r^s \Delta s \\
&= \left( \sum_{s=a}^{T-1} \frac{|f(s)|}{\Gamma(\nu)} \right) (t-a)^{N-1} + \frac{A(t-a)^{N-1}}{\Gamma(\nu)} \left[ \frac{r^s}{r-1} \right]_{s=T}^{s=t-\nu+1}
\end{aligned}$$

$$= \left(\sum_{s=a}^{T-1} \frac{|f(s)|}{\Gamma(\nu)}\right)(t-a)^{N-1} + \frac{A(t-a)^{N-1}}{(r-1)\Gamma(\nu)}[r^{t-\nu+1} - r^T]$$

$$\leq \left(\sum_{s=a}^{T-1} \frac{|f(s)|}{\Gamma(\nu)}\right)(t-a)^{N-1} + \frac{A(t-a)^{N-1}r^{1-\nu}}{(r-1)\Gamma(\nu)}r^t$$

$$= B(t-a)^{N-1} + C(t-a)^{N-1}r^t,$$

where $B$ and $C$ are constants. But for any fixed $\epsilon > 0$ we get by applying L'Hôpital's rule, that

$$\lim_{t\to\infty} \frac{B(t-a)^{N-1} + C(t-a)^{N-1}r^t}{(r+\epsilon)^t} = 0.$$

Therefore, $\Delta_a^{-\nu}f : \mathbb{N}_{a+\nu} \to \mathbb{R}$ is of exponential order $r+\epsilon$ for each fixed $\epsilon > 0$. By Remark 2.64, we also have $\Delta_a^{-\nu}f : \mathbb{N}_{a+\nu-N} \to \mathbb{R}$ is of exponential order $r+\epsilon$ for each fixed $\epsilon > 0$.

Finally, we show $\Delta_a^{\nu}f : \mathbb{N}_{a+N-\nu} \to \mathbb{R}$, where $N-1 < \nu \leq N$, is of exponential order $r+\epsilon$ for each fixed $\epsilon > 0$. Since

$$\Delta_a^{\nu}f(t) = \Delta^N \Delta_a^{-(N-\nu)}f(t)$$

and by the first part of the proof, $\Delta_a^{-(N-\nu)}f(t)$ is of exponential order $r+\epsilon$, we have by Exercise 2.2 that $\Delta_a^{\nu}f$ is of exponential order $r+\epsilon$. □

**Corollary 2.66.** *Suppose that $f : \mathbb{N}_a \to \mathbb{R}$ is of exponential order $r \geq 1$ and let $\nu > 0$ be given with $N-1 < \nu \leq N$. Then*

$$\mathcal{L}_{a+\nu}\left\{\Delta_a^{-\nu}f\right\}(s), \quad \mathcal{L}_{a+\nu-N}\left\{\Delta_a^{-\nu}f\right\}(s), \quad \text{and} \quad \mathcal{L}_{a+N-\nu}\left\{\Delta_a^{\nu}f\right\}(s)$$

*converge for all $|s+1| > r$.*

*Proof.* Suppose $f, r$, and $\nu$ are as in the statement of this corollary and fix $s_0$ so that $|s_0+1| > r$. Then there is an $\epsilon_0 > 0$ so that $|s_0+1| > r+\epsilon_0$. Since we know by Theorem 2.65 that $\Delta_a^{-\nu}f : \mathbb{N}_{a+\nu} \to \mathbb{R}$, $\Delta_a^{-\nu}f : \mathbb{N}_{a+\nu-N} \to \mathbb{R}$, and $\Delta_a^{\nu}f : \mathbb{N}_{a+N-\nu} \to \mathbb{R}$ are of exponential order $r+\epsilon_0$, it follows from Theorem 2.4 that $\mathcal{L}_{a+\nu}\left\{\Delta_a^{-\nu}f\right\}(s_0)$, $\mathcal{L}_{a+\nu-N}\left\{\Delta_a^{-\nu}f\right\}(s_0)$, and $\mathcal{L}_{a+N-\nu}\left\{\Delta_a^{\nu}f\right\}(s_0)$ converge. Since $|s_0+1| > r$ is arbitrary, we have that

$$\mathcal{L}_{a+\nu}\left\{\Delta_a^{-\nu}f\right\}(s), \quad \mathcal{L}_{a+\nu-N}\left\{\Delta_a^{-\nu}f\right\}(s), \quad \text{and} \quad \mathcal{L}_{a+N-\nu}\left\{\Delta_a^{\nu}f\right\}(s)$$

all converge for all $|s+1| > r$. □

## 2.8 The Laplace Transform of Fractional Operators

With Corollary 2.66 in hand to insure the correct domain of convergence for the Laplace transform of any fractional operator, we may now safely develop formulas for applying the Laplace transform to fractional operators. This is the content of the next theorem.

**Theorem 2.67.** *Suppose $f : \mathbb{N}_a \to \mathbb{R}$ is of exponential order $r \geq 1$, and let $\nu > 0$ be given with $N - 1 < \nu \leq N$. Then for $|s+1| > r$,*

$$\mathcal{L}_{a+\nu}\left\{\Delta_a^{-\nu} f\right\}(s) = \frac{(s+1)^\nu}{s^\nu} F_a(s), \tag{2.49}$$

*and*

$$\mathcal{L}_{a+\nu-N}\left\{\Delta_a^{-\nu} f\right\}(s) = \frac{(s+1)^{\nu-N}}{s^\nu} F_a(s). \tag{2.50}$$

*Proof.* Since $f : \mathbb{N}_a \to \mathbb{R}$ is of exponential order $r \geq 1$, $F_a(s)$ exists for $|s+1| > r$ and by Corollary 2.66 both $\mathcal{L}_{a+\nu}\left\{\Delta_a^{-\nu} f\right\}(s)$ and $\mathcal{L}_{a+\nu-N}\left\{\Delta_a^{-\nu} f\right\}(s)$ exist for $|s+1| > r$. First, we find a relationship between the left-hand sides of equations (2.49) and (2.50). Using (2.37), we get

$$\begin{aligned}
&\mathcal{L}_{a+\nu-N}\left\{\Delta_a^{-\nu} f\right\}(s) \\
&\qquad = \frac{1}{(s+1)^N} \mathcal{L}_{a+\nu}\left\{\Delta_a^{-\nu} f\right\}(s) + \sum_{k=0}^{N-1} \frac{\Delta_a^{-\nu} f(a+\nu-N+k)}{(s+1)^{k+1}} \\
&\qquad = \frac{1}{(s+1)^N} \mathcal{L}_{a+\nu}\left\{\Delta_a^{-\nu} f\right\}(s),
\end{aligned} \tag{2.51}$$

using the fact that $\Delta_a^{-\nu} f(a+\nu-N+k) = 0$ for $0 \leq k \leq N-1$, by our convention on sums.

To see that (2.49) holds, note that

$$\begin{aligned}
&\mathcal{L}_{a+\nu}\left\{\Delta_a^{-\nu} f\right\}(s) \\
&= \sum_{k=0}^{\infty} \frac{\Delta_a^{-\nu} f(a+k+\nu)}{(s+1)^{k+1}} \\
&= \sum_{k=0}^{\infty} \frac{1}{(s+1)^{k+1}} \sum_{r=a}^{k+a} h_{\nu-1}(a+k+\nu, \sigma(r)) f(r) \\
&= \sum_{k=0}^{\infty} \frac{1}{(s+1)^{k+1}} \sum_{r=a}^{k+a} f(r) h_{\nu-1}\left((a+k+1) - \sigma(r) + a, a - (\nu-1)\right)
\end{aligned}$$

$$
\begin{aligned}
&= \sum_{k=0}^{\infty} \frac{(f * h_{\nu-1}(t, a-(\nu-1)))\,(a+1+k)}{(s+1)^{k+1}}, && \text{by (2.42)}\\
&= \mathcal{L}_{a+1}\left\{f * h_{\nu-1}(t, a-(\nu-1))\right\}(s)\\
&= (s+1)\,\mathcal{L}_a\left\{f * h_{\nu-1}(t, a-(\nu-1))\right\}(s), && \text{using (2.38) and (2.42)}\\
&= (s+1)\,F_a(s)\,\mathcal{L}_a\left\{h_{\nu-1}(t, a-(\nu-1))\right\}(s), && \text{by (2.43)}\\
&= \frac{(s+1)^{\nu}}{s^{\nu}} F_a(s), \text{ applying (2.38), since } r \geq 1
\end{aligned}
$$

proving (2.49). Finally, using (2.51) and (2.49), we get

$$
\begin{aligned}
\mathcal{L}_{a+\nu-N}\left\{\Delta_a^{-\nu} f\right\}(s) &= \frac{1}{(s+1)^N} \mathcal{L}_{a+\nu}\left\{\Delta_a^{-\nu} f\right\}(s)\\
&= \frac{(s+1)^{\nu-N}}{s^{\nu}} F_a(s),
\end{aligned}
$$

for $|s+1| > r$, proving (2.50). □

*Example 2.68.* Find $\mathcal{L}_{2+\pi+e}\{\Delta_{5+\pi}^{-e} f\}(s)$ given that

$$
f(t) = (t-5)^{\underline{\pi}}, \quad t \in \mathbb{N}_{5+\pi}.
$$

First note that

$$
f(t) = \Gamma(\pi+1) h_\pi(t, 5), \quad t \in \mathbb{N}_{5+\pi},
$$

and hence using (2.39) we have that

$$
F_{5+\pi}(s) = \Gamma(\pi+1)\mathcal{L}_{5+\pi}\{h_\pi(t,5)\}(s) = \Gamma(\pi+1)\frac{(s+1)^\pi}{s^{\pi+1}}
$$

for $|s+1| > 1$.

Then using (2.50) gives us

$$
\begin{aligned}
\mathcal{L}_{2+\pi+e}\left\{\Delta_{5+\pi}^{-e} f\right\}(s) &= \mathcal{L}_{(5+\pi)+e-3}\{\Delta_{5+\pi}^{-e} f\}(s)\\
&= \frac{(s+1)^{e-3}}{s^e}\left(\Gamma(\pi+1)\frac{(s+1)^\pi}{s^{\pi+1}}\right)\\
&= \Gamma(\pi+1)\frac{(s+1)^{\pi+e-3}}{s^{\pi+e+1}}
\end{aligned}
$$

for $|s+1| > 1$.

*Remark 2.69.* Note that when $\nu = N$ in (2.50), the correct well-known formula (2.46) for $N = 1$, is obtained. This holds true for the Laplace transform of a fractional difference as well, as the following theorem shows (Holm [123]).

**Theorem 2.70.** *Suppose $f : \mathbb{N}_a \to \mathbb{R}$ is of exponential order $r \geq 1$, and let $\nu > 0$ be given with $N - 1 < \nu \leq N$. Then for $|s+1| > r$*

$$\mathcal{L}_{a+N-\nu}\left\{\Delta_a^\nu f\right\}(s) = s^\nu (s+1)^{N-\nu} F_a(s) - \sum_{j=0}^{N-1} s^j \Delta_a^{\nu-1-j} f(a+N-\nu). \tag{2.52}$$

*Proof.* Let $f, r, \nu$, and $N$ be given as in the statement of the theorem. By Exercise 2.28 we have that (2.52) holds when $\nu = N$. Hence we assume $N-1 < \nu < N$. To see this, consider

$$\begin{aligned}
&\mathcal{L}_{a+N-\nu}\left\{\Delta_a^\nu f\right\}(s)\\
&= \mathcal{L}_{a+N-\nu}\left\{\Delta^N \Delta_a^{-(N-\nu)} f\right\}(s)\\
&= s^N \mathcal{L}_{a+N-\nu}\left\{\Delta_a^{-(N-\nu)} f\right\}(s)\\
&\quad - \sum_{j=0}^{N-1} s^j \Delta^{N-1-j} \Delta_a^{-(N-\nu)} f(a+N-\nu)\\
&= s^N \frac{(s+1)^{N-\nu}}{s^{N-\nu}} F_a(s)\\
&\quad - \sum_{j=0}^{N-1} s^j \Delta^{N-1-j} \Delta_a^{-(N-\nu)} f(a+N-\nu)\\
&= s^\nu (s+1)^{N-\nu} F_a(s) - \sum_{j=0}^{N-1} s^j \Delta_a^{\nu-1-j} f(a+N-\nu).
\end{aligned}$$

This completes the proof. □

## 2.9 Power Rule and Composition Rule

In this section (see Atici and Eloe [34], Holm [123, 125]), we present a number of properties and formulas concerning fractional sum and difference operators are developed. These include composition rules and fractional power rules, whose proofs employ a variety of tools, none of which involves the Laplace transform. However, some of these results may also be proved using the Laplace transform. The following are two previously known results for which the Laplace transform provides a significantly shorter and cleaner proof than the original ones found in [34, 123].

**Theorem 2.71 (Power Rule).** *Let* $\nu, \mu > 0$ *be given. Then for* $t \in \mathbb{N}_{a+\mu+\nu}$,

$$\Delta_{a+\mu}^{-\nu} (t-a)^{\underline{\mu}} = \frac{\Gamma(\mu+1)}{\Gamma(\mu+1+\nu)} (t-a)^{\underline{\mu+\nu}}$$

*or equivalently*

$$\Delta_{a+\mu}^{-\nu} h_\mu(t,a) = h_{\mu+\nu}(t,a).$$

*Proof.* Applying Remark 2.57 together with Lemma 2.63, we conclude that for each $\epsilon > 0$, $(t-a)^{\underline{\mu}}$ is of exponential order $1+\epsilon$ and therefore we have that $\Delta_{a+\mu}^{-\nu} (t-a)^{\underline{\mu}}$ is of exponential order $1+2\epsilon$. Thus, after employing an argument similar to that given in Corollary 2.66, we conclude that both $\mathcal{L}_{a+\mu}\{(t-a)^{\underline{\mu}}\}$ and $\mathcal{L}_{a+\mu+\nu}\left\{\Delta_{a+\mu}^{-\nu} (t-a)^{\underline{\mu}}\right\}$ converge for $|s+1| > 1$. Hence, for $|s+1| > 1$, we have

$$\begin{aligned}
&\mathcal{L}_{a+\mu+\nu}\left\{\Delta_{a+\mu}^{-\nu} (t-a)^{\underline{\mu}}\right\}(s) \\
&\quad = \frac{(s+1)^\nu}{s^\nu}\mathcal{L}_{a+\mu}\{(t-a)^{\underline{\mu}}\}(s), \quad \text{using (2.49)} \\
&\quad = \frac{(s+1)^\nu}{s^\nu}\Gamma(\mu+1)\mathcal{L}_{a+\mu}\{h_\mu(t,a)\}(s) \\
&\quad = \frac{(s+1)^\nu}{s^\nu}\Gamma(\mu+1)\frac{(s+1)^\mu}{s^{\mu+1}}, \quad \text{applying (2.39)} \\
&\quad = \Gamma(\mu+1)\frac{(s+1)^{\mu+\nu}}{s^{\mu+\nu+1}} \\
&\quad = \Gamma(\mu+1)\mathcal{L}_{a+\mu+\nu}\{h_{\mu+\nu}(t,a)\}(s) \\
&\quad = \mathcal{L}_{a+\mu+\nu}\left\{\frac{\Gamma(\mu+1)}{\Gamma(\mu+\nu+1)}(t-a)^{\underline{\mu+\nu}}\right\}(s).
\end{aligned}$$

Since the Laplace transform is injective, it follows that

$$\Delta_{a+\mu}^{-\nu} (t-a)^{\underline{\mu}} = \frac{\Gamma(\mu+1)}{\Gamma(\mu+1+\nu)} (t-a)^{\underline{\mu+\nu}}, \text{ for } t \in \mathbb{N}_{a+\mu+\nu}.$$

This completes the proof. □

**Theorem 2.72.** *Suppose that* $f : \mathbb{N}_a \to \mathbb{R}$ *is of exponential order* $r \geq 1$*, and let* $\nu, \mu > 0$ *be given. Then*

$$\Delta_{a+\mu}^{-\nu}\Delta_a^{-\mu}f(t) = \Delta_a^{-\nu-\mu}f(t) = \Delta_{a+\nu}^{-\mu}\Delta_a^{-\nu}f(t), \text{ for all } t \in \mathbb{N}_{a+\mu+\nu}.$$

*Proof.* Let $f, r, \nu$, and $\mu$ be given as in the statement of the theorem. It follows from Corollary 2.66 that each of

$$\mathcal{L}_{a+\mu+\nu}\left\{\Delta_{a+\mu}^{-\nu}\Delta_a^{-\mu}f\right\}, \; \mathcal{L}_{a+\mu}\left\{\Delta_a^{-\mu}f\right\} \text{ and } \mathcal{L}_{a+(\nu+\mu)}\left\{\Delta_a^{-(\nu+\mu)}f\right\}$$

exists for $|s+1| > r$. Therefore, we may apply (2.49) multiple times to write for $|s+1| > r$,

$$\begin{aligned}
\mathcal{L}_{a+\mu+\nu}\left\{\Delta_{a+\mu}^{-\nu}\Delta_a^{-\mu}f\right\}(s) &= \frac{(s+1)^\nu}{s^\nu}\mathcal{L}_{a+\mu}\left\{\Delta_a^{-\mu}f\right\}(s) \\
&= \frac{(s+1)^\nu}{s^\nu}\frac{(s+1)^\mu}{s^\mu}\mathcal{L}_a\{f\}(s) \\
&= \frac{(s+1)^{\nu+\mu}}{s^{\nu+\mu}}\mathcal{L}_a\{f\}(s) \\
&= \mathcal{L}_{a+(\nu+\mu)}\left\{\Delta_a^{-(\nu+\mu)}f\right\}(s) \\
&= \mathcal{L}_{a+\mu+\nu}\left\{\Delta_a^{-\nu-\mu}f\right\}(s).
\end{aligned}$$

The result follows from symmetry and the fact that the operator $\mathcal{L}_{a+\mu+\nu}$ is injective (see Theorem 2.7). □

## 2.10 The Laplace Transform Method

The tools developed in the previous sections of this chapter enable us to solve a general fractional initial value problem using the Laplace transform. The initial value problem (2.53) below is identical to that studied and solved using the composition rules in Holm [123, 125]. In Theorem 2.76 below, we present only that part of the proof involving the Laplace transform method.

**Theorem 2.73.** *Assume* $f : \mathbb{N}_a \to \mathbb{R}$ *is of exponential order* $r \geq 1$ *and* $\nu > 0$ *with* $N - 1 < \nu \leq N$*. Then the unique solution of the IVP*

$$\Delta_{a+\nu-N}^{\nu}y(t) = f(t), \quad t \in \mathbb{N}_a$$

$$\Delta^i y(a+\nu-N) = 0, \quad 0 \leq i \leq N-1,$$

*is given by*

$$y(t) = \Delta_a^{-\nu} f(t) = \int_a^t h_{\nu-1}(t, \sigma(k)) f(k) \Delta k,$$

*for* $t \in \mathbb{N}_{a+\nu-N}$.

*Proof.* Since

$$\Delta_{a+\nu-N}^{\nu} y(t) = f(t), \quad t \in \mathbb{N}_a,$$

we have that

$$\mathcal{L}_a\{\Delta_{a+\nu-N}^{\nu} y\}(s) = F_a(s)$$

for $|s+1| > r$. Assume for the moment that the Laplace transform (based at $a + \nu - N$) of the solution of the given IVP converges for $|s+1| > r$. It follows from (2.52) that

$$\begin{aligned}
\mathcal{L}_a\{\Delta_{a+\nu-N}^{\nu} y\}(s) &= s^{\nu}(s+1)^{N-\nu} Y_{a+\nu-N}(s) - \sum_{j=0}^{N-1} s^j \Delta_a^{\nu-1-j} y(a) \\
&= s^{\nu}(s+1)^{N-\nu} Y_{a+\nu-N}(s),
\end{aligned}$$

where we have used the initial conditions. It follows that

$$\begin{aligned}
\mathcal{L}_{a+\nu-N}\{y\}(s) &= Y_{a+\nu-N}(s) \\
&= \frac{(s+1)^{\nu-N}}{s^{\nu}} F_a(s) \\
&= \mathcal{L}_{a+\nu-N}\{\Delta_a^{-\nu} f\}(s), \quad \text{by (2.50)}.
\end{aligned}$$

It then follows from the uniqueness theorem for Laplace transforms, Theorem 2.7, that

$$y(t) = \Delta_a^{-\nu} f(t), \quad t \in \mathbb{N}_{a+\nu-N}.$$

From this we now know that $y$ is of exponential order $r$ and hence the above arguments hold and the proof is complete. □

Using Theorem 2.73 and Theorem 2.43 it is easy to prove the following result.

**Theorem 2.74.** *Assume* $f : \mathbb{N}_a \to \mathbb{R}$ *is of exponential order* $r \geq 1$ *and* $\nu > 0$ *with* $N - 1 < \nu \leq N$. *Then a general solution of the nonhomogeneous equation*

$$\Delta_{a+\nu-N}^{\nu} y(t) = f(t), \quad t \in \mathbb{N}_a$$

*is given by*

$$y(t) = \sum_{k=1}^{N} c_k (t-a)^{\underline{\nu-k}} + \Delta_a^{-\nu} f(t)$$

*for* $t \in \mathbb{N}_{a+\nu-N}$.

*Example 2.75.* Solve the IVP

$$\Delta_{a-\frac{1}{2}}^{\frac{1}{2}} y(t) = h_{\frac{1}{2}}(t,a), \quad t \in \mathbb{N}_a$$

$$y\left(a - \frac{1}{2}\right) = \frac{1}{2}.$$

Note this IVP is of the form of the IVP in Theorem 2.74, where

$$\nu = \frac{1}{2}, \quad N = 1, \quad a + N - \nu = a - \frac{1}{2}, \quad f(t) = h_{\frac{1}{2}}(t,a).$$

From Theorem 2.74 a general solution of the fractional equation $\Delta_{a-\frac{1}{2}}^{\frac{1}{2}} y(t) = h_{\frac{1}{2}}(t,a)$ is given by

$$\begin{aligned} y(t) &= c_1 (t-a)^{\underline{\nu-1}} + \Delta_a^{-\frac{1}{2}} h_{\frac{1}{2}}(t,a) \\ &= c_1 (t-a)^{\underline{-\frac{1}{2}}} + (t-a). \end{aligned}$$

Applying the initial condition we get $c_1 = \frac{1}{\sqrt{\pi}}$. Hence the solution of the given IVP in this example is given by

$$y(t) = \frac{1}{\sqrt{\pi}} (t-a)^{\underline{-\frac{1}{2}}} + (t-a)$$

for $t \in \mathbb{N}_{a-\frac{1}{2}}$.

The following theorem appears in Ahrendt et al. [3].

**Theorem 2.76.** *Suppose that* $f : \mathbb{N}_a \to \mathbb{R}$ *is of exponential order* $r \geq 1$, *and let* $\nu > 0$ *be given with* $N - 1 < \nu \leq N$. *The unique solution to the fractional initial value problem*

$$\begin{cases} \Delta_{a+\nu-N}^{\nu} y(t) = f(t), & t \in \mathbb{N}_a \\ \Delta^i y(a+\nu-N) = A_i, & i \in \{0, 1, \cdots, N-1\}; A_i \in \mathbb{R} \end{cases} \tag{2.53}$$

*is given by*

$$y(t) = \sum_{i=0}^{N-1} \alpha_i (t-a)^{\underline{i+\nu-N}} + \Delta_a^{-\nu} f(t), \text{ for } t \in \mathbb{N}_{a+\nu-N},$$

*where*

$$\alpha_i := \sum_{p=0}^{i} \sum_{k=0}^{i-p} \frac{(-1)^k}{i!} (i-k)^{\underline{N-\nu}} \binom{i}{p} \binom{i-p}{k} A_p,$$

*for* $i \in \{0, 1, \cdots, N-1\}$.

*Proof.* Since $f$ is of exponential order $r$, we know that $F_a(s) = \mathcal{L}_a \{f\}(s)$ exists for $|s+1| > r$. So, applying the Laplace transform to both sides of the difference equation in (2.53), we have for $|s+1| > r$

$$\mathcal{L}_a \left\{ \Delta_{a+\nu-N}^{\nu} y \right\}(s) = F_a(s).$$

Using (2.52), we get

$$s^{\nu} (s+1)^{N-\nu} Y_{a+\nu-N}(s) - \sum_{j=0}^{N-1} s^j \Delta_{a+\nu-N}^{\nu-j-1} y(a) = F_a(s).$$

This implies that

$$Y_{a+\nu-N}(s) = \frac{F_a(s)}{s^{\nu} (s+1)^{N-\nu}} + \sum_{j=0}^{N-1} \frac{\Delta_{a+\nu-N}^{\nu-j-1} y(a)}{s^{\nu-j} (s+1)^{N-\nu}}.$$

From (2.50), we have immediately that

$$\frac{F_a(s)}{s^{\nu} (s+1)^{N-\nu}} = \mathcal{L}_{a+\nu-N} \left\{ \Delta_a^{-\nu} f \right\}(s).$$

Considering next the terms in the summation, we have for each fixed $j \in \{0, \cdots, N-1\}$,

$$\begin{aligned}
\frac{1}{s^{\nu-j} (s+1)^{N-\nu}} &= \frac{1}{(s+1)^{N-j-1}} \frac{(s+1)^{\nu-j-1}}{s^{\nu-j}} \\
&= \frac{1}{(s+1)^{N-j-1}} \mathcal{L}_{a+\nu-j-1} \left\{ h_{\nu-j-1}(t,a) \right\}(s), && \text{by (2.39)} \\
&= \mathcal{L}_{a+\nu-N} \left\{ h_{\nu-j-1}(t,a) \right\}(s)
\end{aligned}$$

$$-\sum_{k=0}^{N-j-2}\frac{h_{\nu-j-1}(k+a+\nu-N,a)}{(s+1)^{k+1}}, \text{ by (2.37)}$$
$$=\mathcal{L}_{a+\nu-N}\left\{h_{\nu-j-1}(t,a)\right\}(s),$$

since

$$\begin{aligned} h_{\nu-j-1}(k+a+\nu-N,a) &= \frac{(k+\nu-N)^{\underline{\nu-j-1}}}{\Gamma(\nu-j)} \\ &= \frac{\Gamma(k+\nu-N+1)}{\Gamma(k-(N-j-2))\,\Gamma(\nu-j)} \\ &= 0, \end{aligned}$$

for $k\in\{0,\cdots,N-j-2\}$. It follows that for $|s+1|>r$,

$$\begin{aligned} &\mathcal{L}_{a+\nu-N}\{y\}(s) \\ &= \mathcal{L}_{a+\nu-N}\left\{\Delta_a^{-\nu}f\right\}(s)+\sum_{j=0}^{N-1}\Delta_{a+\nu-N}^{\nu-j-1}y(a)\mathcal{L}_{a+\nu-N}\left\{h_{\nu-j-1}(t,a)\right\}(s) \\ &= \mathcal{L}_{a+\nu-N}\left\{\sum_{j=0}^{N-1}\Delta_{a+\nu-N}^{\nu-j-1}y(a)h_{\nu-j-1}(t,a)+\Delta_a^{-\nu}f\right\}(s). \end{aligned}$$

Since the Laplace transform is injective, we conclude that for $t\in\mathbb{N}_{a+\nu-N}$,

$$\begin{aligned} y(t) &= \sum_{j=0}^{N-1}\Delta_{a+\nu-N}^{\nu-j-1}y(a)h_{\nu-j-1}(t,a)+\Delta_a^{-\nu}f(t) \\ &= \sum_{j=0}^{N-1}\frac{\Delta_{a+\nu-N}^{\nu-j-1}y(a)}{\Gamma(\nu-j)}(t-a)^{\underline{\nu-j-1}}+\Delta_a^{-\nu}f(t) \\ &= \sum_{i=0}^{N-1}\left(\frac{\Delta_{a+\nu-N}^{i+\nu-N}y(a)}{\Gamma(i+\nu-N+1)}\right)(t-a)^{\underline{i+\nu-N}}+\Delta_a^{-\nu}f(t). \end{aligned}$$

Moreover, Holm [125] showed that

$$\frac{\Delta_{a+\nu-N}^{i+\nu-N}y(a)}{\Gamma(i+\nu-N+1)}=\sum_{p=0}^{i}\sum_{k=0}^{i-p}\frac{(-1)^k}{i!}(i-k)^{\underline{N-\nu}}\binom{i}{p}\binom{i-p}{k}\Delta^i y(a+\nu-N),$$

for $i\in\{0,1,\cdots,N-1\}$, concluding the proof. □

Theorem 2.76 shows how we can solve the general IVP (2.53) using the discrete Laplace transform method. We offer a brief example.

*Example 2.77.* Consider the IVP given by

$$\begin{cases} \Delta^{\pi}_{\pi-4}y(t) = \pi^4 t^{\underline{2}}, \; t \in \mathbb{N}_0 \\ y(\pi-4) = 2, \; \Delta y(\pi-4) = 3, \; \Delta^2 y(\pi-4) = 5, \; \Delta^3 y(\pi-4) = 7. \end{cases} \tag{2.54}$$

Note that (2.54) is a specific case of (2.53) from Theorem 2.76, with

$$a = 0, \quad \nu = \pi, \quad N = 4, f(t) = \pi^4 t^{\underline{2}}$$
$$A_0 = 2, A_1 = 3, A_2 = 5\ A_3 = 7.$$

After applying the discrete Laplace transform method as described in Theorem 2.76, we have

$$\begin{aligned} y(t) &= \sum_{i=0}^{3} \alpha_i t^{\underline{i+\pi-4}} + \Delta_0^{-\pi}\left(\pi^4 t^{\underline{2}}\right) \\ &= \sum_{i=0}^{3} \alpha_i t^{\underline{i+\pi-4}} + \Delta_2^{-\pi}\left(\pi^4 t^{\underline{2}}\right), \text{ since } t^{\underline{2}} = t\,(t-1), \\ &\approx 0.303 t^{\underline{\pi-4}} + 5.040 t^{\underline{\pi-3}} + 6.977 t^{\underline{\pi-2}} + 4.876 t^{\underline{\pi-1}} + 3.272 t^{\underline{\pi+2}}, \end{aligned}$$

where in this last step, we calculated

$$\alpha_i = \sum_{p=0}^{i} \sum_{k=0}^{i-p} \frac{(-1)^k}{i!} (i-k)^{\underline{4-\pi}} \binom{i}{p} \binom{i-p}{k} A_p, \text{ for } i = 0, 1, 2, 3,$$

for the first four terms and applied the power rule (Theorem 2.71) on the last term.

## 2.11 Exercises

**2.1.** Show that $f : \mathbb{N}_a \to \mathbb{R}$ is of exponential order $r = 1$ iff $f$ is bounded on $\mathbb{N}_a$.

**2.2.** Prove that if $f : \mathbb{N}_a \to \mathbb{R}$ is of exponential order $r > 0$, then $\Delta^n f : \mathbb{N}_a \to \mathbb{R}$ is also of exponential order $r$ for $n \in \mathbb{N}_0$.

**2.3.** Show that if $f : \mathbb{N}_a \to \mathbb{R}$ is of exponential order $r > 1$, then $h(t) := \int_a^t f(\tau)\Delta\tau, t \in \mathbb{N}_a$ is also of exponential order $r$.

**2.4.** Show that $h_0(t, a)$ is of exponential order 1 and for each $n \geq 0$, $h_n(t, a)$ is of exponential order $1 + \epsilon$ for all $\epsilon > 0$.

**2.5.** Prove formula (i) in Theorem 2.8, that is

$$\mathcal{L}_a\{\cosh_p(t,a)\}(s) = \frac{s}{s^2 - p^2}$$

for $|s+1| > \max\{|1+p|, |1-p|\}$.

**2.6.** Prove formula (ii) in Theorem 2.9, that is

$$\mathcal{L}_a\{\sin_p(t,a)\}(s) = \frac{p}{s^2 + p^2}$$

for $|s+1| > \max\{|1+ip|, |1-ip|\}$.

**2.7.** Prove formula (ii) in Theorem 2.10, that is

$$\mathcal{L}_a\{e_\alpha(t,a)\sinh_{\frac{\beta}{1+\alpha}}(t,a)\}(s) = \frac{\beta}{(s-\alpha)^2 - \beta^2},$$

for $|s+1| > \max\{|1+\alpha+\beta|, |1+\alpha-\beta|\}$.

**2.8.** Prove Theorem 2.11.

**2.9.** For each of the following find $y(t)$ given that

(i) $Y_a(s) = \frac{14-s}{s^2+2s-8}$;
(ii) $Y_0(s) = \frac{2s^2}{s^2-\sqrt{2}s+1}$.

**2.10.** Use Laplace transforms to solve the following IVPs

(i)

$$y(t+2) - 7y(t+1) + 12y(t) = 0, \quad t \in \mathbb{N}_0;$$
$$y(0) = 2, \quad y(1) = 4.$$

(ii)

$$y(t+1) - 2y(t) = 3^t, \quad t \in \mathbb{N}_0;$$
$$y(0) = 5.$$

(iii)

$$y(t+2) - 6y(t+1) + 8y(t) = 20(4)^t, \quad t \in \mathbb{N}_0$$
$$y(0) = 0, \quad y(1) = 4.$$

**2.11.** Use Laplace transforms to solve the IVP

$$u(t+1)+v(t)=0$$
$$-u(t)+v(t+1)=0$$
$$u(0)=1, \quad v(0)=0.$$

**2.12.** Solve each of the following IVPs:

(i)

$$\Delta y(t)-2y(t)=\delta_4(t), \quad t\in\mathbb{N}_0;$$
$$y(0)=2,$$

(ii)

$$\Delta y(t)-5y(t)=3u_{60}(t), \quad t\in\mathbb{N}_0$$
$$y(0)=4, \quad t\in\mathbb{N}_0.$$

**2.13.** Solve the following summation equations using Laplace transforms:

(i) $y(t)=2+4\sum_{r=0}^{t-1}3^{t-r-1}y(r), \quad t\in\mathbb{N}_0;$
(ii) $y(t)=3\cdot 5^t-4\sum_{r=0}^{t-1}5^{t-r-1}y(r), \quad t\in\mathbb{N}_0;$
(iii) $y(t)=t+\sum_{r=0}^{t-1}y(r), \quad t\in\mathbb{N}_0;$
(iv) $y(t)=2^{t-a}+\sum_{r=a}^{t-1}4^{t-r-1}y(r), \quad t\in\mathbb{N}_a.$

**2.14.** Use Laplace transforms to solve each of the following:

(i) $y(t)=3^t+\sum_{m=0}^{t-1}3^{k-m-1}y_m, \quad t\in\mathbb{N}_0;$
(ii) $y(t)=3^t+\sum_{m=0}^{t-1}4^{k-m-1}y_m, \quad t\in\mathbb{N}_0.$

**2.15.** Show that

(i) $\Delta_a^{-\nu}f(a+\nu)=f(a);$
(ii) $\Delta_a^{-\nu}f(a+\nu+1)=\nu f(a)+f(a+1).$

**2.16.** Complete the proof of Theorem 2.27.

**2.17.** Work each of the following:

(i) Use the definition of the $\nu$-th fractional sum (Definition 2.25) to find $\Delta_a^{-\frac{1}{3}}1;$
(ii) Use the definition of the fractional difference (Definition 2.29) and part (2.32) to find $\Delta_a^{\frac{2}{3}}1.$

**2.18.** Show that the following hold:

(i) $\Delta_{a+\mu}^{-\nu}(t-a)^{\underline{\mu}}=\mu^{\underline{-\nu}}(t-a)^{\underline{\mu+\nu}}, \quad t\in\mathbb{N}_{a+\mu+\nu};$
(ii) $\Delta_{a+\mu}^{\nu}(t-a)^{\underline{\mu}}=\mu^{\underline{\nu}}(t-a)^{\underline{\mu-\nu}}, \quad t\in\mathbb{N}_{a+\mu+N-\nu}.$

**2.19.** Verify that (2.12) holds.

**2.20.** Show that $h_\mu(t, t-\mu+k) = 0$ for $k \in \mathbb{N}_1$, $\mu - k + 1 \notin \{0, -1, -2, \cdots\}$.

**2.21.** Evaluate each of the following using Theorem 2.38 and Theorem 2.40

(i) $\Delta_{\frac{3}{2}}^{-1}(t-1)^{\underline{\frac{1}{2}}}, \quad t \in \mathbb{N}_{\frac{5}{2}}$;
(ii) $\Delta_{4}^{-.7}(t-1.7)^{\underline{2.3}}, \quad t \in \mathbb{N}_{4.7}$;
(iii) $\Delta_{5.5}^{.5}(t-3)^{\underline{2.5}}, \quad t \in \mathbb{N}_5$;
(iv) $\Delta_{3}^{\frac{1}{2}} t(t-1)(t-2), \quad t \in \mathbb{N}_{\frac{5}{2}}$.

**2.22.** Prove that part (ii) of Theorem 2.42, follows from Theorem 2.40.

**2.23.** Prove (2.26).

**2.24.** Solve each of the following IVPs:

(i) $\Delta_{-0.3}^{2.7} x(t) = t^{\underline{2}}, \quad t \in \mathbb{N}_0$
$x(-0.3) = x(0.7) = x(1.7) = 0$;
(ii) $\Delta_{-0.4}^{1.6} x(t) = t^{\underline{4}}, \quad t \in \mathbb{N}_0$
$x(-0.4) = x(0.6) = 0$;
(iii) $\Delta_{-0.1}^{0.9} x(t) = t^{\underline{5}}, \quad t \in \mathbb{N}_0$
$x(-0.1) = 0$.

**2.25.** Use Theorems 2.54 and 2.58 to show that $\mathcal{L}_a\{h_1(t,a)\} = \frac{1}{s^2}$. Evaluate the convolution product $1 * 1$ and show directly (do not use the convolution theorem) that $\mathcal{L}_a\{1 * 1\}(s) = \mathcal{L}_a\{1\}(s)\, \mathcal{L}_a\{1\}(s)$.

**2.26.** Assume $p \in \mathcal{R}$ and $p \neq 0$. Using the definition of the convolution product (Definition 2.59), find

$$[h_1(t,a) * e_p(t,a)](t).$$

**2.27.** Assume $p, q \in \mathcal{R}$ and $p \neq q$. Using the definition of the convolution product (Definition 2.59), find

$$[e_p(t,a) * e_q(t,a)](t).$$

**2.28.** For $N$ a positive integer, use the definition of the Laplace transform to prove that (2.4) holds (that is, (2.52) holds when $\nu = N$).

# Chapter 3
# Nabla Fractional Calculus

## 3.1 Introduction

As mentioned in the previous chapter and as demonstrated on numerous occasions, the disadvantage of the discrete delta fractional calculus is the shifting of domains when one goes from the domain of the function to the domain of its delta fractional difference. This problem is not as great with the fractional nabla difference as noted by Atici and Eloe. In this chapter we study the discrete fractional nabla calculus. We then define the corresponding nabla Laplace transform motivated by a particularly general definition of the delta Laplace transform that was first defined in a very general way by Bohner and Peterson [62]. Several properties of this nabla Laplace transform are then derived. Fractional nabla Taylor monomials are defined and formulas for their nabla Laplace transforms are presented. Then the discrete nabla version of the Mittag–Leffler function and its nabla Laplace transform is obtained. Finally, a variation of constants formula for an initial value problem for a $\nu$-th, $0 < \nu < 1$, order nabla fractional difference equation is given along with some applications. Much of the work in this chapter comes from the results in Hein et al. [119], Holm [123–125], Brackins [64], Ahrendt et al. [3, 4], and Baoguo et al. [49, 52].

## 3.2 Preliminary Definitions

We first introduce some notation and state elementary results concerning the nabla calculus, which we will use in this chapter. As in Chaps. 1 and 2 for $a \in \mathbb{R}$, the sets $\mathbb{N}_a$ and $\mathbb{N}_a^b$, where $b - a$ is a positive integer, are defined by

$$\mathbb{N}_a := \{a, a+1, a+2, \dots\}, \quad \mathbb{N}_a^b := \{a, a+1, a+2, \dots, b\}.$$



C. Goodrich, A.C. Peterson, *Discrete Fractional Calculus*,
DOI 10.1007/978-3-319-25562-0_3

For an arbitrary function $f : \mathbb{N}_a \to \mathbb{R}$ we define the **nabla operator (backwards difference operator)**, $\nabla$, by

$$(\nabla f)(t) := f(t) - f(t-1), \quad t \in \mathbb{N}_{a+1}.$$

For convenience, we adopt the convention that $\nabla f(t) := (\nabla f)(t)$. Sometimes it is useful to use the relation

$$\nabla f(t) = \Delta f(t-1) \tag{3.1}$$

to get results for the nabla calculus from the delta calculus and vice versa. Since many readers will be interested only in the nabla calculus, we want this chapter to be self-contained. So we will not use the formula (3.1) in this chapter. The operator $\nabla^n$ is defined recursively by $\nabla^n f(t) := \nabla\big(\nabla^{n-1} f(t)\big)$ for $t \in \mathbb{N}_{a+n}$, $n \in \mathbb{N}_1$, where $\nabla^0$ is the identity operator defined by $\nabla^0 f(t) = f(t)$. We define the **backward jump operator**, $\rho : \mathbb{N}_{a+1} \to \mathbb{N}_a$, by

$$\rho(t) = t - 1.$$

Also we let $f^\rho$ denote the composition function $f \circ \rho$. It is easy (Exercise 3.1) to see that if $f : \mathbb{N} \to \mathbb{R}$ and $\nabla f(t) = 0$ for $t \in \mathbb{N}_{a+1}$, then

$$f(t) = C, \quad t \in \mathbb{N}_a, \quad \text{where } C \text{ is a constant.}$$

The following theorem gives several properties of the nabla difference operator.

**Theorem 3.1.** *Assume $f, g : \mathbb{N}_a \to \mathbb{R}$ and $\alpha, \beta \in \mathbb{R}$. Then for $t \in \mathbb{N}_{a+1}$,*

(i) $\nabla \alpha = 0$;
(ii) $\nabla \alpha f(t) = \alpha \nabla f(t)$;
(iii) $\nabla\left(f(t) + g(t)\right) = \nabla f(t) + \nabla g(t)$;
(iv) *if* $\alpha \neq 0$, *then* $\nabla \alpha^{t+\beta} = \frac{\alpha-1}{\alpha}\alpha^{t+\beta}$;
(v) $\nabla\left(f(t)g(t)\right) = f(\rho(t))\nabla g(t) + \nabla f(t) g(t)$;
(vi) $\nabla\left(\frac{f(t)}{g(t)}\right) = \frac{g(t)\nabla f(t) - f(t)\nabla g(t)}{g(t)g(\rho(t))}$, *if* $g(t) \neq 0$, $t \in \mathbb{N}_{a+1}$.

*Proof.* We will just prove (iv) and (v) and leave the proof of the other parts to the reader. To see that (iv) holds assume that $\alpha \neq 0$ and note that

$$\begin{aligned}\nabla \alpha^{t+\beta} &= \alpha^{t+\beta} - \alpha^{t-1+\beta} \\ &= [\alpha - 1]\alpha^{t-1+\beta} \\ &= \frac{\alpha - 1}{\alpha}\alpha^{t+\beta}.\end{aligned}$$

Next we prove the product rule (v). For $t \in \mathbb{N}_{a+1}$, consider

$$\begin{aligned}\nabla[f(t)g(t)] &= f(t)g(t) - f(t-1)g(t-1)\\ &= f(t-1)[g(t) - g(t-1)] + [f(t) - f(t-1)]g(t)\\ &= f(\rho(t))\nabla g(t) + \nabla f(t)g(t),\end{aligned}$$

which is the desired result. □

Next we define the rising function.

**Definition 3.2.** Assume $n$ is a positive integer and $t \in \mathbb{R}$. Then we define the **rising function**, $t^{\overline{n}}$, read "$t$ to the $n$ rising,"by

$$t^{\overline{n}} := t(t+1)\cdots(t+n-1).$$

Readers familiar with the Pochhammer function may recognize this notation in its alternative form, $(k)_n$. See Knuth [139].

The rising function is defined this way so that the following power rule holds.

**Theorem 3.3 (Nabla Power Rule).** *For $n \in \mathbb{N}_1$, $\alpha \in \mathbb{R}$,*

$$\nabla(t+\alpha)^{\overline{n}} = n\,(t+\alpha)^{\overline{n-1}},$$

*for $t \in \mathbb{R}$.*

*Proof.* We simply write

$$\begin{aligned}\nabla(t+\alpha)^{\overline{n}} &= (t+\alpha)^{\overline{n}} - (t-1+\alpha)^{\overline{n}}\\ &= [(t+\alpha)(t+\alpha+1)\cdots(t+\alpha+n-1)]\\ &\quad - [(t+\alpha-1)(t+\alpha)\cdots(t+\alpha+n-2)]\\ &= (t+\alpha)(t+\alpha+1)\cdots(t+\alpha+n-2)\\ &\quad \cdot[(t+\alpha+n-1) - (t+\alpha-1)]\\ &= n\,(t+\alpha)^{\overline{n-1}}.\end{aligned}$$

This completes the proof. □

Note that for $n \in \mathbb{N}_1$,

$$\begin{aligned}t^{\overline{n}} &:= t(t+1)\cdots(t+n-1)\\ &= (t+n-1)(t+n-2)\cdots(t+1)\cdot t\\ &= \frac{(t+n-1)(t+n-2)\cdots t\cdot\Gamma(t)}{\Gamma(t)}\\ &= \frac{\Gamma(t+n)}{\Gamma(t)}, \quad t \notin \{0,-1,-2,\cdots\},\end{aligned}$$

where $\Gamma$ is the gamma function (Definition 1.6). Motivated by this we next define the **(generalized) rising function** as follows.

**Definition 3.4.** The (generalized) rising function is defined by

$$t^{\overline{r}} = \frac{\Gamma(t+r)}{\Gamma(t)}, \tag{3.2}$$

for those values of $t$ and $r$ so that the right-hand side of equation (3.2) is sensible. Also, we use the convention that if $t$ is a nonpositive integer, but $t + r$ is not a nonpositive integer, then $t^{\overline{r}} := 0$.

We then get the following generalized power rules.

**Theorem 3.5 (Generalized Nabla Power Rules).** *The formulas*

$$\nabla(t+\alpha)^{\overline{r}} = r\,(t+\alpha)^{\overline{r-1}}, \tag{3.3}$$

*and*

$$\nabla(\alpha - t)^{\overline{r}} = -r(\alpha - \rho(t))^{\overline{r-1}}, \tag{3.4}$$

*hold for those values of $t$, $r$, and $\alpha$ so that the expressions in equations* (3.3) *and* (3.4) *are sensible. In particular,* $t^{\overline{0}} = 1, t \neq 0, -1, -2, \cdots$.

*Proof.* Consider that

$$\begin{aligned}
\nabla(t+\alpha)^{\overline{r}} &= (t+\alpha)^{\overline{r}} - (t-1+\alpha)^{\overline{r}} \\
&= \frac{\Gamma(t+\alpha+r)}{\Gamma(t+\alpha)} - \frac{\Gamma(t+\alpha+r-1)}{\Gamma(t+\alpha-1)} \\
&= [(t+\alpha+r-1) - (t+\alpha-1)]\frac{\Gamma(t+\alpha+r-1)}{\Gamma(t+\alpha)} \\
&= r\frac{\Gamma(t+\alpha+r-1)}{\Gamma(t+\alpha)} \\
&= r(t+\alpha)^{\overline{r-1}}.
\end{aligned}$$

Hence, (3.3) holds. Next we prove (3.4). To see this, note that

$$\begin{aligned}
\nabla(\alpha - t)^{\overline{r}} &= (\alpha - t)^{\overline{r}} - (\alpha - t + 1)^{\overline{r}} \\
&= \frac{\Gamma(\alpha - t + r)}{\Gamma(\alpha - t)} - \frac{\Gamma(\alpha - t + 1 + r)}{\Gamma(\alpha - t + 1)} \\
&= [(\alpha - t) - (\alpha - t + r)]\frac{\Gamma(\alpha - t + r)}{\Gamma(\alpha - t + 1)} \\
&= -r\,\frac{\Gamma(\alpha - t + r)}{\Gamma(\alpha - t + 1)}
\end{aligned}$$

$$= -r\frac{\Gamma(\alpha - \rho(t) + r - 1)}{\Gamma(\alpha - \rho(t))}$$

$$= -r(\alpha - \rho(t))^{\overline{r-1}}.$$

This completes the proof. □

## 3.3 Nabla Exponential Function

In this section we want to study the nabla exponential function that plays a similar role in the nabla calculus that the exponential function $e^{pt}$ does in the continuous calculus. Motivated by the fact that when $p$ is a constant, $x(t) = e^{pt}$ is the unique solution of the initial value problem

$$x' = px, \quad x(0) = 1,$$

we define the nabla exponential function, $E_p(t, s)$ based at $s \in \mathbb{N}_a$, where the function $p$ is in the set of (nabla) regressive functions

$$\mathcal{R} := \{p : \mathbb{N}_{a+1} \to \mathbb{R} : \quad 1 - p(t) \neq 0, \quad \text{for} \quad t \in \mathbb{N}_{a+1}\},$$

to be the unique solution of the initial value problem

$$\nabla y(t) = p(t)y(t), \quad t \in \mathbb{N}_{a+1} \tag{3.5}$$

$$y(s) = 1. \tag{3.6}$$

After reading the proof of the next theorem one sees why this IVP has a unique solution. In the next theorem we give a formula for the exponential function $E_p(t, s)$.

**Theorem 3.6.** *Assume $p \in \mathcal{R}$ and $s \in \mathbb{N}_a$. Then*

$$E_p(t, s) = \begin{cases} \prod_{\tau=s+1}^{t} \frac{1}{1-p(\tau)}, & t \in \mathbb{N}_s \\ \prod_{\tau=t+1}^{s} [1 - p(\tau)], & t \in \mathbb{N}_a^{s-1}. \end{cases} \tag{3.7}$$

*Here it is understood that $\prod_{\tau=t+1}^{t} h(\tau) = 1$ for any function $h$.*

*Proof.* First we find a formula for $E_p(t, s)$ for $t \geq s + 1$ by solving the IVP (3.5), (3.6) by iteration. Solving the nabla difference equation (3.5) for $y(t)$ we obtain

$$y(t) = \frac{1}{1 - p(t)} y(t - 1), \quad t \in \mathbb{N}_{a+1}. \tag{3.8}$$

Letting $t = s + 1$ in (3.8) we get

$$y(s+1) = \frac{1}{1 - p(s+1)} y(s) = \frac{1}{1 - p(s+1)}.$$

Then letting $t = s + 2$ in (3.8) we obtain

$$y(s+2) = \frac{1}{1 - p(s+2)} y(s+1) = \frac{1}{[1 - p(s+1)]\,[1 - p(s+2)]}.$$

Proceeding in this matter we get by mathematical induction that

$$E_p(t, a) = \prod_{\tau=s+1}^{t} \frac{1}{1 - p(\tau)},$$

for $t \in \mathbb{N}_{s+1}$. By our convention on products we get

$$E_p(s, s) = \prod_{\tau=s+1}^{s} [1 - p(\tau)] = 1$$

as desired. Now assume $a \le t < s$. Solving the nabla difference equation (3.5) for $y(t-1)$ we obtain

$$y(t-1) = [1 - p(t)]y(t), \quad t \in \mathbb{N}_{a+1}. \tag{3.9}$$

Letting $t = s$ in (3.9) we get

$$y(s-1) = [1 - p(s)]y(s) = [1 - p(s)].$$

If $s - 2 \ge a$, we obtain by letting $t = s - 1$ in (3.9)

$$y(s-2) = [1 - p(s-1)]y(s-1) = [1 - p(s)]\,[1 - p(s-1)]\,.$$

By mathematical induction we arrive at

$$E_p(t, s) = \prod_{\tau=t+1}^{s} [1 - p(\tau)], \quad \text{for} \quad t \in \mathbb{N}_a^s.$$

Hence, $E_p(t, s)$ is given by (3.7). □

Theorem 3.6 gives us the following example.

*Example 3.7.* If $s \in \mathbb{N}_a$ and $p(t) \equiv p_0$, where $p_0 \neq 1$ is a constant, then

$$E_p(t, s) = (1 - p_0)^{s-t}, \quad t \in \mathbb{N}_a.$$

We now set out to prove properties of the exponential function $E_p(t,s)$. To motivate some of these properties, consider, for $p,q \in \mathcal{R}$, the product

$$\begin{aligned} E_p(t,a)E_q(t,a) &= \prod_{\tau=a+1}^{t} \frac{1}{1-p(\tau)} \prod_{\tau=a+1}^{t} \frac{1}{1-q(\tau)} \\ &= \prod_{\tau=a+1}^{t} \frac{1}{[1-p(\tau)]\,[1-q(\tau)]} \\ &= \prod_{\tau=a+1}^{t} \frac{1}{1-[p(\tau)+q(\tau)-p(\tau)q(\tau)]} \\ &= \prod_{\tau=a+1}^{t} \frac{1}{1-(p \boxplus q)(\tau)} \quad \text{if } (p \boxplus q)(t) := p(t)+q(t)-p(t)q(t) \\ &= E_{p\boxplus q}(t,a) \end{aligned}$$

for $t \in \mathbb{N}_a$.

Hence, we deduce that the nabla exponential function satisfies the law of exponents

$$E_p(t,a)E_q(t,a) = E_{p\boxplus q}(t,a), \quad t \in \mathbb{N}_a,$$

if we define the box plus addition $\boxplus$ on $\mathcal{R}$ by

$$(p \boxplus q)(t) := p(t)+q(t)-p(t)q(t), \quad t \in \mathbb{N}_{a+1}.$$

We now give an important result concerning the box plus addition $\boxplus$.

**Theorem 3.8.** *If we define the* **box plus addition**, $\boxplus$, *on* $\mathcal{R}$ *by*

$$p \boxplus q := p+q-pq,$$

*then* $\mathcal{R}$, $\boxplus$ *is an Abelian group.*

*Proof.* First, to see that the closure property is satisfied, note that if $p,q \in \mathcal{R}$, then $1-p(t) \neq 0$ and $1-q(t) \neq 0$ for $t \in \mathbb{N}_{a+1}$. It follows that

$$1-(p \boxplus q)(t) = 1-[p(t)+q(t)-p(t)q(t)] = (1-p(t))(1-q(t)) \neq 0,$$

for $t \in \mathbb{N}_{a+1}$, and hence the function $p \boxplus q \in \mathcal{R}$.

Next, notice that the zero function, 0, is in $\mathcal{R}$, since the regressivity condition $1-0=1\neq 0$ holds. Also

$$0 \boxplus p = 0+p-0\cdot p = p, \quad \text{for all } p \in \mathcal{R},$$

so the zero function 0 is the identity element in $\mathcal{R}$.

We now show that every element in $\mathcal{R}$ has an additive inverse let $p \in \mathcal{R}$. So, set $q = \frac{-p}{1-p}$ and note that since

$$1 - q(t) = 1 - \frac{-p(t)}{1-p(t)} = \frac{1}{1-p(t)} \neq 0, \quad t \in \mathbb{N}_{a+1}$$

we have that $q \in \mathcal{R}$, and we also have that

$$p \boxplus q = p \boxplus \frac{-p}{1-p} = p + \frac{-p}{1-p} - \frac{-p^2}{1-p} = 0,$$

so $q$ is the additive inverse of $p$. For $p \in \mathcal{R}$, we use the following notation for the additive inverse of $p$:

$$\boxminus p := \frac{-p}{1-p}. \tag{3.10}$$

The fact that the addition $\boxplus$ is associative and commutative is Exercise 3.4. □

We can now define box minus subtraction, $\boxminus$, on $\mathcal{R}$ in a standard manner as follows.

**Definition 3.9.** We define box minus subtraction on $\mathcal{R}$ by

$$p \boxminus q := p \boxplus [\boxminus q].$$

By Exercise 3.5 we have that if $p, q \in \mathcal{R}$, then

$$(p \boxminus q)(t) = \frac{p(t) - q(t)}{1 - q(t)}, \quad t \in \mathbb{N}_a.$$

In addition, we define the set of **(nabla) positively regressive functions**, $\mathcal{R}^+$, by

$$\mathcal{R}^+ = \{p : \mathbb{N}_{a+1} :\to \mathbb{R}, \quad \text{such that} \quad 1 - p(t) > 0 \quad \text{for} \quad t \in \mathbb{N}_{a+1}\}.$$

The proof of the following theorem is left as an exercise (see Exercise 3.8).

**Theorem 3.10.** *The set of positively regressive functions, $\mathcal{R}^+$, with the addition $\boxplus$, is a subgroup of $\mathcal{R}$.*

In the next theorem we give several properties of the exponential function $E_p(t, s)$.

**Theorem 3.11.** *Assume $p, q \in \mathcal{R}$ and $s, r \in \mathbb{N}_a$. Then*

(i) $E_0(t, s) = 1, \quad t \in \mathbb{N}_a$;
(ii) $E_p(t, s) \neq 0, \quad t \in \mathbb{N}_a$;
(iii) *if $p \in \mathcal{R}^+$, then* $E_p(t, s) > 0, \quad t \in \mathbb{N}_a$;

(iv) $\nabla E_p(t,s) = p(t)E_p(t,s)$, $t \in \mathbb{N}_{a+1}$, *and* $E_p(t,t) = 1$, $t \in \mathbb{N}_a$;
(v) $E_p(\rho(t),s) = [1-p(t)]E_p(t,s)$, $t \in \mathbb{N}_{a+1}$;
(vi) $E_p(t,s)E_p(s,r) = E_p(t,r)$, $t \in \mathbb{N}_a$;
(vii) $E_p(t,s)E_q(t,s) = E_{p\boxplus q}(t,s)$, $t \in \mathbb{N}_a$;
(viii) $E_{\boxminus p}(t,s) = \frac{1}{E_p(t,s)}$, $t \in \mathbb{N}_a$;
(ix) $\frac{E_p(t,s)}{E_q(t,s)} = E_{p\boxminus q}(t,s)$, $t \in \mathbb{N}_a$.

*Proof.* Using Example 3.7, we have that

$$E_0(t,s) = (1-0)^{s-t} = 1$$

and thus (i) holds.

To see that (ii) holds, note that since $p \in \mathcal{R}$, it follows that $1 - p(t) \neq 0$, and hence we have that for $t \in \mathbb{N}_s$

$$E_p(t,s) = \prod_{\tau=s+1}^{t} \frac{1}{1-p(\tau)} \neq 0$$

and for $t \in \mathbb{N}_a^{s-1}$

$$E_p(t,s) = \prod_{\tau=t+1}^{s} [1-p(\tau)] \neq 0.$$

Hence, (ii) holds. The proof of (iii) is similar to the proof of (ii), whereas property (iv) follows from the definition of $E_p(t,s)$.

Since, for $t \in \mathbb{N}_s$,

$$\begin{aligned} E_p(\rho(t),s) &= \prod_{\tau=s+1}^{t-1} \frac{1}{1-p(\tau)} \\ &= [1-p(t)] \prod_{\tau=s+1}^{t} \frac{1}{1-p(\tau)} \\ &= [1-p(t)]E_p(t,s) \end{aligned}$$

we have that (v) holds for $t \in \mathbb{N}_{s+1}$. Next assume $t \in \mathbb{N}_{a+1}^{s-1}$. Then

$$\begin{aligned} E_p(\rho(t),s) &= \prod_{\tau=\rho(t)+1}^{s} [1-p(\tau)] \\ &= \prod_{\tau=t}^{s} [1-p(\tau)] \end{aligned}$$

$$= [1 - p(t)] \prod_{\tau=t+1}^{s} [1 - p(\tau)]$$
$$= [1 - p(t)]E_p(t, s).$$

Hence, (v) holds for $t \in \mathbb{N}_{a+1}^{s-1}$. It is easy to check that $E_p(\rho(s), s) = [1 - p(s)]\, E_p(s, s)$. This completes the proof of (v).

We will just show that (vi) holds when $s \geq r \geq a$. First consider the case $t \in \mathbb{N}_s$. In this case

$$E_p(t, s)E_p(s, r) = \prod_{\tau=s+1}^{t} \frac{1}{1 - p(\tau)} \prod_{\tau=r+1}^{s} \frac{1}{1 - p(\tau)}$$
$$= \prod_{\tau=r+1}^{t} \frac{1}{1 - p(\tau)}$$
$$= E_p(t, r).$$

Next, consider the case $t \in \mathbb{N}_r^{s-1}$. Then

$$E_p(t, s)E_p(s, r) = \prod_{\tau=t+1}^{s} [1 - p(\tau)] \prod_{\tau=r+1}^{s} \frac{1}{1 - p(\tau)}$$
$$= \prod_{\tau=r+1}^{t} \frac{1}{1 - p(\tau)}$$
$$= E_p(t, r).$$

Finally, consider the case $t \in \mathbb{N}_a^{r-1}$. Then

$$E_p(t, s)E_p(s, r) = \prod_{\tau=t+1}^{s} [1 - p(\tau)] \prod_{\tau=r+1}^{s} \frac{1}{1 - p(\tau)}$$
$$= \prod_{\tau=r+1.}^{t} [1 - p(\tau)]$$
$$= E_p(t, r).$$

This completes the proof of (vi) for the special case $s \geq r \geq a$. The case $a \leq s \leq r$ is left to the reader (Exercise 3.9). The proof of the law of exponents (vii) is Exercise 3.10. To see that (viii) holds, note that for $t \in \mathbb{N}_s$

$$E_{\boxminus p}(t,s) = \prod_{\tau=s+1}^{t} \frac{1}{1-(\boxminus p)(\tau)}$$
$$= \prod_{\tau=s+1}^{t} [1-p(\tau)]$$
$$= \frac{1}{E_p(t,s)}.$$

Also, if $t \in \mathbb{N}_a^{s-1}$

$$E_{\boxminus p}(t,s) = \prod_{\tau=t+1}^{s} [1-(\boxminus p)(\tau)]$$
$$= \prod_{\tau=t+1}^{s} \frac{1}{1-p(\tau)}$$
$$= \frac{1}{E_p(t,s)}.$$

Hence (viii) holds for $t \in \mathbb{N}_a$. Finally, using (viii) and then (vii), we have that

$$\frac{E_p(t,s)}{E_q(t,s)} = E_p(t,s)E_{\boxminus q}(t,s) = E_{p\boxplus[\boxminus q]}(t,s) = E_{p\boxminus q}(t,s),$$

from which it follows that (ix) holds. □

Next we define the **scalar box dot multiplication**, $\boxdot$.

**Definition 3.12.** For $\alpha \in \mathbb{R}$, $p \in \mathcal{R}^+$ the scalar box dot multiplication, $\alpha \boxdot p$, is defined by

$$\alpha \boxdot p = 1 - (1-p)^\alpha.$$

It follows that for $\alpha \in \mathbb{R}$, $p \in \mathcal{R}^+$

$$1 - (\alpha \boxdot p)(t) = 1 - \{1 - [1-p(t)]^\alpha\}$$
$$= [1-p(t)]^\alpha > 0$$

for $t \in \mathbb{N}_{a+1}$. Hence $\alpha \boxdot p \in \mathcal{R}^+$.

Now we can prove the following law of exponents.

**Theorem 3.13.** *If $\alpha \in \mathbb{R}$ and $p \in \mathcal{R}^+$, then*

$$E_p^\alpha(t,a) = E_{\alpha \boxdot p}(t,a)$$

*for $t \in \mathbb{N}_a$.*

*Proof.* Consider that, for $t \in \mathbb{N}_a$,

$$\begin{aligned} E_p^\alpha(t,a) &= \left[\prod_{\tau=a+1}^{t} \frac{1}{1+p(\tau)}\right]^\alpha \\ &= \prod_{\tau=a+1}^{t} \frac{1}{[1+p(\tau)]^\alpha} \\ &= \prod_{\tau=a+1}^{t} \frac{1}{1-[1-(1-p(\tau))^\alpha]} \\ &= \prod_{\tau=a+1}^{t} \frac{1}{1-[\alpha \boxdot p](\tau)} \\ &= E_{\alpha \boxdot p}(t,a). \end{aligned}$$

This completes the proof. □

**Theorem 3.14.** *The set of positively regressive functions $\mathcal{R}^+$, with the addition $\boxplus$ and the scalar multiplication $\boxdot$, is a vector space.*

*Proof.* From Theorem 3.10 we know that $\mathcal{R}^+$ with the addition $\boxplus$ is an Abelian group. The four remaining nontrivial properties of a vector space are the following:

(i) $1 \boxdot p = p$;
(ii) $\alpha \boxdot (p \boxplus q) = (\alpha \boxdot p) \boxplus (\alpha \boxdot q)$;
(iii) $\alpha \boxdot (\beta \boxdot p) = (\alpha\beta) \boxdot p$;
(iv) $(\alpha + \beta) \boxdot p = (\alpha \boxdot p) \boxplus (\beta \boxdot p)$,

where $\alpha, \beta \in \mathbb{R}$ and $p, q \in \mathcal{R}^+$. We will prove properties (i)–(iii) and leave property (iv) as an exercise (Exercise 3.12).

Property (i) follows immediately from the following:

$$1 \boxdot p = 1 - (1-p)^1 = p.$$

To prove (ii) consider

$$\begin{aligned} &(\alpha \boxdot p) \boxplus (\alpha \boxdot q) \\ &\quad = \alpha \boxdot p + \alpha \boxdot q - (\alpha \boxdot p)(\alpha \boxdot q) \end{aligned}$$

$$= [1-(1-p)^\alpha] + [1-(1-q)^\alpha] - [1-(1-p)^\alpha][1-(1-q)^\alpha]$$
$$= 1-(1-p)^\alpha(1-q)^\alpha$$
$$= 1-(1-p-q+pq)^\alpha$$
$$= 1-(1-p \boxplus q)^\alpha$$
$$= \alpha \boxdot (p \boxplus q).$$

Hence, (ii) holds. Finally, consider

$$\begin{aligned}\alpha \boxdot (\beta \boxplus p) &= 1-(1-\beta \boxdot p)^\alpha \\ &= 1-\left[1-\left[1-(1-p)^\beta\right]\right]^\alpha \\ &= 1-(1-p)^{\alpha\beta} \\ &= (\alpha\beta) \boxdot p.\end{aligned}$$

Hence, property (iii) holds. □

## 3.4 Nabla Trigonometric Functions

In this section we introduce the discrete nabla hyperbolic sine and cosine functions, the discrete sine and cosine functions and give some of their properties. First we define the nabla hyperbolic sine and cosine functions.

**Definition 3.15.** Assume $p, -p \in \mathcal{R}$. Then the generalized nabla hyperbolic sine and cosine functions are defined as follows:

$$\mathrm{Cosh}_p(t,a) := \frac{E_p(t,a) + E_{-p}(t,a)}{2}, \quad \mathrm{Sinh}_p(t,a) := \frac{E_p(t,a) - E_{-p}(t,a)}{2}$$

for $t \in \mathbb{N}_a$.

The following theorem gives various properties of the nabla hyperbolic sine and cosine functions.

**Theorem 3.16.** *Assume $p, -p \in \mathcal{R}$. Then*

(i) $Cosh_p(a,a) = 1, \quad Sinh_p(a,a) = 0;$
(ii) $Cosh_p^2(t,a) - Sinh_p^2(t,a) = E_{p^2}(t,a), \quad t \in \mathbb{N}_a;$
(iii) $\nabla Cosh_p(t,a) = p(t)\, Sinh_p(t,a), \quad t \in \mathbb{N}_{a+1};$
(iv) $\nabla Sinh_p(t,a) = p(t)\, Cosh_p(t,a), \quad t \in \mathbb{N}_{a+1};$
(v) $Cosh_{-p}(t,a) = Cosh_p(t,a), \quad t \in \mathbb{N}_a;$
(vi) $Sinh_{-p}(t,a) = -Sinh_p(t,a), \quad t \in \mathbb{N}_a.$

*Proof.* Parts (i), (v), (vi) follow immediately from the definitions of the nabla hyperbolic sine and cosine functions. To see that (ii) holds, note that

$$\begin{aligned}
&\text{Cosh}_p^2(t,a) - \text{Sinh}_p^2(t,a) \\
&= \frac{\left[E_p(t,a) + E_{-p}(t,a)\right]^2 - \left[E_p(t,a) - E_{-p}(t,a)\right]^2}{4} \\
&= E_p(t,a)E_{-p}(t,a) \\
&= E_{p\boxplus(-p)}(t,a) \\
&= E_{p^2}(t,a).
\end{aligned}$$

To see that (iii) holds, consider

$$\begin{aligned}
\nabla \text{Cosh}_p(t,a) &= \frac{1}{2}\nabla E_p(t,a) + \frac{1}{2}\nabla E_{-p}(t,a) \\
&= \frac{1}{2}\left[pE_p(t,a) - pE_{-p}(t,a)\right] \\
&= p\,\frac{E_p(t,a) - E_{-p}(t,a)}{2} \\
&= p\,\text{Sinh}_p(t,a).
\end{aligned}$$

The proof of (iv) is similar (Exercise 3.13). □

Next, we define the nabla sine and cosine functions.

**Definition 3.17.** Assume $ip, -ip \in \mathcal{R}$. Then we define the nabla sine and cosine functions as follows:

$$\text{Cos}_p(t,a) = \frac{E_{ip}(t,a) + E_{-ip}(t,a)}{2}, \quad \text{Sin}_p(t,a) = \frac{E_{ip}(t,a) - E_{-ip}(t,a)}{2i}$$

for $t \in \mathbb{N}_a$.

Using the definitions of $\text{Cos}_p(t,a)$ and $\text{Sin}_p(t,a)$ we get immediately Euler's formula

$$E_{ip}(t,a) = \text{Cos}_p(t,a) + i\text{Sin}_p(t,a), \quad t \in \mathbb{N}_a \tag{3.11}$$

provided $ip, -ip \in \mathcal{R}$.

The following theorem gives some relationships between the nabla trigonometric functions and the nabla hyperbolic trigonometric functions.

**Theorem 3.18.** *Assume p is a constant. Then the following hold:*

(i) $Sin_{ip}(t,a) = iSinh_p(t,a), \quad if\ p \neq \pm 1;$
(ii) $Cos_{ip}(t,a) = Cosh_p(t,a), \quad if\ p \neq \pm 1;$

(iii) $Sinh_{ip}(t,a) = iSin_p(t,a)$, *if* $p \neq \pm i$;
(iv) $Cosh_{ip}(t,a) = Cos_p(t,a)$, *if* $p \neq \pm i$,

*for* $t \in \mathbb{N}_a$.

*Proof.* To see that (i) holds, note that

$$\begin{aligned}
\mathrm{Sin}_{ip}(t,a) &= \frac{1}{2i}[E_{i^2p}(t,a) - E_{-i^2p}(t,a)] \\
&= \frac{1}{2i}[E_{-p}(t,a) - E_p(t,a)] \\
&= i\frac{E_p(t,a) - E_{-p}(t,a)}{2} \\
&= i\ \mathrm{Sinh}_p(t,a).
\end{aligned}$$

The proof of parts (ii), (iii), and (iv) are similar. □

The following theorem gives various properties of the generalized sine and cosine functions.

**Theorem 3.19.** *Assume* $ip, -ip \in \mathcal{R}$. *Then*

(i) $Cos_p(a,a) = 1, \quad Sin_p(a,a) = 0$;
(ii) $Cos_p^2(t,a) + Sin_p^2(t,a) = E_{-p^2}(t,a), \quad t \in \mathbb{N}_a$;
(iii) $\nabla Cos_p(t,a) = -p(t)\ Sin_p(t,a), \quad t \in \mathbb{N}_{a+1}$;
(iv) $\nabla Sin_p(t,a) = p(t)\ Cos_p(t,a), \quad t \in \mathbb{N}_{a+1}$;
(v) $Cos_{-p}(t,a) = Cos_p(t,a), \quad t \in \mathbb{N}_a$;
(vi) $Sin_{-p}(t,a) = -Sin_p(t,a), \quad t \in \mathbb{N}_a$.

*Proof.* The proof of this theorem follows from Theorems 3.16 and 3.18. □

## 3.5 Second Order Linear Equations with Constant Coefficients

The nonhomogeneous second order linear nabla difference equation is given by

$$\nabla^2 y(t) + p(t)\nabla y(t) + q(t)y(t) = f(t), \quad t \in \mathbb{N}_{a+2}, \tag{3.12}$$

where we assume $p, g, f : \mathbb{N}_{a+2} \to \mathbb{R}$ and $1 + p(t) + q(t) \neq 0$ for $t \in \mathbb{N}_{a+2}$. In this section we will see that we can easily solve the corresponding second order linear homogeneous nabla difference equation with constant coefficients

$$\nabla^2 y(t) + p\nabla y(t) + qy(t) = 0, \quad t \in \mathbb{N}_{a+2}, \tag{3.13}$$

where we assume the constants $p, q \in \mathbb{R}$ satisfy $1 + p + q \neq 0$.

First we prove an existence-uniqueness theorem for solutions of initial value problems (IVPs) for (3.12).

**Theorem 3.20.** *Assume that* $p, q, f : \mathbb{N}_{a+2} \to \mathbb{R}$, $1 + p(t) + q(t) \neq 0$, $t \in \mathbb{N}_{a+2}$, $A, B \in \mathbb{R}$, *and* $t_0 \in \mathbb{N}_{a+1}$. *Then the IVP*

$$\nabla^2 y(t) + p(t)\nabla y(t) + q(t)y(t) = f(t), \quad t \in \mathbb{N}_{a+2}, \tag{3.14}$$

$$y(t_0 - 1) = A, \quad y(t_0) = B, \tag{3.15}$$

*where* $t_0 \in \mathbb{N}_{a+1}$ *and* $A, B \in \mathbb{R}$ *has a unique solution* $y(t)$ *on* $\mathbb{N}_a$.

*Proof.* Expanding equation (3.14) we have by first solving for $y(t)$ and then solving for $y(t-2)$ that, since $1 + p(t) + q(t) \neq 0$,

$$\begin{aligned} y(t) = & \frac{2 + p(t)}{1 + p(t) + q(t)} y(t-1) \\ & - \frac{1}{1 + p(t) + q(t)} y(t-2) + \frac{f(t)}{1 + p(t) + q(t)} \end{aligned} \tag{3.16}$$

and

$$y(t-2) = -[1 + p(t) + q(t)]y(t) + [2 + p(t)]y(t-1) + f(t). \tag{3.17}$$

If we let $t = t_0 + 1$ in (3.16), then equation (3.14) holds at $t = t_0 + 1$ iff

$$\begin{aligned} y(t_0 + 1) = & \frac{[2 + p(t_0 + 1)]B}{1 + p(t_0 + 1) + q(t_0 + 1)} - \frac{A}{1 + p(t_0 + 1) + q(t_0 + 1)} \\ & + \frac{f(t_0 + 1)}{1 + p(t_0 + 1) + q(t_0 + 1)}. \end{aligned}$$

Hence, the solution of the IVP (3.14), (3.15) is uniquely determined at $t_0 + 1$. But using the equation (3.16) evaluated at $t = t_0 + 2$, we have that the unique values of the solution at $t_0$ and $t_0 + 1$ uniquely determine the value of the solution at $t_0 + 2$. By induction we get that the solution of the IVP (3.14), (3.15) is uniquely determined on $\mathbb{N}_{t_0 - 1}$. On the other hand if $t_0 \geq a + 2$, then using equation (3.17) with $t = t_0$, we have that

$$y(t_0 - 2) = -[1 + p(t_0) + q(t_0)]B + [2 + p(t_0)]A + f(t_0).$$

Hence the solution of the IVP (3.14), (3.15) is uniquely determined at $t_0 - 2$. Similarly, if $t_0 - 3 \geq a$, then the value of the solution at $t_0 - 2$ and at $t_0 - 1$ uniquely determines the value of the solution at $t_0 - 3$. Proceeding in this manner we have by mathematical induction that the solution of the IVP (3.14), (3.15) is uniquely determined on $\mathbb{N}_a^{t_0 - 1}$. Hence the result follows. □

*Remark 3.21.* Note that the so-called initial conditions in (3.15)

$$y(t_0 - 1) = A, \quad y(t_0) = B$$

hold iff the equations $y(t_0) = C$, $\nabla y(t_0) = D := B - A$ are satisfied. Because of this we also say that

$$y(t_0) = C, \quad \nabla y(t_0) = D$$

are initial conditions for solutions of equation (3.12). In particular, Theorem 3.20 holds if we replace the conditions (3.15) by the conditions

$$y(t_0) = C, \quad \nabla y(t_0) = D.$$

*Remark 3.22.* From Exercise 3.21 we see that if $1 + p(t) + q(t) \neq 0$, $t \in \mathbb{N}_{a+2}$, then the general solution of the linear homogeneous equation

$$\nabla^2 y(t) + p(t)\nabla y(t) + q(t)y(t) = 0$$

is given by

$$y(t) = c_1 y_1(t) + c_2 y_2(t), \quad t \in \mathbb{N}_a,$$

where $y_1(t)$, $y_2(t)$ are any two linearly independent solutions of (3.13) on $\mathbb{N}_a$.

Next we show we can solve the second order linear nabla difference equation with constant coefficients (3.13). We say the equation

$$\lambda^2 + p\lambda + q = 0$$

is the characteristic equation of the nabla linear difference equation (3.13) and the solutions of this characteristic equation are called the characteristic values of (3.13).

**Theorem 3.23 (Distinct Roots).** *Assume* $1 + p + q \neq 0$ *and* $\lambda_1 \neq \lambda_2$ *(possibly complex) are the characteristic values of* (3.13). *Then*

$$y(t) = c_1 E_{\lambda_1}(t, a) + c_2 E_{\lambda_2}(t, a)$$

*is a general solution of* (3.13) *on* $\mathbb{N}_a$.

*Proof.* Since $\lambda_1$, $\lambda_2$ satisfy the characteristic equation for (3.13), we have that the characteristic polynomial for (3.13) is given by

$$(\lambda - \lambda_1)(\lambda - \lambda_2) = \lambda^2 - (\lambda_1 + \lambda_2)\lambda + \lambda_1\lambda_2$$

and hence

$$p = -\lambda_1 - \lambda_2, \quad q = \lambda_1\lambda_2.$$

Since

$$1 + p + q = 1 + (-\lambda_1 - \lambda_2) + \lambda_1\lambda_2 = (1 - \lambda_1)(1 - \lambda_2) \neq 0,$$

we have that $\lambda_1, \lambda_2 \neq 1$ and hence $E_{\lambda_1}(t,a)$ and $E_{\lambda_2}(t,a)$ are well defined. Next note that

$$\begin{aligned}\nabla^2 E_{\lambda_i}(t,a) + p\,\nabla E_{\lambda_i}(t,a) + q\,E_{\lambda_i}(t,a)\\ &= [\lambda_i^2 + p\lambda_i + q]E_{\lambda_i}(t,a)\\ &= 0,\end{aligned}$$

for $i = 1, 2$. Hence $E_{\lambda_i}(t,a)$, $i = 1, 2$ are solutions of (3.13). Since $\lambda_1 \neq \lambda_2$, these two solutions are linearly independent on $\mathbb{N}_a$, and by Remark 3.22,

$$y(t) = c_1 E_{\lambda_1}(t,a) + c_2 E_{\lambda_2}(t,a)$$

is a general solution of (3.13) on $\mathbb{N}_a$. □

*Example 3.24.* Solve the nabla linear difference equation

$$\nabla^2 y(t) + 2\nabla y(t) - 8y(t) = 0. \quad t \in \mathbb{N}_{a+2}.$$

The characteristic equation is

$$\lambda^2 + 2\lambda - 8 = (\lambda - 2)(\lambda + 4) = 0$$

and the characteristic roots are

$$\lambda_1 = 2, \quad \lambda_2 = -4.$$

Note that $1 + p + q = -5 \neq 0$, so we can apply Theorem 3.23. Then we have that

$$\begin{aligned}y(t) &= c_1 E_{\lambda_1}(t,a) + c_2 E_{\lambda_2}(t,a)\\ &= c_1 E_2(t,a) + c_2 E_{-4}(t,a)\\ &= c_1(-1)^{a-t} + c_2 5^{a-t}\end{aligned}$$

is a general solution on $\mathbb{N}_a$.

Usually, we want to find all real-valued solutions of (3.13). When a characteristic value $\lambda_1$ of (3.13) is complex, $E_{\lambda_1}(t,a)$ is a complex-valued solution. In the next theorem we show how to use this complex-valued solution to find two linearly independent real-valued solutions on $\mathbb{N}_a$.

**Theorem 3.25 (Complex Roots).** *Assume the characteristic values of* (3.13) *are* $\lambda = \alpha \pm i\beta$, $\beta > 0$ *and* $\alpha \neq 1$. *Then a general solution of* (3.13) *is given by*

$$y(t) = c_1 E_\alpha(t,a) Cos_\gamma(t,a) + c_2 E_\alpha(t,a) Sin_\gamma(t,a),$$

*where* $\gamma := \frac{\beta}{1-\alpha}$.

*Proof.* Since the characteristic roots are $\lambda = \alpha \pm i\beta$, $\beta > 0$, we have that the characteristic equation is given by

$$\lambda^2 - 2\alpha\lambda + \alpha^2 + \beta^2 = 0.$$

It follows that $p = -2\alpha$ and $q = \alpha^2 + \beta^2$, and hence

$$1 + p + q = (1-\alpha)^2 + \beta^2 \neq 0.$$

Hence, Remark 3.22 applies. By the proof of Theorem 3.23, we have that $y(t) = E_{\alpha+i\beta}(t,a)$ is a complex-valued solution of (3.13). Using

$$\alpha + i\beta = \alpha \boxplus i\frac{\beta}{1-\alpha} = \alpha \boxplus i\gamma,$$

where $\gamma = \frac{\beta}{1-\alpha}$, $\alpha \neq 1$, we get that

$$y(t) = E_{\alpha+i\beta}(t,a) = E_{\alpha \boxplus i\gamma}(t,a) = E_\alpha(t,a) E_{i\gamma}(t,a)$$

is a nontrivial solution. It follows from Euler's formula (3.11) that

$$\begin{aligned} y(t) &= E_\alpha(t,a) E_{i\gamma}(t,a) \\ &= E_\alpha(t,a)[\mathrm{Cos}_\gamma(t,a) + i\mathrm{Sin}_\gamma(t,a)] \\ &= y_1(t) + i y_2(t) \end{aligned}$$

is a solution of (3.13). But since $p$ and $q$ are real, we have that the real part, $y_1(t) = E_\alpha(t,a)\mathrm{Cos}_\gamma(t,a)$, and the imaginary part, $y_2(t) = E_\alpha(t,a)\mathrm{Sin}_\gamma(t,a)$, of $y(t)$ are solutions of (3.13). But $y_1(t), y_2(t)$ are linearly independent on $\mathbb{N}_a$, so we get that

$$y(t) = c_1 E_\alpha(t,a)\mathrm{Cos}_\gamma(t,a) + c_2 E_\alpha(t,a)\mathrm{Sin}_\gamma(t,a)$$

is a general solution of (3.13) on $\mathbb{N}_a$. □

*Example 3.26.* Solve the nabla difference equation

$$\nabla^2 y(t) + 2\nabla y(t) + 2y(t) = 0, \quad t \in \mathbb{N}_{a+2}. \tag{3.18}$$

The characteristic equation is

$$\lambda^2 + 2\lambda + 2 = 0,$$

and so, the characteristic roots are $\lambda = -1 \pm i$. Note that $1 + p + q = 5 \neq 0$. So, applying Theorem 3.25, we find that

$$y(t) = c_1 E_{-1}(t,a)\mathrm{Cos}_{\frac{1}{2}}(t,a) + c_2 E_{-1}(t,a)\mathrm{Sin}_{\frac{1}{2}}(t,a)$$

is a general solution of (3.18) on $\mathbb{N}_a$.

The previous theorem (Theorem 3.25) excluded the case when the characteristic roots of (3.13) are $1 \pm i\beta$, where $\beta > 0$. The next theorem considers this case.

**Theorem 3.27.** *If the characteristic values of* (3.13) *are* $1 \pm i\beta$, *where* $\beta > 0$, *then a general solution of* (3.13) *is given by*

$$y(t) = c_1 \beta^{a-t} \cos\left[\frac{\pi}{2}(t-a)\right] + c_2 \beta^{a-t} \sin\left[\frac{\pi}{2}(t-a)\right],$$

$t \in \mathbb{N}_a$.

*Proof.* Since $1 - i\beta$ is a characteristic value of (3.13), we have that $y(t) = E_{1-i\beta}(t,a)$ is a complex-valued solution of (3.13). Now

$$\begin{aligned}
y(t) &= E_{1-i\beta}(t,a) \\
&= (i\beta)^{a-t} \\
&= \left(\beta e^{i\frac{\pi}{2}}\right)^{a-t} \\
&= \beta^{a-t} e^{i\frac{\pi}{2}(a-t)} \\
&= \beta^{a-t}\left\{\cos\left[\frac{\pi}{2}(a-t)\right] + i\sin\left[\frac{\pi}{2}(a-t)\right]\right\} \\
&= \beta^{a-t}\cos\left[\frac{\pi}{2}(t-a)\right] - i\beta^{a-t}\sin\left[\frac{\pi}{2}(t-a)\right].
\end{aligned}$$

It follows that

$$y_1(t) = \beta^{a-t}\cos\left[\frac{\pi}{2}(t-a)\right], \quad y_2(t) = \beta^{a-t}\sin\left[\frac{\pi}{2}(t-a)\right]$$

are solutions of (3.13). Since these solutions are linearly independent on $\mathbb{N}_a$, we have that

$$y(t) = c_1\beta^{a-t}\cos\left[\frac{\pi}{2}(t-a)\right] + c_2\beta^{a-t}\sin\left[\frac{\pi}{2}(t-a)\right]$$

is a general solution of (3.13). □

*Example 3.28.* Solve the nabla linear difference equation

$$\nabla^2 y(t) - 2\nabla y(t) + 5y(t) = 0, \quad t \in \mathbb{N}_2.$$

The characteristic equation is $\lambda^2 - 2\lambda + 5 = 0$, so the characteristic roots are $\lambda = 1 \pm 2i$. It follows from Theorem 3.27 that

$$y(t) = c_1 2^{-t}\cos\left(\frac{\pi}{2}t\right) + c_2 2^{-t}\sin\left(\frac{\pi}{2}t\right),$$

for $t \in \mathbb{N}_0$.

**Theorem 3.29 (Double Root).** *Assume $\lambda_1 = \lambda_2 = r \neq 1$ is a double root of the characteristic equation. Then*

$$y(t) = c_1 E_r(t,a) + c_2(t-a)E_r(t,a)$$

*is a general solution of* (3.13).

*Proof.* Since $\lambda_1 = r$ is a double root of the characteristic equation, we have that $\lambda^2 - 2r\lambda + r^2 = 0$ is the characteristic equation. It follows that $p = -2r$ and $q = r^2$. Therefore

$$1 + p + q = 1 - 2r + r^2 = (1-r)^2 \neq 0$$

since $r \neq 1$. Hence, Remark 3.22 applies. Since $r \neq 1$ is a characteristic root, we have that $y_1(t) = E_r(t,a)$ is a nontrivial solution of (3.13). From Exercise 3.14, we have that $y_2(t) = (t-a)E_r(t,a)$ is a second solution of (3.13) on $\mathbb{N}_a$. Since these two solutions are linearly independent on $\mathbb{N}_a$, we have from Remark 3.22 that

$$y(t) = c_1 E_r(t,a) + c_2(t-a)E_r(t,a)$$

is a general solution of (3.13). □

*Example 3.30.* Solve the nabla difference equation

$$\nabla^2 y(t) + 12\nabla y(t) + 36y(t) = 0, \quad t \in \mathbb{N}_{a+2}.$$

The corresponding characteristic equation is

$$\lambda^2 + 12\lambda + 36 = (\lambda + 6)^2 = 0.$$

Hence $r = -6 \neq 1$ is a double root, so by Theorem 3.29 a general solution is given by

$$\begin{aligned} y(t) &= c_1 E_{-6}(t,a) + c_2(t-a)E_{-6}(t,a) \\ &= c_1 7^{a-t} + c_2(t-a)7^{a-t} \end{aligned}$$

for $t \in \mathbb{N}_a$.

## 3.6 Discrete Nabla Integral

In this section we define the nabla definite and indefinite integral, give several of their properties, and present a nabla fundamental theorem of calculus.

**Definition 3.31.** Assume $f : \mathbb{N}_{a+1} \to \mathbb{R}$ and $b \in \mathbb{N}_a$. Then the nabla integral of $f$ from $a$ to $b$ is defined by

$$\int_a^b f(t)\nabla t := \sum_{t=a+1}^{b} f(t), \quad t \in \mathbb{N}_a$$

with the convention that

$$\int_a^a f(t)\nabla t = \sum_{t=a+1}^{a} f(t) := 0.$$

Note that even if $f$ had the domain $\mathbb{N}_a$ instead of $\mathbb{N}_{a+1}$ the value of the integral $\int_a^b f(t)\nabla t$ does not depend on the value of $f$ at $a$. Also note if $f : \mathbb{N}_{a+1} \to \mathbb{R}$, then $F(t) := \int_a^t f(\tau)\nabla\tau$ is defined on $\mathbb{N}_a$ with $F(a) = 0$.

The following theorem gives some important properties of this nabla integral.

**Theorem 3.32.** *Assume $f, g : \mathbb{N}_{a+1} \to \mathbb{R}$, $b, c, d \in \mathbb{N}_a$, $b \leq c \leq d$, and $\alpha \in \mathbb{R}$. Then*

(i) $\int_b^c \alpha f(t)\nabla t = \alpha \int_b^c f(t)\nabla t$;
(ii) $\int_b^c (f(t) + g(t))\nabla t = \int_b^c f(t)\nabla t + \int_b^c g(t)\nabla t$;
(iii) $\int_b^b f(t)\nabla t = 0$;
(iv) $\int_b^d f(t)\nabla t = \int_b^c f(t)\nabla t + \int_c^d f(t)\nabla t$;
(v) $|\int_b^c f(t)\nabla t| \leq \int_b^c |f(t)|\nabla t$;
(vi) *if* $F(t) := \int_b^t f(s)\nabla s$, *for* $t \in \mathbb{N}_b^c$, *then* $\nabla F(t) = f(t)$, $t \in \mathbb{N}_{b+1}^c$;
(vii) *if* $f(t) \geq g(t)$ *for* $t \in \mathbb{N}_{b+1}^c$, *then* $\int_b^c f(t)\nabla t \geq \int_b^c g(t)\nabla t$.

*Proof.* To see that (vi) holds, assume

$$F(t) = \int_b^t f(s)\nabla s, \quad t \in \mathbb{N}_b^c.$$

Then, for $t \in \mathbb{N}^c_{b+1}$, we have that

$$\nabla F(t) = \nabla\left(\int_b^t f(s)\nabla s\right)$$
$$= \nabla\left(\sum_{s=b+1}^{t} f(s)\right)$$
$$= \sum_{s=b+1}^{t} f(s) - \sum_{s=b+1}^{t-1} f(s)$$
$$= f(t).$$

Hence property (vi) holds. All the other properties of the nabla integral in this theorem hold since the corresponding properties for the summations hold. □

**Definition 3.33.** Assume $f : \mathbb{N}^b_{a+1} \to \mathbb{R}$. We say $F : \mathbb{N}^b_a \to \mathbb{R}$ is a **nabla antidifference** of $f(t)$ on $\mathbb{N}^b_a$ provided

$$\nabla F(t) = f(t), \quad t \in \mathbb{N}^b_{a+1}.$$

If $f : \mathbb{N}^b_{a+1} \to \mathbb{R}$, then if we define $F$ by

$$F(t) := \int_a^t f(s)\nabla s, \quad t \in \mathbb{N}^b_a$$

we have from part (vi) of Theorem 3.32 that $\nabla F(t) = f(t)$, for $t \in \mathbb{N}^b_{a+1}$, that is, $F(t)$ is a nabla antidifference of $f(t)$ on $\mathbb{N}^b_a$. Next we show that if $f : \mathbb{N}^b_{a+1} \to \mathbb{R}$, then $f(t)$ has infinitely many antidifferences on $\mathbb{N}^b_a$.

**Theorem 3.34.** *If $f : \mathbb{N}^b_{a+1} \to \mathbb{R}$ and $G(t)$ is a nabla antidifference of $f(t)$ on $\mathbb{N}^b_a$, then $F(t) = G(t) + C$, where $C$ is a constant, is a general nabla antidifference of $f(t)$ on $\mathbb{N}^b_a$.*

*Proof.* Assume $G(t)$ is a nabla antidifference of $f(t)$ on $\mathbb{N}^b_a$. Let $F(t) := G(t) + C$, $t \in \mathbb{N}^b_a$, where $C$ is a constant. Then

$$\nabla F(t) = \nabla G(t) = f(t), \quad t \in \mathbb{N}^b_{a+1},$$

and so, $F(t)$ is a antidifference of $f(t)$ on $\mathbb{N}^b_a$.

Conversely, assume $F(t)$ is a nabla antidifference of $f(t)$ on $\mathbb{N}^b_a$. Then

$$\nabla(F(t) - G(t)) = \nabla F(t) - \nabla G(t) = f(t) - f(t) = 0$$

for $t \in \mathbb{N}_{a+1}^b$. This implies $F(t) - G(t) = C$, for $t \in \mathbb{N}_a^b$, where $C$ is a constant. Hence

$$F(t) := G(t) + C, \quad t \in \mathbb{N}_a^b.$$

This completes the proof. □

**Definition 3.35.** If $f : \mathbb{N}_a \to \mathbb{R}$, then the nabla indefinite integral of $f$ is defined by

$$\int f(t)\nabla t = F(t) + C,$$

where $F(t)$ is a nabla antidifference of $f(t)$ and $C$ is an arbitrary constant.

Since any formula for a nabla derivative gives us a formula for an indefinite integral, we have the following theorem.

**Theorem 3.36.** *The following hold:*

(i) $\int \alpha^{t+\beta}\nabla t = \frac{\alpha}{\alpha-1}\alpha^{t+\beta} + C, \quad \alpha \neq 1;$
(ii) $\int (t-\alpha)^{\overline{r}}\nabla t = \frac{1}{r+1}(t-\alpha)^{\overline{r+1}} + C, \quad r \neq -1;$
(iii) $\int (\alpha-\rho(t))^{\overline{r}}\nabla t = -\frac{1}{r+1}(\alpha-t)^{\overline{r+1}} + C, \quad r \neq -1;$
(iv) $\int p(t)\, E_p(t,a)\nabla t = E_p(t,a) + C, \quad \text{if} \quad p \in \mathcal{R};$
(v) $\int p(t)Cosh_p(t,a)\nabla t = Sinh_p(t,a) + C, \quad \text{if} \quad \pm p \in \mathcal{R};$
(vi) $\int p(t)Sinh_p(t,a)\nabla t = Cosh_p(t,a) + C, \quad \text{if} \quad \pm p \in \mathcal{R};$
(vii) $\int p(t)Cos_p(t,a)\nabla t = Sin_p(t,a) + C, \quad \text{if} \quad \pm ip \in \mathcal{R};$
(viii) $\int p(t)Sin_p(t,a)\nabla t = -Cos_p(t,a) + C, \quad \text{if} \quad \pm ip \in \mathcal{R},$

*where $C$ is an arbitrary constant.*

*Proof.* The formula

$$\int \alpha^{t+\beta}\nabla t = \frac{\alpha}{\alpha-1}\alpha^{t+\beta} + C, \quad \alpha \neq 1,$$

is clear when $\alpha = 0$, and for $\alpha \neq 0$ it follows from part (iv) of Theorem 3.1. Parts (ii) and (iii) of this theorem follow from the power rules (3.3) and (3.4), respectively. Part (iv) of this theorem follows from part (iv) of Theorem 3.11. Parts (v) and (vi) of this theorem follow from parts (iv) and (iii) of Theorem 3.16, respectively. Finally, parts (vii) and (viii) of this theorem follow from parts (iv) and (iii) of Theorem 3.19, respectively. □

We now state and prove the fundamental theorem for the nabla calculus.

**Theorem 3.37 (Fundamental Theorem of Nabla Calculus).** *We assume $f : \mathbb{N}_{a+1}^b \to \mathbb{R}$ and $F$ is any nabla antidifference of $f$ on $\mathbb{N}_a^b$. Then*

$$\int_a^b f(t)\nabla t = F(t)\Big|_a^b := F(b) - F(a).$$

*Proof.* By Theorem 3.32, (vi), we have that $G$ defined by $G(t) := \int_a^t f(s)\nabla s$, for $t \in \mathbb{N}_a^b$, is a nabla antidifference of $f$ on $\mathbb{N}_a^b$. Since $F$ is a nabla antidifference of $f$ on $\mathbb{N}_a^b$, it follows from Theorem 3.34 that $F(t) = G(t) + C$, $t \in \mathbb{N}_a^b$, for some constant $C$. Hence,

$$\begin{aligned} F(t)\Big|_a^b &= F(b) - F(a) \\ &= [G(b) + C] - [G(a) + C] \\ &= G(b) - G(a) \\ &= \int_a^b f(s)\nabla s - \int_a^a f(s)\nabla s \\ &= \int_a^b f(s)\nabla s. \end{aligned}$$

This completes the proof. □

*Example 3.38.* Assume $p \neq 0, 1$ is a constant. Use the integration formula

$$\int_a^t E_p(t,a)\nabla t = \frac{1}{p} E_p(t,a)\Big|_a^t$$

to evaluate the integral $\int_0^4 f(t)\nabla t$, where $f(t) := (-3)^{-t}$, $t \in \mathbb{N}_0$. We calculate

$$\begin{aligned} \int_0^4 (-3)^{-t}\nabla t &= \int_0^4 (1-4)^{0-t}\nabla t \\ &= \int_0^4 E_4(t,0)\nabla t \\ &= \frac{1}{4} E_4(t,0)\Big|_0^4 \\ &= \frac{1}{4}[E_4(4,0) - E_4(0,0)] \\ &= \frac{1}{4}[(-3)^{-4} - 1] \\ &= -\frac{20}{81}. \end{aligned}$$

Check this answer by using part (i) in Theorem 3.36.

Using the product rule (part (v) in Theorem 3.1) we can prove the following integration by parts formulas.

**Theorem 3.39 (Integration by Parts).** *Given two functions $u, v : \mathbb{N}_a \to \mathbb{R}$ and $b, c \in \mathbb{N}_a$, $b < c$, we have the integration by parts formulas:*

$$\int_b^c u(t)\nabla v(t)\nabla t = u(t)v(t)\Big|_b^c - \int_b^c v(\rho(t))\nabla u(t)\nabla t, \tag{3.19}$$

$$\int_b^c u(\rho(t))\nabla v(t)\nabla t = u(t)v(t)\Big|_b^c - \int_b^c v(t)\nabla u(t)\nabla t. \tag{3.20}$$

*Example 3.40.* Given $f(t) = (t-1)3^{1-t}$ for $t \in \mathbb{N}_1$, evaluate the integral $\int_1^t f(\tau)\nabla\tau$. Note that

$$\int_1^t f(\tau)\nabla\tau = \int_1^t (\tau - 1)E_{-2}(\tau, 1)\nabla\tau.$$

To set up to use the integration by parts formula (3.19), set

$$u(\tau) = \tau - 1, \quad \nabla v(\tau) = E_{-2}(\tau, 1).$$

It follows that

$$\nabla u(\tau) = 1, \quad v(\tau) = -\frac{1}{2}E_{-2}(\tau, 1), \quad v(\rho(\tau)) = -\frac{3}{2}E_{-2}(\tau, 1).$$

Hence, using the integration by parts formula (3.19), we get

$$\begin{aligned}
\int_1^t f(\tau)\nabla\tau &= \int_1^t (\tau - 1)E_{-2}(\tau, 1)\nabla\tau \\
&= -\frac{1}{2}(\tau - 1)E_{-2}(\tau, 1)\Big|_{\tau=1}^{\tau=t} + \frac{3}{2}\int_1^t E_{-2}(\tau, 1)\nabla\tau \\
&-\frac{1}{2}(t-1)E_{-2}(t, 1) - \frac{3}{4}E_{-2}(\tau, 1)\Big|_{\tau=1}^{\tau=t} \\
&= -\frac{1}{2}(t-1)E_{-2}(t, 1) - \frac{3}{4}E_{-2}(t, 1) + \frac{3}{4} \\
&= -\frac{3}{2}t\left(\frac{1}{3}\right)^t - \frac{1}{4}\left(\frac{1}{3}\right)^t + \frac{3}{4},
\end{aligned}$$

for $t \in \mathbb{N}_1$.

## 3.7 First Order Linear Difference Equations

In this section we show how to solve the first order nabla linear equation

$$\nabla y(t) = p(t)y(t) + q(t), \quad t \in \mathbb{N}_{a+1}, \tag{3.21}$$

where we assume $p, q : \mathbb{N}_{a+1} \to \mathbb{R}$ and $p \in \mathcal{R}$. At the end of this section we will then show how to use the fact that we can solve the first order nabla linear equation (3.21) to solve certain nabla second order linear equations with variable coefficients (3.13) by the method of factoring.

We begin by using one of the following nabla Leibniz's formulas to find a variation of constants formula for (3.21).

**Theorem 3.41 (Nabla Leibniz Formulas).** *Assume $f : \mathbb{N}_a \times \mathbb{N}_{a+1} \to \mathbb{R}$. Then*

$$\nabla\left(\int_a^t f(t,\tau)\nabla\tau\right) = \int_a^t \nabla_t f(t,\tau)\nabla\tau + f(\rho(t),t), \tag{3.22}$$

$t \in \mathbb{N}_{a+1}$. *Also*

$$\nabla\left(\int_a^t f(t,\tau)\nabla\tau\right) = \int_a^{t-1} \nabla_t f(t,\tau)\nabla\tau + f(t,t), \tag{3.23}$$

*for $t \in \mathbb{N}_{a+1}$*

*Proof.* The proof of (3.22) follows from the following:

$$\begin{aligned}
\nabla\left(\int_a^t f(t,\tau)\nabla\tau\right) &= \int_a^t f(t,\tau)\nabla\tau - \int_a^{t-1} f(t-1,\tau)\nabla\tau \\
&= \int_a^t [f(t,\tau) - f(t-1,\tau)]\nabla\tau + \int_{t-1}^t f(t-1,\tau)\nabla\tau \\
&= \int_a^t \nabla_t f(t,\tau)\nabla\tau + f(\rho(t),t)
\end{aligned}$$

for $t \in \mathbb{N}_{a+1}$. The proof of (3.23) is Exercise 3.22. □

**Theorem 3.42 (Variation of Constants Formula).** *Assume $p, q : \mathbb{N}_{a+1} \to \mathbb{R}$ and $p \in \mathcal{R}$. Then the unique solution of the IVP*

$$\begin{aligned}
\nabla y(t) &= p(t)y(t) + q(t), \quad t \in \mathbb{N}_{a+1} \\
y(a) &= A
\end{aligned}$$

*is given by*

$$y(t) = AE_p(t,a) + \int_a^t E_p(t,\rho(s))q(s)\nabla s, \quad t \in \mathbb{N}_a.$$

*Proof.* The proof of uniqueness is left to the reader. Let

$$y(t) := AE_p(t,a) + \int_a^t E_p(t,\rho(s))q(s)\nabla s, \quad t \in \mathbb{N}_a.$$

Using the nabla Leibniz formula (3.22), we obtain

$$\begin{aligned}\nabla y(t) &= Ap(t)E_p(t,a) + \int_a^t p(t)E_p(t,\rho(s))q(s)\nabla s + E_p(\rho(t),\rho(t))q(t)\\ &= p(t)\left[AE_p(t,a) + \int_a^t E_p(t,\rho(s))q(s)\nabla s\right] + q(t)\\ &= p(t)y(t) + q(t)\end{aligned}$$

for $t \in \mathbb{N}_{a+1}$. We also see that $y(a) = A$. And this completes the proof. □

*Example 3.43.* Assuming $r \in \mathcal{R}$, solve the IVP

$$\nabla y(t) = r(t)y(t) + E_r(t,a), \quad t \in \mathbb{N}_{a+1} \tag{3.24}$$

$$y(a) = 0. \tag{3.25}$$

Using the variation of constants formula in Theorem 3.42, we have

$$\begin{aligned}y(t) &= \int_a^t E_r(t,\rho(s))E_r(s,a)\nabla s\\ &= E_r(t,a)\int_a^t E_r(a,\rho(s))E_r(s,a)\nabla s\\ &= E_r(t,a)\int_a^t \frac{E_r(s,a)}{E_r(\rho(s),a)}\nabla s\\ &= E_r(t,a)\int_a^t \frac{E_r(s,a)}{[1-r(s)]E_r(s,a)}\nabla s\\ &= E_r(t,a)\int_a^t \frac{1}{1-r(s)}\nabla s.\end{aligned}$$

If we further assume $r(t) = r \neq 1$ is a constant, then we obtain that the function $\frac{1}{1-r}(t-a)E_r(t,a)$ is the solution of the IVP

$$\nabla y(t) = ry(t) + E_r(t,a), \quad y(a) = 0.$$

A general solution of the linear equation (3.21) is given by adding a general solution of the corresponding homogeneous equation $\nabla y(t) = p(t)y(t)$ to a particular solution to the nonhomogeneous difference equation (3.21). Hence,

$$y(t) = cE_p(t,a) + \int_0^t E_p(t,\rho(s))q(s)\nabla s$$

is a general solution of (3.21). We use this fact in the following example.

*Example 3.44.* Find a general solution of the linear difference equation

$$\nabla y(t) = (\boxminus 3)y(t) + 3t, \quad t \in \mathbb{N}_1. \tag{3.26}$$

Note that the constant function $p(t) := \boxminus 3$ is a regressive function on $\mathbb{N}_1$. Hence, the general solution of (3.26) is given by

$$\begin{aligned}
y(t) &= cE_p(t,a) + \int_a^t E_p(t,\rho(s))q(s)\nabla s \\
&= cE_{\boxminus 3}(t,0) + 3\int_0^t sE_{\boxminus 3}(t,\rho(s))\nabla s \\
&= cE_{\boxminus 3}(t,0) + 3\int_0^t sE_3(\rho(s),t)\nabla s \\
&= cE_{\boxminus 3}(t,0) - 6\int_0^t sE_3(s,t)\nabla s,
\end{aligned}$$

for $t \in \mathbb{N}_0$. Integrating by parts we get

$$\begin{aligned}
y(t) &= cE_{\ominus 3}(t,0) - 2sE_3(s,t)\Big|_{s=0}^{t} + 2\int_0^t E_3(\rho(s),t)\nabla s \\
&= cE_{\boxminus 3}(t,0) - 2t - 4\int_0^t E_3(s,t)\nabla s \\
&= cE_{\boxminus 3}(t,0) - 2t - \frac{4}{3}E_3(s,t)\Big|_0^t \\
&= cE_{\boxminus 3}(t,0) - 2t - \frac{4}{3} + \frac{4}{3}E_3(0,t) \\
&= \alpha E_{\boxminus 3}(t,0) - 2t - \frac{4}{3} \\
&= \alpha(-2)^t - 2t - \frac{4}{3}
\end{aligned}$$

for $t \in \mathbb{N}_0$.

*Example 3.45.* Assuming $r \neq 1$, use the method of factoring to solve the nabla difference equation

$$\nabla^2 y(t) - 2r\nabla y(t) + r^2 y(t) = 0, \quad t \in \mathbb{N}_a. \tag{3.27}$$

A factored form of (3.27) is

$$(\nabla - rI)(\nabla - rI)y(t) = 0, \quad t \in \mathbb{N}_a. \tag{3.28}$$

It follows from (3.28) that any solution of $(\nabla - rI)y(t) = 0$ is a solution of (3.27). Hence $y_1(t) = E_r(t,a)$ is a solution of (3.27). It also follows from the factored equation (3.28) that the solution $y(t)$ of the IVP

$$(\nabla - rI)y(t) = E_r(t,a), \quad y(a) = 0$$

is a solution of (3.27). Hence, by the variation of constants formula in Theorem 3.42,

$$\begin{aligned}
y(t) &= \int_a^t E_r(t,\rho(s))E_r(s,a)\nabla s = E_r(t,a)\int_a^t E_r(a,\rho(s))E_r(s,a)\nabla s \\
&= E_r(t,a)\int_a^t E_{\ominus r}(\rho(s),a)E_r(s,a)\nabla s \\
&= E_r(t,a)\int_a^t [1-\ominus r]E_{\ominus r}(s,a)E_r(s,a)\nabla s \\
&= E_r(t,a)\int_a^t [1-\ominus r]\nabla s = E_r(t,a)\int_a^t \frac{1}{1-r}\nabla s \\
&= \frac{1}{1-r}(t-a)E_r(t,a)
\end{aligned}$$

is a solution of (3.27). But this implies that $y_2(t) = (t-a)E_r(t,a)$ is a solution of (3.27). Since $y_1(t)$ and $y_2(t)$ are linearly independent on $\mathbb{N}_a$,

$$y(t) = c_1 E_r(t,a) + c_2(t-a)E_r(t,a)$$

is a general solution of (3.27) on $\mathbb{N}_a$.

## 3.8 Nabla Taylor's Theorem

In this section we want to prove the nabla version of Taylor's Theorem. To do this we first study the nabla Taylor monomials and give some of their important properties. These nabla Taylor monomials will appear in the nabla Taylor's Theorem. We then will find nabla Taylor series expansions for the nabla exponential, hyperbolic, and trigonometric functions. Finally, as a special case of our Taylor's theorem we will obtain a variation of constants formula for $\nabla^n y(t) = h(t)$.

**Definition 3.46.** We define the **nabla Taylor monomials**, $H_n(t, a)$, $n \in \mathbb{N}_0$, by $H_0(t, a) = 1$, for $t \in \mathbb{N}_a$, and

$$H_n(t, a) = \frac{(t-a)^{\overline{n}}}{n!}, \quad t \in \mathbb{N}_{a-n+1}, \quad n \in \mathbb{N}_1.$$

**Theorem 3.47.** *The nabla Taylor monomials satisfy the following:*

(i) $H_n(t, a) = 0, \quad a - n + 1 \le t \le a, \quad n \in \mathbb{N}_1;$
(ii) $\nabla H_{n+1}(t, a) = H_n(t, a), \quad t \in \mathbb{N}_{a-n+1}, \quad n \in \mathbb{N}_0;$
(iii) $\int_a^t H_n(\tau, a)\nabla\tau = H_{n+1}(t, a), \quad t \in \mathbb{N}_a, \quad n \in \mathbb{N}_0;$
(iv) $\int_a^t H_n(t, \rho(s))\nabla s = H_{n+1}(t, a), \quad t \in \mathbb{N}_a, \quad n \in \mathbb{N}_0.$

*Proof.* Part (i) of this theorem follows from the definition (Definition 3.46) of the nabla Taylor monomials. By the first power rule (3.3), it follows that

$$\begin{aligned} \nabla H_{n+1}(t, a) &= \nabla \frac{(t-a)^{\overline{n+1}}}{(n+1)!} \\ &= \frac{(t-a)^{\overline{n}}}{n!} \\ &= H_n(t, a), \end{aligned}$$

and so, we have that part (ii) of this theorem holds. Part (iii) follows from parts (ii) and (i). Finally, to see that (iv) holds we use the integration formula in part (iii) in Theorem 3.36 to get

$$\begin{aligned} \int_a^t H_n(t, \rho(s))\nabla s &= \frac{1}{n!}\int_a^t (t - \rho(s))^{\overline{n}}\nabla s \\ &= \frac{-1}{(n+1)!}(t-s)^{\overline{n+1}}\Big|_{s=a}^{s=t} \\ &= \frac{(t-a)^{\overline{n+1}}}{(n+1)!} \\ &= H_{n+1}(t, a). \end{aligned}$$

This completes the proof. □

Now we state and prove the nabla Taylor's Theorem.

**Theorem 3.48 (Nabla Taylor's Formula).** *Assume* $f : \mathbb{N}_{a-n} \to \mathbb{R}$, *where* $n \in \mathbb{N}_0$. *Then*

$$f(t) = p_n(t) + R_n(t), \quad t \in \mathbb{N}_{a-n},$$

*where the n-th degree nabla Taylor polynomial, $p_n(t)$, is given by*

$$p_n(t) := \sum_{k=0}^{n} \nabla^k f(a) \frac{(t-a)^{\overline{k}}}{k!} = \sum_{k=0}^{n} \nabla^k f(a) H_k(t,a)$$

*and the Taylor remainder, $R_n(t)$, is given by*

$$R_n(t) = \int_a^t \frac{(t-\rho(s))^{\overline{n}}}{n!} \nabla^{n+1} f(s) \nabla s = \int_a^t H_n(t,\rho(s)) \nabla^{n+1} f(s) \nabla s,$$

*for $t \in \mathbb{N}_{a-n}$. (By convention we assume $R_n(t) = 0$ for $a - n \leq t < a$.)*

*Proof.* We will use the second integration by parts formula in Theorem 3.39, namely (3.20), to evaluate the integral in the definition of $R_n(t)$. To do this we set

$$u(\rho(s)) = H_n(t,\rho(s)), \quad \nabla v(s) = \nabla^{n+1} f(s).$$

Then it follows that

$$u(s) = H_n(t,s), \quad v(s) = \nabla^n f(s).$$

Using part (iv) of Theorem 3.47, we get

$$\nabla u(s) = -H_{n-1}(t,\rho(s)).$$

Hence we get from the second integration by parts formula (3.20) that

$$\begin{aligned} R_n(t) &= \int_a^t H_n(t,\rho(s)) \nabla^{n+1} f(s) \nabla s \\ &= H_n(t,s) \nabla^n f(s) \Big|_{s=a}^{s=t} + \int_a^t H_{n-1}(t,\rho(s)) \nabla^n f(s) \nabla s \\ &= -\nabla^n f(a) H_n(t,a) + \int_a^t H_{n-1}(t,\rho(s)) \nabla^n f(s) \nabla s. \end{aligned}$$

Again, using the second integration by parts formula (3.20), we have that

$$\begin{aligned} R_n(t) =& -\nabla^n f(a) H_n(t,a) + H_{n-1}(t,s) \nabla^{n-1} f(s) \Big|_{s=a}^{s=t} \\ &+ \int_a^t H_{n-2}(t,\rho(s)) \nabla^{n-1} f(s) \nabla s \\ =& -\nabla^n f(a) H_n(t,a) - \nabla^{n-1} f(a) H_{n-1}(t,a) \\ &+ \int_a^t H_{n-2}(t,\rho(s)) \nabla^{n-1} f(s) \nabla s. \end{aligned}$$

By induction on $n$ we obtain

$$\begin{aligned} R_n(t) &= -\sum_{k=1}^{n} \nabla^k f(a) H_k(t,a) + \int_a^t H_0(t,\rho(s)) \nabla f(s) \nabla s \\ &= -\sum_{k=1}^{n} \nabla^k f(a) H_k(t,a) + f(t) - f(a) H_0(t,a) \\ &= -\sum_{k=0}^{n} \nabla^k f(a) H_k(t,a) + f(t) \\ &= -p_n(t) + f(t). \end{aligned}$$

Solving for $f(t)$ we get the desired result. □

We next define the formal nabla power series of a function at a point.

**Definition 3.49.** Let $a \in \mathbb{R}$ and let

$$\mathbb{Z}_a := \{\ldots, a-2, a-1, a, a+1, a+2, \ldots\}.$$

If $f : \mathbb{Z}_a \to \mathbb{R}$, then we call

$$\sum_{k=0}^{\infty} \nabla^k f(a) \frac{(t-a)^{\overline{k}}}{k!} = \sum_{k=0}^{\infty} \nabla^k f(a) H_k(t,a)$$

the (formal) nabla Taylor series of $f$ at $t = a$

The following theorem gives some convergence results for nabla Taylor series for various functions.

**Theorem 3.50.** *Assume $|p| < 1$ is a constant. Then the following hold:*

(i) $E_p(t,a) = \sum_{n=0}^{\infty} p^n H_n(t,a)$;
(ii) $Sin_p(t,a) = \sum_{n=0}^{\infty} (-1)^n p^{2n+1} H_{2n+1}(t,a)$;
(iii) $Cos_p(t,a) = \sum_{n=0}^{\infty} (-1)^n p^{2n} H_{2n}(t,a)$;
(iv) $Cosh_p(t,a) = \sum_{n=0}^{\infty} p^{2n} H_{2n}(t,a)$;
(v) $Sinh_p(t,a) = \sum_{n=0}^{\infty} p^{2n+1} H_{2n+1}(t,a)$,

*for $t \in \mathbb{N}_a$.*

*Proof.* First we prove part (i). Since $\nabla^n E_p(t,a) = p^n E_p(t,a)$ for $n \in \mathbb{N}_0$, we have that the Taylor series for $E_p(t,a)$ is given by

$$\sum_{n=0}^{\infty} \nabla^n E_p(a,a) H_n(t,a) = \sum_{n=0}^{\infty} p^n H_n(t,a).$$

To show that the above Taylor series converges to $E_p(t,a)$ when $|p| < 1$ is a constant, for each $t \in \mathbb{N}_a$, it suffices to show that the remainder term, $R_n(t)$, in Taylor's Formula satisfies

$$\lim_{n\to\infty} R_n(t) = 0$$

when $|p| < 1$, for each fixed $t \in \mathbb{N}_a$,

So fix $t \in \mathbb{N}_a$ and consider

$$\begin{aligned}|R_n(t)| &= \left|\int_a^t H_n(t,\rho(s))\nabla^{n+1}E_p(s,a)\nabla s\right| \\ &= \left|\int_a^t H_n(t,\rho(s))p^{n+1}E_p(s,a)\nabla s\right|.\end{aligned}$$

Since $t$ is fixed, there is a constant $C$ such that

$$|E_p(s,a)| \leq C, \quad a \leq s \leq t.$$

Hence,

$$\begin{aligned}|R_n(t)| &\leq C\int_a^t H_n(t,\rho(s))|p|^{n+1}\nabla s \\ &= C|p|^{n+1}\int_a^t H_n(t,\rho(s))\nabla s \\ &= C|p|^{n+1}H_{n+1}(t,a) \quad \text{by Theorem 3.47, (iv)} \\ &= C|p|^{n+1}\frac{(t-a)^{\overline{n+1}}}{(n+1)!}.\end{aligned}$$

By the ratio test, if $|p| < 1$, the series

$$\sum_{n=0}^{\infty}\frac{|p|^{n+1}(t-a)^{\overline{n+1}}}{(n+1)!}$$

converges. It follows that if $|p| < 1$, then by the $n$-th term test

$$\lim_{n\to\infty}\frac{|p|^{n+1}(t-a)^{\overline{n+1}}}{(n+1)!} = 0.$$

This implies that if $|p| < 1$, then for each fixed $t \in \mathbb{N}_a$

$$\lim_{n\to\infty} R_n(t) = 0,$$

and hence if $|p| < 1$,

$$E_p(t,a) = \sum_{n=0}^{\infty} p^n H_n(t,a)$$

for all $t \in \mathbb{N}_a$. Since the functions $\mathrm{Sin}_p(t,a)$, $\mathrm{Cos}_p(t,a)$, $\mathrm{Sinh}_p(t,a)$, and $\mathrm{Cosh}_p(t,a)$ are defined in terms of $E_p(t,a)$, parts (ii)–(v) follow easily from part (i). □

We now see that the integer order variation of constants formula follows from Taylor's formula.

**Theorem 3.51 (Integer Order Variation of Constants Formula).** *Assume $h : \mathbb{N}_{a+1} \to \mathbb{R}$ and $n \in \mathbb{N}_1$. Then the solution of the IVP*

$$\begin{aligned} \nabla^n y(t) &= h(t), \quad t \in \mathbb{N}_{a+1} \\ \nabla^k y(a) &= C_k, \quad 0 \le k \le n-1, \end{aligned} \tag{3.29}$$

*where $C_k$, $0 \le k \le n-1$, are given constants, is given by the variation of constants formula*

$$y(t) = \sum_{k=0}^{n-1} C_k H_k(t,a) + \int_a^t H_{n-1}(t,\rho(s))h(s)\nabla s, \quad t \in \mathbb{N}_{a-n+1}.$$

*Proof.* It is easy to see that the given IVP has a unique solution $y$ that is defined on $\mathbb{N}_{a-n+1}$. By Taylor's formula (see Theorem 3.48) with $n$ replaced by $n-1$ we get that

$$\begin{aligned} y(t) &= \sum_{k=0}^{n-1} \nabla^k y(a) H_k(t,a) + \int_a^t H_{n-1}(t,\rho(s))\nabla^n y(s)\nabla s \\ &= \sum_{k=0}^{n-1} C_k H_k(t,a) + \int_a^t H_{n-1}(t,\rho(s))h(s)\nabla s, \end{aligned}$$

$t \in \mathbb{N}_{a-n+1}$. □

We immediately get the following special case of Theorem 3.51. This special case, which we label Corollary 3.52, is also called a variation of constants formula.

**Corollary 3.52 (Integer Order Variation of Constants Formula).** *Assume the function $h : \mathbb{N}_{a+1} \to \mathbb{R}$ and $n \in \mathbb{N}_0$. Then the solution of the IVP*

$$\begin{aligned} \nabla^n y(t) &= h(t), \quad t \in \mathbb{N}_{a+1} \\ \nabla^k y(a) &= 0, \quad 0 \le k \le n-1 \end{aligned} \tag{3.30}$$

*is given by the variation of constants formula.*

$$y(t) = \int_a^t H_{n-1}(t, \rho(s))h(s)\nabla s, \quad t \in \mathbb{N}_{a-n+1}.$$

*Example 3.53.* Use the variation of constants formula to solve the IVP

$$\nabla^2 y(t) = (-2)^{a-t}, \quad t \in \mathbb{N}_{a+1}$$
$$y(a) = 2, \quad \nabla y(a) = 1.$$

By the variation of constants formula in Theorem 3.51 the solution of this IVP is given by

$$\begin{aligned}
y(t) &= C_0 H_0(t,a) + C_1 H_1(t,a) + \int_a^t H_1(t,\rho(s))(-2)^{a-s}\nabla s \\
&= 2H_0(t,a) + H_1(t,a) + \int_a^t H_1(t,\rho(s))E_3(s,a)\nabla s \\
&= 2H_0(t,a) + H_1(t,a) + \frac{1}{3}H_1(t,s)E_3(s,a)\Big|_{s=a}^{t} + \frac{1}{3}\int_a^t E_3(s,a)\nabla s \\
&= 2H_0(t,a) + H_1(t,a) - \frac{1}{3}H_1(t,a) + \frac{1}{9}E_3(s,a)\Big|_a^t \\
&= 2H_0(t,a) + H_1(t,a) - \frac{1}{3}H_1(t,a) + \frac{1}{9}E_3(t,a) - \frac{1}{9} \\
&= 2 + \frac{2}{3}H_1(t,a) + \frac{1}{9}(-2)^{a-t} - \frac{1}{9} \\
&= \frac{17}{9} + \frac{2}{3}(t-a) + \frac{1}{9}(-2)^{a-t},
\end{aligned}$$

for $t \in \mathbb{N}_{a-1}$.

## 3.9 Fractional Sums and Differences

With the relevant preliminaries established, we are now ready to develop what we mean by fractional nabla differences and fractional nabla sums. We first give the motivation for how we define nabla integral sums.

In the previous section (see Corollary 3.52) we saw that

$$y(t) = \int_a^t H_{n-1}(t, \rho(s))f(s)\nabla s$$

is the unique solution of the nabla difference equation $\nabla^n y(t) = f(t), t \in \mathbb{N}_{a+1}$ satisfying the initial conditions $\nabla^i y(a) = 0, 0 \le i \le n-1$, for $t \in \mathbb{N}_a$. Integrating $n$ times both sides of $\nabla^n y(t) = f(t)$ and using the initial conditions $\nabla^i y(a) = 0$, $0 \le i \le n-1$, we get by uniqueness

$$\int_a^t \int_a^{\tau_1} \cdots \int_a^{\tau_{n-1}} f(\tau_n)\nabla\tau_n \cdots \nabla\tau_2\nabla\tau_1$$
$$= \int_a^t H_{n-1}(t, \rho(s))f(s)\nabla s. \qquad (3.31)$$

The formula (3.31) can also be easily proved by repeated integration by parts. Motivated by this we define the nabla integral order sum as in the following definition.

**Definition 3.54 (Integral Order Sum).** Let $f : \mathbb{N}_{a+1} \to \mathbb{R}$ be given and $n \in \mathbb{N}_1$. Then

$$\nabla_a^{-n} f(t) := \int_a^t H_{n-1}(t, \rho(s))f(s)\nabla s, \quad t \in \mathbb{N}_a.$$

Also, we define $\nabla^{-0} f(t) := f(t)$.

Note that the function $\nabla_a^{-n} f$ depends on the values of $f$ at all the points $a+1 \le s \le t$, unlike the positive integer nabla difference $\nabla^n f(t)$, which just depends on the values of $f$ at the $n+1$ points $t-n \le s \le t$. Another interesting observation is that we could think of $\nabla_a^{-n} f(t)$ as defined on $\mathbb{N}_{a-n+1}$, from which we obtain that $f(t) = 0, \quad a+n-1 \le t \le a$ by our convention that the nabla integral from a point to a smaller point is zero (see Definition 3.31). The following example appears in Hein et al. [119].

*Example 3.55.* Use the definition (Definition 3.54) of the fractional sum to find $\nabla_a^{-2} E_p(t, a)$, where $p \ne 0, 1$ is a constant. By definition we obtain, using the second integration by parts formula (3.20),

$$\begin{aligned}
\nabla_a^{-2} E_p(t, a) &= \int_a^t H_1(t, \rho(s))E_p(s, a)\nabla s \\
&= \frac{1}{p}E_p(s, a)H_1(t, s)\Big|_{s=a}^t + \frac{1}{p}\int_a^t E_p(s, a)\nabla s \\
&= -\frac{1}{p}H_1(t, a) + \frac{1}{p^2}E_p(s, a)\Big|_{s=a}^t \\
&= -\frac{1}{p}H_1(t, a) + \frac{1}{p^2}E_p(t, a) - \frac{1}{p^2} \\
&= -\frac{1}{p}(t-a) + \frac{1}{p^2}(1-p)^{a-t} - \frac{1}{p^2}.
\end{aligned}$$

Note that if $n \in \mathbb{N}_1$, then

$$H_n(t,a) = \frac{(t-a)^{\overline{n}}}{n!} = \frac{(t-a)^{\overline{n}}}{\Gamma(n+1)}.$$

Motivated by this we define the fractional $\mu$-th order nabla Taylor monomial as follows.

**Definition 3.56.** Let $\mu \neq -1, -2, -3, \cdots$. Then we define the $\mu$-th order nabla fractional Taylor monomial, $H_\mu(t,a)$, by

$$H_\mu(t,a) = \frac{(t-a)^{\overline{\mu}}}{\Gamma(\mu+1)},$$

whenever the right-hand side of this equation is sensible.

In the next theorem we collect some of the properties of fractional nabla Taylor monomials.

**Theorem 3.57.** *The following hold:*

(i) $H_\mu(a,a) = 0$;
(ii) $\nabla H_\mu(t,a) = H_{\mu-1}(t,a)$;
(iii) $\int_a^t H_\mu(s,a)\nabla s = H_{\mu+1}(t,a)$;
(iv) $\int_a^t H_\mu(t,\rho(s))\nabla s = H_{\mu+1}(t,a)$;
(v) *for* $k \in \mathbb{N}_1$, $H_{-k}(t,a) = 0$, $t \in \mathbb{N}_a$,

*provided the expressions in this theorem are well defined.*

*Proof.* Part (i) follows immediately from the definition of $H_\mu(t,a)$. The proofs of parts (ii)–(iii) of this theorem are the same as the proof of Theorem 3.47, where we used the fractional power rules instead of the integer power rules. Finally, part (v) follows since

$$H_{-k}(t,a) = \frac{(t-a)^{\overline{-k}}}{\Gamma(-k+1)} = 0$$

by our earlier convention when the denominator is undefined but the numerator is defined. □

Now we can define the fractional nabla sum in terms of the nabla fractional Taylor monomial as follows.

**Definition 3.58 (Nabla Fractional Sum).** Let $f : \mathbb{N}_{a+1} \to \mathbb{R}$ be given and assume $\mu > 0$. Then

$$\nabla_a^{-\mu} f(t) := \int_a^t H_{\mu-1}(t,\rho(s)) f(s)\nabla s,$$

for $t \in \mathbb{N}_a$, where by convention $\nabla_a^{-\mu} f(a) = 0$.

The following example appears in Hein et al. [119].

*Example 3.59.* Use the definition (Definition 3.58) of the fractional sum to find $\nabla_a^{-\mu}1$. By definition

$$\begin{aligned}\nabla_a^{-\mu}1 &= \int_a^t H_{\mu-1}(t,\rho(s)) \cdot 1\nabla s \\ &= \int_a^t H_{\mu-1}(t,\rho(s))\nabla s \\ &= H_\mu(t,a), \quad t \in \mathbb{N}_a\end{aligned}$$

by part (iv) of Theorem 3.57.

For those readers that have read Chap. 2 we gave a relationship between a certain delta fractional sum and a certain nabla fractional sum. This formula is sometimes useful for obtaining results for the nabla fractional calculus from the delta fractional calculus. Since we want this chapter to be self-contained we will not use this formula in this chapter.

**Theorem 3.60.** *Assume $f : \mathbb{N}_a \to \mathbb{R}$ and $\nu > 0$. Then*

$$\Delta_a^{-\nu}f(t+\nu) = \nabla_a^{-\nu}f(t) + H_{\nu-1}(t,\rho(a))f(a),$$

*for $t \in \mathbb{N}_a$. In particular, if $f(a) = 0$, then*

$$\Delta_a^{-\nu}f(t+\nu) = \nabla_a^{-\nu}f(t),$$

*for $t \in \mathbb{N}_a$.*

*Proof.* Note that $f : \mathbb{N}_a \to \mathbb{R}$ implies $\Delta_a^{-\nu}f(t+\nu)$ is defined for $t \in \mathbb{N}_a$. Using the definition of the $\nu$-th order fractional sum (Definition 3.58) we find that

$$\begin{aligned}\Delta_a^{-\nu}f(t+\nu) &= \int_a^{t+1} h_{\nu-1}(t+\nu,\sigma(\tau))f(\tau)\nabla\tau \\ &= \sum_{\tau=a}^{t} h_{\nu-1}(t+\nu,\sigma(\tau))f(\tau) \\ &= \sum_{\tau=a}^{t} \frac{(t+\nu-\rho(\tau))^{\underline{\nu-1}}}{\Gamma(\nu)}f(\tau) \\ &= \sum_{\tau=a}^{t} \frac{\Gamma(t+\nu-\tau)}{\Gamma(\nu)\Gamma(t+\nu-\tau)} \\ &= \sum_{\tau=a}^{t} \frac{(t-\tau+1)^{\overline{\nu-1}}}{\Gamma(\nu)}f(\tau) \\ &= \nabla_a^{-\nu}f(t)\end{aligned}$$

for $t \in \mathbb{N}_a$. □

We next define the nabla fractional difference (nabla Riemann–Liouville fractional difference) in terms of a nabla fractional sum. The Caputo fractional sum (Definition 3.117) will be considered in Sect. 3.18.

**Definition 3.61 (Nabla Fractional Difference).** Let $f : \mathbb{N}_{a+1} \to \mathbb{R}$, $\nu \in \mathbb{R}^+$ and choose $N$ such that $N - 1 < \nu \leq N$. Then we define the $\nu$-th order nabla fractional difference, $\nabla_a^\nu f(t)$, by

$$\nabla_a^\nu f(t) := \nabla^N \nabla_a^{-(N-\nu)} f(t) \quad \text{for} \quad t \in \mathbb{N}_{a+N}.$$

We now have a definition for both fractional sums and fractional differences; however, they may still be unified to a similar form. We will show here that the traditional definition of a fractional difference can be rewritten in a form similar to the definition for a fractional sum. The following result appears in Ahrendt et al. [3].

**Theorem 3.62.** *Assume $f : \mathbb{N}_a \to \mathbb{R}$, $\nu > 0$, $\nu \notin \mathbb{N}_1$, and choose $N \in \mathbb{N}_1$ such that $N - 1 < \nu < N$. Then*

$$\nabla_a^\nu f(t) = \int_a^t H_{-\nu-1}(t, \rho(\tau)) f(\tau) \nabla\tau, \tag{3.32}$$

*for $t \in \mathbb{N}_{a+1}$.*

*Proof.* Note that

$$\begin{aligned}
\nabla_a^\nu f(t) &= \nabla^N \nabla_a^{-(N-\nu)} f(t) \\
&= \nabla^N \left( \int_a^t H_{N-\nu-1}(t, \rho(\tau)) f(\tau) \nabla\tau \right) \\
&= \nabla^{N-1} \nabla \left( \int_a^t H_{N-\nu-1}(t, \rho(\tau)) f(\tau) \nabla\tau \right) \\
&= \nabla^{N-1} \left( \int_a^t H_{N-\nu-2}(t, \rho(\tau)) f(\tau) \nabla\tau + H_{N-\nu-1}(\rho(t), \rho(t)) f(t) \right) \\
&= \nabla^{N-1} \int_a^t H_{N-\nu-2}(t, \rho(\tau)) f(\tau) \nabla\tau.
\end{aligned}$$

By applying Leibniz's Rule $N - 1$ more times, we deduce that

$$\nabla_a^\nu f(t) = \int_a^t H_{-\nu-1}(t, \rho(\tau)) f(\tau) \nabla\tau,$$

which is the desired result. □

In the following theorem we show that the nabla fractional difference, for each fixed $t \in \mathbb{N}_a$, is a continuous function of $\nu$ for $\nu > 0$. The following theorem appears in Ahrendt et al. [3].

**Theorem 3.63 (Continuity of the Nabla Fractional Difference).** *Assume $f : \mathbb{N}_a \to \mathbb{R}$. Then the fractional difference $\nabla_a^\nu f$ is continuous with respect to $\nu$ for $\nu > 0$.*

*Proof.* It is sufficient for this proof to show that for $f : \mathbb{N}_a \to \mathbb{R}$, $N-1 < \nu \leq N$, and $m \in \mathbb{N}_0$, the following hold:

$$\nabla_a^\nu f(a+N+m) \text{ is continuous with respect to } \nu \text{ on } (N-1,N), \tag{3.33}$$

$$\nabla_a^\nu f(a+N+m) \to \nabla^N f(a+N+m) \text{ as } \nu \to N^-, \tag{3.34}$$

and

$$\nabla_a^\nu f(a+N+m) \to \nabla^{N-1} f(a+N+m) \text{ as } \nu \to (N-1)^+. \tag{3.35}$$

Let $\nu$ be fixed such that $N-1 < \nu < N$. We now show that (3.33) holds. To see this note that we have the following:

$$\begin{aligned}
\nabla_a^\nu f(a+N+m) &= \int_a^t H_{-\nu-1}(t,\rho(\tau))\nabla\tau \\
&= \frac{1}{\Gamma(-\nu)} \sum_{\tau=a+1}^{t} (t-\rho(\tau))^{\overline{-\nu-1}}\bigg|_{t=a+N+m} f(\tau) \\
&= \frac{1}{\Gamma(-\nu)} \sum_{\tau=a+1}^{a+N+m} (a+N+m-\rho(\tau))^{\overline{-\nu-1}} f(\tau) \\
&= \sum_{\tau=a+1}^{a+N+m} \frac{\Gamma(a+N+m-\tau+1-\nu-1)}{\Gamma(a+N+m-\tau+1)\Gamma(-\nu)} f(\tau) \\
&= \sum_{\tau=a+1}^{a+N+m} \frac{(a+N+m-\tau-\nu-1)\cdots(-\nu)\Gamma(-\nu)}{(a+N+m-\tau)!\Gamma(-\nu)} f(\tau) \\
&= \sum_{\tau=a+1}^{a+N+m-1} \frac{(a+N+m-\tau-\nu-1)\cdots(-\nu)}{(a+N+m-\tau)!} f(\tau) + f(a+N+m).
\end{aligned}$$

Letting $i := a+N+m-\tau$, we get

$$\nabla_a^\nu f(a+N+m)$$

$$= \sum_{i=1}^{N+m} \frac{(i-1-\nu)\cdots(1-\nu)(-\nu)}{i!} f(a+N+M-i) + f(a+N+M).$$

This shows that the $\nu$-th order fractional difference is continuous on $N-1 < \nu < N$, showing (3.33) holds.

Now we consider the case $\nu \to N^-$ in order to show that (3.34) holds:

$$\lim_{\nu \to N^-} \nabla_a^\nu f(a+N+m)$$

$$= \lim_{\nu \to N^-} \sum_{i=1}^{N+m} \frac{(i-1-\nu)\cdots(1-\nu)(-\nu)}{i!} f(a+N+M-i)$$

$$+ f(a+N+M)$$

$$= \sum_{i=1}^{N+m} \left( \frac{(i-1-N)\cdots(-N)}{i!} f(a+N+m-i) \right) + f(a+N+m)$$

$$= \sum_{i=0}^{N+m} \left( (-1)^i \frac{(N+1-i)\cdots(N)}{i!} f(a+N+m-i) \right)$$

$$= \sum_{i=0}^{N+m} (-1)^i \binom{N}{i} f(a+N+m-i)$$

$$= \sum_{i=0}^{N} (-1)^i \binom{N}{i} f(a+m+N-i)$$

$$= \nabla^N f(a+N+m).$$

Finally we want to show (3.35) holds. So we write

$$\lim_{\nu \to (N-1)^+} \nabla_a^\nu f(a+N+m)$$

$$= \lim_{\nu \to (N-1)^+} \sum_{i=1}^{N+m} \frac{(i-1-\nu)\cdots(1-\nu)(-\nu)}{i!} f(a+N+M-i)$$

$$+ f(a+N+M)$$

$$= \sum_{i=1}^{N+m} \frac{(i-N)(i-N-1)\cdots(1-N)}{i!} f(a+N+m-i)$$

$$+ f(a+N+m)$$

$$= \sum_{i=0}^{N+m} (-1)^i \frac{(N-i)(N+1-i)\cdots(N-1)}{i!} f(a+N+m-i)$$

$$= \sum_{i=0}^{N+m} (-1)^i \binom{N-1}{i} f(a+m+1+N-1-i)$$

$$= \nabla^{N-1} f(a+N+m).$$

This completes the proof. □

To prove various properties for the nabla fractional sums and differences it is convenient to develop the theory of the nabla Laplace transform, which we do in the next section.

## 3.10 Nabla Laplace Transforms

Having established the necessary preliminaries, we are now ready to discuss an important application of this material: the Laplace transform. The Laplace transform, as in the standard calculus, will provide us with an elegant way to solve initial value problems for a fractional nabla difference equation. In this section, we will lay the groundwork for this method, prove the basic properties, and establish a means in which to solve various initial value (nabla) fractional difference equations. We begin this section by defining the nabla Laplace transform operator $\mathcal{L}_a$ (based at $a$) as follows:

**Definition 3.64.** Assume $f : \mathbb{N}_{a+1} \to \mathbb{R}$. Then the nabla Laplace transform of $f$ is defined by

$$\mathcal{L}_a\{f\}(s) = \int_a^\infty E_{\boxminus s}(\rho(t), a) f(t) \nabla t,$$

for those values of $s \neq 1$ such that this improper integral converges.

In the following theorem we give another formula for the Laplace transform, which is often more convenient to use.

**Theorem 3.65.** *Assume* $f : \mathbb{N}_{a+1} \to \mathbb{R}$. *Then*

$$\mathcal{L}_a\{f\}(s) = \sum_{k=1}^{\infty} (1-s)^{k-1} f(a+k), \tag{3.36}$$

*for those values of $s$ such that this infinite series converges.*

*Proof.* Assume $f : \mathbb{N}_{a+1} \to \mathbb{R}$. Then

$$\begin{aligned}
\mathcal{L}_a\{f\}(s) &= \int_a^\infty E_{\boxminus s}(\rho(t), a) f(t) \nabla t \\
&= \int_a^\infty [1 - \boxminus s]^{a-t+1} f(t) \nabla t \\
&= \int_a^\infty \left(\frac{1}{1-s}\right)^{a-t+1} f(t) \nabla t \\
&= \int_a^\infty (1-s)^{t-a-1} f(t) \nabla t \\
&= \sum_{t=a+1}^{\infty} (1-s)^{t-a-1} f(t) \\
&= \sum_{k=1}^{\infty} (1-s)^{k-1} f(a+k),
\end{aligned}$$

for those values of $s$ such that this infinite series converges. □

In the definition of the nabla Laplace transform we assumed $s \neq 1$ because we do not define $E_{\boxminus 1}(t, a)$. But the formula of the nabla Laplace transform (3.36) is well defined when $s = 1$. From now on we will always include $s = 1$ in the domain of convergence for the nabla Laplace transform although in the proofs we will often assume $s \neq 1$. In fact the formula (3.36) for any $f : \mathbb{N}_{a+1} \to \mathbb{R}$ gives us that

$$\mathcal{L}_a\{f\}(1) = f(a+1).$$

*Example 3.66.* We use the last theorem to find $\mathcal{L}_a\{1\}(s)$. By Theorem 3.65 we obtain

$$\begin{aligned}
\mathcal{L}_a\{1\}(s) &= \sum_{k=1}^{\infty} (1-s)^{k-1} 1 \\
&= \sum_{k=0}^{\infty} (1-s)^k \\
&= \frac{1}{1-(1-s)}, \quad \text{for } |1-s| < 1 \\
&= \frac{1}{s}.
\end{aligned}$$

That is

$$\mathcal{L}_a\{1\}(s) = \frac{1}{s}, \quad |s-1| < 1.$$

**Theorem 3.67.** *For all nonnegative integers n, we have that*

$$\mathcal{L}_a\{H_n(\cdot,a)\}(s) = \frac{1}{s^{n+1}}, \quad \textit{for} \quad |s-1| < 1.$$

*Proof.* The proof is by induction on $n$. The result is true for $n = 0$ by the previous example. Suppose now that $\mathcal{L}_a\{H_n(\cdot,a)\}(s) = \frac{1}{s^{n+1}}$ for some fixed $n \geq 0$ and $|s-1| < 1$. Then consider

$$\mathcal{L}_a\{H_{n+1}(\cdot,a)\}(s) = \int_a^\infty E_{\boxminus s}(\rho(t),a)H_{n+1}(t,a)\nabla t.$$

We will apply the first integration by parts formula (3.19) with

$$u(t) = H_{n+1}(t,a), \quad \text{and} \quad \nabla v(t) = E_{\boxminus s}(\rho(t),a) = -\frac{1}{s} \boxminus sE_{\boxminus s}(t,a).$$

It follows that

$$\nabla u(t) = H_n(t,a), \quad v(\rho(t)) = -\frac{1}{s}E_{\boxminus s}(\rho(t),a).$$

Hence by the integration by parts formula (3.19)

$$\begin{aligned}
\mathcal{L}_a\{H_{n+1}(\cdot,a)\}(s) &= \int_a^\infty E_{\boxminus s}(\rho(t),a)H_{n+1}(t,a)\nabla t \\
&= -\frac{1}{s}E_{\boxminus s}(t,a)H_{n+1}(t,a)\Big|_a^\infty + \frac{1}{s}\int_a^\infty E_{\boxminus s}(\rho(t),a)H_n(t,a)\nabla t \\
&= -\frac{1}{s}(1-s)^{t-a}H_{n+1}(t,a)\Big|_a^\infty + \frac{1}{s}\mathcal{L}\{H_n(\cdot,a)\}(s).
\end{aligned}$$

Using the nabla form of L'Hôpital's rule (Exercise 3.19) we calculate

$$\begin{aligned}
\lim_{t\to\infty} |(1-s)^{t-a}H_{n+1}(t,a)| &= \lim_{t\to\infty} \frac{H_{n+1}(t,a)}{|1-s|^{a-t}} \\
&= \lim_{t\to\infty} \frac{H_n(t,a)}{[1-|1-s|]|1-s|^{a-t}} \\
&= \lim_{t\to\infty} \frac{H_{n-1}(t,a)}{[1-|1-s|]^2|1-s|^{a-t}} \\
&= \cdots \\
&= \lim_{t\to\infty} \frac{H_0(t,a)}{[1-|1-s|]^{n+1}|1-s|^{a-t}} \\
&= 0,
\end{aligned}$$

since $|s-1|<1$. Thus we have that

$$\mathcal{L}_a\{H_{n+1}(\cdot,a)\}(s)=\frac{1}{s^{n+2}},\quad |s-1|<1$$

completing the proof. □

**Definition 3.68.** A function $f:\mathbb{N}_{a+1}\to\mathbb{R}$ is said to be of **exponential order** $r>0$ if there exist a constant $M>0$ and a number $T\in\mathbb{N}_{a+1}$ such that

$$|f(t)|\le Mr^t,\quad \text{for all } t\in\mathbb{N}_T.$$

**Theorem 3.69.** *For $n\in\mathbb{N}_1$, the Taylor monomials $H_n(t,a)$ are of exponential order $1+\epsilon$ for all $\epsilon>0$. Also, $H_0(t,a)$ is of exponential order 1.*

*Proof.* Since

$$|H_0(t,a)|=1\cdot 1^t,\quad t\in\mathbb{N}_{a+1},$$

$H_0(t,a)$ is of exponential order 1. Next, assume $n\in\mathbb{N}_1$ and $\epsilon>0$ is fixed. Using repeated applications of the nabla L'Hôpital's rule, we get

$$\begin{aligned}\lim_{t\to\infty}\frac{H_n(t,a)}{(1+\epsilon)^t}&=\lim_{t\to\infty}\frac{H_{n-1}(t,a)}{\frac{\epsilon}{1+\epsilon}(1+\epsilon)^t}\\&=\lim_{t\to\infty}\frac{H_{n-2}(t,a)}{(\frac{\epsilon}{1+\epsilon})^2(1+\epsilon)^t}\\&\cdots\\&=\lim_{t\to\infty}\frac{H_0(t,a)}{(\frac{\epsilon}{1+\epsilon})^n(1+\epsilon)^t}\\&=0.\end{aligned}$$

It follows from this that each $H_n(t,a)$, $n\in\mathbb{N}_1$, is of exponential order $1+\epsilon$ for all $\epsilon>0$. □

**Theorem 3.70 (Existence of Nabla Laplace Transform).** *If $f:\mathbb{N}_{a+1}\to\mathbb{R}$ is a function of exponential order $r>0$, then its Laplace transform exists for $|s-1|<\frac{1}{r}$.*

*Proof.* Let $f$ be a function of exponential order $r$. Then there is a constant $M>0$ and a number $T\in\mathbb{N}_{a+1}$ such that $|f(t)|\le Mr^t$ for all $t\in\mathbb{N}_T$. Pick $K$ so that $T=a+K$, then we have that

$$|f(a+k)|\le Mr^{a+k},\quad k\in\mathbb{N}_K.$$

We now show that

$$\mathcal{L}_a\{f\}(s) = \sum_{k=1}^{\infty}(1-s)^{k-1}f(a+k)$$

converges for $|s-1| < \frac{1}{r}$. To see this, consider

$$\begin{aligned}\sum_{k=K}^{\infty}|(1-s)^{k-1}f(a+k)| &= \sum_{k=K}^{\infty}|1-s|^{k-1}|f(a+k)| \\ &\leq \sum_{k=K}^{\infty}|1-s|^{k-1}Mr^{a+k} \\ &= Mr^{a+1}\sum_{k=K}^{\infty}[r|s-1|]^{k-1},\end{aligned}$$

which converges since $r|s-1| < 1$. It follows that $\mathcal{L}_a\{f\}(s)$ converges absolutely for $|s-1| < \frac{1}{r}$. □

**Theorem 3.71.** *The Laplace transform of the Taylor monomial, $H_n(t,a)$, $n \in \mathbb{N}_0$, exists for $|s-1| < 1$.*

*Proof.* The proof of this theorem follows from Theorems 3.69 and 3.70. □

Similarly, by Exercise 3.30 each of the functions $E_p(t,a)$, $\mathrm{Cosh}_p(t,a)$, $\mathrm{Sinh}_p(t,a)$, $\mathrm{Cos}_p(t,a)$, and $\mathrm{Sin}_p(t,a)$ is of exponential order $|1+p|$, and hence by Theorem 3.70 their Laplace transforms exist for $|s-1| < \frac{1}{|1+p|}$.

**Theorem 3.72 (Uniqueness Theorem).** *Assume $f, g : \mathbb{N}_{a+1} \to \mathbb{R}$. Then $f(t) = g(t)$, $t \in \mathbb{N}_{a+1}$, if and only if*

$$\mathcal{L}_a\{f\}(s) = \mathcal{L}_a\{g\}(s), \quad \textit{for} \quad |s-1| < r$$

*for some $r > 0$.*

*Proof.* Since $\mathcal{L}_a$ is a linear operator it suffices to show that $f(t) = 0$ for $t \in \mathbb{N}_{a+1}$ if and only if $\mathcal{L}_a\{f\}(s) = 0$ for $|s-1| < r$ for some $r > 0$. If $f(t) = 0$ for $t \in \mathbb{N}_{a+1}$, then trivially $\mathcal{L}_a\{f\}(s) = 0$ for all $s \in \mathbb{C}$. Conversely, assume that $\mathcal{L}_a\{f\}(s) = 0$ for $|s-1| < r$ for some $r > 0$. In this case we have that

$$\sum_{k=1}^{\infty}f(a+k)(1-s)^{k-1} = 0, \quad |s-1| < r.$$

This implies that

$$f(t) = 0, \quad t \in \mathbb{N}_{a+1}.$$

This completes the proof. □

## 3.11 Fractional Taylor Monomials

To find the formula for the Laplace transform of a fractional nabla Taylor monomial we will use the following lemma which appears in Hein et al [119].

**Lemma 3.73.** *For $\nu \in \mathbb{C}\backslash\mathbb{Z}$ and $n \geq 0$, we have that*

$$(1+\nu)^{\overline{n}} = \frac{(-1)^n\Gamma(-\nu)}{\Gamma(-\nu-n)}. \tag{3.37}$$

*Proof.* The proof of (3.37) is by induction for $n \in \mathbb{N}_0$. For $n = 0$ (3.37) clearly holds. Assume (3.37) is true for some fixed $n \geq 0$. Then,

$$\begin{aligned}(1+\nu)^{\overline{n+1}} &= (1+\nu)^{\overline{n}}(\nu+n+1)\\ &= \frac{(-1)^n\Gamma(-\nu)(\nu+n+1)}{\Gamma(-\nu-n)}, \quad \text{by the induction hypothesis}\\ &= \frac{(-1)^{n+1}\Gamma(-\nu)}{\Gamma(-\nu-(n+1))}.\end{aligned}$$

The result follows. □

We now determine the Laplace transform of the fractional nabla Taylor monomial.

**Theorem 3.74.** *For $\nu$ not an integer, we have that*

$$\mathcal{L}_a\{H_\nu(\cdot,a)\}(s) = \frac{1}{s^{\nu+1}}, \quad for \quad |s-1|<1.$$

*Proof.* Consider for $|s-1|<1$, $|s|^p > 1$

$$\begin{aligned}\mathcal{L}_a\{H_\nu(\cdot,a)\}(s) &= \sum_{k=1}^{\infty}(1-s)^{k-1}H_\nu(a+k,a) = \sum_{k=1}^{\infty}(1-s)^{k-1}\frac{k^{\overline{\nu}}}{\Gamma(\nu+1)}\\ &= \sum_{k=1}^{\infty}(1-s)^{k-1}\frac{\Gamma(k+\nu)}{\Gamma(k)\Gamma(\nu+1)} = \sum_{k=0}^{\infty}(1-s)^{k}\frac{\Gamma(k+1+\nu)}{\Gamma(k+1)\Gamma(\nu+1)}\\ &= \sum_{k=0}^{\infty}(1-s)^{k}\frac{(1+\nu)^{\overline{k}}}{\Gamma(k+1)}\\ &= \sum_{k=0}^{\infty}(-1)^k(1-s)^{k}\frac{\Gamma(-\nu)}{\Gamma(k+1)\Gamma(-\nu-k)} \qquad \text{(by Lemma 3.73)}\\ &= \sum_{k=0}^{\infty}(-1)^k(1-s)^{k}\frac{[-(\nu+1)]^{\underline{k}}}{\Gamma(k+1)}\end{aligned}$$

$$= \sum_{k=0}^{\infty}(-1)^k \binom{-(\nu+1)}{k}(1-s)^k$$

$$= [1-(1-s)]^{-(\nu+1)} \qquad \text{(by the Generalized Binomial Theorem)}$$

$$= \frac{1}{s^{\nu+1}}.$$

This completes the proof. □

Combining Theorems 3.67 and 3.74, we get the following corollary:

**Corollary 3.75.** *For $\nu \in \mathbb{C}\backslash\{-1,-2,-3,\ldots\}$, we have that*

$$\mathcal{L}_a\{H_\nu(\cdot,a)\}(s) = \frac{1}{s^{\nu+1}}, \quad \textit{for} \quad |1-s| < 1.$$

**Theorem 3.76.** *The following hold:*

(i) $\mathcal{L}_a\{E_p(\cdot,a)\}(s) = \frac{1}{s-p}, \quad p \neq 1;$
(ii) $\mathcal{L}_a\{Cosh_p(\cdot,a)\}(s) = \frac{s}{s^2-p^2}, \quad p \neq \pm 1;$
(iii) $\mathcal{L}_a\{Sinh_p(\cdot,a)\}(s) = \frac{p}{s^2-p^2}, \quad p \neq \pm 1;$
(iv) $\mathcal{L}_a\{Cos_p(\cdot,a)\}(s) = \frac{s}{s^2+p^2}, \quad p \neq \pm i;$
(v) $\mathcal{L}_a\{Sin_p(\cdot,a)\}(s) = \frac{p}{s^2+p^2}, \quad p \neq \pm i;$

*where (i) holds for $|s-1| < |1-p|$, (ii) and (iii) hold for $|s-1| < min\{|1-p|, |1+p|\}$, and (iv) and (v) hold for $|s-1| < min\{|1-ip|, |1+ip|\}$.*

*Proof.* To see that (i) holds, note that

$$\mathcal{L}_a\{E_p(\cdot,a)\}(s) = \sum_{k=1}^{\infty}(1-s)^{k-1}E_p(a+k,a)$$

$$= \sum_{k=1}^{\infty}(1-s)^{k-1}(1-p)^{-k};$$

$$= \frac{1}{1-p}\sum_{k=1}^{\infty}\left(\frac{1-s}{1-p}\right)^{k-1}$$

$$= \frac{1}{1-p}\frac{1}{1-\frac{1-s}{1-p}}, \quad \text{for} \quad \left|\frac{1-s}{1-p}\right| < 1$$

$$= \frac{1}{s-p}$$

for $|s-1| < |1-p|$. To see that (ii) holds, note that for $|s-1| < \min\{|1-p|, |1+p|\}$

$$\begin{aligned}\mathcal{L}_a\{\mathrm{Cosh}_p(\cdot, a)\}(s) &= \frac{1}{2}\mathcal{L}_a\{E_p(\cdot, a)\}(s) + \frac{1}{2}\mathcal{L}_a\{E_{-p}(\cdot, a)\}(s)\\ &= \frac{1}{2(s-p)} + \frac{1}{2(s+p)}\\ &= \frac{s}{s^2-p^2}.\end{aligned}$$

To see that (iv) holds, note that

$$\begin{aligned}\mathcal{L}_a\{\mathrm{Cos}_p(\cdot, a)\}(s) &= \mathcal{L}_a\{Cosh_{ip}(\cdot, a)\}(s)\\ &= \frac{s}{s^2-(ip)^2}\\ &= \frac{s}{s^2+p^2}\end{aligned}$$

for $|s-1| < \min\{|1-ip|, |1+ip|\}$. The proofs of parts (iii) and (v) are left as an exercise (Exercise 3.31). □

## 3.12 Convolution

We are now ready to investigate one of the most important properties in solving initial-value fractional nabla difference equations: convolution. This definition is motivated by the desire to express the fractional nabla sums and fractional nabla differences as convolutions of arbitrary functions and Taylor monomials. As a consequence, the resulting properties that stem from this definition are, in fact, consistent with the standard convolution. Many of the results in this section appear in Hein et al. [119] and Ahrendt et al. [3].

**Definition 3.77.** For $f, g : \mathbb{N}_{a+1} \to \mathbb{R}$, we define the nabla convolution product of $f$ and $g$ by

$$(f * g)(t) := \int_a^t f(t-\rho(\tau)+a)g(\tau)\nabla\tau, \quad t \in \mathbb{N}_{a+1}.$$

*Example 3.78.* Use the definition of the nabla convolution product to find $1 * \mathrm{Sin}_p(\cdot, a)$, $p \neq 0, \pm i$. By Definition 3.77,

$$\begin{aligned}(1 * \mathrm{Sin}_p(\cdot, a))(t) &= \int_a^t 1 \cdot \mathrm{Sin}_p(\tau, a)\nabla\tau\\ &= \int_a^t \mathrm{Sin}_p(\tau, a)\nabla\tau\end{aligned}$$

$$= -\frac{1}{p}\mathrm{Cos}_p(\tau, a)\Big|_a^t$$
$$= \frac{1}{p} - \frac{1}{p}\mathrm{Cos}_p(t, a),$$

for $t \in \mathbb{N}_{a+1}$.

*Example 3.79.* Use the definition of the (nabla) convolution product to find

$$\left(E_p(\cdot, a) * E_q(\cdot, a)\right)(t), \quad p, q \neq 1, \quad p \neq q.$$

Assume $q \neq 0$. By Definition 3.77 and using the second integration by parts formula, we have that

$$\left(E_p(\cdot, a) * E_q(\cdot, a)\right)(t)$$
$$= \int_a^t E_p(t - \rho(\tau) + a, a)E_q(\tau, a)\nabla\tau$$
$$= \frac{1}{q}E_p(t - \tau + a, a)E_q(\tau, a)\Big|_{\tau=a}^{\tau=t} + \frac{p}{q}\int_a^t E_p(t - \rho(\tau) + a, a)E_q(\tau, a)\nabla\tau$$
$$= \frac{1}{q}E_q(t, a) - \frac{1}{q}E_p(t, a) + \frac{p}{q}\left(E_p(\cdot, a) * E_q(\cdot, a)\right)(t).$$

Solving for $\left(E_p(\cdot, a) * E_q(\cdot, a)\right)(t)$, we obtain

$$\left(E_p(\cdot, a) * E_q(\cdot, a)\right)(t) = \frac{1}{p - q}E_p(t, a) + \frac{1}{q - p}E_q(t, a)$$

for $t \in \mathbb{N}_{a+1}$. We leave it to the reader to show that this last formula is also valid if $q = 0$.

**Theorem 3.80.** *Assume* $\nu \in \mathbb{R}\backslash\{0, -1, -2, \dots\}$ *and* $f : \mathbb{N}_{a+1} \to \mathbb{R}$. *Then*

$$\nabla_a^{-\nu} f(t) = (H_{\nu-1}(\cdot, a) * f)(t),$$

*for* $t \in \mathbb{N}_{a+1}$.

*Proof.* The result follows from the following:

$$(H_{\nu-1}(\cdot, a) * f)(t) = \int_a^t H_{\nu-1}(t - \rho(\tau) + a, a)f(\tau)\nabla\tau$$
$$= \int_a^t \frac{(t - \rho(\tau) + a - a)^{\overline{\nu-1}}}{\Gamma(\nu)}f(\tau)\nabla\tau$$

$$= \int_a^t \frac{(t-\rho(\tau))^{\overline{\nu-1}}}{\Gamma(\nu)} f(\tau)\nabla\tau$$
$$= \int_a^t H_{\nu-1}(t,\rho(\tau)) f(\tau)\nabla\tau$$
$$= \nabla_a^{-\nu} f(t),$$

for $t \in \mathbb{N}_{a+1}$. □

**Theorem 3.81 (Nabla Convolution Theorem).** *Assume $f, g : \mathbb{N}_{a+1} \to \mathbb{R}$ and their nabla Laplace transforms converge for $|s-1| < r$ for some $r > 0$. Then*

$$\mathcal{L}_a\{f * g\}(s) = \mathcal{L}_a\{f\}(s)\mathcal{L}_a\{g\}(s),$$

*for $|s-1| < r$.*

*Proof.* The following proves our result:

$$\begin{aligned}
\mathcal{L}_a\{f * g\}(s) &= \sum_{k=1}^{\infty} (1-s)^{k-1} (f * g)(a+k) \\
&= \sum_{k=1}^{\infty} (1-s)^{k-1} \int_a^{a+k} f(a+k-\rho(\tau)+a) g(\tau)\nabla\tau \\
&= \sum_{k=1}^{\infty} (1-s)^{k-1} \sum_{\tau=a+1}^{a+k} f(a+k-\rho(\tau)+a) g(\tau) \\
&= \sum_{k=1}^{\infty} \sum_{\tau=1}^{k} (1-s)^{k-1} f(k-\rho(\tau)+a) g(\tau+a) \\
&= \sum_{\tau=1}^{\infty} \sum_{k=\tau}^{\infty} (1-s)^{k-1} f(k-\rho(\tau)+a) g(a+\tau) \\
&= \left( \sum_{\tau=1}^{\infty} (1-s)^{\tau-1} g(a+\tau) \right) \left( \sum_{k=1}^{\infty} (1-s)^{k-1} f(a+k) \right) \\
&= \mathcal{L}_a\{g\}(s)\mathcal{L}_a\{f\}(s)
\end{aligned}$$

for $|s-1| < r$. □

With the above result and the uniqueness of the Laplace transform, it follows that the convolution product is commutative and associative (see Exercise 3.32).

We next establish properties of the Laplace transform that will be useful in solving initial value problems for integer nabla difference equations.

**Theorem 3.82 (Transformation of Fractional Sums).** *Assume $\nu > 0$ and the nabla Laplace transform of $f : \mathbb{N}_{a+1} \to \mathbb{R}$ converges for $|s-1| < r$ for some $r > 0$. Then*

$$\mathcal{L}_a\{\nabla_a^{-\nu} f\}(s) = \frac{1}{s^\nu}\mathcal{L}_a\{f\}(s)$$

*for $|s-1| < \min\{1, r\}$.*

*Proof.* The result follows since

$$\begin{aligned}\mathcal{L}_a\{\nabla_a^{-\nu} f\}(s) &= \mathcal{L}_a\{H_{\nu-1}(\cdot,a) * f\}(s)\\ &= \mathcal{L}_a\{H_{\nu-1}(\cdot,a)\}(s)\mathcal{L}_a\{f\}(s)\\ &= \frac{1}{s^\nu}\mathcal{L}_a\{f\}(s),\end{aligned}$$

for $|s-1| < \min\{1, r\}$. □

Assuming that $\nu$ is a positive integer, this result is consistent with the formula in the continuous case for the Laplace transform of the $n$-th iterated integral of a function. We want to establish similar properties for fractional differences; however, we will first establish integer-order difference properties.

**Theorem 3.83 (Transform of Nabla Difference).** *Assume $f : \mathbb{N}_a \to \mathbb{R}$ is of exponential order $r > 0$. Then*

$$\mathcal{L}_a\{\nabla f\}(s) = s\mathcal{L}_a\{f\}(s) - f(a)$$

*for $|s-1| < r$.*

*Proof.* Note that since we are assuming that $f : \mathbb{N}_a \to \mathbb{R}$, we have that $\nabla f : \mathbb{N}_{a+1} \to \mathbb{R}$ and so we can consider $\mathcal{L}_a\{\nabla f\}(s)$. Since $f : \mathbb{N}_a \to \mathbb{R}$ is of exponential order $r > 0$, it follows that $\nabla f : \mathbb{N}_{a+1} \to \mathbb{R}$ is of exponential order $r > 0$. It follows that for $|s-1| < r$,

$$\begin{aligned}\mathcal{L}_a\{\nabla f\}(s) &= \sum_{k=1}^{\infty}(1-s)^{k-1}\nabla f(a+k)\\ &= \sum_{k=1}^{\infty}(1-s)^{k-1}\left[f(a+k) - f(a+k-1)\right]\\ &= \mathcal{L}_a\{f\}(s) - \sum_{k=1}^{\infty}(1-s)^{k-1}f(a+k-1)\\ &= \mathcal{L}_a\{f\}(s) - \sum_{k=0}^{\infty}(1-s)^{k}f(a+k)\end{aligned}$$

$$= \mathcal{L}_{a+1}\{f\}(s) - f(a) - (1-s)\sum_{k=1}^{\infty}(1-s)^{k-1}f(a+k)$$
$$= \mathcal{L}_a\{f\}(s) - f(a) - (1-s)\mathcal{L}_a\{f\}(s)$$
$$= s\mathcal{L}_a\{f\}(s) - f(a).$$

This completes the proof. □

The following is a simple example where we will use the Nabla Convolution Theorem and Theorem 3.83 to solve an initial value problem.

*Example 3.84.* Use the Nabla Convolution Theorem to help you solve the IVP

$$\nabla y(t) - 3y(t) = E_4(t,a), \quad t \in \mathbb{N}_{a+1}$$
$$y(a) = 0.$$

If $y(t)$ is the solution of this IVP and its Laplace transform, $Y_a(s)$, exists, then we have that

$$sY_a(s) - y(a) - 3Y_a(s) = \frac{1}{s-4}.$$

Using the initial condition and solving for $Y_a(s)$ we obtain

$$Y_a(s) = \frac{1}{s-4}\frac{1}{s-3}.$$

Using the Nabla Convolution Theorem we see that

$$y(t) = (E_4(\cdot,a) * E_3(\cdot,a))(t).$$

Using Example 3.79 we find that

$$y(t) = E_4(t,a) - E_3(t,a)$$
$$= (-3)^{a-t} - (-2)^{a-t}$$

is the solution of our given IVP on $\mathbb{N}_a$. Of course, in this simple example one could also use partial fractions to find $y(t)$.

We can then generalize this result for an arbitrary number of nabla differences.

**Theorem 3.85 (Transform of $n$-th-Order Nabla Difference).** *Assume $f : \mathbb{N}_{a-n+1} \to \mathbb{R}$ is of exponential order $r > 0$. Then*

$$\mathcal{L}_a\{\nabla^n f\}(s) = s^n \mathcal{L}_a\{f\}(s) - \sum_{k=1}^{n} s^{n-k}\nabla^{k-1}f(a). \tag{3.38}$$

*for $|s-1| < r$, for each $n \in \mathbb{N}_1$.*

*Proof.* Note that since $f$ is of exponential order $r > 0$, $\nabla^n f$ is of exponential order $r > 0$ for each $n \in \mathbb{N}_1$. Hence $\mathcal{L}_a\{\nabla^n f\}(s)$ converges for $|s-1| < r$ for each $n \in \mathbb{N}_1$. The proof of (3.38) is by induction for $n \in \mathbb{N}_1$. The base case $n = 1$ follows from Theorem 3.83. Now assume $n \geq 1$ and (3.38) holds for $|s-1| < r$. Then, using Theorem 3.83, we have that

$$\begin{aligned}
\mathcal{L}_a\{\nabla^{n+1} f\}(s) &= \mathcal{L}_a\{\nabla\left[\nabla^n f\right]\}(s) \\
&= s\mathcal{L}_a\{\nabla^n f\}(s) - \nabla^n f(a) \\
&= s\left[s^n \mathcal{L}_a\{f\}(s) - \sum_{k=1}^{n} s^{n-k}\nabla^{k-1} f(a)\right] - \nabla^n f(a) \\
&= s^{n+1}\mathcal{L}_a\{f\}(s) - \sum_{k=1}^{n+1} s^{(n+1)-k}\nabla^{k-1} f(a).
\end{aligned}$$

Hence, (3.38) holds when $n$ is replaced by $n+1$ and the proof is complete. □

*Example 3.86.* Solve the IVP

$$\begin{aligned}
\nabla^2 y(t) - 6\nabla y(t) + 8y(t) &= 0, \quad t \in \mathbb{N}_{a+1} \\
y(a) = 1, \quad \nabla y(a) &= -1.
\end{aligned}$$

If $y(t)$ is the solution of this IVP and we let $Y_a(s) := \mathcal{L}_a\{y\}(s)$, we have that

$$\left[s^2 Y_a(s) - sy(a) - \nabla y(a)\right] - 6\left[sY_a(s) - y(a)\right] + 8Y_a(s) = 0.$$

Using the initial conditions we have

$$\left[s^2 Y_a(s) - s + 1\right] - 6\left[sY_a(s) - 1\right] + 8Y_a(s) = 0.$$

Solving for $Y_a(s)$ we have that

$$\begin{aligned}
Y_a(s) &= \frac{s-7}{s^2 - 6s + 8} \\
&= \frac{s-7}{(s-2)(s-4)} \\
&= \frac{5}{2}\frac{1}{s-2} - \frac{3}{2}\frac{1}{s-4}.
\end{aligned}$$

It follows that

$$y(t) = \frac{5}{2}E_2(t,a) - \frac{3}{2}E_4(t,a)$$

$$= \frac{5}{2}(-1)^{a-t} - \frac{3}{2}(-3)^{a-t}$$

for $t \in \mathbb{N}_{a-1}$.

## 3.13 Further Properties of the Nabla Laplace Transform

In this section we want to find the Laplace transform of a $\nu$-th order fractional difference of a function, where $0 < \nu < 1$.

**Theorem 3.87.** *Assume $f : \mathbb{N}_{a+1} \to \mathbb{R}$ is of exponential order $r > 0$ and $0 < \nu < 1$. Then*

$$\mathcal{L}_a\{\nabla_a^\nu f\}(s) = s^\nu \mathcal{L}_a\{f\}(s)$$

*for $|s-1| < r$.*

*Proof.* Using Theorems 3.82 and 3.83 we have that

$$\begin{aligned}\mathcal{L}_a\{\nabla_a^\nu f\}(s) &= \mathcal{L}_a\{\nabla\nabla_a^{-(1-\nu)} f\}(s)\\ &= s\mathcal{L}_a\{\nabla_a^{-(1-\nu)} f\}(s) - \nabla_a^{-(1-\nu)} f(a)\\ &= \frac{s}{s^{1-\nu}} s^\nu \mathcal{L}_a\{f\}(s)\\ & s^\nu \mathcal{L}_a\{f\}(s)\end{aligned}$$

for $|s-1| < 1$. □

Next we state and prove a useful lemma (see Hein et al. [119] for $n = 1$ and see Ahrendt et al. [3] for general $n$).

**Lemma 3.88 (Shifting Base Lemma).** *Given $f : \mathbb{N}_{a+1} \to \mathbb{R}$ and $n \in \mathbb{N}_1$, we have that*

$$\mathcal{L}_{a+n}\{f\}(s) = \left(\frac{1}{1-s}\right)^n \mathcal{L}_a\{f\}(s) - \sum_{k=1}^{n} \frac{f(a+k)}{(1-s)^{n-k+1}}.$$

*Proof.* Consider

$$\begin{aligned}\mathcal{L}_{a+n}\{f\}(s) &= \sum_{k=1}^{\infty}(1-s)^{k-1} f(a+n+k)\\ &= \sum_{k=n+1}^{\infty} (1-s)^{k-n-1} f(a+k)\end{aligned}$$

$$= \frac{1}{(1-s)^n}\mathcal{L}_a\{f\}(s) - \sum_{k=1}^{n} \frac{f(a+k)}{(1-s)^{n-k+1}},$$

which is what we wanted to prove. □

With this, we are ready to provide the general form of the Laplace transform of a $\nu$-th order fractional difference of a function $f$, where $0 < \nu < 1$.

**Theorem 3.89.** *Given* $f : \mathbb{N}_{a+1} \to \mathbb{R}$ *and* $0 < \nu < 1$. *Then we have*

$$\mathcal{L}_{a+1}\{\nabla_a^\nu f\}(s) = s^\nu \mathcal{L}_{a+1}\{f\}(s) - \frac{1-s^\nu}{1-s} f(a+1).$$

*Proof.* Consider

$$\begin{aligned}
&\mathcal{L}_{a+1}\{\nabla_a^\nu f\}(s)\\
&\quad = \mathcal{L}_{a+1}\{\nabla\nabla_a^{-(1-\nu)} f\}(s)\\
&\quad = s\mathcal{L}_{a+1}\{\nabla_a^{-(1-\nu)} f\}(s) - \nabla_a^{-(1-\nu)} f(a+1), \quad \text{by Theorem 3.83}\\
&\quad = s\mathcal{L}_{a+1}\{\nabla_a^{-(1-\nu)} f\}(s) - f(a+1), \quad \text{by Exercise 3.27.}
\end{aligned}$$

From this and Lemma 3.88, we have that

$$\begin{aligned}
&\mathcal{L}_{a+1}\{\nabla_a^\nu f\}(s)\\
&\quad = s\left(\frac{1}{1-s}\mathcal{L}_a\{\nabla_a^{-(1-\nu)} f\}(s) - \frac{1}{1-s}\nabla_a^{-(1-\nu)} f(a+1)\right) - f(a+1)\\
&\quad = \frac{s^\nu}{1-s}\mathcal{L}_a\{f\}(s) - \frac{1}{1-s} f(a+1) \quad \text{(by Theorem 3.82).}
\end{aligned}$$

Applying Lemma 3.88 again we obtain

$$\begin{aligned}
&\mathcal{L}_{a+1}\{\nabla_a^\nu f\}(s)\\
&\quad = s^\nu\left(\mathcal{L}_{a+1}\{f\}(s) + \frac{1}{1-s} f(a+1)\right) - \frac{1}{1-s} f(a+1),
\end{aligned}$$

which is the desired result. □

The following theorem was proved by Jia Baoguo.

**Theorem 3.90.** *Let $f : \mathbb{N}_{a-N+1} \to \mathbb{R}$ and $N-1 < \nu < N$ be given. Then we have*

$$\begin{aligned}\mathcal{L}_{a+1}\{\nabla_a^\nu f\}(s) =& s^\nu \mathcal{L}_{a+1}\{f\}(s) + \frac{s^\nu - s^{N-1}}{1-s} f(a+1) \\ & - \sum_{k=2}^{N} s^{N-k} \nabla^{k-1} f(a+1).\end{aligned}$$

*Proof.* We first calculate

$$\begin{aligned}
&\mathcal{L}_{a+1}\{\nabla_a^\nu f\}(s) \\
&= \mathcal{L}_{a+1}\{\nabla^N \nabla_a^{-(N-\nu)} f\}(s) \\
&\overset{\text{Theorem 3.85}}{=} s^N \mathcal{L}_{a+1}(\nabla_a^{-(N-\nu)} f)(s) - \sum_{k=1}^{N} s^{N-k} \nabla^{k-1} \nabla_a^{-(N-\nu)} f(a+1) \\
&\overset{\text{Lemma 3.88}}{=} \frac{s^N}{1-s} \mathcal{L}_a\{\nabla_a^{-(N-\nu)} f\}(s) - s^N \frac{\nabla_a^{-(N-\nu)} f(a+1)}{1-s} \\
&\quad - \sum_{k=1}^{N} s^{N-k} \nabla^{k-1} \nabla_a^{-(N-\nu)} f(a+1) \\
&\overset{\text{Theorem 3.82}}{=} \frac{s^N}{1-s} \cdot \frac{1}{s^{N-\nu}} \mathcal{L}_a\{f\}(s) - s^N \frac{\nabla_a^{-(N-\nu)} f(a+1)}{1-s} \\
&\quad - \sum_{k=1}^{N} s^{N-k} \nabla^{k-1} \nabla_a^{-(N-\nu)} f(a+1) \\
&\overset{\text{Theorem 3.88}}{=} \frac{s^\nu}{1-s} [(1-s)\mathcal{L}_{a+1}\{f\}(s) + f(a+1)] - s^N \frac{\nabla_a^{-(N-\nu)} f(a+1)}{1-s} \\
&\quad - \sum_{k=1}^{N} s^{N-k} \nabla^{k-1} \nabla_a^{-(N-\nu)} f(a+1).
\end{aligned}$$

Since

$$\begin{aligned}\nabla_a^{-(N-\nu)} f(a+1) &= \int_a^{a+1} H_{N-\nu-1}(a+1, a) f(s) \nabla s \\ &= H_{N-\nu}(a+1, a) f(a+1) = f(a+1),\end{aligned}$$

we have

$$\mathcal{L}_{a+1}\{\nabla_a^\nu f\}(s)$$

$$= s^\nu \mathcal{L}_{a+1}\{f\}(s) + \frac{s^\nu}{1-s}f(a+1) - \frac{s^N f(a+1)}{1-s}$$

$$-\sum_{k=1}^{N} s^{N-k}\nabla^{k-1}f(a+1)$$

$$= s^\nu \mathcal{L}_{a+1}\{f\}(s) + \frac{s^\nu - s^{N-1}}{1-s}f(a+1) - \sum_{k=2}^{N} s^{N-k}\nabla^{k-1}f(a+1).$$

□

*Remark 3.91.* When $N = 1$, Theorem 3.90 becomes the Theorem 3.89. When $N = 2$, we can get the following Corollary.

**Corollary 3.92.** *Let $f : \mathbb{N}_a \to \mathbb{R}$ and $1 < \nu < 2$ be given. Then we have*

$$\mathcal{L}_{a+1}\{\nabla_a^\nu f\}(s) = s^\nu \mathcal{L}_{a+1}\{f\}(s) + \frac{s^\nu - s}{1-s}f(a+1) - \nabla f(a+1)$$

$$= s^\nu \mathcal{L}_{a+1}\{f\}(s) + \frac{s^\nu - 1}{1-s}f(a+1) + f(a).$$

## 3.14 Generalized Power Rules

We now see that with the use of the Laplace transform it is very easy to prove the following generalized power rules.

**Theorem 3.93 (Generalized Power Rules).** *Let $\nu \in \mathbb{R}^+$ and $\mu \in \mathbb{R}$ such that $\mu$, $\nu + \mu$, and $\mu - \nu$ are nonnegative integers. Then we have that*

(i) $\nabla_a^{-\nu} H_\mu(t,a) = H_{\mu+\nu}(t,a)$;
(ii) $\nabla_a^{\nu} H_\mu(t,a) = H_{\mu-\nu}(t,a)$;
(iii) $\nabla_a^{-\nu}(t-u)^{\overline{\mu}} = \frac{\Gamma(\mu+1)}{\Gamma(\mu+\nu+1)}(t \quad a)^{\overline{\mu+\nu}}$;
(iv) $\nabla_a^{\nu}(t-a)^{\overline{\mu}} = \frac{\Gamma(\mu+1)}{\Gamma(\mu-\nu+1)}(t-a)^{\overline{\mu-\nu}}$;

*for $t \in \mathbb{N}_a$.*

*Proof.* To see that (i) holds, note that

$$\mathcal{L}_a\{\nabla_a^{-\nu} H_\mu(\cdot,a)\}(s) = \frac{1}{s^\nu}\mathcal{L}_a\{H_\mu(\cdot,a)\}(s)$$

$$= \frac{1}{s^{\nu+\mu+1}}$$

$$= \mathcal{L}_a\{H_{\mu+\nu}(\cdot,a)\}(s).$$

Hence, by the uniqueness theorem (Theorem 3.72) we have that

$$\nabla_a^{-\nu} H_\mu(t,a) = H_{\mu+\nu}(t,a), \quad t \in \mathbb{N}_{a+1}.$$

Also, this last equation holds for $t = a$ and hence (i) holds. To see that (i) implies (iii), note that

$$\begin{aligned}
\nabla_a^{-\nu}(t-a)^{\overline{\mu}} &= \Gamma(\mu+1)\nabla_a^{-\nu} H_\mu(t,a) \\
&= \Gamma(\mu+1) H_{\mu+\nu}(t,a) \\
&= \frac{\Gamma(\mu+1)}{\Gamma(\mu+\nu+1)}(t-a)^{\overline{\mu+\nu}}.
\end{aligned}$$

To show that part (ii) holds we will first show that

$$\mathcal{L}_a\{\nabla_a^{-\nu} H_\mu(\cdot,a)\}(s) = \mathcal{L}_a\{H_{\mu-\nu}(\cdot,a)\}(s), \tag{3.39}$$

for $|s-1| < 1$. On the one hand, using Lemma 3.88 with $n = 1$ we have that

$$\begin{aligned}
\mathcal{L}_{a+1}\{H_{\mu-\nu}(\cdot,a)\}(s) &= \frac{1}{1-s}\mathcal{L}_a\{H_{\mu-\nu}(\cdot,a)\}(s) - \frac{H_{\mu-\nu}(a+1,a)}{1-s} \\
&= \frac{1}{1-s}\frac{1}{s^{\mu-\nu+1}} - \frac{1}{1-s}.
\end{aligned} \tag{3.40}$$

On the other hand, using Theorem 3.89 we have that

$$\begin{aligned}
\mathcal{L}_{a+1}\{\nabla_a^\nu H_\mu(\cdot,a)\}(s) &= s^\nu \mathcal{L}_{a+1}\{H_\mu(\cdot,a)\}(s) - \frac{1-s^\nu}{1-s} H_\mu(a+1,a) \\
&= s^\nu\left[\frac{1}{1-s}\mathcal{L}_a\{H_\mu(\cdot,a)\}(s) - \frac{1}{1-s}\right] - \frac{1-s^\nu}{1-s} \\
&= \frac{s^\nu}{1-s}\left[\frac{1}{s^{\mu+1}} - 1\right] - \frac{1-s^\nu}{1-s} \\
&= \frac{1}{1-s}\frac{1}{s^{\mu-\nu+1}} - \frac{1}{1-s}.
\end{aligned} \tag{3.41}$$

From (3.40) and (3.41) we get that (3.39) holds. Hence, by the uniqueness theorem (Theorem 3.72)

$$\nabla_a^\nu H_\mu(t,a) = H_{\mu-\nu}(t,a)$$

for $t \in \mathbb{N}_{a+1}$. But this last equation also holds for $t = a$. Thus, part (ii) holds for $t \in \mathbb{N}_a$. The proof of (iv) is left to the reader (Exercise 3.34). □

Next we consider the fractional difference equation

$$\nabla_a^\nu x(t) = f(t), \quad t \in \mathbb{N}_{a+N}, \tag{3.42}$$

where $N-1 < \nu < N, N \in \mathbb{N}_1$. First we prove the following existence-uniqueness theorem for the fractional difference equation 3.42.

**Theorem 3.94.** *Assume $f : \mathbb{N}_{a+N} \to \mathbb{R}$ and $N-1 < \nu < N, N \in \mathbb{N}_1$. Then the IVP*

$$\nabla_a^\nu x(t) = f(t), \; t \in \mathbb{N}_{a+N}$$
$$x(a+k) = c_k, \; 1 \leq k \leq N,$$

*where $c_k$, for $1 \leq k \leq N$, are given constants, has a unique solution, which is defined on $\mathbb{N}_{a+1}$.*

*Proof.* First note that if we write the fractional equation $\nabla_a^\nu x(t) = h(t)$ in expanded form we have that

$$\sum_{\tau=a}^{t-1} H_{-\nu-1}(t, \rho(\tau))x(\tau) + x(t) = f(t).$$

It follows that the given IVP is equivalent to the summation equation

$$x(t) = f(t) - \sum_{\tau=a}^{t-1} H_{-\nu-1}(t, \rho(\tau))x(\tau) \tag{3.43}$$

$$x(a+k) = c_k, \quad 1 \leq k \leq N. \tag{3.44}$$

Letting $t = a+N+1$ in this summation IVP we have that $x(t)$ solves our IVP at $t = a+N+1$ iff

$$x(a+N+1) = f(a+) - \sum_{\tau=a}^{t-1} c_k H_{-\nu-1}(t, \rho(\tau))x(\tau).$$

□

**Theorem 3.95.** *Assume $\nu > 0$ and $N-1 < \nu \leq N$. Then a general solution of $\nabla_a^\nu x(t) = 0$ is given by*

$$x(t) = c_1 H_{\nu-1}(t,a) + c_2 H_{\nu-2}(t,a) + \cdots + c_N H_{\nu-N}(t,a)$$

*for $t \in \mathbb{N}_a$.*

*Proof.* For $1 \leq k \leq N$, we have from (3.58) that

$$\begin{aligned}
\nabla_a^\nu H_{\nu-k}(t,a) &= \nabla^k \nabla_a^{\nu-k} H_{\nu-k}(t,a) \\
&= \nabla^k H_0(t,a) \quad \text{(by Theorem 3.93)} \\
&= \nabla^k 1 \\
&= 0
\end{aligned}$$

for $t \in \mathbb{N}_a$. Since these $N$ solutions are linearly independent on $\mathbb{N}_a$ we have that

$$x(t) = c_1 H_{\nu-1}(t,a) + c_2 H_{\nu-2}(t,a) + \cdots + c_N H_{\nu-N}(t,a)$$

is a general solution of $\nabla_a^\nu x(t) = 0$ on $\mathbb{N}_a$. □

The next theorem relates fractional Taylor monomials based at values that differ by a positive integer. This result is in Hein et al. [119] and Ahrendt et al. [3].

**Theorem 3.96.** *For $\nu \in \mathbb{R}\backslash\{-1,-2,\ldots\}$ and $N, m \in \mathbb{N}$,*

$$H_{\nu-N}(t, a+m) = \sum_{k=0}^{m} \binom{m}{k} (-1)^k H_{\nu-N-k}(t,a).$$

*Proof.* The proof is by induction on $m$ for $m \geq 1$. Consider the base case $m = 1$

$$\begin{aligned}
&H_{\nu-N}(t,a) - H_{\nu-N-1}(t,a)\\
&= \frac{(t-a)^{\overline{\nu-N}}}{\Gamma(\nu-N+1)} - \frac{(t-a)^{\overline{\nu-N-1}}}{\Gamma(\nu-N)}\\
&= \frac{\Gamma(t-a+\nu-N)}{\Gamma(\nu-N+1)\Gamma(t-a)} - \frac{\Gamma(t-a+\nu-N-1)}{\Gamma(\nu-N)\Gamma(t-a)}\\
&= \frac{\Gamma(t-a+\nu-N-1)}{\Gamma(t-a)\Gamma(\nu-N+1)}\big[(t-a+\nu-N-1)-(\nu-N)\big]\\
&= \frac{(t-\rho(a))\Gamma(t-a+\nu-N-1)}{\Gamma(t-a)\Gamma(\nu-N+1)}\\
&= \frac{\Gamma(t-a+\nu-N-1)}{\Gamma(t-\rho(a))\Gamma(\nu-N+1)}\\
&= \frac{[t-(a+1)]^{\overline{\nu-N}}}{\Gamma(\nu-N+1)}\\
&= H_{\nu-N}(t,a+1).
\end{aligned}$$

Hence the base case, $m = 1$, holds. Now assume $m \geq 1$ is fixed and

$$H_{\nu-N}(t, a+m) = \sum_{k=0}^{m} \binom{m}{k} (-1)^k H_{\nu-N-k}(t,a).$$

From the base case with the number $a$ replaced by the number $a+m$ we have that

$$H_{\nu-N}(t,a+m+1) = H_{\nu-N}(t,a+m) - H_{\nu-N-1}(t,a+m).$$

Applying the induction hypothesis to both terms on the right side of this equation gives

$$\begin{aligned}
&H_{\nu-N}(t, a+m+1)\\
&= \sum_{k=0}^{m}\binom{m}{k}(-1)^k H_{\nu-N-k}(t,a) - \sum_{k=0}^{m}\binom{m}{k}(-1)^k H_{\nu-N-1-k}(t,a)\\
&= \sum_{k=0}^{m}\binom{m}{k}(-1)^k H_{\nu-N-k}(t,a) - \sum_{k=1}^{m+1}\binom{m}{k-1}(-1)^{k-1} H_{\nu-N-k}(t,a)\\
&= \sum_{k=0}^{m+1}\binom{m}{k}(-1)^k H_{\nu-N-k}(t,a) - \binom{m}{m+1}(-1)^{m+1} H_{\nu-N-m-1}(t,a)\\
&\quad - \sum_{k=0}^{m+1}\binom{m}{k-1}(-1)^{k-1} H_{\nu-N-k}(t,a) + \binom{m}{-1}(-1)^{-1} H_{\nu-N}(t,a)\\
&= \sum_{k=0}^{m+1}\left(\binom{m}{k}+\binom{m}{k-1}\right)(-1)^k H_{\nu-N-k}(t,a)\\
&= \sum_{k=0}^{m+1}\binom{m+1}{k}(-1)^k H_{\nu-N-k}(t,a).
\end{aligned}$$

This completes the proof. □

## 3.15 Mittag–Leffler Function

In this section we define the nabla Mittag–Leffler function, which is useful for solving certain IVPs. First we give an alternate proof of part (i) of Theorem 3.50.

**Theorem 3.97.** *For $|p| < 1$, we have that $E_p(t,a) = \sum_{k=0}^{\infty} p^k H_k(t,a)$ for $t \in \mathbb{N}_a$.*

*Proof.* We will show that $E_p(t,a)$ and $\sum_{k=0}^{\infty} p^k H_k(t,a)$ have the same Laplace transform. In order to ensure convergence, we restrict the transform domain such that $|s| < |p|$, $|1-s| < 1$, and $|1-s| < |1-p|$. First, we determine the Laplace transform of the exponential function as follows:

$$\begin{aligned}
\mathcal{L}_a\{E_p(\cdot,a)\}(s) &= \sum_{k=1}^{\infty}(1-s)^{k-1}(1-p)^{-k}\\
&= \frac{1}{1-p}\sum_{k=0}^{\infty}\left(\frac{1-s}{1-p}\right)^k = \frac{1}{s-p}.
\end{aligned}$$

Next, we have

$$\mathcal{L}_a\left\{\sum_{k=0}^{\infty} p^k H_k(\cdot, a)\right\}(s) = \sum_{k=0}^{\infty} p^k \mathcal{L}_a\{H_k(\cdot, a)\}(s)$$

$$= \frac{1}{s}\sum_{k=0}^{\infty}\left(\frac{p}{s}\right)^k = \frac{1}{s-p}.$$

Finally, $E_p(a, a) = 1$ by definition, and $\sum_{k=0}^{\infty} p^k H_k(a, a) = 1$ since $p^0 H_0(a, a) = 1$ and $p^k H_k(a, a) = 0$ for $k \geq 1$. Therefore, we obtain the desired result on $\mathbb{N}_a$. □

Next we define the nabla Mittag–Leffler function, which is a generalization of the exponential function $E_p(t, a)$.

**Definition 3.98 (Mittag–Leffler Function).** For $|p| < 1, \alpha > 0, \beta \in \mathbb{R}$, we define the nabla Mittag–Leffler function by

$$E_{p,\alpha,\beta}(t, a) := \sum_{k=0}^{\infty} p^k H_{\alpha k+\beta}(t, a), \quad t \in \mathbb{N}_a.$$

*Remark 3.99.* Since $H_0(t, a) = 1$, we have that $E_{0,\nu,0}(t, a) = 1$ and $E_{p,\nu,0}(a, a) = 1$. Also note that $E_{p,1,0}(t, a) = E_p(t, a)$, for $|p| < 1$.

**Theorem 3.100.** *Assume $|p| < 1$, $\alpha > 0$, $\beta \in \mathbb{R}$. Then*

$$\nabla^{\nu}_{\rho(a)} E_{p,\alpha,\beta}(t, \rho(a)) = E_{p,\alpha,\beta-\nu}(t, \rho(a)) \tag{3.45}$$

*for $t \in \mathbb{N}_a$.*

*Proof.* Since

$$\nabla^{\nu}_{\rho(a)} E_{p,\alpha,\beta}(t, \rho(a)) = \nabla^{\nu}_{\rho(a)}\left(\sum_{k=0}^{\infty} p^k H_{\alpha k+\beta}(t, \rho(a))\right)$$

$$= \sum_{k=0}^{\infty} p^k \nabla^{\nu}_{\rho(a)} H_{\alpha k+\beta}(t, \rho(a))$$

$$= \sum_{k=0}^{\infty} p^k H_{\alpha k+\beta-\nu}(t, \rho(a))$$

$$= E_{p,\alpha,\beta-\nu}(t, \rho(a))$$

we have that (3.45) holds for $t \in \mathbb{N}_a$. □

**Theorem 3.101.** *Assume $N - 1 < \nu \leq N$, $N \in \mathbb{N}$ and $|c| < 1$. Then*

$$E_{-c,\nu,\nu-i}(t, \rho(a)) \qquad 1 \leq i \leq N$$

*are N linearly independent solutions on* $\mathbb{N}_a$ *of*

$$\nabla^{\nu}_{\rho(a)}x(t) + cx(t) = 0, \; t \in \mathbb{N}_{a+N}.$$

*In particular, a general solution of the fractional equation* $\nabla^{\nu}_{\rho(a)}x(t) + cx(t) = 0$ *is given by*

$$x(t) = c_1 E_{-c,\nu,\nu-1}(t, \rho(a)) + c_2 E_{-c,\nu,\nu-2}(t, \rho(a)) + \cdots + c_N E_{-c,\nu,\nu-N}(t, \rho(a)),$$

*for* $t \in \mathbb{N}_a$.

*Proof.* If $c = 0$, then this result follows from Theorem 3.95. Now assume $c \neq 0$. Fix $1 \leq i \leq N$ and consider for $t \in N_{a+1}$,

$$\begin{aligned}
\nabla^{\nu}_{\rho(a)} E_{-c,\nu,\nu-i}(t, \rho(a)) &= E_{-c,\nu,-i}(t, \rho(a)), \quad \text{by (3.45)} \\
&= \sum_{k=0}^{\infty} (-c)^k H_{\nu k - i}(t, \rho(a)) \\
&= \sum_{k=1}^{\infty} (-c)^k H_{\nu k - i}(t, \rho(a)) \\
&= \sum_{k=0}^{\infty} (-c)^{k+1} H_{\nu(k+1)-i}(t, \rho(a)) \\
&= -c \sum_{k=0}^{\infty} (-c)^k H_{\nu k + (\nu - i)}(t, \rho(a)) \\
&= -cE_{-c,\nu,\nu-i}(t, \rho(a)).
\end{aligned}$$

Hence, for each $1 \leq i \leq N$, $E_{-c,\nu,\nu-i}(t, \rho(a))$ is a solution of $\nabla^{\nu}_{\rho(a)}x(t) + cx(t) = 0$ on $\mathbb{N}_a$. It follows that a general solution of $\nabla^{\nu}_{\rho(a)}x(t) + cx(t) = 0$ is given by

$$x(t) = c_1 E_{-c,\nu,\nu-1}(t, \rho(a)) + c_2 E_{-c,\nu,\nu-2}(t, \rho(a)) + \cdots + c_N E_{-c,\nu,\nu-N}(t, \rho(a))$$

for $t \in \mathbb{N}_a$. □

The following example was suggested by Jia Baoguo.

*Example 3.102.* Consider the second order nabla difference equation

$$\nabla^2 x(t) + cx(t) = 0, \qquad 0 < c < 1. \tag{3.46}$$

From Definition 3.98, Theorem 3.50, and Theorem 3.101, we have the following are two solutions of (3.46):

$$\begin{aligned} E_{-c,2,1}(t,a) &= \sum_{k=0}^{\infty}(-c)^k H_{2k+1}(t,a) \\ &= \frac{1}{\sqrt{c}}\sum_{k=0}^{\infty}(-1)^k(\sqrt{c})^{2k+1}H_{2k+1}(t,a) \\ &= \frac{1}{\sqrt{c}}\mathrm{Sin}_{\sqrt{c}}(t,a). \end{aligned} \tag{3.47}$$

and

$$\begin{aligned} E_{-c,2,0}(t,a) &= \sum_{k=0}^{\infty}(-c)^k H_{2k}(t,a) \\ &= \sum_{k=0}^{\infty}(-1)^k(\sqrt{c})^{2k}H_{2k+1}(t,a) \\ &= \mathrm{Cos}_{\sqrt{c}}(t,a). \end{aligned} \tag{3.48}$$

The characteristic values of the equation (3.46) are $\lambda_{1,2} = \pm\sqrt{c}i$. So the solutions of (3.46) are $x_1(x,a) = E_{\sqrt{ci}}(t,a)$ and $x_2(x,a) = E_{-\sqrt{ci}}(t,a)$. So

$$\begin{aligned} E_{-\sqrt{ci}} = (1+\sqrt{ci})^{a-t} &= (1+c)^{\frac{a-t}{2}}[\cos\theta + i\sin\theta]^{a-t} \\ &= (1+c)^{\frac{a-t}{2}}[\cos(a-t)\theta + i\sin(a-t)\theta] \end{aligned} \tag{3.49}$$

and

$$E_{\sqrt{ci}} = (1-\sqrt{ci})^{a-t} = (1+c)^{\frac{a-t}{2}}[\cos(a-t)\theta - i\sin(a-t)\theta], \tag{3.50}$$

where $\cos\theta = \frac{1}{\sqrt{1+c}}$, $\sin\theta = \frac{\sqrt{c}}{\sqrt{1+c}}$.

From the definitions (see Definition 3.17) of $\mathrm{Cos}_{\sqrt{c}}(t,a)$ and $\mathrm{Sin}_{\sqrt{c}}(t,a)$, we have

$$\begin{aligned} \mathrm{Cos}_{\sqrt{c}}(t,a) &= \frac{E_{\sqrt{ci}}(t,a) + E_{-\sqrt{ci}}(t,a)}{2} \\ &= (1+c)^{\frac{a-t}{2}}\cos(a-t)\theta \end{aligned} \tag{3.51}$$

and

$$\begin{aligned} \mathrm{Sin}_{\sqrt{c}}(t,a) &= \frac{E_{\sqrt{ci}}(t,a) - E_{-\sqrt{ci}}(t,a)}{2i} \\ &= -(1+c)^{\frac{a-t}{2}}\sin(a-t)\theta. \end{aligned} \tag{3.52}$$

Consequently, using (3.47), (3.48), (3.51), and (3.52), we find that

$$E_{-c,2,1}(t,a) = -\frac{1}{\sqrt{c}}(1+c)^{\frac{a-t}{2}} \sin(a-t)\theta$$

and that

$$E_{-c,2,0}(t,a) = (1+c)^{\frac{a-t}{2}} \cos(a-t)\theta.$$

Thus, from Theorem 3.101, the general solution of the equation (3.46) is given by

$$x(t) = (1+c)^{\frac{a-t}{2}} [c_1 \sin(a-t)\theta + c_2 \cos(a-t)\theta].$$

Finally, the real part

$$y_1(t,a) = (1+c)^{\frac{a-t}{2}} \cos(a-t)\theta$$

and the imaginary part

$$y_2(t,a) = (1+c)^{\frac{a-t}{2}} \sin(a-t)\theta$$

are solutions of (3.46).

We will now determine the Laplace transform of the Mittag–Leffler function.

**Theorem 3.103.** *Assume* $|p| < 1$ *a constant,* $\alpha > 0$*, and* $\beta \in \mathbb{R}$*. Then*

$$\mathcal{L}_a\{E_{p,\alpha,\beta}(\cdot,a)\}(s) = \frac{s^{\alpha-\beta-1}}{s^\alpha - p}.$$

*for* $|1-s| < 1$, $|s^\alpha| > |p|$.

*Proof.* Note that

$$\begin{aligned}\mathcal{L}_a\{E_{p,\alpha,\beta}(\cdot,a)\}(s) &= \sum_{k=0}^{\infty} p^k \mathcal{L}_a\{H_{\alpha k+\beta}(\cdot,a)\}(s)\\ &= \frac{1}{s^{\beta+1}} \sum_{k=0}^{\infty} \left(\frac{p}{s^\alpha}\right)^k\\ &= \frac{s^{\alpha-\beta-1}}{s^\alpha - p}.\end{aligned}$$

This completes the proof. □

## 3.16 Solutions to Initial Value Problems

We will now consider a $\nu$-th order fractional nabla initial value problem and give a formula for its solution in case $0 < \nu < 1$.

**Theorem 3.104 (Fractional Variation of Constants Formula [119]).** *Assume* $f : \mathbb{N}_a \to \mathbb{R}$, $|c| < 1$ *and* $0 < \nu < 1$. *Then the unique solution of the fractional initial value problem*

$$\begin{cases} \nabla^{\nu}_{\rho(a)} x(t) + cx(t) = f(t), & t \in \mathbb{N}_{a+1}, \\ x(a) = A, \quad A \in \mathbb{R} \end{cases} \tag{3.53}$$

*is given by*

$$x(t) = \left[E_{-c,\nu,\nu-1}(\cdot, \rho(a)) * f(\cdot)\right](t) + \left[A(c+1) - f(a)\right]E_{-c,\nu,\nu-1}(t, \rho(a)). \tag{3.54}$$

*Proof.* We begin by taking the Laplace transform based at $a$ of both sides of the fractional equation in (3.53) to get that

$$\mathcal{L}_a\{\nabla^{\nu}_{\rho(a)} x\}(s) + c\mathcal{L}_a\{x\}(s) = \mathcal{L}_a\{f\}(s).$$

Applying Theorem 3.89 and using the initial condition, we have that

$$(s^{\nu} + c)\mathcal{L}_a\{x\}(s) - A\left(\frac{1 - s^{\nu}}{1 - s}\right) = \mathcal{L}_a\{f\}(s).$$

Using Lemma 3.88 implies that

$$(s^{\nu} + c)\underbrace{\left[\frac{1}{1-s}\mathcal{L}_{\rho(a)}\{x\}(s) - \frac{1}{1-s}x(a)\right]}_{=\mathcal{L}_a\{x\}(s)} - A\left(\frac{1 - s^{\nu}}{1 - s}\right)$$
$$= \underbrace{\frac{1}{1-s}\mathcal{L}_{\rho(a)}\{f\}(s) - \frac{1}{1-s}f(a)}_{=\mathcal{L}_a\{f\}(s)}.$$

Multiplying both sides of the preceding equality by $(1 - s)$ and then solving for $\mathcal{L}_{\rho(a)}\{x\}(s)$ we obtain

$$\mathcal{L}_{\rho(a)}\{x\}(s) = \frac{1}{s^{\nu} + c}\mathcal{L}_{\rho(a)}\{f\}(s) + [A(c+1) - f(a)]\frac{1}{s^{\nu} + c}.$$

Since

$$\mathcal{L}_{\rho(a)}\{E_{-c,\nu,\nu-1}(\cdot,\rho(a))\}(s) = \frac{1}{s^\nu + c},$$

we have by the convolution theorem that

$$x(t) = \big(E_{-c,\nu,\nu-1}(\cdot,\rho(a)) * f(\cdot)\big)(t) + [A(c+1) - f(a)]\, E_{-c,\nu,\nu-1}(t,\rho(a)) \tag{3.55}$$

for $t \in \mathbb{N}_{a+1}$. □

Letting $c = 0$ in the above fractional initial value problem, we get the following corollary.

**Corollary 3.105.** *Let $f : \mathbb{N}_a \to \mathbb{R}$ and $0 < \nu < 1$. Then the unique solution of the fractional IVP*

$$\begin{cases} \nabla^\nu_{\rho(a)} x(t) = f(t), & t \in \mathbb{N}_{a+1} \\ x(a) = A, & A \in \mathbb{R} \end{cases}$$

*is given by*

$$x(t) = \nabla^{-\nu}_{\rho(a)} f(t) + \big(A - f(a)\big) H_{\nu-1}(t,\rho(a)). \tag{3.56}$$

*Proof.* First, we observe that

$$E_{0,\nu,\nu-1}(t,\rho(a)) = H_{\nu-1}(t,\rho(a)).$$

Finally, we have $\big(E_{0,\nu,\nu-1}(\cdot,\rho(a)) * f(\cdot)\big)(t) = [H_{\nu-1}(\cdot,\rho(a)) * f(\cdot)]\,(t) = \nabla_{\rho(a)} a^{-\nu} f(t)$ by Theorem 3.80. From this, the stated solution to the initial value problem follows. □

*Example 3.106.* Use Corollary 3.105 to solve the IVP

$$\nabla^{\frac{1}{2}}_{\rho(0)} x(t) = t, \quad t \in \mathbb{N}_1$$
$$x(0) = 2.$$

By the variation of constants formula (3.56), the solution of this IVP is given by

$$\begin{aligned} x(t) &= \nabla^{-\frac{1}{2}}_{\rho(0)} H_1(t,0) + (2-1) H_{-\frac{1}{2}}(t,\rho(0)) \\ &= \nabla^{-\frac{1}{2}}_{\rho(0)} H_1(t,0) + \frac{(t-\rho(0))^{\overline{-\frac{1}{2}}}}{\Gamma(\frac{1}{2})} \end{aligned}$$

$$= \int_{\rho(0)}^{0} H_{-\frac{1}{2}}(t,\rho(s)) s \nabla s + \int_{0}^{t} H_{-\frac{1}{2}}(t,\rho(s)) s \nabla s + \frac{(t-\rho(0))^{\overline{-\frac{1}{2}}}}{\Gamma(\frac{1}{2})}$$

$$= \int_{0}^{t} H_{-\frac{1}{2}}(t,\rho(s)) s \nabla s + \frac{(t-\rho(0))^{\overline{-\frac{1}{2}}}}{\Gamma(\frac{1}{2})}$$

$$= \nabla_0^{-\frac{1}{2}} H_1(t,0) + \frac{(t-\rho(0))^{\overline{-\frac{1}{2}}}}{\Gamma(\frac{1}{2})}$$

$$= \nabla_0^{-\frac{1}{2}} H_1(t,0) + \frac{(t+1)^{\overline{-\frac{1}{2}}}}{\Gamma(\frac{1}{2})}$$

$$= H_{\frac{3}{2}}(t,0) + \frac{1}{\sqrt{\pi}}(t+1)^{\overline{-\frac{1}{2}}}$$

$$= \frac{t^{\overline{\frac{3}{2}}}}{\Gamma(\frac{5}{2})} + \frac{1}{\sqrt{\pi}}(t+1)^{\overline{-\frac{1}{2}}}$$

$$= \frac{4}{3\sqrt{\pi}} t^{\overline{\frac{3}{2}}} + \frac{1}{\sqrt{\pi}}(t+1)^{\overline{-\frac{1}{2}}},$$

where we used Theorem 3.93 in step seven.

## 3.17 Nabla Fractional Composition Rules

In this section we prove several composition rules for nabla fractional sums and differences. Most of these results can be found in Ahrendt et al. [3]. First we prove the following formula for the composition of two fractional sums.

**Theorem 3.107.** *Assume* $f : \mathbb{N}_{a+1} \to \mathbb{R}$, *and* $\nu, \mu > 0$. *Then*

$$\nabla_a^{-\nu} \nabla_a^{-\mu} f(t) = \nabla_a^{-\nu-\mu} f(t), \quad t \in \mathbb{N}_a.$$

*Proof.* Note that

$$\begin{aligned} \mathcal{L}_a\{\nabla_a^{-\nu} \nabla_a^{-\mu} f\}(s) &= \tfrac{1}{s^\nu} \mathcal{L}_a\{\nabla_a^{-\mu} f\}(s) \\ &= \tfrac{1}{s^{\nu+\mu}} \mathcal{L}\{f\}(s) \\ &= \mathcal{L}_a\{\nabla_a^{-\nu-\mu} f\}(s). \end{aligned}$$

By the uniqueness theorem for Laplace transforms (Theorem 3.72), we have

$$\nabla_a^{-\nu}\nabla_a^{-\mu}f(t) = \nabla_a^{-\nu-\mu}f(t)$$

for $t \in \mathbb{N}_{a+1}$. Also

$$\nabla_a^{-\nu}\nabla_a^{-\mu}f(a) = 0 = \nabla_a^{-\nu-\mu}f(a).$$

□

Next we prove a theorem concerning the composition of an integer-order difference with a fractional sum and with a fractional difference. This result was first proved in this generality by Ahrendt et al. [3].

**Lemma 3.108.** *Let $k \in \mathbb{N}_0$, $\mu > 0$, and choose $N \in \mathbb{N}$ such that $N - 1 < \mu \leq N$. Then*

$$\nabla^k\nabla_a^{-\mu}f(t) = \nabla_a^{k-\mu}f(t), \tag{3.57}$$

*and*

$$\nabla^k\nabla_a^{\mu}f(t) = \nabla_a^{k+\mu}f(t). \tag{3.58}$$

*for $t \in \mathbb{N}_{a+k}$.*

*Proof.* Assume $k \in \mathbb{N}_0$, $\mu > 0$, and choose $N \in \mathbb{N}_1$ such that $N - 1 < \mu \leq N$. First we prove (3.57) for $\mu = N$. To see this first note that

$$\begin{aligned}\nabla\nabla_a^{-1}f(t) &= \nabla\int_a^t H_0(t,\rho(\tau))f(\tau)\nabla\tau\\ &= \nabla\int_a^t f(\tau)\nabla\tau\\ &= f(t), \quad t \in \mathbb{N}_{a+1}.\end{aligned}$$

So, then, for the case of $\mu = N$ we have

$$\begin{aligned}\nabla^k\nabla_a^{-N}f(t) &= \nabla^{k-1}[\nabla\nabla_a^{-1}(\nabla_a^{-(N-1)}f(t))]\\ &= \nabla^{k-1}\nabla^{-(N-1)}f(t)\\ &= \nabla^{k-2}\nabla^{-(N-2)}f(t)\\ &\cdots\\ &= \nabla^{-(N-k)}f(t)\\ &= \nabla^{k-N}f(t), \quad t \in \mathbb{N}_{a+k}.\end{aligned}$$

Hence (3.57) holds for $\mu = N$. Now we consider (3.58). It is trivial to prove (3.58) when $\mu = N$, so we assume $N-1 < \mu < N$. First we will show that (3.58) holds for the base case

$$\nabla\nabla_a^{\mu} f(t) = \nabla_a^{1+\mu} f(t), \quad t \in \mathbb{N}_{a+1}.$$

This follows from the following:

$$\begin{aligned}
&\nabla\nabla_a^{\mu} f(t) \\
&= \nabla\left(\int_a^t H_{-\mu-1}(t,\rho(\tau))f(\tau)\nabla\tau\right) \\
&= \int_a^t H_{-\mu-2}(t,\rho(\tau))f(\tau)\nabla\tau + H_{-\mu-1}(\rho(t),\rho(t))f(t) \quad \text{(by (3.22))} \\
&= \int_a^t H_{-\mu-2}(t,\rho(\tau))f(\tau)\nabla\tau \\
&= \nabla_a^{1+\mu} f(t).
\end{aligned}$$

Then, for any $k \in \mathbb{N}_0$,

$$\begin{aligned}
\nabla^k\nabla_a^{\mu} f(t) &= \nabla^{k-1}(\nabla\nabla_a^{\mu} f(t)) \\
&= \nabla^{k-1}\nabla_a^{1+\mu} f(t) \\
&= \nabla^{k-2}\nabla_a^{2+\mu} f(t) \\
&\cdots \\
&= \nabla_a^{k+\mu} f(t), \quad t \in \mathbb{N}_{a+k},
\end{aligned}$$

which shows (3.58) holds for this case. In case $N-1 < \mu < N$, noticing that $\nabla_a^{k+\mu} f(t)$ can be obtained from $\nabla_a^{k-\mu} f(t)$ by replacing $\mu$ by $-\mu$, we obtain by a similar argument that (3.57) holds for the case $N-1 < \mu < N$. And this completes the proof. □

**Theorem 3.109.** *Assume* $f : \mathbb{N}_a \to \mathbb{R}$ *and* $\nu, \mu > 0$. *Then*

$$\nabla_a^{\nu}\nabla_a^{-\mu} f(t) = \nabla_a^{\nu-\mu} f(t).$$

*Proof.* Let $\nu, \mu > 0$ be given, and $N \in \mathbb{N}$ such that $N-1 < \nu \le N$. Then we have

$$\begin{aligned}
\nabla_a^{\nu}\nabla_a^{-\mu} f(t) &= \nabla^N\nabla_a^{-(N-\nu)}\nabla_a^{-\mu} f(t) \\
&= \nabla^N\nabla_a^{-(N-\nu)-\mu} f(t) \qquad \text{by Theorem 3.107}
\end{aligned}$$

$$= \nabla_a^{N-N+\nu-\mu} f(t) \qquad \text{by Lemma 3.108}$$
$$= \nabla_a^{\nu-\mu} f(t).$$

This completes the proof. □

The following theorem for $N = 1$ appears in Hein et al. [119] and for general $N$ appears in Ahrendt et al. [3].

**Theorem 3.110.** *Assume the nabla Laplace transform of $f : \mathbb{N}_{a+1} \to \mathbb{R}$ exists for $|s-1| < r$, $r > 0$, $\nu > 0$, and pick $N \in \mathbb{N}_1$ so that $N - 1 < \nu \leq N$. Then*

$$\mathcal{L}_{a+N}\{\nabla_a^\nu f\}(s) = s^\nu \mathcal{L}_{a+N}\{f\}(s) + \sum_{k=0}^{N-1} \Bigg[ \frac{s^\nu}{(1-s)^{N-k}} f(a+k+1) - \frac{s^N}{(1-s)^{N-k}} \nabla_a^{-(N-\nu)} f(a+k+1) - \nabla^{N-k-1} \nabla_a^{-(N-\nu)} f(a+N) s^k \Bigg],$$

*for $|s-1| < r$.*

*Proof.* Consider

$$\begin{aligned}
\mathcal{L}_{a+N}\{\nabla_a^\nu f\}(s) &= \mathcal{L}_{a+N}\{\nabla^N \nabla_a^{-(N-\nu)} f\}(s) \\
&= s^N \mathcal{L}_{a+N}\{\nabla_a^{-(N-\nu)} f\}(s) - \sum_{k=1}^{N} s^{N-k} \nabla^{k-1} \nabla_a^{-(N-\nu)} f(a+N) \quad \text{by (3.85)} \\
&= s^N \left[ \frac{1}{(1-s)^N} \mathcal{L}_a\{\nabla_a^{-(N-\nu)} f\}(s) - \sum_{k=1}^{N} \frac{\nabla_a^{-(N-\nu)} f(a+k)}{(1-s)^{N-k+1}} \right] \\
&\quad - \sum_{k=0}^{N-1} s^k \nabla^{N-k-1} \nabla_a^{-(N-\nu)} f(a+N) \quad \text{by Lemma 3.88} \\
&= \frac{s^\nu}{(1-s)^N} \mathcal{L}_a\{f\}(s) - s^N \sum_{k=1}^{N} \frac{\nabla_a^{-(N-\nu)} f(a+k)}{(1-s)^{N-k+1}} \\
&\quad - \sum_{k=0}^{N-1} s^k \nabla^{N-k-1} \nabla_a^{-(N-\nu)} f(a+N) \\
&= s^\nu \left[ \mathcal{L}_{a+N}\{f\}(s) + \sum_{k=1}^{N} \frac{f(a+k)}{(1-s)^{n-k+1}} \right] - s^N \sum_{k=1}^{N} \frac{\nabla_a^{-(N-\nu)} f(a+k)}{(1-s)^{N-k+1}}
\end{aligned}$$

$$
-\sum_{k=0}^{N-1} s^k \nabla^{N-k-1} \nabla_a^{-(N-\nu)} f(a+N) \quad \text{by Lemma 3.88}
$$

$$
= s^\nu \mathcal{L}_{a+N}\{f\}(s) + \sum_{k=0}^{N-1} \left[ \frac{s^\nu}{(1-s)^{N-k}} f(a+k+1) \right.
$$

$$
\left. \frac{s^N}{(1-s)^{N-k}} \nabla_a^{-(N-\nu)} f(a+k+1) - \nabla^{N-k-1} \nabla_a^{-(N-\nu)} f(a+N) s^k \right],
$$

which is what we wanted to prove. □

**Theorem 3.111.** *Assume* $f : \mathbb{N}_a \to \mathbb{R}$, $\nu > 0$ *and* $k \in \mathbb{N}_0$. *Then*

$$
\nabla_{a+k}^{-\nu} \nabla^k f(t) = \nabla_{a+k}^{k-\nu} f(t) - \sum_{j=0}^{k-1} \nabla^j f(a+k) H_{\nu-k+j}(t, a+k).
$$

*Proof.* Integrating by parts on two different occasions below we get

$$
\begin{aligned}
\nabla_{a+k}^{-\nu} \nabla^k f(t) &= \int_{a+k}^t H_{\nu-1}(t, \rho(\tau)) \nabla^k f(\tau) \nabla\tau \\
&= \int_{a+k}^t H_{\nu-1}(t, \rho(\tau)) \nabla \nabla^{k-1} f(\tau) \nabla\tau \\
&= \nabla^{k-1} f(\tau) H_{\nu-1}(t, \tau) \Big|_{a+k}^t + \int_{a+k}^t H_{\nu-2}(t, \rho(\tau)) \nabla^{k-1} f(\tau) \\
&= -\nabla^{k-1} f(a+k) H_{\nu-1}(t, a+k) + H_{\nu-1}(t, a+k) \nabla^{k-1} f(\tau) \\
&= \nabla_{a+k}^{-(\nu-1)} \nabla^{k-1} f(t) - \nabla^{k-1} f(a+k) H_{\nu-1}(t, a+k) \\
&= \nabla [\nabla_{a+k}^{-\nu} \nabla^{k-1} f(t)] - \nabla^{k-1} f(a+k) H_{\nu-1}(t, a+k) \\
&= \nabla_{a+k}^{-(\nu-2)} \nabla^{k-2} f(t) - \nabla^{k-2} f(a+k) H_{\nu-2}(t, a+k) \\
&\quad - \nabla^{k-1} f(a+k) H_{\nu-1}(t, a+k).
\end{aligned}
$$

Integrating by parts $k-2$ more times gives

$$
\nabla_{a+k}^{-\nu} \nabla^k f(t) = \nabla_{a+k}^{k-\nu} f(t) - \sum_{j=0}^{k-1} \nabla^j f(a+k) H_{\nu-k+j}(t, a+k),
$$

which was what we wanted to prove. □

**Theorem 3.112.** *Let $\nu > 0$ and $k \in \mathbb{N}_0$ be given and choose $N \in \mathbb{N}$ such that $N-1 < \nu \leq N$. Then*

$$\nabla^{\nu}_{a+k}\nabla^k f(t) = \nabla^{k+\nu}_{a+k} f(t) - \sum_{j=0}^{k-1} \nabla^j f(a+k) H_{-\nu-k+j}(t, a+k).$$

*Proof.* Consider

$$\begin{aligned}
\nabla^{\nu}_{a+k}\nabla^k f(t) &= \nabla^N(\nabla^{-(N-\nu)}_{a+k}\nabla^k f(t)) \\
&= \nabla^N\left(\nabla^{k-(N-\nu)}_{a+k} f(t) - \sum_{j=0}^{k-1} \nabla^j f(a+k) H_{N-\nu-k+j}(t, a+k)\right) \\
&= \nabla^{k+\nu}_{a+k} f(t) - \sum_{j=0}^{k-1} \nabla^j f(a+k) H_{N-\nu-k+j}(t, a) \\
&= \nabla^{k+\nu}_{a+k} f(t) - \sum_{j=0}^{k-1} \nabla^j f(a+k) \nabla^{N-1} H_{N-\nu-k+j}(t, a) \\
&= \nabla^{k+\nu}_{a+k} f(t) - \sum_{j=0}^{k-1} \nabla^j f(a+k) \nabla^{N-1} H_{N-\nu-k+j-1}(t, a).
\end{aligned}$$

Taking the difference inside the summation $N-1$ more times, we get

$$\nabla^{\nu}_{a+k}\nabla^k f(t) = \nabla^{k+\nu}_{a+k} f(t) - \sum_{j=0}^{k-1} \nabla^j f(a+k) H_{-\nu-k+j}(t, a+k),$$

which is what we wanted to prove. □

**Theorem 3.113.** *Assume $1 < \nu \leq 2$. Then the unique solution of the fractional IVP*

$$\begin{aligned}
\nabla^{\nu}_a x(t) &= 0, \quad t \in \mathbb{N}_{a+2} \\
x(a+2) &= A_0, \\
\nabla x(a+2) &= A_1,
\end{aligned}$$

*where $A_0, A_1 \in \mathbb{R}$, is given by*

$$\begin{aligned}
x(t) &= [(2-\nu)A_0 + (\nu-1)A_1]h_{\nu-1}(t, a) \\
&\quad + [(\nu-1)A_0 - \nu A_1]h_{\nu-2}(t, a), \qquad (3.59)
\end{aligned}$$

*for $t \in \mathbb{N}_{a+1}$.*

*Proof.* Let $x(t)$ be the solution of the IVP (3.59) $\nabla_a^\nu x(t) = 0$. Then

$$x(t) = c_1 H_{\nu-1}(t,a) + c_2 H_{\nu-2}(t,a).$$

Using the IC's we get that

$$x(a+1) = x(a+2) - \nabla x(a+2) = A_0 - A_1.$$

It follows from this that

$$x(a+1) = c_1 H_{\nu-1}(a+1,a) + c_2 H_{\nu-2}(a+1,a) = A_0 - A_1.$$

Since $H_{\nu-1}(a+1,a) = H_{\nu-2}(a+1,a) = 1$, we have that

$$c_1 + c_2 = A_0 - A_1.$$

Since $\nabla x(t) = c_1 H_{\nu-2}(t,a) + c_2 H_{\nu-3}(t,a)$, we get that

$$\begin{aligned}\nabla x(a+2) &= c_1 H_{\nu-2}(a+2,a) + c_2 H_{\nu-3}(a+2,a)\\ &= c_1(\nu-1) + c_2(\nu-2)\\ &= A_1.\end{aligned}$$

Solving the system

$$\begin{aligned}c_1 + c_2 &= A_0 - A_1\\ (\nu-1)c_1 + (\nu-2)c_2 &= A_1\end{aligned}$$

we get

$$c_1 = (2-\nu)A_0 + (\nu-1)A_1, \qquad c_2 = (\nu-1)A_0 - \nu A_1.$$

Hence,

$$x(t) = [(2-\nu)A_0 + (\nu-1)A_1]h_{\nu-1}(t,a) + [(\nu-1)A_0 - \nu A_1]h_{\nu-2}(t,a),$$

for $t \in \mathbb{N}_{a+1}$. □

Next, we look at the nonhomogeneous equation with zero initial conditions.

**Theorem 3.114.** *Let $g : \mathbb{N}_a \to \mathbb{R}$ and $1 < \nu \leq 2$. Then, for $t \in \mathbb{N}_{a+2}$, the fractional initial value problem*

$$\begin{aligned}\nabla_a^\nu x(t) &= g(t), \qquad t \in \mathbb{N}_{a+2}\\ x(a+2) &= 0\\ \nabla x(a+2) &= 0\end{aligned}$$

*has the unique solution*

$$x(t) = \nabla_a^{-\nu} g(t) - [g(a+1) + g(a+2)]h_{\nu-1}(t,a) \\ + g(a+2)h_{\nu-2}(t,a). \tag{3.60}$$

*Proof.* We take the Laplace transform based at $a+2$ of both sides of the equation.

$$\mathcal{L}_{a+2}\{\nabla_a^\nu x\}(s) = \mathcal{L}_{a+2}\{g\}(s).$$

Next, we use Theorem 3.110 and Lemma 3.88 on the left-hand side and the Laplace transform shifting theorem on the right-hand side of the equation.

$$\begin{aligned}
&\frac{s^\nu}{(1-s)^2}\mathcal{L}_a\{x\}(s) \\
&\quad - (\frac{s}{1-s})^2 \nabla_a^{-(2-\nu)} x(a+1) - \nabla\nabla_a^{-(2-\nu)} x(a+2) \\
&\quad - (\frac{s^2}{1-s})\nabla_a^{-(2-\nu)} x(a+2) - s\nabla_a^{-(2-\nu)} x(a+2) \\
&= \frac{1}{(1-s)^2}\mathcal{L}_a\{g\}(s) - \frac{1}{(1-s)^2} g(a+1) - \frac{1}{1-s} g(a+2).
\end{aligned}$$

Using $x(a+2) = 0 = \nabla x(a+2)$, we obtain

$$\begin{aligned}
\frac{s^\nu}{(1-s)^2}\mathcal{L}_a\{x\}(s) = \frac{1}{(1-s)^2}\mathcal{L}_a\{g\}(s) - \frac{1}{(1-s)^2} g(a+1) \\
- \frac{1}{1-s} g(a+2).
\end{aligned}$$

Next, we solve for the Laplace transform of $x(t)$ to obtain

$$\begin{aligned}
\mathcal{L}_a\{x\}(s) &= \frac{1}{s^\nu}\mathcal{L}_a\{g\}(s) - \frac{1}{s^\nu} g(a+1) - \frac{(1-s)}{s^\nu} g(a+2) \\
&= [\mathcal{L}_a\{h_{\nu-1}(t,a)\}(s)\mathcal{L}_a\{g\}(s)] - \frac{1}{s^\nu} g(a+1) \\
&\quad - \frac{1}{s^\nu} g(a+2) + \frac{1}{s^{\nu-1}} g(a+2).
\end{aligned}$$

Finally, we take the inverse Laplace transform and note that

$$\nabla_a^{-\nu} g(t) = h_{\nu-1}(t,a) * g(t),$$

which yields

$$\begin{aligned} x(t) &= [h_{\nu-1}(\cdot,a) * g(\cdot)] - [g(a+1)+g(a+2)]h_{\nu-1}(t,a) \\ &\quad + g(a+2)h_{\nu-2}(t,a) \\ &= \nabla_a^{-\nu} g(t) - [g(a+1)+g(a+2)]h_{\nu-1}(t,a) + g(a+2)h_{\nu-2}(t,a) \end{aligned}$$

for $t \in \mathbb{N}_{a+2}$. □

## 3.18 Monotonicity for the Nabla Case

Most of the results in this section appear in the paper [49]. These results were motivated by the paper by Dahal and Goodrich [67]. The results of Dahal and Goodrich are treated in Sect. 7.2. First, we derive a nabla difference inequality which plays an important role in proving our main result on monotonicity.

**Theorem 3.115.** *Assume that* $f : \mathbb{N}_a \to \mathbb{R}$, $\nabla_a^\nu f(t) \geq 0$, *for each* $t \in \mathbb{N}_{a+1}$, *with* $1 < \nu < 2$. *Then*

$$\begin{aligned} \nabla f(t) &\geq -f(a+1)[H_{-\nu-1}(t,a) + H_{-\nu}(t,a+1)] \\ &\quad - \sum_{\tau=a+2}^{t-1} H_{-\nu}(t,\rho(\tau))\nabla f(\tau) \quad (3.61) \\ &= -f(a+1)\frac{(-\nu+1)^{\overline{t-\rho(a)}}}{(t-\rho(a))!} - \sum_{\tau=a+2}^{t-1} \frac{(-\nu+1)^{\overline{t-\tau}}}{(t-\tau)!}\nabla f(\tau) \quad (3.62) \end{aligned}$$

*for* $t \in \mathbb{N}_{a+1}$, *where for* $t \geq \tau$,

$$H_{-\nu}(t,\rho(\tau)) = \frac{(-\nu+1)^{\overline{t-\tau}}}{(t-\tau)!} < 0. \quad (3.63)$$

*Proof.* Note that

$$\begin{aligned} \nabla_a^\nu f(t) &= \int_a^{a+1} H_{-\nu-1}(t,\rho(\tau))f(\tau)\nabla\tau + \int_{a+1}^t H_{-\nu-1}(t,\rho(\tau))f(\tau)\nabla\tau \\ &= H_{-\nu-1}(t,a)f(a+1) + \int_{a+1}^t H_{-\nu-1}(t,\rho(\tau))f(\tau)\nabla\tau. \quad (3.64) \end{aligned}$$

Integrating by parts and using the power rule

$$\nabla_\tau H_{-\nu}(t,\tau) = -H_{-\nu-1}(t,\rho(\tau))$$

we have that (where we use $H_{-\nu}(t, \rho(t)) = 1$)

$$\begin{aligned}
&\int_{a+1}^{t} H_{-\nu-1}(t, \rho(\tau))f(\tau)\nabla\tau \\
&= -f(\tau)H_{-\nu}(t,\tau)|_{\tau=a+1}^{t} + \int_{a+1}^{t} H_{-\nu}(t, \rho(\tau))\nabla f(\tau)\nabla\tau \\
&= f(a+1)H_{-\nu}(t, a+1) + \sum_{\tau=a+2}^{t} H_{-\nu}(t, \rho(\tau))\nabla f(\tau) \\
&= f(a+1)H_{-\nu}(t, a+1) + \sum_{\tau=a+2}^{t-1} H_{-\nu}(t, \rho(\tau))\nabla f(\tau) \\
&\qquad + H_{-\nu}(t, \rho(t))\nabla f(t) \\
&= f(a+1)H_{-\nu}(t, a+1) + \sum_{\tau=a+2}^{t-1} H_{-\nu}(t, \rho(\tau))\nabla f(\tau) + \nabla f(t). \qquad (3.65)
\end{aligned}$$

Using (3.64) and (3.65), we obtain

$$\begin{aligned}
0 &\leq \Delta_a^{\nu} f(t) \\
&= [H_{-\nu-1}(t, a) + H_{-\nu}(t, a+1)]f(a+1) \\
&\quad + \sum_{\tau=a+2}^{t-1} H_{-\nu}(t, \rho(\tau))\nabla f(\tau) + \nabla f(t),
\end{aligned}$$

for $t \in \mathbb{N}_{a+1}$ by assumption. Solving this last inequality for $\nabla f(t)$ we obtained the desired inequality (3.61). Next we show that (3.63) holds. This follows from the following:

$$\begin{aligned}
&H_{-\nu}(t, \rho(\tau)) \\
&= \frac{(t-\rho(\tau))^{\overline{-\nu}}}{\Gamma(-\nu+1)} \\
&= \frac{(t-\tau+1)^{\overline{-\nu}}}{\Gamma(-\nu+1)} \\
&= \frac{\Gamma(t+1-\nu-\tau)}{\Gamma(t-\tau+1)\Gamma(-\nu+1)} \\
&= \frac{(-\nu+t-\tau)(-\nu+t-\tau-1)\cdots(-\nu+1)}{(t-\tau)!} \quad (\text{since} \quad t-\tau \geq 1)
\end{aligned}$$

$$= \frac{(-\nu+1)^{\overline{t-\tau}}}{(t-\tau)!}$$
$$< 0$$

since $1 < \nu < 2$. Also

$$\begin{aligned}
&-[H_{-\nu-1}(t,a) + H_{-\nu}(t,a+1)] \\
&= -\Big[\frac{(t-a)^{\overline{-\nu-1}}}{\Gamma(-\nu)} + \frac{(t-\rho(a))^{\overline{-\nu}}}{\Gamma(-\nu+1)}\Big] \\
&= -\Big[\frac{\Gamma(-\nu+t-\rho(a))}{\Gamma(t-a)\Gamma(-\nu)} + \frac{\Gamma(-\nu+t-\rho(a))}{\Gamma(t-\rho(a))\Gamma(-\nu+1)}\Big] \\
&= -\frac{(-\nu+t-\rho(a))(-\nu+t-a-2)\cdots(-\nu+2)(-\nu+1)}{(t-\rho(a))!} \\
&= -\frac{\Gamma(-\nu+t-a)}{\Gamma(-\nu+1)(t-\rho(a))!} \\
&= -\frac{(-\nu+1)^{\overline{t-\rho(a)}}}{(t-\rho(a))!} \\
&> 0.
\end{aligned}$$

This completes the proof. □

**Theorem 3.116.** *Assume $f : \mathbb{N}_{a+1} \to \mathbb{R}$, $\nabla_a^\nu f(t) \geq 0$, for each $t \in \mathbb{N}_{a+1}$, with $1 < \nu < 2$. Then $\nabla f(t) \geq 0$, for $t \in \mathbb{N}_{a+2}$.*

*Proof.* We prove that $\nabla f(a+k) \geq 0$, for $k \geq 0$ by the principle of strong induction. Since $\nabla_a^\nu f(a+1) = f(a+1)$, we have by assumption that $f(a+1) \geq 0$. When $t = a+2$, it follows that

$$\begin{aligned}
\nabla_0^\nu f(a+2) &= \int_a^{a+2} H_{-\nu-1}(a+2,\tau-1)f(\tau)\nabla\tau \\
&= f(a+2) - \nu f(a+1) \\
&= \nabla f(a+2) - (\nu-1)f(a+1).
\end{aligned}$$

From our assumption $\nabla_a^\nu f(a+2) \geq 0$ and the fact that $\nabla_a^\nu f(a+1) = f(a+1)$, we have

$$\nabla f(a+2) \geq (\nu-1)f(a+1) \geq 0.$$

Suppose $k \geq 2$ and that $\nabla f(a+i) \geq 0$, for $i = 2,3,4,\cdots,k$. Then from Theorem 3.115, we have $\nabla f(a+k+1) \geq 0$, so this completes the proof. □

## 3.19 Caputo Fractional Difference

In this section we define the $\mu$-th Caputo fractional difference operator and give some of its properties. Many of the results in this section and related results are contained in the papers by Anastassiou [7–13] and the paper by Ahrendt et al. [4].

**Definition 3.117.** Assume $f : \mathbb{N}_{a-N+1} \to \mathbb{R}$ and $\mu > 0$. Then the $\mu$-th Caputo nabla fractional difference of $f$ is defined by

$$\nabla_{a*}^{\mu} f(t) = \nabla_a^{-(N-\mu)} \nabla^N f(t)$$

for $t \in \mathbb{N}_{a+1}$, where $N = \lceil \mu \rceil$.

One nice property of the Caputo nabla fractional difference is that if $\mu \geq 1$ and $C$ is any constant, then

$$\nabla_{a*}^{\mu} C = 0.$$

Note that this is not true for the nabla Riemann–Liouville fractional difference, when $C \neq 0$ and $\mu > 0$ is not an integer (see Exercise 3.28).

The following theorem follows immediately from the definition of the Caputo nabla fractional difference and the definition of the Taylor monomials.

**Theorem 3.118.** *Assume $\mu > 0$ and $N = \lceil \mu \rceil$. Then the nabla Taylor monomials, $H_k(t, a)$, $0 \leq k \leq N - 1$, are $N$ linearly independent solutions of $\nabla_{a*}^{\mu} x = 0$ on $\mathbb{N}_{a-N+1}$.*

The reader should compare this theorem (Theorem 3.118) to Theorem 3.95 which gives the analogue result for the nabla Riemann–Liouville case $\nabla_a^{\mu} x = 0$. That is, $H_{\mu-k}(t, a)$, where $0 \leq k \leq N - 1$, are $N$ linearly independent solutions of $\nabla_a^{\mu} x = 0$.

The following result appears in Anastassiou [7].

**Theorem 3.119 (Nabla Taylor's Theorem with Caputo Differences).** *Assume $f : \mathbb{N}_{a-N+1} \to \mathbb{R}$, $\mu > 0$ and $N - 1 < \mu \leq N$. Then*

$$f(t) = \sum_{k=0}^{N-1} H_k(t, a) \nabla^k f(a) + \int_a^t H_{\mu-1}(t, \rho(\tau)) \nabla_{a*}^{\mu} f(\tau) \nabla \tau,$$

*for $t \in \mathbb{N}_{a-N+1}$.*

*Proof.* By Taylor's Theorem (Theorem 3.48) with $n = N - 1$, we have that

$$f(t) = \sum_{k=0}^{N-1} H_k(t, a) \nabla^k f(a) + \int_a^t H_{N-1}(t, \rho(\tau)) \nabla^N f(\tau) \nabla \tau,$$

for $t \in \mathbb{N}_{a-N+1}$. Hence to complete the proof we just need to show that

$$\int_a^t H_{N-1}(t,\rho(\tau))\nabla^N f(\tau)\nabla\tau = \int_a^t H_{\mu-1}(t,\rho(\tau))\nabla_{a*}^{\mu} f(\tau)\nabla\tau, \tag{3.66}$$

holds for $t \in \mathbb{N}_{a-N+1}$. By convention both integrals in (3.66) are equal to zero for $t \in \mathbb{N}_{a-N+1}^{a}$. Hence it remains to prove that (3.66) holds for $t \in \mathbb{N}_a$. To see that this is true note that

$$\begin{aligned}
\int_a^t H_{\mu-1}(t,\rho(\tau))\nabla_{a*}^{\mu} f(\tau)\nabla\tau &= \nabla_a^{-\mu}\nabla_{a*}^{\mu} f(t) \\
&= \nabla_a^{-\mu}\nabla_a^{-(N-\mu)}\nabla^N f(t) \\
&= \nabla_a^{-\mu-N+\mu}\nabla^N f(t) \quad \text{by Theorem 3.107} \\
&= \nabla_a^{-N}\nabla^N f(t), \\
&= \int_a^t H_{N-1}(t,\rho(\tau))\nabla^N f(\tau)\nabla\tau
\end{aligned}$$

for $t \in \mathbb{N}_a$. □

## 3.20 Nabla Fractional Initial Value Problems

In this section we will consider the nabla fractional initial value problem (IVP)

$$\begin{cases} \nabla_{a*}^{\nu} x(t) = h(t), & t \in \mathbb{N}_{a+1} \\ \nabla^k x(a) = c_k, & 0 \le k \le N-1, \end{cases} \tag{3.67}$$

where we always assume that $a, \nu \in \mathbb{R}$, $\nu > 0$, $N := \lceil \nu \rceil$, $c_k \in \mathbb{R}$ for $0 \le k \le N-1$, and $h : \mathbb{N}_{a+1} \to \mathbb{R}$. In the next theorem we will see that this IVP has a unique solution which is defined on $\mathbb{N}_{a-N+1}$.

**Theorem 3.120.** *The unique solution to the IVP* (3.67) *is given by*

$$x(t) = \sum_{k=0}^{N-1} H_k(t,a)c_k + \nabla_a^{-\nu} h(t),$$

*for $t \in \mathbb{N}_{a-N+1}$, where by convention $\nabla_a^{-\nu} h(t) = 0$ for $a - N + 1 \le t \le a$.*

*Proof.* Define $f : \mathbb{N}_{a-N+1} \to \mathbb{R}$ by

$$\nabla^k f(a) = c_k,$$

for $0 \leq k \leq N-1$ (note that this uniquely defines $f(t)$ for $a-N+1 \leq t \leq a$), and for $t \in \mathbb{N}_{a+1}$ define $f(t)$ recursively by

$$\begin{aligned}\nabla^N f(t) &= h(t) - \int_a^{t-1} H_{N-\nu-1}(t,\rho(\tau))\nabla^N f(\tau)\nabla\tau \\ &= h(t) - \sum_{\tau=a+1}^{t-1} H_{N-\nu-1}(t,\rho(\tau))\nabla^N f(\tau).\end{aligned}$$

So for any $t \in \mathbb{N}_{a+1}$,

$$\nabla^N f(t) + \int_a^{t-1} H_{N-\nu-1}(t,\rho(\tau))\nabla^N f(\tau)\nabla\tau = h(t). \tag{3.68}$$

It follows that

$$\begin{aligned}\nabla_{a*}^{\nu} f(t) &= \nabla_a^{-(N-\nu)}\nabla^N f(t) \\ &= \int_a^t H_{N-\nu-1}(t,\rho(\tau))\nabla^N f(\tau)\nabla\tau \\ &= \int_a^{t-1} H_{N-\nu-1}(t,\rho(\tau))\nabla^N f(\tau)\nabla\tau + \nabla^N f(t) \\ &= h(t) \quad \text{by (3.68)}\end{aligned}$$

for $t \in \mathbb{N}_{a+1}$. Therefore, $f(t)$ solves the IVP (3.67). Conversely, if we suppose that there is a function $f : \mathbb{N}_{a-N+1} \to \mathbb{R}$ that satisfies the IVP, reversing the above steps would lead to the same recursive definition. Therefore the solution to the IVP is uniquely defined. By the Caputo Discrete Taylor's Theorem, $x(t) = f(t)$ is given by

$$\begin{aligned}x(t) &= \sum_{k=0}^{N-1} H_k(t,a)\nabla^k x(a) + \nabla_a^{-\nu}\nabla_{a*}^{\nu} x(t) \\ &= \sum_{k=0}^{N-1} c_k H_k(t,a) + \nabla_a^{-\nu} h(t).\end{aligned}$$

□

The following example appears in [4].

*Example 3.121.* Solve the IVP

$$\begin{cases}\nabla_{0*}^{0.7} x(t) = t, & t \in \mathbb{N}_1 \\ x(0) = 2.\end{cases}$$

Applying the variation of constants formula in Theorem 3.120 we get

$$x(t) = \sum_{k=0}^{0} \frac{t^{\overline{k}}}{k!} 2 + \nabla_0^{-0.7} h(t)$$

$$= 2 + \int_0^t H_{-.3}(t, \rho(s)) s \nabla s$$

for $t \in \mathbb{N}_0$. Integrating by parts, we have

$$x(t) = 2 - sH_{.7}(t,s)\Big|_{s=0}^{s=t} + \int_0^t H_{.7}(t, \rho(s)) \nabla s$$

$$= 2 - H_{1.7}(t,s)\Big|_{s=0}^{s=t}$$

$$= 2 + \frac{1}{\Gamma(2.7)} t^{\overline{1.7}}$$

for $t \in \mathbb{N}_0$.

**Corollary 3.122.** *For $\nu > 0$, $N = \lceil \nu \rceil$, and $h : \mathbb{N}_{a+1} \to \mathbb{R}$, we have that*

$$\nabla_a^{-(N-\nu)} \nabla_a^{N-\nu} h(t) = h(t), \quad t \in \mathbb{N}_{a+1}.$$

*Proof.* Assume $N \neq \nu$; otherwise, the proof is trivial. Let $\nu > 0$, $N = \lceil \nu \rceil$, and $h : \mathbb{N}_{a+1} \to \mathbb{R}$. Let $c_k \in \mathbb{R}$ for $0 \leq k \leq N-1$, and define $f : \mathbb{N}_{a-N+1} \to \mathbb{R}$ in terms of $h$ by

$$f(t) := \sum_{k=0}^{N-1} \frac{(t-a)^{\overline{k}}}{k!} c_k + \nabla_a^{-\nu} h(t).$$

Then by Theorem 3.120, $f(t)$ solves the IVP

$$\begin{cases} \nabla_{a*}^{\nu} f(t) = h(t), & t \in \mathbb{N}_{a+1} \\ \nabla^k f(a) = c_k, & 0 \leq k \leq N-1. \end{cases}$$

With repeated applications of the Leibniz rule (3.22) we get

$$\nabla^N f(t) = \nabla^N \left[ \sum_{k=0}^{N-1} H_k(t,a) c_k + \nabla_a^{-\nu} h(t) \right]$$

$$= \nabla^N \nabla_a^{-\nu} h(t)$$

$$
\begin{aligned}
&= \nabla^N\left[\int_a^t H_{\nu-1}(t,\rho(\tau))h(\tau)\nabla\tau\right]\\
&= \nabla^{N-1}\left[\int_a^t H_{\nu-2}(t,\rho(\tau))h(\tau)\nabla\tau + H_{\nu-1}(\rho(t),\rho(t))h(t)\right]\\
&= \nabla^{N-1}\left[\int_a^t H_{\nu-2}(t,\rho(\tau))h(\tau)\nabla\tau\right]\\
&= \nabla^{N-2}\left[\int_a^t H_{\nu-3}(t,\rho(\tau))h(\tau)\nabla\tau + H_{\nu-2}(\rho(t),\rho(t))h(t)\right]\\
&= \nabla^{N-2}\left[\int_a^t H_{\nu-3}(t,\rho(\tau))h(\tau)\nabla\tau\right]\\
&= \cdots\\
&= \nabla\left[\int_a^t H_{\nu-N}(t,\rho(\tau)h(\tau)\nabla\tau\right]\\
&= \int_a^t H_{\nu-N-1}(t,\rho(\tau))h(\tau)\nabla\tau + H_{\nu-N}(\rho(t),\rho(t))h(t)\\
&= \int_a^t H_{\nu-N-1}(t,\rho(\tau))h(\tau)\nabla\tau\\
&= \int_a^t H_{-(N-\nu)-1}(t,\rho(\tau)h(\tau)\nabla\tau\\
&= \nabla_a^{N-\nu}h(t).
\end{aligned}
$$

It follows that

$$\nabla_a^{-(N-\nu)}\nabla_a^{N-\nu}h(t) = \nabla_a^{-(N-\nu)}\nabla^N f(t) = \nabla_{a*}^{\nu}f(t) = h(t),$$

and the proof is complete. □

## 3.21 Monotonicity (Caputo Case)

Many of the results in this section appear in Baoguo et al. [49]. This work is motivated by the paper by R. Dahal and C. Goodrich [67], where they obtained some interesting monotonicity results for the delta fractional difference operator. These monotonicity results for the delta case will be discussed in Sect. 7.2. In this section, we prove the following corresponding results for Caputo fractional differences.

**Theorem 3.123.** *Assume that* $N-1<\nu<N$, $f:\mathbb{N}_{a-N+1}\to\mathbb{R}$, $\nabla_{a*}^{\nu}f(t)\geq 0$ *for* $t\in\mathbb{N}_{a+1}$ *and* $\nabla^{N-1}f(a)\geq 0$. *Then* $\nabla^{N-1}f(t)\geq 0$ *for* $t\in\mathbb{N}_a$.

**Theorem 3.124.** *Assume $N-1<\nu<N$, $f:\mathbb{N}_{a-N+1}\to\mathbb{R}$, and $\nabla^N f(t)\geq 0$ for $t\in\mathbb{N}_{a+1}$. Then $\nabla^{\nu}_{a*}f(t)\geq 0$, for each $t\in\mathbb{N}_{a+1}$.*

When $N=2$ in Theorem 3.123 we get the important monotonicity result.

**Theorem 3.125.** *Assume that $1<\nu<2$, $f:\mathbb{N}_{\rho(a)}\to\mathbb{R}$, $\nabla^{\nu}_{a*}f(t)\geq 0$ for $t\in\mathbb{N}_{a+1}$ and $f(a)\geq f(\rho(a))$. Then $f(t)$ is an increasing function for $t\in\mathbb{N}_{\rho(a)}$.*

Also the following partial converse of Theorem 3.123 is true.

**Theorem 3.126.** *Assume $0<\nu<1$, $f:\mathbb{N}_{\rho(a)}\to\mathbb{R}$ and $f$ is an increasing function for $t\in\mathbb{N}_a$. Then $\nabla^{\nu}_{a*}f(t)\geq 0$, for each $t\in\mathbb{N}_{a+1}$.*

We also give a counterexample to show that the above assumption $f(a)\geq f(\rho(a))$ in 3.125 is essential. We begin by proving the following theorem.

**Theorem 3.127.** *Assume that $f:\mathbb{N}_{a-N+1}\to\mathbb{R}$, and $\nabla^{\mu}_{a*}f(t)\geq 0$, for each $t\in\mathbb{N}_{a+1}$, with $N-1<\mu<N$. Then*

$$\begin{aligned}&\nabla^{N-1}f(a+k)\\&\quad\geq\sum_{i=1}^{k-1}\left[\frac{(k-i+1)^{\overline{N-\mu-2}}}{\Gamma(N-\mu-1)}\right]\nabla^{N-1}f(a+i-1)\\&\quad+H_{N-\mu-1}(a+k,a)\nabla^{N-1}f(a),\end{aligned}\tag{3.69}$$

*for $k\in\mathbb{N}_1$ (note by our convention on sums the first term on the right-hand side is zero when $k=1$).*

*Proof.* If $t=a+1$, we have that

$$\begin{aligned}0\leq\nabla^{\mu}_{a*}f(a+1)&=\nabla_a^{-(N-\mu)}\nabla^N f(t)\\&=\int_a^{a+1}H_{N-\mu-1}(a+1,\rho(s))\nabla^N f(s)\nabla s\\&=H_{N-\mu-1}(a+1,a)\nabla^N f(a+1)\\&=\nabla^N f(a+1)=\nabla^{N-1}f(a+1)-\nabla^{N-1}f(a),\end{aligned}$$

where we used $H_{N-\mu-1}(a+1,a)=1$. Solving for $\nabla^{N-1}f(a+1)$ we get the inequality

$$\nabla^{N-1}f(a+1)\geq\nabla^{N-1}f(a),$$

which gives us the inequality (3.69) for $t=a+1$. Hence, the inequality (3.69) holds for $t=a+1$.

Next consider the case $t = a+k$ for $k \geq 2$. Taking $t = a+k$, $k \geq 2$ we have that

$$\begin{aligned}
0 &\leq \nabla_{a*}^{\mu} f(t) \\
&= \nabla_a^{-(N-\mu)} \nabla^N f(t) \\
&= \int_a^t H_{N-\mu-1}(t, \rho(s)) \nabla^N f(s) \nabla s \\
&= \int_a^{a+k} H_{N-\mu-1}(a+k, \rho(s)) \nabla^N f(s) \nabla s \\
&= \sum_{i=1}^{k} H_{N-\mu-1}(a+k, a+i-1) \nabla^N f(a+i) \\
&= \sum_{i=1}^{k} H_{N-\mu-1}(a+k, a+i-1) \left[\nabla^{N-1} f(a+i) - \nabla^{N-1} f(a+i-1)\right] \\
&= \sum_{i=1}^{k} H_{N-\mu-1}(a+k, a+i-1) \nabla^{N-1} f(a+i) \\
&\quad - \sum_{i=1}^{k} H_{N-\mu-1}(a+k, a+i-1) \nabla^{N-1} f(a+i-1) \\
&= \nabla^{N-1} f(a+k) + \sum_{i=1}^{k-1} H_{N-\mu-1}(a+k, a+i-1) \nabla^{N-1} f(a+i) \\
&\quad - H_{N-\mu-1}(a+k, a) \nabla^{N-1} f(a) \\
&\quad - \sum_{i=2}^{k} H_{N-\mu-1}(a+k, a+i-1) \nabla^{N-1} f(a+i-1),
\end{aligned}$$

where we used $H_{N-\mu-1}(a+k, a+k-1) = 1$. It follows that

$$\begin{aligned}
0 &\leq \nabla^{N-1} f(a+k) + \sum_{i=1}^{k-1} H_{N-\mu-1}(a+k, a+i-1) \nabla^{N-1} f(a+i) \\
&\quad - H_{N-\mu-1}(a+k, a) \nabla^{N-1} f(a) - \sum_{i=1}^{k-1} H_{N-\mu-1}(a+k, a+i) \nabla^{N-1} f(a+i) \\
&= \nabla^{N-1} f(a+k) - H_{N-\mu-1}(a+k, a) \nabla^{N-1} f(a+i) \\
&\quad - \sum_{i=1}^{k-1} \Big[ H_{N-\mu-1}(a+k, a+i) \\
&\quad - H_{N-\mu-1}(a+k, a+i-1) \Big] \nabla^{N-1} f(a+i).
\end{aligned}$$

Hence,

$$\begin{aligned}
0 \leq &\nabla^{N-1}f(a+k) - H_{N-\mu-1}(a+k,a)\nabla^{N-1}f(a) \\
&- \sum_{i=1}^{k-1} \nabla_s H_{N-\mu-1}(a+k,s)|_{s=a+i}\nabla^{N-1}f(a+i) \\
=&\nabla^{N-1}f(a+k) - H_{N-\mu-1}(a+k,a)\nabla^{N-1}f(a) \\
&+ \sum_{i=1}^{k-1} H_{N-\mu-2}(a+k,a+i-1)\nabla^{N-1}f(a+i) \\
=&\nabla^{N-1}f(a+k) - H_{N-\mu-1}(a+k,a)\nabla^{N-1}f(a) \\
&+ \sum_{i=1}^{k-1} \Big[\frac{(k-i+1)^{\overline{N-\mu-2}}}{\Gamma(N-\mu-1)}\Big]\nabla^{N-1}f(a+i).
\end{aligned}$$

Solving the above inequality for $\nabla^{N-1}f(a+k)$, we obtain the desired inequality (3.69).

Next we consider for $1 \leq i \leq k-1$

$$\begin{aligned}
\frac{(k-i+1)^{\overline{N-\mu-2}}}{\Gamma(N-\mu-1)} &= \frac{\Gamma(N-\mu+k-i-1)}{\Gamma(k-i+1)\Gamma(N-\mu-1)} \\
&= \frac{(N-\mu+k-i-2)\cdots(N-\mu-1)}{(k-i)!} < 0
\end{aligned}$$

since $N < \mu + 1$. Also

$$\begin{aligned}
H_{N-\mu-1}(a+k,a) &= \frac{k^{\overline{N-\mu-1}}}{\Gamma(N-\mu)} \\
&= \frac{\Gamma(N-\mu+k-1)}{\Gamma(k)\Gamma(N-\mu)} \\
&= \frac{(N-\mu+k-2)\cdots(N-\mu)}{(k-1)!} > 0.
\end{aligned}$$

And this completes the proof. □

From Theorem 3.127 we have the following

**Theorem 3.128.** *Assume that* $N-1 < \nu < N$, $f : \mathbb{N}_{a-N+1} \to \mathbb{R}$, $\nabla^{\nu}_{a*}f(t) \geq 0$ *for* $t \in \mathbb{N}_{a+1}$ *and* $\nabla^{N-1}f(a) \geq 0$. *Then* $\nabla^{N-1}f(t) \geq 0$ *for* $t \in \mathbb{N}_a$.

*Proof.* By using the principle of strong induction, we prove that the conclusion of the theorem is correct. By assumption, the result holds for $t = a$. Suppose that $\nabla^{N-1}f(t) \geq 0$, for $t = a, a+1, \ldots, a+k-1$. From Theorem 3.127 and (3.69), we have $\nabla^{N-1}f(a+k) \geq 0$, and the proof is complete. □

Taking $N = 2$ and $N = 3$, we can get the following corollaries.

**Corollary 3.129.** *Assume that* $1 < \nu < 2, f : N_{\rho(a)} \to \mathbb{R}$, $\nabla^{\nu}_{a*}f(t) \geq 0$ *for* $t \in \mathbb{N}_{a+1}$ *and* $f(a) \geq f(\rho(a))$. *Then* $f(t)$ *is increasing for* $t \in \mathbb{N}_{\rho(a)}$.

**Corollary 3.130.** *Assume that* $2 < \nu < 3, f : N_{a-2} \to \mathbb{R}$, $\nabla^{\nu}_{a*}f(t) \geq 0$ *for* $t \in \mathbb{N}_{a+1}$ *and* $\nabla^2 f(a) \geq 0$. *Then* $\nabla f(t)$ *is increasing for* $t \in \mathbb{N}_a$.

One should compare the next result with Theorem 3.128.

**Theorem 3.131.** *Assume that* $N-1 < \nu < N, f : \mathbb{N}_{a-N+1} \to \mathbb{R}$, *and* $\nabla^N f(t) \geq 0$ *for* $t \in \mathbb{N}_{a+1}$. *Then* $\nabla^{\nu}_{a*}f(t) \geq 0$, *for each* $t \in \mathbb{N}_{a+1}$.

*Proof.* Taking $t = a+k$, we have

$$\begin{aligned}
&\nabla^{-\mu}_{a*}f(t)\\
&= \nabla_a^{-(N-\mu)}\nabla^N f(t)\\
&= \int_a^t H_{N-\mu-1}(t,\rho(s))\nabla^N f(s)\nabla s\\
&= \sum_{i=1}^{k} H_{N-\mu-1}(a+k, a+i-1)\nabla^N f(a+i).
\end{aligned} \tag{3.70}$$

Since

$$\begin{aligned}
&H_{N-\mu-1}(a+k, a+i-1)\\
&= \frac{(k-i+1)^{\overline{N-\mu-1}}}{\Gamma(N-\mu)}\\
&= \frac{\Gamma(k+N-i-\mu)}{\Gamma(N-\mu)\Gamma(k-i+1)}\\
&= \frac{(-\mu+k+N-i)\cdots(N-\mu+1)(N-\mu)}{(k-i)!} > 0,
\end{aligned} \tag{3.71}$$

where we used $\mu < N$, from (3.70) and (3.71) we get that $\nabla^{\nu}_{a*}f(t) \geq 0$, for each $t \in \mathbb{N}_{a+1}$, □

Taking $N = 1$ and $N = 2$, we get the following corollaries.

**Corollary 3.132.** *Assume that* $0 < \nu < 1, f : \mathbb{N}_{\rho(a)} \to \mathbb{R}$ *and* $f$ *is an increasing function for* $t \in \mathbb{N}_a$. *Then* $\nabla^{\nu}_{a*}f(t) \geq 0$, *for* $t \in \mathbb{N}_{a+1}$.

**Corollary 3.133.** *Assume that* $1 < \nu < 2$, $f : \mathbb{N}_{\rho(a)} \to \mathbb{R}$ *and* $\nabla^2 f(t) \geq 0$ *for* $t \in \mathbb{N}_{a+1}$. *Then* $\nabla^{\nu}_{a*} f(t) \geq 0$, *for each* $t \in \mathbb{N}_{a+1}$.

In the following, we will give a counterexample to show that the assumption in Corollary 3.129 "$f(a) \geq f(\rho(a))$" is essential. To verify this example we will use the following simple lemma.

**Lemma 3.134.** *Assume* $f \in C^2([a, \infty))$ *and* $f''(t) \geq 0$ *on* $[a, \infty)$. *Then* $\nabla^{\nu}_{a*} f(t) \geq 0$, *for each* $t \in \mathbb{N}_{a+1}$, *with* $1 < \nu < 2$.

*Proof.* By Taylor's Theorem,

$$f(a+i+1) = f(a+i) + f'(a+i) + \frac{f''(\xi^i)}{2}, \quad \xi^i \in [a+i, a+i+1] \tag{3.72}$$

and

$$f(a+i-1) = f(a+i) - f'(a+i) + \frac{f''(\eta^i)}{2}, \quad \eta^i \in [a+i-1, a+i] \tag{3.73}$$

for $i = 0, 1, \ldots, k-1$. Using (3.72) and (3.73), we have

$$\begin{aligned} \nabla^2 f(a+i+1) &= f(a+i+1) - 2f(a+i) + f(a+i-1) \\ &= \frac{f''(\xi^i) + f''(\eta^i)}{2} \\ &\geq 0. \end{aligned} \tag{3.74}$$

From (3.74) and Corollary 3.133, we get that $\nabla^{\nu}_{a*} f(t) \geq 0$, for each $t \in \mathbb{N}_{a+1}$, with $1 < \nu < 2$. □

*Example 3.135.* Let $f(t) = -\sqrt{t}$, $a = 2$. We have $f''(t) \geq 0$, for $t \geq 1$. By Lemma 3.134, we have $\nabla^{\nu}_{a*} f(t) \geq 0$.

Note that $f(\rho(a)) = f(1) = -1 > f(a) = -\sqrt{2}$. Therefore $f(x)$ does not satisfy the assumptions of Corollary 3.129. In fact, $f(x)$ is decreasing, for $t \geq 1$.

Corollary 3.129 could be useful for solving nonlinear fractional equations as the following result shows.

**Corollary 3.136.** *Let* $h : \mathbb{N}_{a+1} \times \mathbb{R} \to \mathbb{R}$ *be a nonnegative, continuous function. Then any solution of the Caputo nabla fractional difference equation*

$$\nabla^{\nu}_{a*} y(t) = h(t, y(t)), \quad t \in \mathbb{N}_{a+1}, \quad 1 < \nu < 2 \tag{3.75}$$

*satisfying* $\nabla y(a) = A \geq 0$ *is increasing on* $\mathbb{N}_{\rho(a)}$.

## 3.22 Asymptotic Behavior and Comparison Theorems

In this section we will determine the asymptotic behavior of solutions of a nabla Caputo fractional equation of the form

$$\nabla_{a*}^{\nu}x(t) = c(t)x(t), \quad t \in \mathbb{N}_{a+1}, \tag{3.76}$$

where $c : \mathbb{N}_{a+1} \to \mathbb{R}$, $0 < \nu < 1$. We will prove important comparison theorems to help us prove our asymptotic results. Most of the results in this section appear in Baoguo et al. [52]. The following lemma will be useful.

**Lemma 3.137.** *Assume that* $c(t) < 1$, $0 < \nu < 1$. *Then any solution of*

$$\nabla_{a*}^{\nu}x(t) = c(t)x(t), \qquad t \in \mathbb{N}_{a+1} \tag{3.77}$$

*satisfying* $x(a) > 0$ *is positive on* $\mathbb{N}_a$.

*Proof.* Using the integrating by parts formula (3.23) and

$$\nabla_s H_{-\nu}(t,s) = -H_{-\nu-1}(t,\rho(s)),$$

we have

$$\begin{aligned}
\nabla_{a*}^{\nu}x(t) &= \nabla_a^{-(1-\nu)}\nabla x(t) \\
&= \int_a^t H_{-\nu}(t,\rho(s))\nabla x(s)\nabla s \\
&= H_{-\nu}(t,s)x(s)|_{s=a}^{t} + \int_a^t H_{-\nu-1}(t,\rho(s))x(s)\nabla s \\
&= -H_{-\nu}(t,a)x(a) + \sum_{s=a+1}^{t} H_{-\nu-1}(t,\rho(s))x(s).
\end{aligned}$$

Taking $t = a + k$, we have

$$\begin{aligned}
\nabla_{a*}^{\nu}x(t) =& \nabla_{a*}^{\nu}x(a+k) \\
=& x(a+k) - \nu x(a+k-1) - \frac{\nu(-\nu+1)}{2!}x(a+k-2) - \cdots \\
&- \frac{\nu(-\nu+1)\cdots(-\nu+k-2)}{(k-1)!}x(a+1) \\
&- \frac{(-\nu+1)\cdots(-\nu+k-1)}{(k-1)!}x(a).
\end{aligned}$$

Using (3.77), we get

$$\begin{aligned}
x(a+k) & \\
&= \frac{1}{1-c(a+k)}\Big[\nu x(a+k-1) + \frac{\nu(-\nu+1)}{2!}x(a+k-2) + \cdots \\
&\quad + \frac{\nu(-\nu+1)\cdots(-\nu+k-2)}{(k-1)!}x(a+1) \\
&\quad + \frac{(-\nu+1)\cdots(-\nu+k-1)}{(k-1)!}x(a)\Big].
\end{aligned}$$

From the strong induction principle, $0 < \nu < 1$, and $x(a) > 0$, it is easy to prove that $x(a+k) > 0$, for $k \in \mathbb{N}_0$. □

The following comparison theorem plays an important role in proving our main results.

**Theorem 3.138.** *Assume $c_2(t) \leq c_1(t) < 1$, $0 < \nu < 1$, and $x(t), y(t)$ are solutions of the equations*

$$\nabla_{a*}^{\nu} x(t) = c_1(t)x(t), \tag{3.78}$$

*and*

$$\nabla_a^{\nu} y(t) = c_2(t)y(t), \tag{3.79}$$

*respectively, for $t \in \mathbb{N}_{a+1}$ satisfying $x(a) \geq y(a) > 0$. Then*

$$x(t) \geq y(t),$$

*for $t \in \mathbb{N}_a$.*

*Proof.* Similar to the proof of Lemma 3.137, taking $t = a + k$, we have

$$\begin{aligned}
x(a+k) & \\
&= \frac{1}{1-c_1(a+k)}\Big[\nu x(a+k-1) + \frac{\nu(-\nu+1)}{2!}x(a+k-2) + \cdots \\
&\quad + \frac{\nu(-\nu+1)\cdots(-\nu+k-2)}{(k-1)!}x(a+1) + \frac{(-\nu+1)\cdots(-\nu+k-1)}{(k-1)!}x(a)\Big]
\end{aligned} \tag{3.80}$$

and

$$\begin{aligned}
y(a+k) & \\
&= \frac{1}{1-c_2(a+k)}\Big[\nu y(a+k-1) + \frac{\nu(-\nu+1)}{2!}y(a+k-2) + \cdots \\
&\quad + \frac{\nu(-\nu+1)\cdots(-\nu+k-2)}{(k-1)!}y(a+1) + \frac{(-\nu+1)\cdots(-\nu+k-1)}{(k-1)!}y(a)\Big].
\end{aligned} \tag{3.81}$$

We will prove $x(a+k) \geq y(a+k) > 0$ for $k \in \mathbb{N}_0$ by using the principle of strong induction. By assumption $x(a) \geq y(a) > 0$ so the base case holds. Now assume that $x(a+i) \geq y(a+i) > 0$, for $i = 0, 1, \ldots, k-1$. Using $c_2(t) \leq c_1(t) < 1$,

$$\frac{\nu(-\nu+1)\cdots(-\nu+i-1)}{i!} > 0,$$

the base case $k = 1$ for $i = 2, 3, \cdots k-1$,

$$\frac{(-\nu+1)(-\nu+2)\cdots(-\nu+k-1)}{(k-1)!} > 0,$$

(3.80), and (3.81) we have

$$x(a+k) \geq y(a+k) > 0.$$

This completes the proof. □

*Remark 3.139.* Since $H_0(t,a) = 1$, we have that $E_{0,\nu,0}(t,a) = 1$ and $E_{p,\nu,0}(a,a) = 1$.

**Lemma 3.140.** *Assume that* $0 < \nu < 1$, $|b| < 1$. *Then*

$$\nabla_{a*}^{\nu} E_{b,\nu,0}(t,a) = bE_{b,\nu,0}(t,a)$$

*for* $t \in \mathbb{N}_{a+1}$.

*Proof.* Integrating by parts, we have

$$\begin{aligned}
&\nabla_{a*}^{\nu} E_{b,\nu,0}(t,a) \\
&= \int_a^t H_{-\nu}(t,\rho(s))\nabla E_{b,\nu,0}(s,a)\nabla s \\
&= [H_{-\nu}(t,s)E_{b,\nu,0}(s,a)]_{s=a}^{t} + \int_a^t H_{-\nu-1}(t,\rho(s))E_{b,\nu,0}(s,a)\nabla s \\
&= -H_{-\nu}(t,a) + \int_a^t H_{-\nu-1}(t,\rho(s))\sum_{k=0}^{\infty} b^k H_{\nu k}(s,a)\nabla s, \qquad (3.82)
\end{aligned}$$

where we used $H_{-\nu}(t,t) = 0$ and $E_{b,\nu,0}(a,a) = 1$. In the following, we first prove that the infinite series

$$H_{-\nu-1}(t,\rho(s))\sum_{k=0}^{\infty} b^k H_{\nu k}(s,a) \qquad (3.83)$$

for each fixed $t$ is uniformly convergent for $s \in [a,t]$.

We will first show that

$$|H_{-\nu-1}(t,\rho(s))| = \left|\frac{\Gamma(-\nu+t-s)}{\Gamma(t-s+1)\Gamma(-\nu)}\right| \leq 1$$

for $a \leq s \leq t$. For $s = t$ we have that

$$|H_{-\nu-1}(t,\rho(t))| = 1.$$

Now assume that $a \leq s < t$. Then

$$\begin{aligned}\left|\frac{\Gamma(-\nu+t-s)}{\Gamma(t-s+1)\Gamma(-\nu)}\right| &= \left|\frac{(t-s-\nu-1)(t-s-\nu-2)\cdots(-\nu)}{(t-s)!}\right| \\ &= \left|\frac{t-s-(\nu+1)}{t-s}\right|\left|\frac{t-s-1-(\nu+1)}{t-s-1}\right|\cdots\left|\frac{-\nu}{1}\right| \\ &\leq 1.\end{aligned}$$

Also consider

$$\begin{aligned}H_{\nu k}(s,a) &= \frac{\Gamma(\nu k+s-a)}{\Gamma(s-a)\Gamma(\nu k+1)} \\ &= \frac{(\nu k+s-a-1)\cdots(\nu k+1)}{(s-a-1)!}.\end{aligned}$$

Note that for large $k$ it follows that

$$\begin{aligned}H_{\nu k}(s,a) &\leq (\nu k+s-a-1)^{s-a-1} \\ &\leq (\nu k+t-a-1)^{t-a-1}\end{aligned}$$

for $a \leq s \leq t$. Applying the Root Test to the infinite series in (3.83) we get that for each fixed $t$

$$\lim_{k\to\infty} \sqrt[k]{|b|^k(\nu k+t-a-1)^{t-a-1}} = |b| < 1.$$

Hence, for each fixed $t$ the infinite series in (3.83) is uniformly convergent for $s \in [a,t]$. So from (3.82), integrating term by term, we obtain, using (3.32) and $\nabla_a^\nu H_{\nu k}(s,a)) = H_{\nu k-\nu}(s,a)$, that

$$\nabla_{a*}^\nu E_{b,\nu,0}(t,a) = -H_{-\nu}(t,a) + \sum_{k=0}^\infty b^k \int_a^t H_{-\nu-1}(t,\rho(s))H_{\nu k}(s,a)\nabla s$$

$$= -H_{-\nu}(t,a) + \sum_{k=0}^{\infty} b^k \nabla_a^\nu H_{\nu k}(t,a)$$

$$= -H_{-\nu}(t,a) + \sum_{k=0}^{\infty} b^k H_{\nu k-\nu}(t,a)$$

$$= \sum_{k=1}^{\infty} b^k H_{\nu k-\nu}(t,a)$$

$$= bE_{b,\nu,0}(t,a),$$

where we also used $H_0(t,a) = 1$. This completes the proof. □

With the aid of Lemma 3.140, we now give a rigorous proof of the following result.

**Lemma 3.141.** *Assume that* $0 < \nu < 1$, $|b| < 1$. *Then* $E_{b,\nu,0}(t,a)$ *is the unique solution of Caputo nabla fractional IVP*

$$\nabla_{a*}^\nu x(t) = bx(t), \qquad t \in \mathbb{N}_{a+1} \tag{3.84}$$
$$x(a) = 1.$$

*Proof.* It is easy to see that the given IVP has a unique solution. If $b = 0$, then

$$E_{0,\nu,0}(t,a) = 1$$

is the solution of the given IVP. For $b \neq 0$ the result follows from Lemma 3.140 and the uniqueness. □

We will see that the following lemma, given in Pudlubny [153], is useful in proving asymptotic properties of certain fractional Taylor monomials and certain nabla Mittag–Leffler functions.

**Lemma 3.142.** *Assume* $\Re(z) > 0$. *Then*

$$\Gamma(z) = \lim_{n\to\infty} \frac{n!n^z}{z(z+1)\cdots(z+n)}.$$

The following lemma is an asymptotic property for certain nabla fractional Taylor monomials.

**Lemma 3.143.** *Assume that* $0 < \nu < 1$. *Then we have*

$$\lim_{t\to\infty} H_{\nu k}(t,a) = \infty, \quad \text{for} \quad k \geq 1,$$

$$\lim_{t\to\infty} H_{\nu k}(t,a) = 1, \quad \text{for} \quad k = 0.$$

*Proof.* Taking $t = a + n$, $n \geq 0$, we have

$$\lim_{t\to\infty} H_{\nu k}(t,a) = \lim_{n\to\infty} H_{\nu k}(a+n,a) = \lim_{n\to\infty} \frac{n^{\overline{\nu k}}}{\Gamma(\nu k+1)}$$
$$= \lim_{n\to\infty} \frac{\Gamma(\nu k+n)}{\Gamma(n)\Gamma(\nu k+1)}$$
$$= \lim_{n\to\infty} \frac{(\nu k+n-1)(\nu k+n-2)\cdots(\nu k+1)}{(n-2)!(n-2)^{\nu k+1}} \cdot \frac{(n-2)^{\nu k+1}}{n-1}. \tag{3.85}$$

Using Lemma 3.142 with $z = \nu k + 1$ and $n$ replaced by $n - 2$, we have

$$\lim_{n\to\infty} \frac{(\nu k+1+n-2)(\nu k+1+n-3)\cdots(\nu k+1)}{(n-2)!(n-2)^{\nu k+1}} = \frac{1}{\Gamma(\nu k+1)},$$

and

$$\lim_{n\to\infty} \frac{(n-2)^{\nu k+1}}{n-1} = \infty, \quad \text{for} \quad k \geq 1,$$

$$\lim_{n\to\infty} \frac{(n-2)^{\nu k+1}}{n-1} = 1, \quad \text{for} \quad k = 0.$$

Using (3.85), we get the desired results. □

**Theorem 3.144.** *Assume* $0 < b_2 \leq c(t) < 1$, $t \in \mathbb{N}_{a+1}$, $0 < \nu < 1$. *Further assume* $x(t)$ *is a solution of Caputo nabla fractional difference equation*

$$\nabla_{a*}^{\nu} x(t) = c(t)x(t), \quad t \in \mathbb{N}_{a+1} \tag{3.86}$$

*satisfying* $x(a) > 0$. *Then*

$$x(t) \geq \frac{x(a)}{2} E_{b_2,\nu,0}(t,a),$$

*for* $t \in \mathbb{N}_{a+1}$.

*Proof.* From Lemma 3.141, we have

$$\nabla_{a*}^{\nu} E_{b_2,\nu,0}(t,a) = b_2 E_{b_2,\nu,0}(t,a)$$

and $E_{b_2,\nu,0}(a,a) = 1$.

In Theorem 3.138, take $c_2(t) = b_2$ . Then $x(t)$ and

$$y(t) = \frac{x(a)}{2} E_{b_2,\nu,0}(t,a)$$

satisfy

$$\nabla_{a*}^{\nu} x(t) = c(t)x(t), \tag{3.87}$$

and

$$\nabla_{a*}^{\nu} y(t) = b_2 y(t), \tag{3.88}$$

respectively, for $t \in \mathbb{N}_{a+1}$ and

$$x(a) > \frac{x(a)}{2} E_{b_2,\nu,0}(a, a) = y(a).$$

From Theorem 3.138, we get that

$$x(t) \geq \frac{x(a)}{2} E_{b_2,\nu,0}(t, a),$$

for $t \in \mathbb{N}_a$. This completes the proof. □

From Lemma 3.143 and the definition of $E_{b_2,\nu,0}(t, a)$, we get the following theorem.

**Theorem 3.145.** *For* $0 < b_2 < 1$, *we have*

$$\lim_{t\to\infty} E_{b_2,\nu,0}(t, a) = +\infty.$$

From Theorem 3.144 and Theorem 3.145, we have the following result holds.

**Theorem 3.146.** *Assume* $0 < \nu < 1$ *and there exists a constant* $b_2$ *such that* $0 < b_2 \leq c(t) < 1$. *Then the solutions of the equation* (3.76) *with* $x(a) > 0$ *satisfy*

$$\lim_{t\to\infty} x(t) = +\infty.$$

Next we consider the case $c(t) \leq b_1 < 0$, $t \in \mathbb{N}_a$. First we prove some preliminary results.

**Lemma 3.147.** *Assume* $f : \mathbb{N}_a \to \mathbb{R}$, $0 < \nu < 1$. *Then*

$$\nabla_a^{-(1-\nu)} \nabla f(t) = \nabla \nabla_a^{-(1-\nu)} f(t) - f(a) H_{-\nu}(t, a). \tag{3.89}$$

*Proof.* Using integration by parts and $H_{-\nu}(t, t) = 0$, we have

$$\begin{aligned}
\nabla_a^{-(1-\nu)} \nabla f(t) &= \int_a^t H_{-\nu}(t, \rho(s)) \nabla f(s) \nabla s \\
&= H_{-\nu}(t, s) f(s)\big|_{s=a}^{t} + \int_a^t H_{-\nu-1}(t, \rho(s)) f(s) \nabla s \\
&= -H_{-\nu}(t, a) f(a) + \int_a^t H_{-\nu-1}(t, \rho(s)) f(s) \nabla s.
\end{aligned} \tag{3.90}$$

Using the composition rule $\nabla_a^\nu \nabla_a^{-\mu} f(t) = \nabla_a^{\nu-\mu} f(t)$, for $\nu, \mu > 0$ in Theorem 3.109, we have

$$\begin{aligned}\nabla\nabla_a^{-(1-\nu)} f(t) &= \nabla_a^\nu f(t)\\ &= \int_a^t H_{-\nu-1}(t,\rho(s)) f(s)\nabla s. \qquad (3.91)\end{aligned}$$

From (3.90) and (3.91), we get that (3.89) holds. □

From Lemma 3.147, it is easy to get the following corollary which will be useful later.

**Corollary 3.148.** *For $0 < \nu < 1$, the following equality holds:*

$$\nabla_a^{-\nu}\nabla f(t) = \nabla\nabla_a^{-\nu} f(t) - H_{\nu-1}(t,a) f(a). \qquad (3.92)$$

*for $t \in \mathbb{N}_a$.*

**Lemma 3.149.** *Assume that $0 < \nu < 1$ and $x(t)$ is a solution of the fractional equation*

$$\nabla_{a*}^\nu x(t) = c(t)x(t), \qquad t \in \mathbb{N}_{a+1} \qquad (3.93)$$

*satisfying $x(a) > 0$. Then $x(t)$ satisfies the integral equation*

$$\begin{aligned}x(t) &= \int_a^t H_{\nu-1}(t,\rho(s)) c(s)x(s)\nabla s + x(a)\\ &= \sum_{s=a+1}^{t} \frac{(t-s+1)^{\overline{\nu-1}}}{\Gamma(\nu)} c(s)x(s) + x(a).\end{aligned}$$

*Proof.* Using Lemma 3.147 and the composition rule

$$\nabla_a^\alpha \nabla_a^{-\beta} f(t) = \nabla_a^{\alpha-\beta} f(t),$$

for $\alpha, \beta > 0$ given in Theorem 3.109, we get

$$\begin{aligned}\nabla_{a*}^\nu x(t) &= \nabla_a^{-(1-\nu)}\nabla x(t)\\ &= \nabla\nabla_a^{-(1-\nu)} x(t) - x(a)H_{-\nu}(t,a)\\ &= \nabla_a^\nu x(t) - x(a)H_{-\nu}(t,a).\end{aligned}$$

From (3.93), we have

$$\nabla_a^\nu x(t) = c(t)x(t) + x(a)H_{-\nu}(t,a).$$

Applying the operator $\nabla_a^{-\nu}$ to each side we obtain

$$\nabla_a^{-\nu}\nabla_a^{\nu}x(t) = \nabla_a^{-\nu}c(t)x(t) + x(a)\nabla_a^{-\nu}H_{-\nu}(t,a),$$

which can be written in the form

$$\nabla_a^{-\nu}\nabla\nabla_a^{-(1-\nu)}x(t) = \nabla_a^{-\nu}c(t)x(t) + x(a)\nabla_a^{-\nu}H_{-\nu}(t,a).$$

Using Corollary 3.148, we obtain

$$\nabla\nabla_a^{-\nu}\nabla_a^{-(1-\nu)}x(t) - \frac{(t-a)^{\overline{\nu-1}}}{\Gamma(\nu)}\nabla_a^{-(1-\nu)}x(t)\Big|_{t=a}$$
$$= \nabla_a^{-\nu}c(t)x(t) + x(a)\nabla_a^{-\nu}H_{-\nu}(t,a).$$

On the other hand, using

$$\nabla_a^{-(1-\nu)}x(t)\Big|_{t=a} = \int_a^a H_{-\nu}(a,\rho(s))x(s)\nabla s = 0,$$

we obtain

$$\nabla\nabla_a^{-\nu}\nabla_a^{-(1-\nu)}x(t) = \nabla_a^{-\nu}c(t)x(t) + x(a)\nabla_a^{-\nu}H_{-\nu}(t,a).$$

By the composition rule, namely Theorem 3.107, it follows both that $\nabla_a^{-\nu}\nabla_a^{-(1-\nu)}x(t) = \nabla_a^{-1}x(t)$ and that $\nabla\nabla_a^{-1}x(t) = x(t)$, from which it follows that

$$x(t) = \nabla_a^{-\nu}c(t)x(t) + x(a)\nabla_a^{-\nu}H_{-\nu}(t,a).$$

Finally, by the power rule $\nabla_a^{-\nu}H_{-\nu}(t,a) = H_0(t,a) = 1$, we obtain

$$\begin{aligned} x(t) &= \nabla_a^{-\nu}c(t)x(t) + x(a) \\ &= \int_a^t H_{\nu-1}(t,\rho(s))c(s)x(s)\nabla s + x(a) \\ &= \sum_{s=a+1}^{t} H_{\nu-1}(t,\rho(s))c(s)x(s) + x(a) \\ &= \sum_{s=a+1}^{t} \frac{(t-s+1)^{\overline{\nu-1}}}{\Gamma(\nu)}c(s)x(s) + x(a). \end{aligned} \tag{3.94}$$

And this completes the proof. □

The following lemma appears in [34].

**Lemma 3.150.** *Assume that* $0 < \nu < 1$, $|b| < 1$. *Then the Mittag–Leffler function* $E_{b,\nu,\nu-1}(t, \rho(a)) = \sum_{k=0}^{\infty} b^k H_{\nu k+\nu-1}(t, \rho(a))$ *is the unique solution of the IVP*

$$\nabla^{\nu}_{\rho(a)} x(t) = bx(t), \qquad t \in \mathbb{N}_{a+1}$$

$$x(a) = \frac{1}{1-b}. \tag{3.95}$$

**Lemma 3.151.** *Assume* $0 < \nu < 1$, $|b| < 1$. *Then any solution of the equation*

$$\nabla^{\nu}_{\rho(a)} x(t) = bx(t), \qquad t \in \mathbb{N}_{a+1} \tag{3.96}$$

*satisfying* $x(a) > 0$ *is positive on* $\mathbb{N}_a$.

*Proof.* From (3.32), we have for $t = a + k$

$$\nabla^{\nu}_{\rho(a)} x(t) = \int_{\rho(a)}^{t} H_{-\nu-1}(t, \rho(s)) x(s) \nabla s$$

$$= \sum_{s=a}^{a+k} H_{-\nu-1}(a+k, s-1) x(s)$$

$$= x(a+k) - \nu x(a+k-1) - \frac{\nu(-\nu+1)}{2} x(a+k-2)$$

$$- \cdots - \frac{\nu(-\nu+1)\cdots(-\nu+k-1)}{k!} x(a).$$

Using (3.92), we have that

$$(1-b)x(a+k)$$

$$= \nu x(a+k-1) + \frac{\nu(-\nu+1)}{2} x(a+k-2)$$

$$+ \cdots + \frac{\nu(-\nu+1)\cdots(-\nu+k-1)}{k!} x(a). \tag{3.97}$$

We will prove $x(a+k) > 0$ for $k \in \mathbb{N}_0$ by using the principle of strong induction. Since $x(a) > 0$ we have that the base case holds. Now assume that $x(a+i) > 0$, for $i = 0, 1, \cdots, k-1$. Since

$$\frac{\nu(-\nu+1)\cdots(-\nu+i-1)}{i!} > 0$$

for $i = 2, 3, \cdots k-1$, from (3.97), we have $x(a+k) > 0$. This completes the proof.

□

**Lemma 3.152.** *Assume that* $0 < \nu < 1$, $-1 < b < 0$. *Then*

$$\lim_{t\to\infty} E_{b,\nu,0}(t,a) = 0.$$

*Proof.* From Lemma 3.150 and Lemma 3.151, we have $E_{b,\nu,\nu-1}(t,\rho(a)) > 0$, for $t \in \mathbb{N}_{a+1}$. So we have

$$\begin{aligned}
\nabla E_{b,\nu,0}(t,a) &= \sum_{k=0}^{\infty} b^k \nabla H_{\nu k}(t,a) \\
&= \sum_{k=0}^{\infty} b^k H_{\nu k-1}(t,a) = \sum_{k=1}^{\infty} b^k H_{\nu k-1}(t,a) \\
&= b\sum_{k=1}^{\infty} b^{k-1} H_{\nu k-1}(t,a) = b\sum_{j=0}^{\infty} b^j H_{\nu j+\nu-1}(t,a) \\
&= bE_{b,\nu,\nu-1}(t,a) = bE_{b,\nu,\nu-1}(t-1,\rho(a)) < 0,
\end{aligned}$$

for $t \in \mathbb{N}_{a+1}$, where we used $H_{-1}(t,a) = 0$. Therefore, $E_{b,\nu,0}(t,a)$ is decreasing for $t \in \mathbb{N}_{a+1}$. From Lemma 3.137, we have $E_{b,\nu,0}(t,a) > 0$ for $t \in \mathbb{N}_{a+1}$. Suppose that

$$\lim_{t\to\infty} E_{b,\nu,0}(t,a) = A \geq 0.$$

In the following, we will prove $A = 0$. If not, $A > 0$. Let $x(t) := E_{b,\nu,0}(t,a) > 0$. From Lemma 3.149, we have

$$\begin{aligned}
x(t) &= \int_a^t H_{\nu-1}(t,\rho(s))bx(s)\nabla s + x(a) \\
&= b[x(t) + \nu x(t-1) + \frac{\nu(\nu+1)}{2!}x(t-2) \\
&\quad + \cdots + H_{\nu-1}(t,a)x(a+1)] + x(a).
\end{aligned}$$

For fixed $k_0 > 0$ and large $t$, we have (since $b < 0$)

$$\begin{aligned}
x(t) \leq b\Big[x(t) + \nu x(t-1) + \frac{\nu(\nu+1)}{2!}x(t-2) \\
+ \cdots + \frac{\nu(\nu+1)\cdots(\nu+k_0-1)}{k_0!}x(t-k_0)\Big] + x(a).
\end{aligned}$$

Letting $t \to \infty$, we get that

$$0 < A \leq bA\left[1+\nu+\frac{\nu(\nu+1)}{2!}+\cdots+\frac{\nu(\nu+1)\cdots(\nu+k_0-1)}{k_0!}\right]+x(a). \quad (3.98)$$

Notice (using mathematical induction in the first step) that

$$\begin{aligned}
&1+\nu+\frac{\nu(\nu+1)}{2!}+\cdots+\frac{\nu(\nu+1)\cdots(\nu+k_0-1)}{k_0!} \\
&= \frac{(\nu+1)(\nu+2)\cdots(\nu+k_0)}{k_0!} \\
&= \frac{(\nu+1)(\nu+2)\cdots(\nu+1+k_0-1)}{(k_0-1)!(k_0-1)^{\nu+1}}\frac{(k_0-1)^{\nu+1}}{k_0} \\
&\to +\infty,
\end{aligned}$$

as $k_0 \to \infty$, where we used (see Lemma 3.142)

$$\frac{1}{\Gamma(\nu+1)} = \lim_{k_0\to\infty}\frac{(\nu+1)(\nu+2)\cdots(\nu+1+k_0-1)}{(k_0-1)!(k_0-1)^{\nu+1}}.$$

So in (3.98), for sufficiently large $k_0$, the right side of (3.98) is negative, but the left side of (3.98) is positive, which is a contradiction. So $A = 0$. This completes the proof. □

**Theorem 3.153.** *Assume $c(t) \leq b_1 < 0$, $0 < \nu < 1$, and $x(t)$ is any solution of the Caputo nabla fractional difference equation*

$$\nabla^{\nu}_{a*}x(t) = c(t)x(t), \quad t \in \mathbb{N}_{a+1} \quad (3.99)$$

*satisfying $x(a) > 0$. Then*

$$x(t) \leq 2x(a)E_{b_1,\nu,0}(t,a),$$

*for $t \in \mathbb{N}_a$.*

*Proof.* Assume that $b_1 > -1$. Otherwise we can choose $0 > b_1' > -1$, $b_1' > b_1$ and replace $b_1$ by $b_1'$. From Lemma 3.141, we have

$$\nabla^{\nu}_{a*}E_{b_1,\nu,0}(t,a) = b_1E_{b_1,\nu,0}(t,a)$$

and $E_{b_1,\nu,0}(a,a) = H_0(a,a) = 1$.

In Theorem 3.138, take $c_2(t) = b_1$. Then it holds that $x(t)$ and $y(t) = 2x(a)E_{b_1,\nu,0}(t,a)$ satisfy

$$\nabla^{\nu}_{a*}x(t) = c(t)x(t), \quad (3.100)$$

and

$$\nabla_{a*}^{\nu} y(t) = b_1 y(t), \tag{3.101}$$

respectively, for $t \in \mathbb{N}_{a+1}$ and

$$x(a) < 2x(a) = 2x(a)E_{b_1,\nu,0}(a,a) = y(a).$$

From Theorem 3.138, we get that

$$x(t) \leq 2x(a)E_{b_1,\nu,0}(t,a),$$

for $t \in \mathbb{N}_a$. This completes the proof. □

From Theorem 3.153 and Lemma 3.152, we get the following result.

**Theorem 3.154.** *Assume* $0 < \nu < 1$ *and there exists a constant* $b_1$ *such that* $c(t) \leq b_1 < 0$. *Then the solutions of the equation* (3.76) *with* $x(a) > 0$ *satisfy*

$$\lim_{t\to\infty} x(t) = 0.$$

Next we consider solutions of the $\nu$-th order Caputo nabla fractional difference equation

$$\nabla_{a*}^{\nu} x(t) = c(t)x(t), \qquad t \in \mathbb{N}_{a+1}, \tag{3.102}$$

satisfying $x(a) < 0$.

By making the transformation $x(t) = -y(t)$ and using Theorem 3.146 and Theorem 3.154, we get the following theorem.

**Theorem 3.155.** *Assume* $0 < \nu < 1$ *and there exists a constant* $b_2$ *such that* $0 < b_2 \leq c(t) < 1$, $t \in \mathbb{N}_{a+1}$. *Then the solutions of the equation* (3.102) *with* $x(a) < 0$ *satisfy*

$$\lim_{t\to\infty} x(t) = -\infty.$$

**Theorem 3.156.** *Assume* $0 < \nu < 1$ *and there exists a constant* $b_1$ *such that* $c(t) \leq b_1 < 0$, $t \in \mathbb{N}_{a+1}$. *Then the solutions of the equation* (3.102) *with* $x(a) < 0$ *satisfy*

$$\lim_{t\to\infty} x(t) = 0.$$

## 3.23 Self-Adjoint Caputo Fractional Difference Equation

Let $\mathcal{D}_a := \{x : \mathbb{N}_a \to \mathbb{R}\}$, and let $L_a : \mathcal{D}_a \to \mathcal{D}_{a+1}$ be defined by

$$(L_a x)(t) := \nabla[p(t+1)\nabla^{\nu}_{a*}x(t+1)] + q(t)x(t), \quad t \in \mathbb{N}_{a+1}, \tag{3.103}$$

where $x \in \mathcal{D}_a, 0 < \nu < 1, p(t) > 0, t \in \mathbb{N}_{a+1}$ and $q : \mathbb{N}_{a+1} \to \mathbb{R}$. Most of the results in this section appear in Ahrendt et al. [4].

**Theorem 3.157.** *The operator $L_a$ in (3.103) is a linear transformation.*

*Proof.* Let $x, y : \mathbb{N}_a \to \mathbb{R}$, and let $\alpha, \beta \in \mathbb{R}$. Then

$$\begin{aligned}
&L_a[\alpha x + \beta y](t)\\
&= \nabla\Big[p(t+1)\nabla^{\nu}_{a*}[\alpha x(t+1) + \beta y(t+1)]\Big] + q(t)[\alpha x(t) + \beta y(t)]\\
&= \nabla\Big[p(t+1)[\alpha\nabla^{\nu}_{a*}x(t+1) + \beta\nabla^{\nu}_{a*}y(t+1)]\Big] + \alpha q(t)x(t) + \beta q(t)y(t)\\
&= \nabla[\alpha p(t+1)\nabla^{\nu}_{a*}x(t+1) + \beta p(t+1)\nabla^{\nu}_{a*}y(t+1)] + \alpha q(t)x(t) + \beta q(t)y(t)\\
&= \alpha\nabla[p(t+1)\nabla^{\nu}_{a*}x(t+1)] + \alpha q(t)x(t) + \beta\nabla[p(t+1)\nabla^{\nu}_{a*}y(t+1)] + \beta q(t)y(t)\\
&= \alpha L_a x(t) + \beta L_a y(t),
\end{aligned}$$

for $t \in \mathbb{N}_{a+1}$. □

**Theorem 3.158 (Existence and Uniqueness for IVPs).** *Let $A, B \in \mathbb{R}$ be given constants and assume $h : \mathbb{N}_{a+1} \to \mathbb{R}$. Then the IVP*

$$\begin{cases} L_a x(t) = h(t), & t \in \mathbb{N}_{a+1} \\ x(a) = A, & \nabla x(a+1) = B, \end{cases} \tag{3.104}$$

*has a unique solution on $\mathbb{N}_a$.*

*Proof.* Let $x : \mathbb{N}_a \to \mathbb{R}$ be defined uniquely by

$$x(a) = A, \quad x(a+1) = A + B,$$

and for $t \in \mathbb{N}_{a+1}$, $x(t)$ satisfies the summation equation

$$\begin{aligned}
x(t+1) = x(t) &- \sum_{\tau=a+1}^{t} \frac{(t-\tau+2)^{\overline{-\nu}}}{\Gamma(1-\nu)}\nabla x(\tau)\\
&+ \frac{1}{p(t+1)}\left[h(t) - q(t)x(t) + p(t)\sum_{\tau=a+1}^{t}\frac{(t-\tau+1)^{\overline{-\nu}}}{\Gamma(1-\nu)}\nabla x(\tau)\right].
\end{aligned}$$

We will show that $x$ solves the IVP (3.104). Clearly the initial conditions are satisfied. Now we show that $x$ is a solution of the nabla Caputo self-adjoint equation on $\mathbb{N}_a$. To see this note that for $t \in \mathbb{N}_{a+1}$, we have from the last equation

$$\nabla x(t+1) + \int_a^t H_{-\nu}(t+1, \rho(\tau))\nabla x(\tau)\nabla\tau$$
$$= \frac{1}{p(t+1)}\left[h(t) - q(t)x(t) + p(t)\nabla_a^{-(1-\nu)}\nabla x(t)\right]. \tag{3.105}$$

But

$$\begin{aligned}
&\nabla_{a*}^{\nu}x(t+1) \\
&= \nabla_a^{-(1-\nu)}\nabla x(t+1) \\
&= \int_a^{t+1} H_{-\nu}(t+1, \rho(\tau))\nabla x(\tau)\nabla\tau \\
&= H_{-\nu}(t+1, t)\nabla x(t+1) + \int_a^t H_{-\nu}(t+1, \rho(\tau))\nabla x(\tau)\nabla\tau \\
&= \nabla x(t+1) + \int_a^t H_{-\nu}(t+1, \rho(\tau))\nabla x(\tau)\nabla\tau.
\end{aligned}$$

Hence, from this last equation and (3.105) we get that

$$\begin{aligned}
p(t+1)\nabla_{a*}^{\nu}x(t+1) &= h(t) - q(t)x(t) + p(t)\nabla_a^{-(1-\nu)}\nabla x(t) \\
&= h(t) - q(t)x(t) + p(t)\nabla_{a*}^{\nu}x(t).
\end{aligned}$$

It follows that

$$\nabla[p(t+1)\nabla_{a*}^{\nu}x(t+1)] + q(t)x(t) = h(t)$$

for $t \in \mathbb{N}_{a+1}$. Reversing the preceding steps shows that if $y(t)$ is a solution to the IVP (3.104), it must be the same solution as $x(t)$. Therefore the IVP (3.104) has a unique solution. □

**Theorem 3.159.** *Let* $0 < \nu < 1$ *and let* $x : \mathbb{N}_a \to \mathbb{R}$. *Then*

$$\nabla_{a*}^{\nu}x(a+1) = \nabla x(a+1).$$

*Proof.* Let $0 < \nu < 1$. Then by the definition of the nabla Caputo fractional difference it holds that

$$\begin{aligned}\nabla_{a*}^{\nu}x(a+1) &= \nabla_a^{-(1-\nu)}\nabla x(a+1)\\ &= \int_a^{a+1} H_{-\nu}(a+1,\rho(\tau))\nabla x(\tau)\nabla\tau\\ &= H_{-\nu}(a+1,a)\nabla x(a+1)\\ &= \nabla x(a+1),\end{aligned}$$

which is what we wanted to prove. □

**Theorem 3.160 (General Solution of the Homogeneous Equation).** *We assume $x_1, x_2$ are linearly independent solutions of $L_a x = 0$ on $\mathbb{N}_a$. Then the general solution to $L_a x = 0$ is given by*

$$x(t) = c_1 x_1(t) + c_2 x_2(t), \quad t \in \mathbb{N}_a$$

*where $c_1, c_2 \in \mathbb{R}$ are arbitrary constants.*

*Proof.* Let $x_1, x_2$ be linearly independent solutions of $L_a x(t) = 0$ on $\mathbb{N}_a$. If we let

$$\alpha := x_1(a), \quad \beta := x_1(a+1), \quad \gamma := x_2(a), \quad \delta := x_2(a+1),$$

then $x_1, x_2$ are the unique solutions to the IVPs

$$\begin{cases} Lx_1 = 0, \\ x_1(a) = \alpha, \\ x_1(a+1) = \beta, \end{cases} \quad \text{and} \quad \begin{cases} Lx_2 = 0, \\ x_2(a) = \gamma, \\ x_2(a+1) = \delta. \end{cases}$$

Since $L_a$ is a linear operator, for any $c_1, c_2 \in \mathbb{R}$, we have

$$L_a[c_1 x_1(t) + c_2 x_2(t)] = c_1 L_a x_1(t) + c_2 L_a x_2(t) = 0,$$

so $x(t) = c_1 x_1(t) + c_2 x_2(t)$ solves $L_a x(t) = 0$. Conversely, suppose $x : \mathbb{N}_a \to \mathbb{R}$ solves $L_a x(t) = 0$. Note that if $A := x(a)$ and $B := x(a+1)$, then $x(t)$ is the unique solution of the IVP

$$\begin{cases} Lx = 0, \\ x(a) = A, \quad x(a+1) = B. \end{cases}$$

It remains to show that the matrix equation

$$\begin{bmatrix} x_1(a) & x_2(a) \\ x_1(a+1) & x_2(a+1) \end{bmatrix} \begin{bmatrix} c_1 \\ c_2 \end{bmatrix} = \begin{bmatrix} A \\ B \end{bmatrix} \tag{3.106}$$

has a unique solution for $c_1, c_2$. Then $x(t)$ and $c_1x_1(t) + c_2x_2(t)$ solve the same IVP, so by Theorem 3.158, every solution to $L_ax(t) = 0$ may be uniquely expressed as a linear combination of $x_1(t)$ and $x_2(t)$. The above matrix equation can be equivalently expressed as

$$\begin{bmatrix} \alpha & \gamma \\ \beta & \delta \end{bmatrix} \begin{bmatrix} c_1 \\ c_2 \end{bmatrix} = \begin{bmatrix} A \\ B \end{bmatrix}.$$

Suppose by way of contradiction that

$$\begin{vmatrix} \alpha & \gamma \\ \beta & \delta \end{vmatrix} = 0.$$

Without loss of generality, there exists a constant $k \in \mathbb{R}$ for which $\alpha = k\gamma$ and $\beta = k\delta$. Then $x_1(a) = k\gamma = kx_2(a)$, and $x_1(a+1) = k\delta = kx_2(a+1)$. Since $kx_2(t)$ solves $L_ax(t) = 0$, we have that $x_1(t)$ and $kx_2(t)$ solve the same IVP. By uniqueness, $x_1(t) = kx_2(t)$. But then $x_1(t)$ and $x_2(t)$ are linearly dependent on $\mathbb{N}_a$, so we have a contradiction. Therefore, the matrix equation (3.106) must have a unique solution.

□

**Corollary 3.161.** *Assume $x_1, x_2$ are linearly independent solutions of $L_ax(t) = 0$ on $\mathbb{N}_a$ and $y_0$ is a particular solution to $L_ax(t) = h(t)$ on $\mathbb{N}_a$ for some $h : \mathbb{N}_{a+1} \to \mathbb{R}$. Then the general solution of $L_ax(t) = h(t)$ is given by*

$$x(t) = c_1x_1(t) + c_2x_2(t) + y_0(t),$$

*where $c_1, c_2 \in \mathbb{R}$ are arbitrary constants, for $t \in \mathbb{N}_a$.*

*Proof.* This proof is left to the reader. □

Next we define the Cauchy function for the Caputo fractional self-adjoint equation, $L_ax = 0$. Later we will see its importance for finding a variation of constants formula for $L_ax = h(t)$ and also its importance for constructing Green's functions for various boundary value problems.

**Definition 3.162.** The **Cauchy function** for $L_ax(t) = 0$ is the real-valued function $x$ with domain $a \leq s \leq t$ such that, for each fixed $s \in \mathbb{N}_a$, $x$ satisfies the IVP

$$\begin{cases} L_sx(t) = 0, & t \in \mathbb{N}_{s+1} \\ x(s,s) = 0, \\ \nabla x(s+1,s) = \frac{1}{p(s+1)}. \end{cases} \tag{3.107}$$

Note by Theorem 3.159, the IVP (3.107) is equivalent to the IVP

$$\begin{cases} L_s x(t) = 0 \quad t \in \mathbb{N}_{s+1}, \\ x(s,s) = 0, \\ \nabla^{\nu}_{s*} x(s+1,s) = \frac{1}{p(s+1)}. \end{cases}$$

*Example 3.163.* We show that the Cauchy function for

$$\nabla[p(t+1)\nabla^{\nu}_{a*} y(t+1)] = 0, \quad t \in \mathbb{N}_{a+1}$$

is given by the formula

$$x(t,s) = \nabla_s^{-\nu}\left(\frac{1}{p(t)}\right) = \int_s^t \frac{H_{\nu-1}(t,\rho(\tau))}{p(\tau)} \nabla\tau. \tag{3.108}$$

for $t \geq s \geq a$. We know for each fixed $s$, the Cauchy function satisfies the equation

$$\nabla[p(t+1)\nabla^{\nu}_{s*} x(t+1,s)] = 0,$$

for $t \geq s \geq a$. It follows that

$$p(t+1)\nabla^{\nu}_{s*} x(t+1,s) = \alpha(s)$$

$$\nabla^{\nu}_{s*} x(t+1,s) = \frac{\alpha(s)}{p(t+1)}.$$

Letting $t = s$ and using the initial condition

$$\nabla x(s+1,s) = \nabla^{\nu}_{s*} x(s+1,s) = \frac{1}{p(s+1)}$$

we get that $\alpha(s) = 1$. Hence we have that

$$\nabla^{\nu}_{s*} x(t+1,s) = \frac{1}{p(t+1)}.$$

By the definition of the Caputo difference, this is equivalent to

$$\nabla_s^{-(1-\nu)} \nabla x(t+1,s) = \frac{1}{p(t+1)}$$

$$\nabla_s^{1-\nu} \nabla_s^{-(1-\nu)} \nabla x(t+1,s) = \nabla_s^{1-\nu}\left(\frac{1}{p(t+1)}\right)$$

$$\nabla x(t+1,s) = \nabla_s^{1-\nu}\left(\frac{1}{p(t+1)}\right).$$

Replacing $t+1$ with $t$ yields

$$\nabla x(t,s) = \nabla_s^{1-\nu}\left(\frac{1}{p(t)}\right).$$

Integrating both sides from $s$ to $t$ and using $x(s,s)=0$ we get

$$\begin{aligned}
x(t,s) &= \int_s^t \nabla_s^{1-\nu}\frac{1}{p(\tau)}\nabla\tau \\
&= \int_s^t \nabla\nabla_s^{-\nu}\frac{1}{p(\tau)}\nabla\tau \\
&= \left[\nabla_s^{-\nu}\frac{1}{p(\tau)}\right]_{\tau=s}^{\tau=t} \\
&= \nabla_s^{-\nu}\left(\frac{1}{p(t)}\right) - \nabla_s^{-\nu}\left(\frac{1}{p}\right)(s) \\
&= \nabla_s^{-\nu}\left(\frac{1}{p(t)}\right) = \int_s^t \frac{H_{\nu-1}(t,\rho(\tau))}{p(\tau)}\nabla\tau.
\end{aligned}$$

*Example 3.164.* Find the Cauchy function for

$$\nabla\nabla_{a*}^{\nu}y(t+1) = 0, \quad t \in \mathbb{N}_{a+1}.$$

From Example 3.163 we have that the Cauchy function is given by

$$\begin{aligned}
x(t,s) &= \nabla_s^{-\nu}\left(\frac{1}{p(t)}\right) = \int_s^t \frac{H_{\nu-1}(t,\rho(\tau))}{p(\tau)}\nabla\tau \\
&= \int_s^t H_{\nu-1}(t,\rho(\tau))\nabla\tau \\
&= -H_\nu(t,\tau)\Big|_{\tau=s}^{\tau=t} \\
&= H_\nu(t,s).
\end{aligned}$$

Note that if $\nu = 1$ we get the well-known result that the Cauchy function for $\nabla^2 x(t+1) = 0$ is given by $x(t,s) = t-s$.

**Theorem 3.165 (Variation of Constants).** *Assume $h : N_{a+1} \to \mathbb{R}$. Then the solution of the IVP*

$$\begin{cases} L_a y(t) = h(t), & t \in \mathbb{N}_{a+1} \\ y(a) = y(a+1) = 0, \end{cases}$$

*is given by*

$$y(t) = \int_a^t x(t,s)h(s)\nabla s, \quad t \in \mathbb{N}_a,$$

*where $x(t,s)$ is the Cauchy function for $L_a x = 0$.*

*Proof.* Let $y(t) = \int_a^t x(t,s)h(s)\nabla s$, $t \in \mathbb{N}_a$. We first note that $y(t)$ satisfies the initial conditions:

$$y(a) = \sum_{s=a+1}^{a} x(a,s)h(s) = 0,$$

$$y(a+1) = \sum_{s=a+1}^{a+1} x(a+1,s)h(s) = x(a+1,a+1)h(a+1) = 0.$$

Next, note that by the Leibniz formula (3.23), we have that

$$\begin{aligned}\nabla y(t) &= \int_a^{t-1} \nabla_t x(t,s)h(s)\nabla s + x(t,t)h(t)\\ &= \int_a^{t-1} \nabla_t x(t,s)h(s)\nabla s. \end{aligned} \tag{3.109}$$

We now show that

$$\nabla_{a*}^{\nu} y(t) = \int_a^{t-1} \nabla_{s*}^{\nu} x(t,s)h(s)\nabla s, \tag{3.110}$$

for $t \in \mathbb{N}_{a+2}$. By the definition of the Caputo fractional difference,

$$\begin{aligned}\nabla_{a*}^{\nu} y(t) &= \nabla_a^{-(1-\nu)} \nabla y(t)\\ &= \int_a^t H_{-\nu}(t,\rho(\tau))\nabla y(\tau)\nabla\tau\\ &= \int_a^t H_{-\nu}(t,\rho(\tau)) \int_a^{\tau-1} \nabla_t x(t,s)h(s)\nabla s\nabla\tau, \quad \text{by (3.109)}\\ &= \int_a^t \int_a^{\tau-1} H_{-\nu}(t,\rho(\tau))\nabla_t x(t,s)h(s)\nabla s\nabla\tau\\ &= \sum_{\tau=a+1}^{t} \sum_{s=a+1}^{\tau-1} H_{-\nu}(t,\rho(\tau))\nabla_t x(t,s)h(s)\end{aligned}$$

$$= \sum_{\tau=a+2}^{t} \sum_{s=a+1}^{\tau-1} H_{-\nu}(t, \rho(\tau)) \nabla_t x(t,s) h(s)$$

$$= \sum_{s=a+1}^{t-1} \sum_{\tau=s+1}^{t} H_{-\nu}(t, \rho(\tau)) \nabla_t x(t,s) h(s)$$

$$= \int_a^{t-1} \int_s^t H_{-\nu}(t, \rho(\tau)) \nabla_t x(t,s) h(s) \nabla \tau \nabla s$$

$$= \int_a^{t-1} \nabla_s^{-(1-\nu)} \nabla_t x(t,s) h(s) \nabla s$$

$$= \int_a^{t-1} \nabla_{s*}^{\nu} x(t,s) h(s) \nabla s$$

for $t \in \mathbb{N}_{a+2}$. Hence (3.110) holds. Then by (3.110) we have that

$$p(t+1) \nabla_{a*}^{\nu} y(t+1) = \int_a^t \left[ p(t+1) \nabla_{s*}^{\nu} x(t+1, s) \right] h(s) \nabla s.$$

Using the Leibniz formula (3.23) we get that

$$\begin{aligned}
&\nabla \left[ p(t+1) \nabla_{a*}^{\nu} y(t+1) \right] \\
&= \int_a^{t-1} \nabla \left[ p(t+1) \nabla_{s*}^{\nu} x(t+1, s) \right] h(s) \nabla s + p(t+1) \nabla_{s*}^{\nu} x(t+1, t) h(t) \\
&= \int_a^{t-1} \nabla \left[ p(t+1) \nabla_{s*}^{\nu} x(t+1, s) \right] h(s) \nabla s + h(t)
\end{aligned}$$

It follows that

$$\begin{aligned}
L_a y(t) &= \nabla [p(t+1) \nabla_{a*}^{\nu} y(t+1)] + q(t) y(t) \\
&= \int_a^{t-1} \nabla \left[ p(t+1) \nabla_{s*}^{\nu} x(t+1, s) \right] h(s) \nabla s + h(t) + \int_a^{t-1} q(t) x(t,s) h(s) \nabla s \\
&= \int_a^{t-1} \left[ \nabla [p(t+1) \nabla_{s*}^{\nu} x(t+1, s)] + q(t) x(t,s) \right] h(s) \nabla s + h(t) \\
&= h(t) + \int_a^{t-1} L_s x(t,s) h(s) \nabla s \\
&= h(t).
\end{aligned}$$

Thus, $y(t)$ solves the given IVP. □

**Theorem 3.166 (Variation of Constants Formula).** *Assume $p, q : \mathbb{N}_{a+1} \to \mathbb{R}$. Then the solution to the IVP*

$$\begin{cases} L_a y(t) = h(t), \\ y(a) = A, \\ \nabla y(a+1) = B, \end{cases}$$

*for $t \in \mathbb{N}_{a+1}$, where $A, B \in \mathbb{R}$ are arbitrary constants, is given by*

$$y(t) = y_0(t) + \int_a^t x(t,s)h(s)\nabla s,$$

*where $y_0(t)$ solves the IVP*

$$\begin{cases} L_a y(t) = 0, \\ y(a) = A, \\ \nabla y(a+1) = B, \end{cases}$$

*for $t \in \mathbb{N}_{a+1}$.*

*Proof.* The proof follows from Theorem 3.165 by linearity. □

**Corollary 3.167.** *Assume $p, h : \mathbb{N}_{a+1} \to \mathbb{R}$. Then the solution of the IVP*

$$\begin{cases} \nabla[p(t+1)\nabla^{\nu}_{a*}y(t+1)] = h(t), \quad t \in \mathbb{N}_{a+1} \\ y(a) = \nabla y(a+1) = 0, \end{cases}$$

*is given by*

$$y(t) = \int_a^t \nabla_s^{-\nu}\left(\frac{1}{p(t)}\right)h(s)\nabla s.$$

*Proof.* From Theorem 3.165, we know that the solution is given by

$$y(t) = \int_a^t x(t,s)h(s)\nabla s,$$

where $x(t,s)$ is the Cauchy function for

$$\nabla[p(t+1)\nabla^{\nu}_{a*}y(t+1)] = 0.$$

By Example 3.163, we know the Cauchy function for the above difference equation is $x(t,s) = \nabla_s^{-\nu}\left(\frac{1}{p(t)}\right)$. Hence the solution of our given IVP is given by

$$y(t) = \int_a^t \nabla_s^{-\nu}\left(\frac{1}{p(t)}\right) h(s)\nabla s,$$

for $t \in \mathbb{N}_a$. □

The following example appears in [4].

*Example 3.168.* Solve the IVP

$$\begin{cases} \nabla\nabla_{0*}^{0.6}x(t+1) = t, & t \in \mathbb{N}_1, \\ x(0) = \nabla x(1) = 0. \end{cases}$$

Note that for this self-adjoint equation, $p(t) \equiv 1$ and $q(t) \equiv 0$. From Example (3.164) we have that $x(t,s) = H_{.6}(t,s)$

Then by Theorem 3.165

$$\begin{aligned} x(t) &= \int_0^t x(t,s)s\nabla s \\ &= \int_0^t H_{.6}(t,s)s\nabla s. \end{aligned}$$

Integrating by parts we get from Exercise 3.37 that

$$x(t) = \frac{1}{\Gamma(3.6)}(t-1)^{\overline{2.6}},$$

for $t \in \mathbb{N}_0$.

## 3.24 Boundary Value Problems

Many of the results in this section appear in Ahrendt et al. [4]. In this section we will consider the nonhomogeneous boundary value problem (BVP)

$$\begin{cases} L_a x(t) = h(t), & t \in \mathbb{N}_{a+1}^{b-1} \\ \alpha x(a) - \beta\nabla x(a+1) = A, \\ \gamma x(b) + \delta\nabla x(b) = B, \end{cases} \tag{3.111}$$

and the corresponding homogeneous BVP

$$\begin{cases} L_a x(t) = 0, \quad t \in \mathbb{N}_{a+1}^{b-1} \\ \alpha x(a) - \beta \nabla x(a+1) = 0, \\ \gamma x(b) + \delta \nabla x(b) = 0, \end{cases} \tag{3.112}$$

where $h : \mathbb{N}_{a+1}^{b-1} \to \mathbb{R}$, $p(t) > 0$, $t \in \mathbb{N}_{a+1}^{b}$, $0 < \nu \leq 1$, and $\alpha, \beta, \gamma, \delta, A, B \in \mathbb{R}$ for which $\alpha^2 + \beta^2 > 0$ and $\gamma^2 + \delta^2 > 0$. Also we always assume $b - a$ is a positive integer and $b - a$ is large enough so that the boundary conditions are not equivalent to initial conditions. The following theorem gives an important relationship between these two BVPs.

**Theorem 3.169.** *The homogeneous BVP* (3.112) *has only the trivial solution iff the nonhomogeneous BVP* (3.111) *has a unique solution.*

*Proof.* Let $x_1, x_2$ be linearly independent solutions to $L_a x(t) = 0$ on $\mathbb{N}_a^b$. By Theorem 3.160, a general solution to $L_a x(t) = 0$ is given by

$$x(t) = c_1 x_1(t) + c_2 x_2(t),$$

where $c_1, c_2 \in \mathbb{R}$ are arbitrary constants. If $x(t)$ solves the homogeneous boundary conditions, then $x(t)$ is the trivial solution if and only if $c_1 = c_2 = 0$. This is true if and only if the system of equations

$$\begin{cases} \alpha[c_1 x_1(a) + c_2 x_2(a)] - \beta \nabla [c_1 x_1(a+1) + c_2 x_2(a+1)] = 0, \\ \gamma[c_1 x_1(b) + c_2 x_2(b)] + \delta \nabla [c_1 x_1(b) + c_2 x_2(b)] = 0, \end{cases}$$

or equivalently,

$$\begin{cases} c_1[\alpha x_1(a) - \beta \nabla x_1(a+1)] + c_2[\alpha x_2(a) - \beta \nabla x_2(a+1)] = 0, \\ c_1[\gamma x_1(b) + \delta \nabla x_1(b)] + c_2[\gamma x_2(b) + \delta \nabla x_2(b)] = 0, \end{cases}$$

has only the trivial solution $c_1 = c_2 = 0$. In other words, $x(t)$ solves (3.112) if and only if

$$D := \begin{vmatrix} \alpha x_1(a) - \beta \nabla x_1(a+1) & \alpha x_2(a) - \beta \nabla x_2(a+1) \\ \gamma x_1(b) + \delta \nabla x_1(b) & \gamma x_2(b) + \delta \nabla x_2(b) \end{vmatrix} \neq 0.$$

Now consider the nonhomogeneous BVP (3.111). By Corollary 3.161, a general solution of the nonhomogeneous equation $L_a y(t) = h(t)$ is given by

$$y(t) = a_1 x_1(t) + a_2 x_2(t) + y_0(t),$$

where $a_1, a_2 \in \mathbb{R}$ are arbitrary constants, and $y_0 : \mathbb{N}_a \to \mathbb{R}$ is a particular solution of the nonhomogeneous equation $L_a y(t) = h(t)$. Then $y(t)$ satisfies the nonhomogeneous boundary conditions in (3.111) if and only if

$$\begin{cases} \alpha[a_1x_1(a) + a_2x_2(a) + y_0(a)] \\ \qquad -\beta\nabla^{\nu}_{a*}[a_1x_1(a+1) + a_2x_2(a+1) + y_0(a+1)] = A, \\ \gamma[a_1x_1(b) + a_2x_2(b) + y_0(b)] + \delta\nabla[a_1x_1(b) + a_2x_2(b) + y_0(b)] = B. \end{cases}$$

This system is equivalent to the system

$$\begin{cases} a_1[\alpha x_1(a) - \beta\nabla x_1(a+1)] + a_2[\alpha x_2(a) - \beta\nabla x_2(a+1)] \\ = A - \alpha y_0(a) + \beta\nabla y_0(a+1), \\ a_1[\gamma x_1(b) + \delta\nabla x_1(b)] + a_2[\gamma x_2(b) + \delta\nabla x_2(b)] \\ = B - \gamma y_0(b) - \delta\nabla y_0(b). \end{cases}$$

Thus, $y(t)$ satisfies the boundary conditions in (3.111) iff $D \neq 0$. Therefore the homogeneous BVP (3.112) has only the trivial solution iff the nonhomogeneous BVP (3.111) has a unique solution. □

In the next theorem we give conditions for which Theorem 3.169 applies.

**Theorem 3.170.** *Let*

$$\rho := \alpha\gamma\nabla_a^{-\nu}\frac{1}{p(b-1)} + \frac{\alpha\delta}{p(b)} + \frac{\beta\gamma}{p(a+1)}.$$

*Then the BVP*

$$\begin{cases} \nabla[p(t+1)\nabla^{\nu}_{a*}x(t+1)] = 0, \quad t \in \mathbb{N}_{a+1}^{b-1}, \\ \alpha x(a) - \beta\nabla^{\nu}_{a*}x(a+1) = 0, \\ \gamma x(b) + \delta\nabla^{\nu}_{a*}x(b) = 0, \end{cases} \tag{3.113}$$

*has only the trivial solution if and only if* $\rho \neq 0$.

*Proof.* Note that $x_1(t) = 1, x_2(t) = \nabla_a^{-\nu}\frac{1}{p(t)}$ are linearly independent solutions to

$$\nabla\{p(t+1)\nabla^{\nu}_{a*}x(t+1)\} = 0.$$

Then a general solution of the difference equation is given by

$$x(t) = c_1x_1(t) + c_2x_2(t) = c_1 + c_2\nabla_a^{-\nu}\frac{1}{p(t)}.$$

The boundary conditions $\alpha x(a) - \beta\nabla^{\nu}_{a*}x(a+1) = 0$, and $\gamma x(b) + \delta\nabla^{\nu}_{a*}x(b) = 0$ give us the linear system

$$c_1\alpha + c_2\left[-\frac{\beta}{p(a+1)}\right] = 0,$$

$$c_1\gamma + c_2\left(\frac{\delta}{p(b)} + \gamma\nabla_a^{-\nu}\frac{1}{p(b)}\right) = 0.$$

The determinant of the coefficients of this system is given by

$$\begin{vmatrix} \alpha & -\frac{\beta}{p(a+1)} \\ \gamma & \frac{\delta}{p(b)} + \gamma\nabla_a^{-\nu}\frac{1}{p(b)} \end{vmatrix} = \alpha\gamma\nabla_a^{-\nu}\frac{1}{p(b-1)} + \frac{\alpha\delta}{p(b)} + \frac{\beta\gamma}{p(a+1)} = \rho.$$

Hence, the BVP has only the trivial solution if and only if $\rho \neq 0$. □

**Corollary 3.171.** *Assume $\alpha$, $\beta$, $\gamma$, and $\delta$ are all greater than or equal to zero with $\alpha^2 + \beta^2 \neq 0$ and $\gamma^2 + \delta^2 \neq 0$. Then the homogeneous BVP*

$$\begin{cases} \nabla[p(t+1)\nabla_{a^*}^{\nu}x(t+1)] = 0, \quad t \in \mathbb{N}_{a+1}^{b-1}, \\ \alpha x(a) - \beta\nabla_{a^*}^{\nu}x(a+1) = 0, \\ \gamma x(b) + \delta\nabla_{a^*}^{\nu}x(b) = 0, \end{cases} \tag{3.114}$$

*has only the trivial solution.*

*Proof.* The hypotheses of this theorem imply that $\rho > 0$. Hence the conclusion follows from Theorem 3.170. □

**Definition 3.172.** Assume the homogeneous BVP (3.112) has only the trivial solution. Then we define the *Green's function,* $G(t,s)$, for the homogeneous BVP 3.112 by

$$G(t,s) := \begin{cases} u(t,s), & a \leq t \leq s \leq b, \\ v(t,s), & a \leq s \leq t \leq b, \end{cases}$$

where for each fixed $s \in \mathbb{N}_{a+1}^{b}$, $u(t,s)$ is the unique solution (guaranteed by Theorem 3.169) of the BVP

$$\begin{cases} L_a u(t) = 0, \quad t \in \mathbb{N}_{a+1}^{b-1} \\ \alpha u(a,s) - \beta\nabla u(a+1,s) = 0, \\ \gamma u(b,s) + \delta\nabla u(b,s) = -[\gamma x(b,s) + \delta\nabla x(b,s)], \end{cases}$$

and

$$v(t,s) := u(t,s) + x(t,s),$$

where $x(t,s)$ is the Cauchy function for $L_a x(t) = 0$.

Note that for each fixed $s \in \mathbb{N}_a^b$, $v(t,s) = u(t,s) + x(t,s)$ is a solution of $L_a x(t) = 0$ and since

$$\begin{aligned}\gamma v(b,s) + \delta\nabla u(b,s) &= \gamma[u(b,s) + x(b-1,s)] + \delta\nabla[u(b,s) + x(b,s)] \\ &= [\gamma u(b,s) + \delta\nabla_t u(b,s)] + [\gamma x(b,s) + \delta\nabla x(b,s)] \\ &= -[\gamma x(b,s) + \delta\nabla_t x(b,s)] + [\gamma x(b,s) + \delta\nabla x(b,s)] \\ &= 0,\end{aligned}$$

we have that for each fixed $s \in \mathbb{N}_a^b$ the function $v(t,s)$ satisfies the homogeneous boundary condition in (3.112) at $t = b$.

**Theorem 3.173 (Green's Function).** *If* (3.112) *has only the trivial solution, then the unique solution to the BVP*

$$\begin{cases} L_a y(t) = h(t), \quad t \in \mathbb{N}_{a+1}^{b-1} \\ \alpha y(a) - \beta\nabla y(a+1) = 0, \\ \gamma y(b) + \delta\nabla y(b) = 0, \end{cases} \tag{3.115}$$

*is given by*

$$y(t) = \int_a^b G(t,s)h(s)\nabla s, \quad t \in \mathbb{N}_a^b,$$

*where* $G(t,s)$ *is the Green's function for the homogeneous BVP* (3.112).

*Proof.* Let

$$\begin{aligned}y(t) := \int_a^b G(t,s)h(s)\nabla s &= \int_a^t G(t,s)h(s)\nabla s + \int_t^b G(t,s)h(s)\nabla s \\ &= \int_a^t v(t,s)h(s)\nabla s + \int_t^b u(t,s)h(s)\nabla s \\ &= \int_a^t [u(t,s) + x(t,s)]h(s)\nabla s + \int_t^b u(t,s)h(s)\nabla s \\ &= \int_a^b u(t,s)h(s)\nabla s + \int_a^t x(t,s)h(s)\nabla s \\ &= \int_a^b u(t,s)h(s)\nabla s + z(t),\end{aligned}$$

where $z(t) := \int_a^t x(t,s)h(s)\nabla s$. Since $x(t,s)$ is the Cauchy function for $L_a x(t) = 0$, it follows that $z$ solves the IVP

$$\begin{cases} L_a z(t) = h(t), & t \in \mathbb{N}_{a+1}^{b-1} \\ z(a) = 0, \\ z(a+1) = 0, \end{cases}$$

for $t \in \mathbb{N}_{a+1}^b$, by Theorem 3.165. Then,

$$\begin{aligned} L_a y(t) &= \int_a^b L_a u(t,s)h(s)\nabla s + L_a z(t) \\ &= 0 + h(t) = h(t), \end{aligned}$$

for $t \in \mathbb{N}_{a+1}^b$. It remains to show that the boundary conditions hold. At $t = a$, we have

$$\begin{aligned} \alpha y(a) - \beta \nabla y(a+1) &= \int_a^b [\alpha u(a,s) - \beta \nabla u(a+1,s)]h(s)\nabla s \\ &\quad + [\alpha z(a) - \beta \nabla z(a+1)] = 0, \end{aligned}$$

and at $t = b$, we have

$$\begin{aligned} &\gamma y(b) + \delta \nabla y(b) \\ &= \gamma z(b) + \int_a^b \gamma u(b,s)h(s)\nabla s + \delta \nabla z(b) + \int_a^b \delta \nabla u(b,s)h(s)\nabla s \\ &= \gamma \int_a^b x(b,s)h(s)\nabla s + \delta \nabla \int_a^b x(b,s)h(s)\nabla s \\ &\quad + \int_a^b [\gamma u(b,s) + \delta \nabla u(b,s)]h(s)\nabla s \\ &= -\int_a^b [\gamma x(b,s) + \delta \nabla x(b,s)]h(s)\nabla s + \int_a^b [\gamma x(b,s) + \delta \nabla x(b,s)]h(s)\nabla s \\ &= 0. \end{aligned}$$

This completes the proof. □

**Corollary 3.174.** *If the homogeneous BVP* (3.112) *has only the trivial solution, then the unique solution of the nonhomogeneous BVP* (3.111) *is given by*

$$y(t) = z(t) + \int_a^b G(t,s)h(s)\nabla s, \quad t \in \mathbb{N}_a^b,$$

*where $z$ is the unique solution to the BVP*

$$\begin{cases} L_a z(t) = 0, & t \in \mathbb{N}_{a+1}^{b-1} \\ \alpha z(a) - \beta \nabla z(a+1) = A, \\ \gamma z(b) + \delta \nabla z(b) = B. \end{cases}$$

*Proof.* This corollary follows directly from Theorem 3.173 by linearity. □

**Theorem 3.175.** *Assume $a, b \in \mathbb{R}$ and $b - a \in \mathbb{N}_2$. Then the Green's function for the BVP*

$$\begin{cases} \nabla \nabla_{a*}^{\nu} x(t+1) = 0, & t \in \mathbb{N}_{a+1}^{b-1} \\ x(a) = x(b) = 0, \end{cases} \tag{3.116}$$

*is given by*

$$G(t,s) = \begin{cases} u(t,s), & a \leq t \leq s \leq b, \\ v(t,s), & a \leq s \leq t \leq b, \end{cases}$$

*where*

$$u(t,s) = -\frac{(b-s)^{\overline{\nu}}(t-a)^{\overline{\nu}}}{\Gamma(1+\nu)(b-a)^{\overline{\nu}}}$$

*and*

$$v(t,s) = u(t,s) + \frac{(t-s)^{\overline{\nu}}}{\Gamma(\nu+1)}.$$

*Proof.* By the definition of the Green's function for the boundary value problem (3.116) we have that

$$G(t,s) = \begin{cases} u(t,s), & a \leq t \leq s \leq b, \\ v(t,s), & a \leq s \leq t \leq b, \end{cases}$$

where $u(t,s)$ for each fixed $s$ solves the BVP

$$\begin{cases} \nabla[\nabla_{a*}^{\nu} u(t+1,s)] = 0, \\ u(a,s) = 0, \\ u(b,s) = -x(b,s), \end{cases}$$

for $t \in \mathbb{N}_{a+1}$, and $v(t,s) = u(t,s) + x(t,s)$. By inspection, we see that $x_1(t) = 1$ is a solution of

$$\nabla[\nabla_{a*}^{\nu} y(t+1)] = 0,$$

for $t \in \mathbb{N}_{a+1}$. Also if $x_2(t) := (\nabla_a^{-\nu} 1)(t)$ we have that

$$\begin{aligned}
\nabla[\nabla_{a*}^{\nu} x_2(t+1)] &= \nabla[\nabla_{a*}^{\nu} \nabla_a^{\nu}(1)] \\
&= \nabla[\nabla_a^{-(N-\nu)} \nabla^N \nabla_a^{\nu}(1)] \\
&= \nabla[\nabla_a^{-(N-\nu)} \nabla_a^{(N-\nu)}(1)] \\
&= (\nabla 1)(t) \\
&= 0,
\end{aligned}$$

so $x_2(t)$ also is a solution $\nabla[\nabla_{a*}^{\nu} y(t+1)] = 0$. Since $x_1(t)$ and $x_2(t)$ are linearly independent, by Theorem 3.160 the general solution is given by

$$y(t) = c_1 + c_2(\nabla_a^{-\nu} 1)(t) = c_1 + c_2 \frac{(t-a)^{\overline{\nu}}}{\Gamma(1+\nu)},$$

and it follows that

$$u(t,s) = c_1(s) + c_2(s) \frac{(t-a)^{\overline{\nu}}}{\Gamma(1+\nu)}.$$

The boundary condition $u(a,s) = 0$ implies that $c_1(s) = 0$. The boundary condition $u(b,s) = -x(b,s)$ then yields

$$-x(b,s) = u(b,s) = c_2(s) \frac{(b-a)^{\overline{\nu}}}{\Gamma(1+\nu)}.$$

From Example 3.163, we know that

$$x(b,s) = (\nabla_s^{-\nu} 1)(b) = \frac{(b-s)^{\overline{\nu}}}{\Gamma(1+\nu)},$$

and thus

$$c_2(s) = -\frac{(b-s)^{\overline{\nu}}}{(b-a)^{\overline{\nu}}}.$$

Hence the Green's function is given by

$$G(t,s) = \begin{cases} -\dfrac{(b-s)^{\overline{\nu}}(t-a)^{\overline{\nu}}}{\Gamma(1+\nu)(b-a)^{\overline{\nu}}}, & a \le t \le s \le b, \\ -\dfrac{(b-s)^{\overline{\nu}}(t-a)^{\overline{\nu}}}{\Gamma(1+\nu)(b-a)^{\overline{\nu}}} + \dfrac{(t-s)^{\overline{\nu}}}{\Gamma(1+\nu)}, & a \le s \le t \le b. \end{cases}$$

□

*Remark 3.176.* Note that in the continuous and integer-order discrete cases, the Green's function is symmetric. This is not necessarily true in the fractional case. By way of counterexample, take $a = 0$, $b = 5$, and $\nu = 0.5$. Then one can show that

$$G(2,3) = u(2,3) = -\frac{(2)^{\overline{0.5}}(2)^{\overline{0.5}}}{\Gamma(1.5)(5)^{\overline{0.5}}} = -\frac{32}{35},$$

but

$$G(3,2) = v(3,2) = -\frac{(3)^{\overline{0.5}}(3)^{\overline{0.5}}}{\Gamma(1.5)(5)^{\overline{0.5}}} + \frac{(1)^{\overline{0.5}}}{\Gamma(1.5)} = -\frac{3}{7}.$$

**Theorem 3.177.** *Assume $a, b \in \mathbb{R}$ and $b - a \in \mathbb{N}_2$. Then the Green's function for the BVP*

$$\begin{cases} \nabla\nabla^{\nu}_{a*}x(t+1) = 0, & t \in \mathbb{N}^{b-1}_{a+1} \\ x(a) = x(b) = 0, \end{cases}$$

*satisfies the inequalities*

(i) $G(t,s) \leq 0$,

(ii) $G(t,s) \geq -\left(\frac{b-a}{4}\right)\left(\frac{\Gamma(b-a+1)}{\Gamma(\nu+1)\Gamma(b-a+\nu)}\right)$,

(iii) $\int_a^b |G(t,s)|\nabla s \leq \frac{(b-a)^2}{4\Gamma(\nu+2)}$,

*for $t \in \mathbb{N}^b_a$, and*

(iv) $\int_a^b |\nabla_t G(t,s)|\nabla s \leq \frac{b-a}{\nu+1}$,

*for $t \in \mathbb{N}^b_{a+1}$.*

*Proof.* First we show that (i) holds. Let $a \leq t \leq s \leq b$. Then

$$G(t,s) = u(t,s) = -\frac{(t-a)^{\overline{\nu}}(b-s)^{\overline{\nu}}}{\Gamma(\nu+1)(b-a)^{\overline{\nu}}} \leq 0.$$

Now let $a \leq s < t \leq b$. Then $G(t,s) = v(t,s)$, so we wish to show that $v(t,s)$ is nonpositive. First, we show that $v(t,s)$ is increasing. Taking the nabla difference with respect to $t$ yields

$$\nabla_t\left[-\frac{(t-a)^{\overline{\nu}}(b-s)^{\overline{\nu}}}{\Gamma(\nu+1)(b-a)^{\overline{\nu}}} + \frac{(t-s)^{\overline{\nu}}}{\Gamma(1+\nu)}\right] = -\frac{(t-a)^{\overline{\nu-1}}(b-s)^{\overline{\nu}}}{\Gamma(\nu)(b-a)^{\overline{\nu}}} + \frac{(t-s)^{\overline{\nu-1}}}{\Gamma(\nu)}.$$

This expression is nonnegative if and only if

$$\frac{(t-a)^{\overline{\nu-1}}(b-s)^{\overline{\nu}}}{\Gamma(\nu)(b-a)^{\overline{\nu}}} \le \frac{(t-s)^{\overline{\nu-1}}}{\Gamma(\nu)}.$$

Since $t-s$ is positive, this is equivalent to

$$\frac{(t-a)^{\overline{\nu-1}}(b-s)^{\overline{\nu}}}{(b-a)^{\overline{\nu}}(t-s)^{\overline{\nu-1}}} \le 1.$$

By the definition of the rising function,

$$\begin{aligned}
&\frac{(t-a)^{\overline{\nu-1}}(b-s)^{\overline{\nu}}}{(b-a)^{\overline{\nu}}(t-s)^{\overline{\nu-1}}}\\
&= \left[\frac{\Gamma(t-a+\nu-1)}{\Gamma(t-a)}\right]\left[\frac{\Gamma(b-s+\nu)}{\Gamma(b-s)}\right]\left[\frac{\Gamma(b-a)}{\Gamma(b-a+\nu)}\right]\left[\frac{\Gamma(t-s)}{\Gamma(t-s+\nu-1)}\right]\\
&= \left[\frac{\Gamma(t-a+\nu-1}{\Gamma(t-s+\nu-1)}\right]\left[\frac{\Gamma(t-s)}{\Gamma(t-a)}\right]\left[\frac{\Gamma(b-a)}{\Gamma(b-s)}\right]\left[\frac{\Gamma(b-s+\nu)}{\Gamma(b-a+\nu)}\right]\\
&= \frac{(t-s+\nu-1)(t-s+\nu)\cdots(t-a+\nu-2)}{(t-s)(t-s+1)\cdots(t-\rho(a))}\\
&\qquad \cdot \frac{(b-s)(b-s+1)\cdots(b-\rho(a))}{(b-s+\nu)(b-s+\nu+1)\cdots(b-a+\nu-1)}\\
&= \frac{(t-s+\nu-1)}{(t-s)}\frac{(t-s+\nu)}{(t-s+1)}\cdots\frac{(t-a+\nu-2)}{(t-\rho(a))}\\
&\qquad \cdot \frac{(b-s)}{(b-s+\nu)}\frac{(b-s+1)}{(b-s+\nu+1)}\cdots\frac{(b-\rho(a))}{(b-a+\nu-1)}\\
&\le 1
\end{aligned}$$

since each factor in the second to last expression is less than or equal to one. Next, we note that $v(t,s)$ at the right endpoint, $t=b$, satisfies

$$v(b,s) = -\frac{(b-a)^{\overline{\nu}}(b-s)^{\overline{\nu}}}{\Gamma(\nu+1)(b-a)^{\overline{\nu}}} + \frac{(b-s)^{\overline{\nu}}}{\Gamma(\nu+1)} = 0.$$

Thus, $v(t,s)$ is nonpositive for $a \le s \le t \le b$. Therefore, for $t \in \mathbb{N}_a^b$, $G(t,s)$ is nonpositive.

Next we show that (ii) holds. Since we know that $v(t,s)$ is always increasing for $a \le s \le t \le b$ and that for $s=t$, $v(t,s)=u(t,s)$, it suffices to show that

$$u(t,s) \ge -\frac{b-a}{4}\left(\frac{\Gamma(b-a+1)}{\Gamma(\nu+1)\Gamma(b-a+\nu)}\right).$$

Let $a \leq t \leq s \leq b$. Then

$$G(t,s) = u(t,s) = -\frac{(t-a)^{\overline{\nu}}(b-s)^{\overline{\nu}}}{\Gamma(\nu+1)(b-a)^{\overline{\nu}}} \geq -\frac{(s-a)^{\overline{\nu}}(b-s)^{\overline{\nu}}}{\Gamma(\nu+1)(b-a)^{\overline{\nu}}}.$$

Note that for $\alpha \in \mathbb{N}_1$ and $0 < \nu < 1$,

$$\alpha^{\overline{\nu}} = \frac{\Gamma(\alpha+\nu)}{\Gamma(\alpha)} \leq \frac{\Gamma(\alpha+1)}{\Gamma(\alpha)} = \alpha^{\overline{1}}.$$

So

$$\begin{aligned}
-\frac{(s-a)^{\overline{\nu}}(b-s)^{\overline{\nu}}}{\Gamma(\nu+1)(b-a)^{\overline{\nu}}} &\geq -\frac{(s-a)^{\overline{1}}(b-s)^{\overline{1}}}{\Gamma(\nu+1)(b-a)^{\overline{\nu}}} \\
&\geq -\frac{(\frac{a+b}{2}-a)(b-\frac{a+b}{2})}{\Gamma(\nu+1)(b-a)^{\overline{\nu}}} \\
&= -\frac{(b-a)(b-a)\Gamma(b-a)}{4\Gamma(\nu+1)\Gamma(b-a+\nu)} \\
&= -\frac{(b-a)\Gamma(b-a+1)}{4\Gamma(\nu+1)\Gamma(b-a+\nu)} \\
&= -\frac{b-a}{4}\left(\frac{\Gamma(b-a+1)}{\Gamma(\nu+1)\Gamma(b-a+\nu)}\right),
\end{aligned}$$

and hence (ii) holds.

Now we show property (iii) holds. Thus, we compute

$$\begin{aligned}
&\int_a^b |G(t,s)|\nabla s \\
&= \int_a^t |v(t,s)|\nabla s + \int_t^b |u(t,s)|\nabla s \\
&= \int_a^t \left| -\frac{(t-a)^{\overline{\nu}}(b-s)^{\overline{\nu}}}{\Gamma(\nu+1)(b-a)^{\overline{\nu}}} + \frac{(t-s)^{\overline{\nu}}}{\Gamma(1+\nu)} \right| \nabla s + \int_t^b \frac{(t-a)^{\overline{\nu}}(b-s)^{\overline{\nu}}}{\Gamma(\nu+1)(b-a)^{\overline{\nu}}}\nabla s \\
&= \int_a^t -\left[ -\frac{(t-a)^{\overline{\nu}}(b-s)^{\overline{\nu}}}{\Gamma(\nu+1)(b-a)^{\overline{\nu}}} + \frac{(t-s)^{\overline{\nu}}}{\Gamma(1+\nu)} \right] \nabla s + \int_t^b \frac{(t-a)^{\overline{\nu}}(b-s)^{\overline{\nu}}}{\Gamma(\nu+1)(b-a)^{\overline{\nu}}}\nabla s \\
&= \int_a^b \frac{(t-a)^{\overline{\nu}}(b-s)^{\overline{\nu}}}{\Gamma(\nu+1)(b-a)^{\overline{\nu}}}\nabla s - \int_a^t \frac{(t-s)^{\overline{\nu}}}{\Gamma(\nu+1)}\nabla s \\
&= -\frac{(t-a)^{\overline{\nu}}(b-s-1)^{\overline{\nu+1}}}{\Gamma(\nu+2)(b-a)^{\overline{\nu}}}\Bigg|_{s=a}^{s=b} + \frac{(t-s-1)^{\overline{\nu+1}}}{\Gamma(\nu+2)}\Bigg|_{s=a}^{s=t}
\end{aligned}$$

$$= \frac{(t-a)^{\overline{\nu}}(b-\rho(a))^{\overline{\nu+1}}}{\Gamma(\nu+2)(b-a)^{\overline{\nu}}} - \frac{(t-\rho(a))^{\overline{\nu+1}}}{\Gamma(\nu+2)}$$

$$= \frac{(t-a)^{\overline{\nu}}(b-\rho(a))(b-a)^{\overline{\nu}}}{\Gamma(\nu+2)(b-a)^{\overline{\nu}}} - \frac{(t-\rho(a))(t-a)^{\overline{\nu}}}{\Gamma(\nu+2)}$$
$$= \frac{(t-a)^{\overline{\nu}}}{\Gamma(\nu+2)}[b-\rho(a)-(t-\rho(a))]$$
$$= \frac{(t-a)^{\overline{\nu}}(b-t)}{\Gamma(\nu+2)}.$$

Hence,

$$\int_a^b |G(t,s)|\nabla s \leq \frac{(t-a)(b-t)}{\Gamma(\nu+2)}$$
$$\leq \frac{(\frac{a+b}{2}-a)(b-\frac{a+b}{2})}{\Gamma(\nu+2)}$$
$$= \frac{(b-a)^2}{4\Gamma(\nu+2)}.$$

Finally, we will show that (iv) holds. First assume that $b-a>1$. Taking the difference with respect to $t$, we have

$$\nabla_t u(t,s) = \nabla_t \frac{-(t-a)^{\overline{\nu}}(b-s)^{\overline{\nu}}}{\Gamma(\nu+1)(b-a)^{\overline{\nu}}} = \frac{-\nu(t-a)^{\overline{\nu-1}}(b-s)^{\overline{\nu}}}{\Gamma(\nu+1)(b-a)^{\overline{\nu}}} \leq 0.$$

For $t \in \mathbb{N}_{a+1}^b$ we compute

$$\int_a^b |\nabla_t G(t,s)|\, \nabla s$$
$$= \int_a^{t-1} |\nabla_t G(t,s)|\, \nabla s + \int_{t-1}^b |\nabla_t G(t,s)|\, \nabla s$$
$$= \int_a^{t-1} |\nabla_t v(t,s)|\, \nabla s + \int_{t-1}^b |\nabla_t u(t,s)|\, \nabla s$$
$$= \int_a^{t-1} \nabla_t \left[\frac{-(t-a)^{\overline{\nu}}(b-s)^{\overline{\nu}}}{\Gamma(\nu+1)(b-a)^{\overline{\nu}}} + \frac{(t-s)^{\overline{\nu}}}{\Gamma(\nu+1)}\right] \nabla s$$
$$+ \int_{t-1}^b \nabla_t \left[\frac{(t-a)^{\overline{\nu}}(b-s)^{\overline{\nu}}}{\Gamma(\nu+1)(b-a)^{\overline{\nu}}}\right] \nabla s$$

$$= \int_a^{t-1} \nabla_t \left[ \frac{-(t-a)^{\overline{\nu}}(b-s)^{\overline{\nu}}}{\Gamma(\nu+1)(b-a)^{\overline{\nu}}} \right] \nabla s + \int_a^{t-1} \nabla_t \left[ \frac{(t-s)^{\overline{\nu}}}{\Gamma(\nu+1)} \right] \nabla s$$

$$+ \int_{t-1}^{b} \nabla_t \left[ \frac{(t-a)^{\overline{\nu}}(b-s)^{\overline{\nu}}}{\Gamma(\nu+1)(b-a)^{\overline{\nu}}} \right] \nabla s$$

$$= \int_a^{t-1} \left[ \frac{-\nu(t-a)^{\overline{\nu-1}}(b-s)^{\overline{\nu}}}{\Gamma(\nu+1)(b-a)^{\overline{\nu}}} \right] \nabla s$$

$$+ \int_a^{t-1} \left[ \frac{\nu(t-s)^{\overline{\nu-1}}}{\Gamma(\nu+1)} \right] \nabla s + \int_{t-1}^{b} \left[ \frac{\nu(t-a)^{\overline{\nu-1}}(b-s)^{\overline{\nu}}}{\Gamma(\nu+1)(b-a)^{\overline{\nu}}} \right] \nabla s$$

$$= \int_a^{t-1} \left[ \frac{-\nu(t-a)^{\overline{\nu-1}}[(b-1)-\rho(s)]^{\overline{\nu}}}{\Gamma(\nu+1)(b-a)^{\overline{\nu}}} \right] \nabla s$$

$$+ \int_a^{t-1} \left[ \frac{\nu[(t-1)-\rho(s)]^{\overline{\nu-1}}}{\Gamma(\nu+1)} \right] \nabla s$$

$$+ \int_{t-1}^{b} \left[ \frac{\nu(t-a)^{\overline{\nu-1}}[(b-1)-\rho(s)]^{\overline{\nu}}}{\Gamma(\nu+1)(b-a)^{\overline{\nu}}} \right] \nabla s$$

$$= \frac{-\nu(t-a)^{\overline{\nu-1}}}{\Gamma(\nu+1)(b-a)^{\overline{\nu}}} \left[ \frac{-1}{\nu+1}(b-s-1)^{\overline{\nu+1}} \right]_{s=a}^{s=t-1}$$

$$+ \frac{\nu}{\Gamma(\nu+1)} \left[ \frac{-1}{\nu}(t-s-1)^{\overline{\nu}} \right]_{s=a}^{s=t-1}$$

$$+ \frac{\nu(t-a)^{\overline{\nu-1}}}{\Gamma(\nu+1)(b-a)^{\overline{\nu}}} \left[ \frac{-1}{\nu+1}(b-s-1)^{\overline{\nu+1}} \right]_{s=t-1}^{s=b}$$

$$= \frac{\nu(t-a)^{\overline{\nu-1}}}{\Gamma(\nu+2)(b-a)^{\overline{\nu}}} \left[ (b-t)^{\overline{\nu+1}} - (b-\rho(a))^{\overline{\nu+1}} \right]$$

$$- \frac{1}{\Gamma(\nu+1)} \left[ (t-t+1-1)^{\overline{\nu}} - (t-\rho(a))^{\overline{\nu}} \right]$$

$$+ \frac{-\nu(t-a)^{\overline{\nu-1}}}{\Gamma(\nu+2)(b-a)^{\overline{\nu}}} \left[ (b-b-1)^{\overline{\nu+1}} - (b-t)^{\overline{\nu}} \right]$$

$$= \frac{2\nu(t-a)^{\overline{\nu-1}}(b-t)^{\overline{\nu+1}}}{\Gamma(\nu+2)(b-a)^{\overline{\nu}}} + \frac{(t-\rho(a))^{\overline{\nu}}}{\Gamma(\nu+1)} - \frac{\nu(t-a)^{\overline{\nu-1}}(b-\rho(a))}{\Gamma(\nu+2)}.$$

Suppose $t = b$. This would imply that

$$\int_a^b |\nabla_t G(t,s)| \nabla s = \frac{2\nu(b-a)^{\overline{\nu-1}}(0)^{\overline{\nu+1}}}{\Gamma(\nu+2)(b-a)^{\overline{\nu}}}$$

$$+\frac{(b-\rho(a))^{\overline{\nu}}}{\Gamma(\nu+1)}-\frac{\nu(b-a)^{\overline{\nu-1}}(b-\rho(a))}{\Gamma(\nu+2)}$$
$$=\frac{(\nu+1)(b-\rho(a))^{\overline{\nu}}}{\Gamma(\nu+2)}-\frac{\nu(b-a)^{\overline{\nu-1}}(b-\rho(a))}{\Gamma(\nu+2)}.$$

For $t=b$ and $b-a=2$, this becomes

$$\begin{aligned}\int_a^b |\nabla_t G(t,s)|\,\nabla s &= \frac{(\nu+1)(1)^{\overline{\nu}}}{\Gamma(\nu+2)}-\frac{\nu(2)^{\overline{\nu-1}}(1)}{\Gamma(\nu+2)}\\ &= \frac{(\nu+1)\Gamma(\nu+1)}{\Gamma(\nu+2)}-\frac{\nu\Gamma(\nu+1)}{\Gamma(\nu+2)}\\ &= 1-\frac{\nu}{\nu+1}\\ &= \frac{1}{\nu+1}\\ &\leq \frac{2}{\nu+1}\\ &= \frac{b-a}{\nu+1}.\end{aligned}$$

On the other hand, for $t=b$ and $b-a=3$, we have

$$\begin{aligned}\int_a^b |\nabla_t G(t,s)|\,\nabla s &= \frac{(\nu+1)(2^{\overline{\nu}})}{\Gamma(\nu+2)}-\frac{\nu(3^{\overline{\nu-1}})(2)}{\Gamma(\nu+2)}\\ &= \frac{(\nu+1)\Gamma(\nu+2)}{\Gamma(\nu+2)}-\frac{2\nu\Gamma(\nu+2)}{\Gamma(\nu+2)\Gamma(3)}\\ &= 1\\ &\leq \frac{b-a}{\nu+1}.\end{aligned}$$

For $t=b$ and $b-a\geq 4$, the result holds since

$$\begin{aligned}&\int_a^b |\nabla_t G(t,s)|\,\nabla s\\ &= \frac{(\nu+1)(b-\rho(a))^{\overline{\nu}}}{\Gamma(\nu+2)}-\frac{\nu(b-a)^{\overline{\nu-1}}(b-\rho(a))}{\Gamma(\nu+2)}\\ &= \frac{(\nu+1)(b-a-2+\nu)\cdots(2+\nu)}{\Gamma(b-\rho(a))}-\frac{\nu\Gamma(b-\rho(a)+\nu)(b-\rho(a))}{\Gamma(2+\nu)\Gamma(b-a)}\end{aligned}$$

$$= \frac{(\nu+1)(b-a-2+\nu)\cdots(2+\nu)}{\Gamma(b-\rho(a))} - \frac{\nu\Gamma(b-\rho(a)+\nu)}{\Gamma(2+\nu)\Gamma(b-\rho(a))}$$
$$= \frac{(\nu+1)(b-a-2+\nu)\cdots(2+\nu)}{(b-a-2)!} - \frac{\nu(b-a-2+\nu)\cdots(2+\nu)}{(b-a-2)!}$$
$$= \frac{(b-a-2+\nu)\cdots(2+\nu)}{(b-a-2)!}$$
$$= \frac{(b-\rho(a))(b-a-2+\nu)\cdots(2+\nu)}{(b-\rho(a))!}$$
$$\leq \frac{(b-\rho(a))(b-\rho(a))\cdots(3)}{(b-\rho(a))!}$$
$$= \frac{\frac{1}{2}(b-\rho(a))!(b-\rho(a))}{(b-\rho(a))!} = \frac{b-\rho(a)}{2} \leq \frac{b-a}{\nu+1}.$$

So the result holds in general when $t = b$. Now, assume $t < b$. If $t = a + 1$, then we have

$$\int_a^b |\nabla_t G(t,s)| \nabla s$$
$$= \frac{2\nu(1^{\overline{\nu-1}})(b-\rho(a))^{\overline{\nu+1}} + (\nu+1)(0^{\overline{\nu}})(b-a)^{\overline{\nu}} - \nu(1^{\overline{\nu-1}})(b-\rho(a))^{\overline{\nu+1}}}{\Gamma(\nu+2)(b-a)^{\overline{\nu}}}$$
$$= \frac{2\nu\Gamma(\nu)(b-\rho(a))(b-a)^{\overline{\nu}} - \nu\Gamma(\nu)(b-\rho(a))(b-a)^{\overline{\nu}}}{\Gamma(\nu+2)(b-a)^{\overline{\nu}}}$$
$$= \frac{2\nu\Gamma(\nu)(b-\rho(a)) - \nu\Gamma(\nu)(b-\rho(a))}{\Gamma(\nu+2)}$$
$$= \frac{\Gamma(\nu+1)(b-\rho(a))}{\Gamma(\nu+2)} = \frac{\Gamma(\nu+1)(b-\rho(a))}{(\nu+1)\Gamma(\nu+1)} \frac{b-\rho(a)}{\nu+1} \leq \frac{b-a}{\nu+1}.$$

If $t = a + 2$, then

$$\int_a^b |\nabla_t G(t,s)| \nabla s$$
$$= \frac{2\nu(2^{\overline{\nu-1}})(b-a-2)^{\overline{\nu+1}} + (\nu+1)(1^{\overline{\nu}})(b-a)^{\overline{\nu}} - \nu(2^{\overline{\nu-1}})(b-\rho(a))^{\overline{\nu+1}}}{\Gamma(\nu+2)(b-a)^{\overline{\nu}}}$$
$$= \frac{2\nu\Gamma(\nu+1)(b-a-2)^{\overline{\nu+1}}}{\Gamma(\nu+2)(b-a)^{\overline{\nu}}} + \frac{\nu(\nu+1)\Gamma(\nu)(b-a)^{\overline{\nu}}}{\Gamma(\nu+2)(b-a)^{\overline{\nu}}}$$
$$- \frac{\nu\Gamma(\nu+1)(b-\rho(a))^{\overline{\nu+1}}}{\Gamma(\nu+2)(b-a)^{\overline{\nu}}}$$

$$
\begin{aligned}
&= \frac{2\nu\Gamma(\nu+1)(b-\rho(a))(b-a-2)}{(\nu+1)(b-\rho(a)+\nu)} + 1 - \frac{\nu(b-\rho(a))}{\nu+1} \\
&\leq \frac{2\nu(b-a-2)}{\nu+1} + 1 - \frac{\nu(b-\rho(a))}{\nu+1} \\
&= \frac{2\nu(b-a-2)}{\nu+1} + \frac{\nu+1}{\nu+1} - \frac{\nu(b-\rho(a))}{\nu+1} \\
&= \frac{\nu(b-a-2)+1}{\nu+1} \leq \frac{b-\rho(a)}{\nu+1} \leq \frac{b-a}{\nu+1}.
\end{aligned}
$$

If $t = a + 3$, then

$$
\begin{aligned}
&\int_a^b |\nabla_t G(t,s)|\, \nabla s \\
&= \frac{2\nu(3^{\overline{\nu-1}})(b-a-3)^{\overline{\nu+1}} + (\nu+1)(2^{\overline{\nu}})(b-a)^{\overline{\nu}} - \nu(3^{\overline{\nu-1}})(b-\rho(a))^{\overline{\nu+1}}}{\Gamma(\nu+2)(b-a)^{\overline{\nu}}} \\
&= \frac{2\nu\Gamma(2+\nu)\Gamma(b-a-2+\nu)\Gamma(b-a)}{\Gamma(3)\Gamma(b-a+\nu)\Gamma(\nu+2)\Gamma(b-a-3)} + (\nu+1) \\
&\quad - \frac{\nu\Gamma(\nu+2)(b-\rho(a))}{\Gamma(\nu+2)\Gamma(3)} \\
&= \frac{\nu(b-\rho(a))(b-a-2)(b-a-3)}{(b-\rho(a)+\nu)(b-a-2+\nu)} + (\nu+1) - \frac{\nu(b-\rho(a))}{2}
\end{aligned}
$$

It follows that

$$
\begin{aligned}
&\int_a^b |\nabla_t G(t,s)|\, \nabla s \\
&\leq \nu(b-a-3) + \nu + 1 - \frac{\nu(b-\rho(a))}{2} \\
&= \frac{2\nu b - 2\nu a - 6\nu + 2\nu + 2 - \nu b + \nu a + \nu}{2} \\
&= \frac{\nu(b-a-3)+2}{2} \\
&\leq \frac{b-a-3+2}{2} \\
&= \frac{b-\rho(a)}{2} \\
&\leq \frac{b-a}{\nu+1}.
\end{aligned}
$$

Now suppose that $t = a + k$, where $k \in \mathbb{N}_4^{b-\rho(a)}$. Then

$$\int_a^b |\nabla_t G(t,s)|\, \nabla s$$

$$= \frac{2\nu(k)^{\overline{\nu-1}}(b-a-k)^{\overline{\nu+1}}}{(b-a)^{\overline{\nu}}\Gamma(\nu+2)} + \frac{(\nu+1)(k-1)^{\overline{\nu}}}{\Gamma(\nu+2)} - \frac{\nu(k)^{\overline{\nu-1}}(b-\rho(a))}{\Gamma(\nu+2)}$$

$$= \frac{2\nu\Gamma(k+\nu-1)\Gamma(b-a-k+\nu+1)\Gamma(b-a)}{\Gamma(k)\Gamma(b-a+\nu)\Gamma(\nu+2)\Gamma(b-a-k)} + \frac{(\nu+1)\Gamma(k-1+\nu)}{\Gamma(\nu+2)\Gamma(k-1)}$$

$$- \frac{\nu\Gamma(k+\nu-1)(b-\rho(a))}{\Gamma(\nu+2)\Gamma(k)}$$

$$= \frac{2\nu(\nu+2)\dots(\nu+k-2)(b-\rho(a))\dots(b-a-k)}{(k-1)!(b-\rho(a)+\nu)\dots(b-a-(k-1)+\nu)}$$

$$+ \frac{(\nu+1)\dots(\nu+k-2)}{(k-2)!}$$

$$- \frac{\nu(\nu+2)\dots(\nu+k-2)(b-\rho(a))}{(k-1)!}$$

Hence,

$$\int_a^b |\nabla_t G(t,s)|\, \nabla s$$

$$\leq \frac{2\nu(\nu+2)\dots(\nu+k-2)(b-a-k)}{(k-1)!} + \frac{(k-1)(\nu+1)\dots(\nu+k-2)}{(k-1)!}$$

$$- \frac{\nu(\nu+2)\dots(\nu+k-2)(b-\rho(a))}{(k-1)!}$$

$$= \frac{\nu(\nu+2)\dots(\nu+k-2)(2b-2a-2k-b+a+1)}{(k-1)!}$$

$$+ \frac{(k-1)(\nu+1)\dots(\nu+k-2)}{(k-1)!}$$

$$= \frac{\nu(\nu+2)\dots(\nu+k-2)(b-a+1-2k)+(k-1)(\nu+1)\dots(\nu+k-2)}{(k-1)!}$$

$$\leq \frac{(1)(3)(4)\dots(k-1)(b-a+1-2k)+(k-1)(2)(3)\dots(k-1)}{(k-1)!}$$

$$= \frac{\frac{1}{2}(k-1)!(b-a+1-2k)+(k-1)(k-1)!}{(k-1)!}$$

$$
\begin{aligned}
&= \frac{(b-a+1-2k)+2(k-1)}{2} \\
&= \frac{b-\rho(a)}{2} \\
&\le \frac{b-a}{\nu+1}.
\end{aligned}
$$

And this completes the proof. □

*Example 3.178.* Use an appropriate Green's function to solve the BVP

$$
\begin{cases}
\nabla\nabla_{0*}^{0.5}x(t+1) = 1, & t \in \mathbb{N}_1^{b-1}, \\
x(0) = 0 = x(b),
\end{cases}
$$

where $b \in \mathbb{N}_2$. By Theorem 3.175 we have that the Green's function for the BVP

$$
\begin{cases}
\nabla\nabla_{0*}^{0.5}x(t+1) = 0, & t \in \mathbb{N}_1, \\
x(0) = 0 = x(b),
\end{cases}
\tag{3.117}
$$

is given by

$$
G(t,s) = \begin{cases}
u(t,s), & 0 \le t \le s \le b, \\
v(t,s), & 0 \le s \le t \le b,
\end{cases}
$$

where

$$
u(t,s) = -\frac{(b-s)^{\overline{0.5}}t^{\overline{0.5}}}{\Gamma(1.5)b^{\overline{0.5}}}, \quad 0 \le t \le s \le b
$$

and

$$
v(t,s) = -\frac{(b-s)^{\overline{0.5}}t^{\overline{0.5}}}{\Gamma(1.5)b^{\overline{0.5}}} + \frac{(t-s)^{\overline{0.5}}}{\Gamma(1.5)}, \quad 0 \le s \le t \le b.
$$

Then the solution of the BVP (3.117) is given by

$$
\begin{aligned}
x(t) &= \int_0^b G(t,s)h(s)\nabla s \\
&= \int_0^t v(t,s)h(s)\nabla s + \int_t^b G(t,s)h(s)\nabla s \\
&= \int_0^t \left[-\frac{(b-s)^{\overline{0.5}}t^{\overline{0.5}}}{\Gamma(1.5)b^{\overline{0.5}}} + \frac{(t-s)^{\overline{0.5}}}{\Gamma(1.5)}\right]\nabla s + \int_t^b \left[-\frac{(b-s)^{\overline{0.5}}t^{\overline{0.5}}}{\Gamma(1.5)b^{\overline{0.5}}}\right]\nabla s
\end{aligned}
$$

$$= -\frac{t^{\overline{0.5}}}{\Gamma(1.5)b^{\overline{0.5}}}\int_0^b (b-s)^{\overline{0.5}}\nabla s + \int_0^t \frac{(t-s)^{\overline{0.5}}}{\Gamma(1.5)}\nabla s$$

$$= -\frac{t^{\overline{0.5}}}{\Gamma(1.5)b^{\overline{0.5}}}\int_0^b [(b-1)-\rho(s)]^{\overline{0.5}}\nabla s + \int_0^t \frac{[(t-1)-\rho(s)]^{\overline{0.5}}}{\Gamma(1.5)}\nabla s$$

$$= \frac{t^{\overline{0.5}}}{\Gamma(1.5)b^{\overline{0.5}}}\frac{[(b-1)-s]^{\overline{1.5}}}{1.5}\Big|_{s=0}^{b} - \frac{[(t-1)-s]^{\overline{1.5}}}{\Gamma(1.5)1.5}\Big|_{s=0}^{t}$$

$$= -\frac{1}{1.5\Gamma(1.5)b^{\overline{.5}}}t^{\overline{.5}} + \frac{1}{1.5\Gamma(1.5)}(t-1)^{\overline{1.5}}$$

$$= -\frac{4(b-1)^{\overline{1.5}}}{3\sqrt{\pi}b^{\overline{0.5}}}t^{\overline{0.5}} + \frac{4}{3\sqrt{\pi}}(t-1)^{\overline{1.5}}$$

for $t \in \mathbb{N}_0^b$.

## 3.25 Exercises

**3.1.** Assume $f : \mathbb{N}_a^b \to \mathbb{R}$. Show that if $\nabla f(t) = 0$ for $t \in \mathbb{N}_{a+1}^b$, then $f(t) = C$ for $t \in \mathbb{N}_a^b$, where $C$ is a constant.

**3.2.** Assume $f, g : \mathbb{N}_a \to \mathbb{R}$. Prove the nabla quotient rule

$$\nabla\left(\frac{f(t)}{g(t)}\right) = \frac{g(t)\nabla f(t) - f(t)\nabla g(t)}{g(t)g(\rho(t))}, \quad g(t) \neq 0, \quad t \in \mathbb{N}_{a+1},$$

which is part (vi) of Theorem 3.1.

**3.3.** Prove that

$$(t-r+1)^{\overline{r}} = t^{\underline{r}}.$$

**3.4.** Prove that the box plus addition, $\boxplus$, on $\mathcal{R}$ is commutative and associative (see Theorem 3.8).

**3.5.** Show that if $p, q \in \mathcal{R}$, then

$$(p \boxminus q)(t) = \frac{p(t) - q(t)}{1 - q(t)}, \quad t \in \mathbb{N}_a.$$

**3.6.** Show that if $p \in \mathcal{R}^+$, then

$$\frac{1}{2} \boxdot p = \frac{p}{1+\sqrt{1-p}}.$$

**3.7.** Prove directly from the definition of the box dot multiplication, $\boxdot$, that if $p \in \mathcal{R}$ and $n \in \mathbb{N}_1$, then

$$n \boxdot p = p \boxplus p \boxplus p \cdots \boxplus p,$$

where the right-hand side of the above expression has $n$ terms.

**3.8.** Show that the set of positively regressive constants $\mathcal{R}^+$ with the addition $\boxplus$ is an Abelian subgroup of $\mathcal{R}$.

**3.9.** Prove part (vi) of Theorem 3.11 for the case $a \le s \le r$. That is, if $p \in \mathcal{R}$ and $a \le s \le r$, then

$$E_p(t,s)E_p(s,r) = E_p(t,r), \quad t \in \mathbb{N}_a.$$

**3.10.** Assume $p, q \in \mathcal{R}$ and $s \in \mathbb{N}_a$. Prove the law of exponents (Theorem 3.11, (vii))

$$E_p(t,s)E_q(t,s) = E_{p\boxplus q}(t,s), \quad t \in \mathbb{N}_a.$$

**3.11.** Prove that if $p, q \in \mathcal{R}$ and

$$E_p(t,a) = E_q(t,a), \quad t \in \mathbb{N}_a,$$

then $p(t) = q(t)$, $t \in \mathbb{N}_{a+1}$.

**3.12.** Show that if $\alpha, \beta \in \mathbb{R}$ and $p \in \mathcal{R}^+$, then

$$(\alpha + \beta) \boxdot p = (\alpha \boxdot p) \boxplus (\beta \boxdot p).$$

**3.13.** Show that if $p, -p \in \mathcal{R}$, then

$$\nabla Sinh_p(t,a) = p(t)Cosh(t,a), \quad t \in \mathbb{N}_a.$$

**3.14.** Show by direct substitution that $y(t) = (t-a)E_r(t,a)$, $r \neq 1$, is a nontrivial solution of the second order linear equation $\nabla^2 y(t) - 2r\nabla y(t) + r^2 y(t) = 0$ on $\mathbb{N}_a$.

**3.15.** Prove part (iii) of Theorem 3.18. That is, if $p \neq \pm i$ is a constant, then

$$\mathrm{Sinh}_{ip}(t,a) = i\ \mathrm{Sin}_p(t,a), \quad t \in \mathbb{N}_a.$$

**3.16.** Solve each of the following nabla difference equations:

(i) $\nabla^2 u(t) - 4\nabla u(t) + 5u(t) = 0, \quad t \in \mathbb{N}_0;$
(ii) $\nabla^2 u(t) - 4\nabla u(t) + 4u(t) = 0, \quad t \in \mathbb{N}_a;$
(iii) $\nabla^2 u(t) - 4\nabla u(t) - 5u(t) = 0, \quad t \in \mathbb{N}_a.$

**3.17.** Solve each of the following nabla linear difference equations:

(i) $\nabla^2 x(t) - 10\nabla x(t) + 25x(t) = 0, \quad t \in \mathbb{N}_0;$
(ii) $\nabla^2 x(t) - 9x(t) = 0, \quad t \in \mathbb{N}_a;$
(iii) $\nabla^2 x(t) + 2\nabla x(t) + 5x(t) = 0, \quad t \in \mathbb{N}_a.$

**3.18.** Solve each of the following nabla linear difference equations:

(i) $\nabla^2 y(t) - 2\nabla y(t) + 2y(t) = 0, \quad t \in \mathbb{N}_a;$
(ii) $\nabla^2 y(t) - 2\nabla y(t) + 10y(t) = 0, \quad t \in \mathbb{N}_a.$

**3.19.** Prove the nabla version of L'Hôpital's rule: If $f, g : \mathbb{N}_a \to \mathbb{R}$, and

$$\lim_{t\to\infty} f(t) = 0 = \lim_{t\to\infty} g(t)$$

and $g(t)\nabla g(t) < 0$ for large $t$, then

$$\lim_{t\to\infty} \frac{f(t)}{g(t)} = \lim_{t\to\infty} \frac{\nabla f(t)}{\nabla g(t)}$$

provided $\lim_{t\to\infty} \frac{\nabla f(t)}{\nabla g(t)}$ exists.

**3.20.** Use the integration formula

$$\int \alpha^{t+\beta} \nabla t = \frac{\alpha}{\alpha - 1}\alpha^{t+\beta} + C$$

to prove the integration formula

$$\int E_p(t, a)\nabla t = \frac{1}{p}E_p(t, a) + C.$$

**3.21.** Show that if $1 + p(t) + q(t) \neq 0$, for $t \in \mathbb{N}_{a+2}$, then the general solution of the linear homogeneous equation

$$\nabla^2 y(t) + p(t)\nabla y(t) + q(t)y(t) = 0$$

is given by

$$y(t) = c_1 y_1(t) + c_2 y_2(t), \quad t \in \mathbb{N}_a,$$

where $y_1(t), y_2(t)$ are any two linearly independent solutions of (3.13) on $\mathbb{N}_a$.

**3.22.** Assume $f : \mathbb{N}_a \times \mathbb{N}_{a+1} \to \mathbb{R}$. Prove the Leibniz formula (3.23). That is,

$$\nabla\left(\int_a^t f(t, \tau)\nabla\tau\right) = \int_a^{t-1} \nabla_t f(t, \tau)\nabla\tau + f(t, t),$$

for $t \in \mathbb{N}_{a+1}$.

**3.23.** Evaluate the nabla integral $\int_0^t f(\tau)\nabla\tau$, when

(i) $f(t) = t(-2)^t, \quad t \in \mathbb{N}_0$;
(ii) $f(t) = H_2(\rho(t), 0)E_3(t, 0), \quad t \in \mathbb{N}_0$.

**3.24.** Use the variation of constants formula in either Corollary 3.52 or Theorem 3.51 to solve each of the following IVPs.

(i)

$$\nabla^2 y(t) = 3^{-t}, \quad t \in \mathbb{N}_1$$
$$y(0) = 0, \quad \nabla y(0) = 0$$

(ii)

$$\nabla^2 y(t) = \mathrm{Sinh}_4(t, 0), \quad t \in \mathbb{N}_1$$
$$y(0) = -1, \quad \nabla y(0) = 1$$

(iii)

$$\nabla^2 y(t) = t - 2, \quad t \in \mathbb{N}_3$$
$$y(2) = 0, \quad \nabla y(2) = 0$$

**3.25.** Show that if $\mu > 0$, then

$$H_\mu(a+1, a) = 1 = H_{-\mu}(a+1, a).$$

**3.26.** Show if $\mu > 0$ is not a positive integer, then

$$H_\mu(t, a) = 0, \quad \text{for } t = a, a-1, a-2, \cdots.$$

Also show that if $\mu$ is a positive integer, then

$$H_\mu(t, a) = 0, \quad \text{for } t \in \mathbb{N}_{a-\mu+1}^{a}.$$

**3.27.** Show that if $f : \mathbb{N}_{a+1} \to \mathbb{R}$ and $\mu > 0$, then

$$\nabla_a^{-\mu} f(a+1) = f(a+1).$$

**3.28.** Use Definition 3.61 to show that if $\mu > 0$ is not an integer and $C$ is a constant, then

$$\nabla_a^\mu C = C H_{-\mu}(t, a), \quad t \in \mathbb{N}_a.$$

On the other hand, if $\mu = k$ is a positive integer, show that $\nabla_a^k C = 0$.

**3.29.** Use the definition (Definition 3.58) of the fractional sum to evaluate each of the following integer sums.

(i) $\nabla_a^{-2}\mathrm{Cosh}_3(t,a), \quad t \in \mathbb{N}_a$;
(ii) $\nabla_a^{-3}H_1(t,a), \quad t \in \mathbb{N}_a$;
(iii) $\nabla_2^{-2}\mathrm{Sin}_4(t,2), \quad t \in \mathbb{N}_2$.

**3.30.** For $p \neq 0, 1$ a constant, show that each of the functions $E_p(t,a)$, $\mathrm{Cosh}_p(t,a)$, $\mathrm{Sinh}_p(t,a)$, $\mathrm{Cos}_p(t,a)$, and $\mathrm{Sin}_p(t,a)$ is of exponential order $\frac{1}{|1-p|}$.

**3.31.** Prove parts (iii) and (v) of Theorem 3.76.

**3.32.** Using the definition (Definition 3.77) of the nabla convolution product, show that the nabla convolution product is commutative—i.e., for all $f, g : \mathbb{N}_{a+1} \to \mathbb{R}$,

$$(f * g)(t) = (g * f)(t), \quad t \in \mathbb{N}_{a+1}.$$

Also show that the nabla convolution product is associative.

**3.33.** Solve each of the following IVPs using the nabla Laplace transform:

(i)

$$\nabla y(t) - 4y(t) = 2E_5(t,a), \quad t \in \mathbb{N}_{a+1}$$
$$y(a) = -2;$$

(ii)

$$\nabla y(t) - 3y(t) = 4, \quad t \in \mathbb{N}_{a+1}$$
$$y(a) = -2;$$

(iii)

$$\nabla^2 y(t) + \nabla y(t) - 6y(t) = 0, \quad t \in \mathbb{N}_{a+2}$$
$$y(a) = 3; \quad y(a+1) = 0$$

(iv)

$$\nabla^2 y(t) - 5\nabla y(t) + 6y(t) = E_4(t,a), \quad t \in \mathbb{N}_{a+2}$$
$$y(a) = 1, \quad y(a+1) = -1.$$

**3.34.** Prove part (iv) of Theorem 3.93

**3.35.** Use Theorem 3.120 to solve each of the following IVPs:

(i)

$$\nabla_{0*}^{\frac{1}{2}}x(t) = 3, \quad t \in \mathbb{N}_1$$
$$x(0) = \pi;$$

(ii)

$$\nabla_{0*}^{\frac{1}{3}}x(t) = t^{\overline{\frac{4}{3}}}, \quad t \in \mathbb{N}_1$$
$$x(0) = 2;$$

(iii)

$$\nabla_{a*}^{\frac{2}{3}}x(t) = t - a, \quad t \in \mathbb{N}_{a+1}$$
$$x(a) = 4.$$

**3.36.** Use Theorem 3.120 to solve each of the following IVPs:

(i)

$$\nabla_{0*}^{1.6}x(t) = 3, \quad t \in \mathbb{N}_1$$
$$x(0) = 2, \quad \nabla x(0) = -1.$$

(ii)

$$\nabla_{0}^{\frac{5}{3}}x(t) = t^{\overline{\frac{4}{3}}}, \quad t \in \mathbb{N}_1$$
$$x(0) = \nabla x(0) = 0.$$

(iii)

$$\nabla_{a*}^{2.7}x(t) = t - a, \quad t \in \mathbb{N}_{a+1}$$
$$x(a) = 0 = \nabla x(a).$$

**3.37.** Show that (see example 3.168)

$$\int_0^t sH_{.6}(t,s)\nabla s = \frac{1}{\Gamma(3.6)}(t-1)^{\overline{2.6}}$$

for $t \in \mathbb{N}_0$.

**3.38.** Solve the following IVPs using Theorem 3.166:

(i)

$$\begin{cases} \nabla\nabla_{0*}^{0.5}x(t+1) = t-a, & t \in \mathbb{N}_{a+1}, \\ x(a) = \nabla x(a) = 0. \end{cases}$$

(ii)

$$\begin{cases} \nabla\nabla_{0*}^{0.3}x(t+1) = t, & t \in \mathbb{N}_1, \\ x(0) = 1, \quad \nabla x(0) = 2. \end{cases}$$

**3.39.** Use an appropriate Green's function to solve the BVP

$$\begin{cases} \nabla\nabla_{0*}^{0.2}x(t+1) = 1, & t \in \mathbb{N}_1^{b-1}, \\ x(0) = 0 = x(b), \end{cases}$$

where $b \in \mathbb{N}_2$.

# Chapter 4
# Quantum Calculus

## 4.1 Introduction

In this chapter we will mainly be concerned with functions defined on the infinite set

$$aq^{\mathbb{N}_0} := \{a, aq, aq^2, \ldots, aq^n, \cdots\}, \quad a > 0, \quad q > 1$$

or the finite set

$$aq^{\mathbb{N}_0^{n_0}} := \{a, aq, aq^2, \ldots, aq^{n_0}\}, \quad a > 0, \quad q > 1,$$

where $n_0$ is a positive integer. We define the jump operator $\sigma$ and the graininess function $\mu$ on $aq^{\mathbb{N}_0}$ by

$$\sigma(t) = qt, \quad \mu(t) = (q-1)t, \quad t \in aq^{\mathbb{N}_0}.$$

This study has important applications in quantum theory (see Kac and Cheung [131]). It is standard to use the letter $q$, since this is the first letter in the word "quantum." Several of the results in this chapter appear in Auch [37], Auch et al. [38], and Baoguo et al. [50].

More specifically, we begin in Sect. 4.2 by defining the quantum difference operator and derive several of its properties. Then in Sect. 4.3 we define an exponential function for the quantum calculus and derive several of its properties. Following the exponential function, we define in Sect. 4.4 appropriate hyperbolic and trigonometric functions for the quantum calculus and derive several of their properties. In Sect. 4.5 we define and prove properties of the nabla integral including integration by parts formulas, and then follow this in Sect. 4.6 by presenting an important variation of parameters formula for certain quantum dynamic equations. In Sect. 4.7 we define an appropriate quantum Laplace transform and derive several



C. Goodrich, A.C. Peterson, *Discrete Fractional Calculus*,
DOI 10.1007/978-3-319-25562-0_4

important quantum Laplace transform formulas. After discussing the Laplace transform, in Sect. 4.8 we define the matrix exponential function and show many of its properties. In Sect. 4.9 Floquet theory for the quantum case is developed. Finally, we conclude this chapter by studying the nabla quantum fractional calculus in Sect. 4.10. Comparison theorems and the asymptotic behavior of solutions of certain $\alpha$-fractional equations with $0 < \alpha < 1$ are given in this section.

## 4.2 The Quantum Difference Operator

We define the quantum difference operator ($q$-difference operator) (Jackson difference operator) as follows.

**Definition 4.1 ($q$-Difference Operator).** If $f : aq^{\mathbb{N}_0} \to \mathbb{R}$, then we define the quantum difference operator, ($q$-difference operator) (Jackson difference operator) $D_q$, of $f$ at $t \in aq^{\mathbb{N}_0}$, by

$$D_q f(t) = \frac{f(\sigma(t)) - f(t)}{\mu(t)}.$$

We also define $D_q^n f(t)$, $n \in \mathbb{N}_1$, recursively by $D_q^n f(t) = D_q D_q^{n-1} f(t)$, where $D_q^0$ is defined to be the identity operator.

The following theorem gives several properties of the quantum difference operator.

**Theorem 4.2.** *Assume $f, g : aq^{\mathbb{N}_0} \to \mathbb{R}$ and $\alpha$ is a constant. Then the following hold:*

(i) $D_q\, \alpha = 0$;
(ii) $D_q \alpha f(t) = \alpha D_q f(t)$;
(iii) $D_q[f(t) + g(t)] = D_q f(t) + D_q g(t)$;
(iv) $D_q[f(t)g(t)] = f(\sigma(t))D_q g(t) + D_q f(t) g(t)$;
(v) $D_q\left(\frac{f(t)}{g(t)}\right) = \frac{g(t)D_q f(t) - f(t)D_q g(t)}{g(t)g(\sigma(t))}$, *if* $g(t)g(\sigma(t)) \neq 0$;
(vi) *if* $D_q f(t) = 0$, $t \in aq^{\mathbb{N}_0}$, *then* $f(t) = C$ *for* $t \in aq^{\mathbb{N}_0}$, *where $C$ is a constant;*
(vii) *if* $D_q f(t) > 0$ *on* $aq^{\mathbb{N}_0}$, *then $f$ is strictly increasing on* $aq^{\mathbb{N}_0}$.

*Proof.* We will just prove the quotient rule (v) and (vi) (the rest of the proof is Exercise 4.1). To see that the quotient rule (v) holds, note that

$$\begin{aligned} D_q\left(\frac{f(t)}{g(t)}\right) &= \frac{\frac{f(qt)}{g(qt)} - \frac{f(t)}{g(t)}}{\mu(t)} \\ &= \frac{f(qt)g(t) - f(t)g(qt)}{\mu(t)g(t)g(qt)} \end{aligned}$$

$$= \frac{f(\sigma(t))g(t) + f(t)g(t) - f(t)g(t) - f(t)g(\sigma(t))}{\mu(t)g(t)g(\sigma(t))}$$
$$= \frac{g(t)[f(\sigma(t)) - f(t)]}{\mu(t)g(t)g(\sigma(t))} - \frac{f(t)[g(\sigma(t)) - g(t)]}{\mu(t)g(t)g(\sigma(t))}$$
$$= \frac{g(t)D_qf(t) - f(t)D_qg(t)}{g(t)g(\sigma(t))}.$$

To see that (vi) holds, assume

$$D_qf(t) = 0, \quad t \in aq^{\mathbb{N}_0}.$$

Then, by the definition of the Jackson difference operator

$$\frac{f(qt) - f(t)}{\mu(t)} = 0, \quad t \in aq^{\mathbb{N}_0}.$$

This implies that $f(qt) = f(t)$ for all $t \in aq^{\mathbb{N}_0}$. It follows that there is a constant $C$ so that $f(t) = C, \quad t \in aq^{\mathbb{N}_0}$. □

Next we give a chain rule formula for functions defined on $aq^{\mathbb{N}_0}$.

**Theorem 4.3 (Chain Rule).** *Assume $n$ is a positive integer and $f : aq^{\mathbb{N}_0} \to \mathbb{R}$. Then*

$$D_q\left(f(q^n t)\right) = q^n D_qf(q^n t), \quad t \in aq^{\mathbb{N}_0}. \tag{4.1}$$

*Proof.* Note that

$$\begin{aligned} D_q\left(f(q^n t)\right) &= \frac{f(q^n\sigma(t)) - f(q^n t)}{\mu(t)} \\ &= q^n \frac{f(\sigma(q^n t)) - f(q^n t)}{q^n\mu(t)} \\ &= q^n \frac{f(\sigma(q^n t)) - f(q^n t)}{\mu(q^n t)} \\ &= q^n D_qf(q^n t) \end{aligned}$$

for $t \in aq^{\mathbb{N}_0}$. □

Let $n \in \mathbb{N}_0$. Then we will use the notation (see [131])

$$[n]_q := \frac{q^n - 1}{q - 1} = 1 + q + q^2 + \cdots + q^{n-1}$$

and

$$[n]_q! := [n]_q\,[n-1]_q \cdots [2]_q\,[1]_q, \quad n \in \mathbb{N}_1,$$

and $[0]_q! := 1$. Next we define the $q$-falling function.

**Definition 4.4.** For $n \in \mathbb{N}$ and $\alpha, \beta \in \mathbb{R}$ the **$q$-falling function** is defined by

$$(\beta - \alpha)_q^{\underline{n}} := \prod_{k=0}^{n-1} (\beta - \alpha q^k),$$

where by our convention on products $(\beta - \alpha)_q^{\underline{0}} := 1$.

Then we define the important Taylor monomials for the $q$-calculus in terms of the $q$-falling functions as follows.

**Definition 4.5.** For each fixed $s \in aq^{\mathbb{N}_0}$, we define the $n$-th, $n \in \mathbb{N}_0$, order Taylor monomial $h_n(t,s)$ based at $s$ by

$$h_n(t,s) = \frac{(t-s)_q^{\underline{n}}}{[n]_q!}.$$

Note that

$$h_0(t,s) = \frac{(t-s)_q^{\underline{0}}}{[0]_q!} := 1, \quad \text{and} \quad h_1(t,s) = t - s.$$

We now state the appropriate power rule for the $q$-calculus.

**Theorem 4.6 (Power Rule).** *For each fixed $s \in aq^{\mathbb{N}_0}$,*

$$D_q h_{n+1}(t,s) = h_n(t,s)$$

$n \in \mathbb{N}_0$, $t \in aq^{\mathbb{N}_0}$.

*Proof.* For each fixed $s \in aq^{\mathbb{N}_0}$, we have that

$$\begin{aligned} D_q h_{n+1}(t,s) &= \frac{h_{n+1}(\sigma(t),s) - h_{n+1}(t,s)}{\mu(t)} \\ &= \frac{(\sigma(t)-s)_q^{\underline{n+1}} - (t-s)_q^{\underline{n+1}}}{\mu(t)\cdot[n+1]_q!} \\ &= \frac{\{q^n(qt-s) - (t-q^n s)\}(t-s)_q^{\underline{n-1}}}{(q-1)t\,[n+1]_q!} \end{aligned}$$

$$= \frac{(t-s)_q^{\underline{n}}}{[n]_q!}$$
$$= h_n(t,s),$$

for $t \in aq^{\mathbb{N}_0}$. □

Note that

$$D_q^i h_n(s,s) = 0 = h_n(q^i s, s), \quad 0 \le i \le n-1, \quad D_q^n h_n(s,s) = 1.$$

The reason we call Theorem 4.6 the "power rule" is because we immediately get the following corollary.

**Corollary 4.7.** *For each fixed* $s \in aq^{\mathbb{N}_0}$,

$$D_q(t-s)_q^{\underline{n}} = [n]_q\,(t-s)_q^{\underline{n-1}}$$

$n \in \mathbb{N}_0$, $t \in aq^{\mathbb{N}_0}$.

The following theorem is also very important.

**Theorem 4.8.** *For any constant* $\alpha \in \mathbb{R}$ *and* $n \in \mathbb{N}_0$

$$D_q h_{n+1}(\alpha, t) = -h_n(\alpha, \sigma(t))$$

*for* $t \in aq^{\mathbb{N}_0}$.

*Proof.* We have that

$$\begin{aligned}
D_q h_{n+1} &= \frac{1}{\mu(t)}[h_{n+1}(\alpha, qt) - h_{n+1}(\alpha, t)] \\
&= \frac{1}{\mu(t)}\left[\frac{(\alpha-qt)_q^{\underline{n+1}}}{[n+1]_q!} - \frac{(\alpha-t)_q^{\underline{n+1}}}{[n+1]_q!}\right] \\
&= \frac{[(\alpha-q^{n+1}t)-(\alpha-t)]}{\mu(t)}\frac{(\alpha-qt)_q^{\underline{n}}}{[n+1]_q!} \\
&= -\frac{[q^{n+1}-1]t(\alpha-qt)_q^{\underline{n}}}{(q-1)t[n+1]_q!} \\
&= -\frac{(\alpha-qt)_q^{\underline{n}}}{[n]_q!} \\
&= -h_n(\alpha, \sigma(t)),
\end{aligned}$$

$t \in aq^{\mathbb{N}_0}$. □

Unlike for most of this chapter, for the rest of this section we assume $0 < q < 1$, which is an important case in quantum theory. For this case we will define a fractional falling function and prove several interesting properties of this $q$-falling function.

We motivate the definition of the $q$-falling function for the fractional case by considering the following for $n \in \mathbb{N}$:

$$
\begin{aligned}
(\beta - \alpha)_q^{\underline{n}} &= \prod_{k=0}^{n-1} (\beta - \alpha q^k) \\
&= \beta^n \prod_{k=0}^{n-1} \left(1 - \frac{\alpha}{\beta} q^k\right) \\
&= \beta^n \frac{\prod_{k=0}^{n-1} (1 - \frac{\alpha}{\beta} q^k) \prod_{r=n}^{\infty} (1 - \frac{\alpha}{\beta} q^r)}{\prod_{r=n}^{\infty} (1 - \frac{\alpha}{\beta} q^r)} \\
&= \beta^n \frac{\prod_{k=0}^{\infty} (1 - \frac{\alpha}{\beta} q^k)}{\prod_{k=0}^{\infty} (1 - \frac{\alpha}{\beta} q^{k+n})}.
\end{aligned}
$$

This leads to the following definition.

**Definition 4.9.** The **fractional q-falling function** for $0 < q < 1$ is defined by

$$
(\beta - \alpha)_q^{\underline{a}} := \beta^a \frac{\prod_{k=0}^{\infty} (1 - \frac{\alpha}{\beta} q^k)}{\prod_{k=0}^{\infty} (1 - \frac{\alpha}{\beta} q^{k+a})}
$$

for $\beta, \alpha, a \in \mathbb{R}$, provided the infinite product in the denominator does not converge to zero.

*Remark 4.10.* Note that in the above definition, both infinite products converge for $q \in (0, 1)$. To see this recall that the infinite product

$$
\prod_{k=0}^{\infty} \left[1 + \left(-\frac{\alpha}{\beta} q^k\right)\right]
$$

converges if and only if the infinite sum $\sum_{k=0}^{\infty} \left(-\frac{\alpha}{\beta} q^k\right)$ converges. Since $\sum_{k=0}^{\infty} \left(-\frac{\alpha}{\beta} q^k\right) = -\frac{\alpha}{\beta} \sum_{k=0}^{\infty} q^k$ is a geometric series with common ratio $q \in (0, 1)$, it converges, and so, the above infinite product converges.

In the next two theorems we prove some interesting properties of this fractional $q$-falling function.

**Theorem 4.11.** *The fractional q-falling function for $\alpha, \gamma, a, \beta \in \mathbb{R}$, $0 < q < 1$, satisfies the following properties:*

(i) $(\beta-\alpha)_q^{\underline{\alpha+\gamma}} = (\beta-\alpha)_q^{\underline{\alpha}}(\beta-q^\alpha\alpha)_q^{\underline{\gamma}}$;
(ii) $(\gamma\beta-\gamma\alpha)_q^{\underline{\alpha}} = \gamma^\alpha(\beta-\alpha)_q^{\underline{\alpha}}$;
(iii) $(t-s)_q^{\underline{\alpha}} = 0, \quad t \geq s, \quad t,s \in aq^{\mathbb{N}_0}, \quad \alpha \notin \mathbb{N}_0$.

*Proof.* To see that part (i) holds note that

$$\begin{aligned}(\beta-\alpha)_q^{\underline{\alpha+\gamma}} &= \beta^{\alpha+\gamma}\frac{\prod_{k=0}^{\infty}(1-\frac{\alpha}{\beta}q^k)}{\prod_{k=0}^{\infty}(1-\frac{\alpha}{\beta}q^{k+\alpha+\gamma})}\\ &= \beta^{\alpha+\gamma}\frac{\prod_{k=0}^{\infty}(1-\frac{\alpha}{\beta}q^k)}{\prod_{k=0}^{\infty}(1-\frac{\alpha}{\beta}q^{k+\alpha+\gamma})}\frac{\prod_{k=0}^{\infty}(1-\frac{\alpha q^\alpha}{\beta}q^k)}{\prod_{k=0}^{\infty}(1-\frac{\alpha}{\beta}q^{k+\alpha})}\\ &= \beta^{\alpha}\frac{\prod_{k=0}^{\infty}(1-\frac{\alpha}{\beta}q^k)}{\prod_{k=0}^{\infty}(1-\frac{\alpha}{\beta}q^{k+\alpha})}\beta^{\gamma}\frac{\prod_{k=0}^{\infty}(1-\frac{\alpha q^\alpha}{\beta}q^k)}{\prod_{k=0}^{\infty}(1-\frac{\alpha}{\beta}q^{k+\alpha+\gamma})}\\ &= (\beta-\alpha)_q^{\underline{\alpha}}(\beta-q^\alpha\alpha)_q^{\underline{\gamma}}.\end{aligned}$$

Part (ii) follows from the following:

$$\begin{aligned}(\gamma\beta-\gamma\alpha)_q^{\underline{\alpha}} &= (\gamma\beta)^\alpha\frac{\prod_{k=0}^{\infty}(1-\frac{\alpha\gamma}{\beta\gamma}q^k)}{\prod_{k=0}^{\infty}(1-\frac{\alpha\gamma}{\beta\gamma}q^{k+\alpha})}\\ &= \gamma^\alpha\beta^\alpha\frac{\prod_{k=0}^{\infty}(1-\frac{\alpha}{\beta}q^k)}{\prod_{k=0}^{\infty}(1-\frac{\alpha}{\beta}q^{k+\alpha})}\\ &= \gamma^\alpha(\beta-\alpha)_q^{\underline{\alpha}}.\end{aligned}$$

Finally, to see that part (iii) holds, note that if $t,s \in aq^{\mathbb{N}_0}$, then there are $m,n \in \mathbb{N}_0$ such that $t = aq^n$, $s = aq^m$, and we have that

$$\begin{aligned}(t-s)_q^{\underline{\alpha}} &= (aq^n-aq^m)_q^{\underline{\alpha}}\\ &= (aq^n)^\alpha\frac{\prod_{k=0}^{\infty}(1-\frac{aq^m}{aq^n}q^k)}{\prod_{k=0}^{\infty}(1-\frac{aq^m}{aq^n}q^{k+\alpha})}\\ &= (aq^n)^\alpha\frac{\prod_{k=0}^{\infty}(1-q^{m-n}q^k)}{\prod_{k=0}^{\infty}(1-q^{m-n}q^{k+\alpha})}\\ &= 0.\end{aligned}$$

This completes the proof. □

**Theorem 4.12.** *For $\alpha, \nu \in \mathbb{R}, n \in \mathbb{N}$ and $q \in (0,1)$, the following equalities hold for $t \in aq^{\mathbb{N}_0}$ :*

(i) $D_q(t-\alpha)_q^{\underline{\nu}} = q^{\nu-1}[\nu]_{1/q}(\sigma(t)-\alpha)_q^{\underline{\nu-1}}$;
(ii) $D_q(\alpha-t)_q^{\underline{\nu}} = -q^{\nu-1}[\nu]_{1/q}(\alpha-t)_q^{\underline{\nu-1}}$.

*Proof.* Part (i) holds since

$$\begin{aligned}
D_q(t-\alpha)_q^{\underline{\nu}} &= D_q t^\nu \frac{\prod_{k=0}^\infty(1-\frac{\alpha}{t}q^k)}{\prod_{k=0}^\infty(1-\frac{\alpha}{t}q^{k+\nu})} \\
&= \frac{\left(\frac{t}{q}\right)^\nu \frac{\prod_{k=0}^\infty(1-\frac{\alpha}{t}q^{k+1})}{\prod_{k=0}^\infty(1-\frac{\alpha}{t}q^{k+\nu+1})} - t^\nu \frac{\prod_{k=0}^\infty(1-\frac{\alpha}{t}q^k)}{\prod_{k=0}^\infty(1-\frac{\alpha}{t}q^{k+\nu})}}{\mu(t)} \\
&= \frac{\left(\frac{t}{q}\right)^\nu \frac{\prod_{k=0}^\infty(1-\frac{\alpha}{t}q^{k+1})}{\prod_{k=0}^\infty(1-\frac{\alpha}{t}q^{k+\nu+1})}\frac{(1-\frac{\alpha}{t}q^\nu)}{(1-\frac{\alpha}{t}q^\nu)} - t^\nu \frac{\prod_{k=0}^\infty(1-\frac{\alpha}{t}q^k)}{\prod_{k=0}^\infty(1-\frac{\nu}{t}q^{k+\nu})}}{\mu(t)} \\
&= t^\nu \frac{\prod_{k=0}^\infty(1-\frac{\alpha}{t}q^{k+1})}{\prod_{k=0}^\infty(1-\frac{\alpha}{t}q^{k+\nu})}\left[\frac{\left(\frac{1}{q}\right)^\nu(1-\frac{\alpha}{t}q^\nu)-(1-\frac{\alpha}{t})}{t\left(\frac{1}{q}-1\right)}\right] \\
&= t^{\nu-1} \frac{\prod_{k=0}^\infty(1-\frac{\alpha}{t}q^{k+1})}{\prod_{k=0}^\infty(1-\frac{\alpha}{t}q^{k+\nu})}\left[\frac{\left(\frac{1}{q}\right)^\nu-\frac{\alpha}{t}-1+\frac{\alpha}{t}}{\frac{1}{q}-1}\right] \\
&= q^{\nu-1}\left(\frac{t}{q}\right)^{\nu-1} \frac{\prod_{k=0}^\infty(1-\frac{\alpha}{t}q^{k+1})}{\prod_{k=0}^\infty(1-\frac{\alpha}{t}q^{k+\nu})}\left[\frac{\frac{1}{q^\nu}-1}{\frac{1}{q}-1}\right] \\
&= q^{\nu-1}[\nu]_{1/q}(\sigma(t)-\alpha)_q^{\underline{\nu-1}}.
\end{aligned}$$

To see that (ii) holds, note that

$$\begin{aligned}
D_q(\alpha-t)_q^{\underline{\nu}} &= D_q\alpha^\nu \frac{\prod_{k=0}^\infty(1-\frac{t}{\alpha}q^k)}{\prod_{k=0}^\infty(1-\frac{t}{\alpha}q^{k+\nu})} \\
&= \frac{\alpha^\nu \frac{\prod_{k=0}^\infty(1-\frac{t}{\alpha}q^{k-1})}{\prod_{k=0}^\infty(1-\frac{t}{\alpha}q^{k+\nu-1})} - \alpha^\nu \frac{\prod_{k=0}^\infty(1-\frac{t}{\alpha}q^k)}{\prod_{k=0}^\infty(1-\frac{t}{\alpha}q^{k+\nu})}}{\mu(t)} \\
&= \frac{\alpha^\nu \frac{\prod_{k=0}^\infty(1-\frac{t}{\alpha}q^{k-1})}{\prod_{k=0}^\infty(1-\frac{t}{\alpha}q^{k+\nu-1})} - \alpha^\nu \frac{\prod_{k=0}^\infty(1-\frac{t}{\alpha}q^k)}{\prod_{k=0}^\infty(1-\frac{t}{\alpha}q^{k+\nu})}\frac{(1-\frac{t}{\alpha}q^{\nu-1})}{(1-\frac{t}{\alpha}q^{\nu-1})}}{\mu(t)} \\
&= \alpha^\nu \frac{\prod_{k=0}^\infty(1-\frac{t}{\alpha}q^k)}{\prod_{k=0}^\infty(1-\frac{t}{\alpha}q^{k+\nu-1})}\left[\frac{(1-\frac{t}{\alpha}p)-(1-\frac{t}{\alpha}q^{\nu-1})}{t\left(\frac{1}{q}-1\right)}\right]
\end{aligned}$$

$$= \alpha^{\nu} \frac{\prod_{k=0}^{\infty}(1 - \frac{t}{\alpha}q^k)}{\prod_{k=0}^{\infty}(1 - \frac{t}{\alpha}q^{k+\nu-1})} \cdot \frac{t}{\alpha} \left[ \frac{q^{\nu-1}(1 - \frac{1}{q^{\alpha}})}{t(\frac{1}{q} - 1)} \right]$$

$$= \alpha^{\nu-1} \frac{\prod_{k=0}^{\infty}(1 - \frac{t}{\alpha}q^k)}{\prod_{k=0}^{\infty}(1 - \frac{t}{\alpha}q^{k+\nu-1})} (-q^{\nu-1})[\nu]_{1/q}$$

$$= -q^{\nu-1}[\nu]_{1/q}(\alpha - t)_q^{\underline{\nu-1}}.$$

□

## 4.3 The Quantum Exponential

In this section we introduce the quantum exponential and prove several of its important properties. First we define the set of $q$-regressive functions by

$$\mathcal{R}_q \equiv \left\{ f : aq^{\mathbb{N}_0} \to \mathbb{R} : 1 + \mu(t)p(t) \neq 0, \ \ t \in aq^{\mathbb{N}_0} \right\}.$$

**Definition 4.13.** For $p \in \mathcal{R}_q$ we define the $q$-exponential function $e_p(\cdot, a)$ to be the unique solution of the IVP

$$D_q x = p(t)x, \quad x(a) = 1.$$

**Theorem 4.14.** *If $p \in \mathcal{R}_q$, then*

$$e_p(t, a) = \prod_{s=a}^{t/q} [1 + \mu(s)p(s)], \quad t \in aq^{\mathbb{N}_0}.$$

*Here it is understood that the above product is 1 when $t = a$.*

*Proof.* Let $x(t)$ be the solution of the IVP

$$D_q x = p(t)x, \quad x(a) = 1.$$

Then

$$D_q x(t) = \frac{x(qt) - x(t)}{\mu(t)} = p(t)x(t).$$

Solving for $x(qt)$ we get

$$x(qt) = [1 + \mu(t)p(t)]x(t), \quad t \in aq^{\mathbb{N}_0}.$$

Iterating this last equation we get the equations:

$$x(aq) = [1 + \mu(a)p(a)]x(a) = [1 + \mu(a)p(a)]$$
$$x(aq^2) = [1 + \mu(aq)p(aq)]x(aq) = [1 + \mu(a)p(a)][1 + \mu(aq)p(aq)]$$
$$\vdots$$
$$x(aq^n) = \prod_{k=0}^{n-1} \left[1 + \mu\left(aq^k\right) p\left(aq^k\right)\right].$$

Thus,

$$e_p(t, a) = \prod_{s=a}^{t/q} [1 + \mu(s)p(s)].$$

And this completes the proof. □

*Remark 4.15.* Sometimes we will be interested in the exponential function $e_p(t, s)$ based at $s \in \mathbb{N}_a$ instead of based at $a$ which is defined as follows:

$$e_p(t, s) = \begin{cases} \prod_{\tau=s}^{\frac{t}{q}} [1 + p(\tau)\mu(\tau)], & t \in \mathbb{N}_{qs} \\ 1, & t = s \\ \prod_{\tau=t}^{\frac{s}{q}} \frac{1}{[1+p(\tau)\mu(\tau)]}, & t \in \mathbb{N}_a^{\frac{s}{q}}. \end{cases}$$

Note that $e_p(t, s)$ is for each fixed $s \in aq^{\mathbb{N}_0}$ the solution of the IVP

$$D_q x(t) = p(t)x(t), \quad x(s) = 1$$

on $\mathbb{N}_a$.

We next motivate how to define an addition on the set of $q$-regressive functions $\mathcal{R}_q$ in order to get an important law of exponents for our quantum exponential. To see this, assume $p, r \in \mathcal{R}_q$ and consider the product

$$\begin{aligned} e_p(t, a)e_r(t, a) &= \prod_{s=a}^{\frac{t}{q}} [1 + \mu(s)p(s)] \prod_{s=a}^{\frac{t}{q}} [1 + \mu(s)r(s)] \\ &= \prod_{s=a}^{t/q} [1 + \mu(s)p(s)][1 + \mu(s)r(s)] \\ &= \prod_{s=a}^{t/q} \{1 + \mu(s)p(s) + \mu(s)r(s) + \mu^2(s)p(s)r(s)\} \end{aligned}$$

$$= \prod_{s=a}^{t/q} \{1 + \mu(s)[p(s) + r(s) + \mu(s)p(s)r(s)]\}$$

$$= \prod_{s=a}^{t/q} [1 + \mu(s)(p \oplus r)(s))]$$

$$= e_{p \oplus r}(t, a)$$

if we define the circle plus addition, $\oplus$, on the set of $q$-regressive functions $\mathcal{R}_q$ by

$$(p \oplus r)(t) := p(t) + r(t) + \mu(t)p(t)r(t), \quad t \in aq^{\mathbb{N}_0}.$$

Similar to the proof of Theorem 1.16, we can prove the following theorem.

**Theorem 4.16.** *The set of regressive functions $\mathcal{R}_q$ with the addition $\oplus$ is an Abelian group.*

*Proof.* First we show that the closure property holds. Let $p, r \in \mathcal{R}_q$. Then

$$1 + \mu(t)p(t) \neq 0, \quad \text{and} \quad 1 + \mu(t)r(t) \neq 0,$$

for $t \in aq^{\mathbb{N}_0}$. We want to show that $1 + \mu(t)(p \oplus r)(t) \neq 0$. To see this consider

$$\begin{aligned} 1 + \mu(t)(p \oplus r)(t) &= 1 + \mu(t)[p(t) + r(t) + \mu(t)p(t)r(t)] \\ &= 1 + \mu(t)p(t) + \mu(t)r(t) + \mu^2(t)p(t)r(t) \\ &= [1 + \mu(t)p(t)][1 + \mu(t)r(t)] \\ &\neq 0. \end{aligned}$$

Thus, $p \oplus r \in \mathcal{R}_q$.

Next we show that the zero function, 0, is the identity element (by context one will know when 0 is the zero function or the number 0) in $\mathcal{R}_q$. We see that $0 \in \mathcal{R}_q$ since $1 + \mu(t)(0) = 1 \neq 0$. Also,

$$(0 \oplus p)(t) = 0 + p(t) + \mu(t)(0)p(t) = p(t),$$

so the zero function 0 is the identity element in $\mathcal{R}_q$.

Next we show that every element in $\mathcal{R}_q$ has an additive inverse. If $p \in \mathcal{R}_q$, then we set $r = \frac{-p}{1+\mu p}$. Then we have that

$$1 + \mu(t)r(t) = 1 + \frac{-\mu(t)p(t)}{1 + \mu(t)p(t)} = \frac{1}{1 + \mu(t)p(t)} \neq 0, \quad \text{for} \quad t \in aq^{\mathbb{N}_0},$$

so $r \in \mathcal{R}_q$. We also have that

$$\begin{aligned}(p \oplus r)(t) &= p(t) \oplus \frac{-p(t)}{1+\mu(t)p(t)} \\ &= p(t) + \frac{-p(t)}{1+\mu(t)p(t)} - \frac{\mu(t)p^2(t)}{1+\mu(t)p(t)} \\ &= \frac{p(t)[1+\mu(t)p(t)] - p(t) - \mu(t)p^2(t)}{1+\mu(t)p(t)} \\ &= 0, \quad \text{for} \quad t \in aq^{\mathbb{N}_0},\end{aligned}$$

which shows that $r$ is the additive inverse of $p$.

Next we show that the associative law holds. Let $p, \ell, r \in \mathcal{R}_q$. Then

$$\begin{aligned}&[(p \oplus \ell) \oplus r)](t) \\ &= (p(t) + \ell(t) + \mu(t)p(t)\ell(t)) \oplus r(t) \\ &= (p(t) + \ell(t) + \mu(t)p(t)\ell(t)) + r(t) + \mu(t)[p(t) + \ell(t) + \mu(t)p(t)\ell(t)]r(t) \\ &= p(t) + q(t) + r(t) + \mu(t)p(t)\ell(t) \\ &\quad + \mu(t)p(t)r(t) + \mu(t)\ell(t)r(t) + \mu^2(t)p(t)\ell(t)r(t) \\ &= p(t) + \ell(t) + r(t) + \mu(t)\ell(t)r(t) \\ &\quad + p(t)\ell(t)\mu(t) + p(t)r(t)\mu(t) + \mu^2(t)p(t)\ell(t)r(t) \\ &= p(t) + [\ell(t) + r(t) + \mu(t)\ell(t)r(t)] + p(t)[\ell(t) + r(t) + \mu(t)\ell(t)r(t)]\mu(t) \\ &= [p \oplus (\ell \oplus r)](t), \quad \text{for} \quad t \in aq^{\mathbb{N}_0}.\end{aligned}$$

Hence, $\mathcal{R}_q$, $\oplus$ is a group.

Finally,

$$\begin{aligned}(p \oplus r)(t) &= p(t) + r(t) + \mu(t)p(t)r(t) \\ &= r(t) + p(t) + \mu(t)r(t)p(t) \\ &= (r \oplus p)(t), \quad \text{for} \quad t \in aq^{\mathbb{N}_0},\end{aligned}$$

so our $\oplus$ addition is commutative. Therefore, we have that $\mathcal{R}_q$, $\oplus$ is an Abelian group. □

For $p \in \mathcal{R}_q$, we use the following notation to denote the additive inverse of $p$:

$$\ominus p = \frac{-p}{1+\mu p}$$

and we then define the circle minus subtraction, $\ominus$, on $\mathcal{R}_q$ by

$$p \ominus r = p \oplus (\ominus r).$$

It follows that

$$p \ominus r = \frac{p-r}{1+\mu r}.$$

**Definition 4.17.** The set of positively $q$-regressive functions, $\mathcal{R}_q^+$, is defined by

$$\mathcal{R}_q^+ := \{p \in \mathcal{R}_q : 1 + \mu(t)p(t) > 0, \quad t \in aq^{\mathbb{N}_0}\}.$$

Note that a positively $q$-regressive function need not be positive on $aq^{\mathbb{N}_0}$.

**Theorem 4.18.** *Assume $p, r \in \mathcal{R}_q$ and $t, s, \tau \in aq^{\mathbb{N}_0}$. Then*

(i) $e_0(t, a) = 1$ *and* $e_p(t, t) = 1$;
(ii) $e_p(t, a) \neq 0$, $t \in aq^{\mathbb{N}_0}$;
(iii) *if* $p \in \mathcal{R}_q^+$, *then* $e_p(t, a) > 0$, $t \in aq^{\mathbb{N}_0}$;
(iv) $D_q e_p(t, s) = p(t)e_p(t, s)$ *and* $e_p(s, s) = 1$;
(v) $e_p(\sigma(t), a)) = [1 + \mu(t)p(t)]e_p(t, a)$, $t \in aq^{\mathbb{N}_0}$;
(vi) $e_p(t, s)e_p(s, \tau) = e_p(t, \tau)$;
(vii) $e_p(t, a)e_r(t, a) = e_{p\oplus r}(t, a)$, $t \in aq^{\mathbb{N}_0}$;
(viii) $e_{\ominus p}(t, a) = \frac{1}{e_p(t,a)}$, $t \in aq^{\mathbb{N}_0}$;
(ix) $\frac{e_p(t,a)}{e_r(t,a)} = e_{p\ominus r}(t, a)$, $t \in aq^{\mathbb{N}_0}$;
(x) *if $p, r$ are (complex) constants with $|p| < |r|$, then we have that* $\lim_{t\to\infty} e_{p\ominus r}(t, a) = 0$.

*Proof.* Since $e_0(t, a) = \prod_{s=a}^{t/q}(1 + 0) = 1$ and $e_p(t, t) = 1$ by our convention on products, we get that (i) holds.

To see that (ii) holds, note that $p \in \mathcal{R}_q$ implies $1 + \mu(t)p(t) \neq 0$ for all $t \in aq^{\mathbb{N}_0}$. Hence,

$$e_p(t, a) = \prod_{s=a}^{t/q}[1 + \mu(s)p(s)] \neq 0,$$

for all $t \in aq^{\mathbb{N}_0}$.

To see that (iii) holds, note that since $p \in \mathcal{R}_q$, we have that $1 + \mu(t)p(t) > 0$, for $\in aq^{\mathbb{N}_0}$, and hence

$$e_p(t, a) = \prod_{s=a}^{t/q}[1 + \mu(s)p(s)] > 0, \quad t \in aq^{\mathbb{N}_0}.$$

Property (iv) follows directly from the definition (Definition 4.13) of $e_p(t, a)$.

To see that (v) holds, the equation

$$D_q e_p(t, a) = p(t)e_p(t, a), \quad t \in aq^{\mathbb{N}_0}$$

implies that

$$p(t)e_p(t,a) = \frac{e_p(\sigma(t),a) - e_p(t,a)}{\mu(t)}.$$

Solving this last equation for $e_p(\sigma(t),a)$ we get the desired result

$$e_p(\sigma(t),a) = [1 + \mu(t)p(t)]e_p(t,a).$$

To see that (vi) holds for the case $t \geq s \geq \tau$ (the other cases are left to the reader), note that

$$\begin{aligned} e_p(t,s)e_p(s,\tau) &= \prod_{\zeta=s}^{t/q}[1 + \mu(\zeta)p(\zeta)] \prod_{\zeta=\tau}^{s/q}[1 + \mu(\zeta)p(\zeta)] \\ &= \prod_{\zeta=\tau}^{t/q}[1 + \mu(\zeta)p(\zeta)] \\ &= e_p(t,\tau). \end{aligned}$$

To see that (vii) holds, we calculate

$$\begin{aligned} e_p(t,a)e_r(t,a) &= \prod_{s=a}^{t/q}[1 + \mu(s)p(s)] \prod_{s=a}^{t/q}[1 + \mu(s)r(s)] \\ &= \prod_{s=a}^{t/q}[1 + \mu(s)p(s)][1 + \mu(s)r(s)] \\ &= \prod_{s=a}^{t/q}[1 + \mu(s)p(s) + \mu(s)r(s) + \mu^2(s)p(s)r(s)] \\ &= \prod_{s=a}^{t/q}\{1 + \mu(s)[p(s) + r(s) + \mu(s)p(s)r(s)]\} \\ &= \prod_{s=a}^{t/q}[1 + \mu(s)(p \oplus r)(s)] \\ &= e_{p\oplus r}(t,a). \end{aligned}$$

And this proves the claim. On the other hand, to see that (viii) holds, we simply write.

$$
\begin{aligned}
e_{\ominus p}(t,a) &= \prod_{s=a}^{t/q}\left[1+\frac{-p(s)}{1+\mu(s)p(s)}\mu(s)\right]\\
&= \prod_{s=a}^{t/q}\frac{1}{1+\mu(s)p(s)}\\
&= \frac{1}{\prod_{s=a}^{t/q}[1+\mu(s)p(s)]} = \frac{1}{e_p(t,a)}.
\end{aligned}
$$

And see that (ix) holds, note that

$$
\begin{aligned}
\frac{e_p(t,a)}{e_r(t,a)} &= \frac{\prod_{s=a}^{t/q}[1+\mu(s)p(s)]}{\prod_{s=a}^{t/q}[1+\mu(s)r(s)]}.\\
&= \prod_{s=a}^{t/q}\frac{1+\mu(s)p(s)}{1+\mu(s)r(s)}\\
&= \prod_{s=a}^{t/q}\frac{(1+\mu(s)p(s))+\mu(s)r(s)-\mu(s)r(s)}{(1+\mu(s)r(s))}\\
&= \prod_{s=a}^{t/q}\left[1+\frac{\mu(s)p(s)-\mu(s)r(s)}{1+\mu(s)r(s)}\right]\\
&= \prod_{s=a}^{t/q}\left[1+\mu(s)\frac{p(s)-r(s)}{1+\mu(s)r(s)}\right]\\
&= \prod_{s=a}^{t/q}[1+\mu(s)(p\oplus(\ominus r))(s)] \quad \text{by Exercise 4.38}\\
&= \prod_{s=a}^{t/q}[1+\mu(s)(p\ominus r)(s)]\\
&= e_{p\ominus r}(t,a).
\end{aligned}
$$

Finally we show that (x) holds. First observe that

$$
\lim_{t\to\infty}\left|e_{p\ominus r}(t,a)\right| = \lim_{t\to\infty}\left|\frac{\prod_{s=a}^{t/q}(1+\mu(s)p)}{\prod_{s=a}^{t/q}(1+\mu(s)r)}\right| = \prod_{s=a}^{\infty}\left|\frac{(1+\mu(s)p)}{(1+\mu(s)r)}\right|
$$

provided the limit exists. To see that this limit exists consider

$$\lim_{t\to\infty}\left|\frac{(1+\mu(t)p)}{(1+\mu(t)r)}\right| = \frac{|p|}{|r|} < 1.$$

It follows that there is a $t_0$ such that for $t \geq t_0$

$$\left|\frac{(1+\mu(t)p)}{(1+\mu(t)r)}\right| \leq \delta_0,$$

for some constant $\delta_0$ such that $\frac{|p|}{|r|} \leq \delta_0 < 1$. Hence,

$$\begin{aligned}
0 \leq \lim_{t\to\infty}\left|e_{p\ominus r}(t,a)\right| &= \prod_{s=a}^{t_0/q}\left|\frac{(1+\mu(s)p)}{(1+\mu(s)r)}\right| \prod_{r=t_0}^{\infty}\left|\frac{(1+\mu(r)p)}{(1+\mu(r)r)}\right| \\
&\leq \prod_{s=a}^{t_0/q}\left|\frac{(1+\mu(s)p)}{(1+\mu(s)r)}\right| \prod_{r=t_0}^{\infty}\delta_0 \\
&= 0.
\end{aligned}$$

This completes the proof. □

Next we define the $q$-scalar dot multiplication, $\odot$, on $\mathcal{R}_q^+$. To motivate this definition, note that if $n$ is a positive integer and $p \in \mathcal{R}$, then

$$\begin{aligned}
\overbrace{(p \oplus p \oplus p \cdots \oplus p)}^{n \text{ terms}}(t) &= \sum_{k=1}^{n}\binom{n}{k}\mu^{k-1}(t)p^k(t) \\
&= \frac{1}{\mu(t)}\sum_{k=1}^{n}\binom{n}{k}\mu^k(t)p^k(t) \\
&= \frac{1}{\mu(t)}\left[\sum_{k=0}^{n}\binom{n}{k}\mu^k(t)p^k(t) - 1\right] \\
&= \frac{[1+\mu(t)p(t)]^n - 1}{\mu(t)}.
\end{aligned}$$

With this in mind we make the following definition.

**Definition 4.19.** We define the $q$-scalar dot multiplication, $\odot$, on $\mathcal{R}_q^+$ by

$$(\alpha \odot p)(t) = \frac{[1+\mu(t)p(t)]^\alpha - 1}{\mu(t)}, \quad t \in aq^{\mathbb{N}_0}.$$

**Lemma 4.20.** *If $p, r \in \mathcal{R}_q$ and $e_p(t, a) = e_r(t, a)$, then*

$$p(t) = r(t), \quad t \in aq^{\mathbb{N}_0}.$$

**Theorem 4.21.** *If $\alpha \in \mathbb{R}$ and $p \in \mathcal{R}_q^+$, then*

$$e_p^{\alpha}(t, a) = e_{\alpha \odot p}(t, a)$$

*for $t \in aq^{\mathbb{N}_0}$.*

*Proof.* We have that

$$\begin{aligned}
e_p^{\alpha}(t, a) &= \left( \prod_{s=a}^{t/q} [1 + \mu(s) p(s)] \right)^{\alpha} \\
&= \prod_{s=a}^{t/q} [1 + \mu(s) p(s)]^{\alpha} \\
&= \prod_{s=a}^{t/q} \left( 1 + \mu(s) \frac{(1 + \mu(s) p(s))^{\alpha} - 1}{\mu(s)} \right) \\
&= \prod_{s=a}^{t/q} \{1 + \mu(s)[\alpha \odot p](s)\} \\
&= e_{\alpha \odot p}(t, a),
\end{aligned}$$

for $t \in aq^{\mathbb{N}_0}$. □

Then, similar to the proof of Theorem 1.23, we get the following result.

**Theorem 4.22.** *The positively q-regressive functions $\mathcal{R}_q^+$ with the addition $\oplus$ and the scalar multiplication $\odot$ is a vector space.*

*Proof.* Since we have already proved that $\mathcal{R}_q^+$ with $\oplus$ is an Abelian group, we have that $\mathcal{R}_q^+$ satisfies many of the properties of a vector space. We now prove some of the remaining properties. First note that

$$\begin{aligned}
e_{\alpha \odot (\beta \odot p)}(t, a) &= [e_{\beta \odot p}(t, a)]^{\alpha} \\
&= [e_p^{\beta}(t, a)]^{\alpha} \\
&= [e_p(t, a)]^{\alpha\beta} \\
&= e_{(\alpha\beta) \odot p}(t, a).
\end{aligned}$$

Therefore, by Lemma 4.20,

$$\alpha \odot (\beta \odot p) = (\alpha\beta) \odot p.$$

Next, we show a distributive property. Let $\alpha \in \mathbb{R}$ and $p, r \in \mathcal{R}_q$. Then

$$\begin{aligned} e_{\alpha\odot(p\oplus r)}(t,a) &= e^{\alpha}_{p\oplus r}(t,a) \\ &= [e_p(t,a)e_r(t,a)]^{\alpha} \\ &= e^{\alpha}_p(t,a)e^{\alpha}_r(t,a) \\ &= e_{\alpha\odot p}(t,a)e_{\alpha\odot r}(t,a) \\ &= e_{(\alpha\odot p)\oplus(\alpha\odot r)}(t,a). \end{aligned}$$

Therefore, by Lemma 4.20,

$$\alpha \odot (p \oplus r) = (\alpha \odot p) \oplus (\alpha \odot q).$$

Similarly one can show that

$$(\alpha + \beta) \odot p = (\alpha \odot p) \oplus (\beta \odot p).$$

Finally, we show that we have a scalar multiplicative identity by writing

$$(1 \odot p)(t) = \frac{1 + \mu(t)p(t) - 1}{\mu(t)} = \frac{\mu(t)p(t)}{\mu(t)} = p(t).$$

So we have that the constant function 1 is our multiplicative identity. □

## 4.4 Quantum Hyperbolic and Trigonometric Functions

In this section we introduce both the quantum hyperbolic sine and cosine functions and the quantum sine and cosine functions and give some of their properties. First, we define the quantum hyperbolic sine and cosine functions.

**Definition 4.23.** Assume $\pm p \in \mathcal{R}_q$, then the quantum hyperbolic sine and cosine functions are defined as follows:

$$\cosh_p(t,a) := \frac{e_p(t,a) + e_{-p}(t,a)}{2}, \quad \sinh_p(t,a) := \frac{e_p(t,a) - e_{-p}(t,a)}{2}$$

for $t \in aq^{\mathbb{N}_0}$.

The following theorem gives various properties of the quantum hyperbolic sine and cosine functions.

**Theorem 4.24.** *Assume* $\pm p \in \mathcal{R}_q$. *Then*

(i) $\cosh_p(a,a) = 1, \quad \sinh_p(a,a) = 0$;
(ii) $\cosh_p^2(t,a) - \sinh_p^2(t,a) = e_{-\mu p^2}(t,a), \quad t \in aq^{\mathbb{N}_0}$;
(iii) $D_q \cosh_p(t,a) = p \sinh_p(t,a), \quad t \in aq^{\mathbb{N}_0}$;
(iv) $D_q \sinh_p(t,a) = p \cosh_p(t,a), \quad t \in aq^{\mathbb{N}_0}$;
(v) $\cosh_{-p}(t,a) = \cosh_p(t,a), \quad t \in aq^{\mathbb{N}_0}$;
(vi) $\sinh_{-p}(t,a) = -\sinh_p(t,a), \quad t \in aq^{\mathbb{N}_0}$.

*Proof.* To see that (ii) holds note that

$$\begin{aligned}
&\cosh_p^2(t,a) - \sinh_p^2(t,a)\\
&= \frac{(e_p(t,a)+e_{-p}(t,a))^2 + (e_p(t,a)-e_{-p}(t,a))^2}{4}\\
&= e_p(t,a)e_{-p}(t,a)\\
&= e_{p\oplus(-p)}(t,a)\\
&= e_{-\mu p^2}(t,a).
\end{aligned}$$

To see that (iii) holds, consider

$$\begin{aligned}
D_q \cosh_p(t,a) &= \frac{1}{2}D_q e_p(t,a) + \frac{1}{2}D_q e_{-p}(t,a)\\
&= \frac{1}{2}[p(t)e_p(t,a) - p(t)e_{-p}(t,a)]\\
&= p \sinh_p(t,a).
\end{aligned}$$

The proof of (iv) is similar. Also (v) and (vi) hold. □

Next, we define the quantum sine and cosine functions.

**Definition 4.25.** For $\pm ip \in \mathcal{R}_q(\mathbb{C})$, we define the generalized sine and cosine functions as follows:

$$\cos_p(t,a) = \frac{e_{ip}(t,a)+e_{-ip}(t,a)}{2}, \quad \sin_p(t,a) = \frac{e_{ip}(t,a)-e_{-ip}(t,a)}{2i}$$

for $t \in aq^{\mathbb{N}_0}$.

The following theorem gives some relationships between the generalized trigonometric functions and the hyperbolic trigonometric functions.

**Theorem 4.26.** *Assume* $\pm p \in \mathcal{R}_q(\mathbb{C})$. *Then for* $t \in aq^{\mathbb{N}_0}$

(i) $\sin_{ip}(t,a) = i\sinh_p(t,a)$;
(ii) $\cos_{ip}(t,a) = \cosh_p(t,a)$;
(iii) $\sinh_{ip}(t,a) = i\sin_p(t,a)$;
(iv) $\cosh_{ip}(t,a) = \cos_p(t,a)$.

*Proof.* To see that (i) holds note that

$$\begin{aligned}\sin_{ip}(t,a) &= \frac{1}{2i}[e_{i^2p}(t,a) - e_{-i^2p}(t,a)]\\ &= i\frac{e_p(t,a) - e_{-p}(t,a)}{2}\\ &= i\sinh_p(t,a).\end{aligned}$$

The proofs of (ii)–(iv) are similar. □

The following theorem gives various properties of the generalized sine and cosine functions.

**Theorem 4.27.** *Assume* $\pm ip \in \mathcal{R}_q(\mathbb{C})$. *Then*

(i) $\cos_p(a,a) = 1, \quad \sin_p(a,a) = 0$;
(ii) $\cos_p^2(t,a) + \sin_p^2(t,a) = e_{\mu p^2}(t,a), \quad t \in aq^{\mathbb{N}_0}$;
(iii) $D_q \cos_p(t,a) = p \ \sin_p(t,a), \quad t \in aq^{\mathbb{N}_0}$;
(iv) $D_q \sin_p(t,a) = p \ \cos_p(t,a), \quad t \in aq^{\mathbb{N}_0}$;
(v) $\cos_{-p}(t,a) = \cos_p(t,a), \quad t \in aq^{\mathbb{N}_0}$;
(vi) $\sin_{-p}(t,a) = -\sin_p(t,a), \quad t \in aq^{\mathbb{N}_0}$.

*Proof.* The proof of this theorem follows from Theorems 4.24 and 4.26. □

## 4.5 The Quantum Integral

We now define the quantum integration of a function defined on $aq^{\mathbb{N}_0}$.

**Definition 4.28.** Assume $f : aq^{\mathbb{N}_0} \to \mathbb{R}$ and $c, d \in aq^{\mathbb{N}_0}$. Then we define the quantum integral ($q$-integral) (Jackson integral) of $f$ from $c$ to $d$ by

$$\int_c^d f(t)D_qt = \begin{cases} \sum_{t=c}^{d/q} f(t)\mu(t), & \text{if } d > c\\ 0, & \text{if } d = c\\ -\sum_{t=d}^{c/q} f(t)\mu(t), & \text{if } d < c.\end{cases}$$

Using well-known properties of sums and the above definition of the integral one can easily prove the following elementary properties of the $q$-integral $\int_c^d f(t)D_qt$.

**Theorem 4.29.** *Assume* $f, g : aq^{\mathbb{N}_0} \to \mathbb{R}$ *and* $c, d, e \in aq^{\mathbb{N}_0}$. *Then*

(i) $\int_c^d \alpha f(t)D_qt = \alpha \int_c^d f(t)D_qt$;
(ii) $\int_c^d [f(t) + g(t)]D_qt = \int_c^d f(t)D_qt + \int_c^d g(t)D_qt$;
(iii) $\int_c^c \alpha f(t)D_qt = 0$;
(iv) $\int_c^d f(t)D_qt = -\int_d^c f(t)D_qt$;

(v) $\int_c^e f(t)D_qt = \int_c^d f(t)D_qt + \int_d^e f(t)D_qt$;
(vi) *if* $c \leq d$ *and* $f(t) \leq g(t)$ *for* $t \in \{c, qc, \cdots, d/q\}$, *then it follows that* $\int_c^d f(t)D_qt \leq \int_c^d g(t)D_qt$.

**Definition 4.30.** Assume $\{c, qc, \ldots, d\} \subset aq^{\mathbb{N}_0}$ and $D_qF(t) = f(t)$ for $t \in \{c, qc, \ldots, d/q\}$. Then we say $F$ is a $q$-antidifference of $f$ on $\{c, qc, \cdots, d\}$.

The following theorem shows that every function $f : aq^{\mathbb{N}_0} \to \mathbb{R}$ has an antidifference on $aq^{\mathbb{N}_0}$.

**Theorem 4.31.** *Assume that* $f : aq^{\mathbb{N}_0} \to \mathbb{R}$. *Then* $F(t) := \int_c^t f(s)D_qs$ *is a* $q$-*antidifference of* $f(t)$ *on* $aq^{\mathbb{N}_0}$; *that is,*

$$D_qF(t) = D_q \int_c^t f(s)D_qt = f(t), \quad t \in aq^{\mathbb{N}_0}.$$

*Proof.* Assume $f : aq^{\mathbb{N}_0} \to \mathbb{R}$ and let

$$F(t) := \int_c^t f(s)D_qs.$$

Then for $t \in aq^{\mathbb{N}_0}$

$$\begin{aligned} D_qf(t) &= D_q \int_c^t f(s)D_qs \\ &= \frac{F(qt) - F(t)}{\mu(t)} \\ &= \frac{\int_c^{qt} f(s)D_qs - \int_c^t f(s)D_qs}{\mu(t)} \\ &= \frac{\int_t^{qt} f(s)D_qs}{\mu(t)} \\ &= \frac{f(t)\mu(t)}{\mu(t)} \\ &= f(t). \end{aligned}$$

Hence, $F(t)$ is a $q$-antidifference of $f(t)$ on $aq^{\mathbb{N}_0}$. □

**Theorem 4.32.** *If* $f : aq^{\mathbb{N}_0^n} \to \mathbb{R}$ *and* $G(t)$ *is an* $q$-*antidifference of* $f(t)$ *on* $aq^{\mathbb{N}_0^n}$, *then* $F(t) = G(t) + C$ *is a general* $q$-*antidifference of* $f(t)$ *on* $aq^{\mathbb{N}_0^n}$.

*Proof.* Assume $G(t)$ is an $q$-antidifference of $f(t)$ on $aq^{\mathbb{N}_0^n}$. Let $F(t) = G(t) + C$, where $C$ is constant and $t \in aq^{\mathbb{N}_0^n}$. Then on $aq^{\mathbb{N}_0^n}$

$$D_qF(t) = D_qG(t) = f(t), \quad t \in aq^{\mathbb{N}_0^{n-1}},$$

and so, $F(t)$ is an $q$-antiderivative of $f(t)$ on $aq^{\mathbb{N}_0^n}$. Conversely, assume $F(t)$ is an $q$-antidifference of $f(t)$ on $aq^{\mathbb{N}_0^n}$. Then

$$D_q(F(t) - G(t)) = D_qF(t) - D_qG(t) = f(t) - f(t) = 0$$

for $t \in aq^{\mathbb{N}_0^{n-1}}$. This implies $F(t) - G(t) = C$, for $t \in aq^{\mathbb{N}_0^n}$. Hence,

$$F(t) = G(t) + C,$$

for $t \in aq^{\mathbb{N}_0^n}$. □

**Theorem 4.33 (Fundamental Theorem of $q$-Calculus).** *Assume $f : aq^{\mathbb{N}_0^n} \to \mathbb{R}$ and $F(t)$ is any $q$-antidifference of $f(t)$ on $aq^{\mathbb{N}_0^n}$. Then*

$$\int_a^{aq^n} f(t)D_qt = \int_a^{aq^n} D_qF(t)D_qt = F(t)\Big|_a^{aq^n}.$$

*Proof.* Assume $F(t)$ is any $q$-antidifference of $f(t)$ on $aq^{\mathbb{N}_0^n}$. Let

$$G(t) := \int_a^t f(s)D_qs, \quad t \in aq^{\mathbb{N}_0^n}.$$

By Theorem 4.31, $G(t)$ is an $q$-antidifference of $f(t)$. Hence by the previous theorem, $F(t) = G(t) + C$, where $C$ is a constant. Then

$$\begin{aligned} F(t)\Big|_a^{aq^n} &= F(aq^n) - F(a) = [G(aq^n) + C] - [G(a) + C] \\ &= G(aq^n) - G(a). \end{aligned}$$

By Theorem 4.29, part (iii), $G(a) = 0$ and hence

$$F(t)\Big|_a^{aq^n} = \int_a^{aq^n} f(t)D_qt.$$

This completes the proof. □

**Theorem 4.34 (Integration by Parts).** *Given two functions $u, v : aq^{\mathbb{N}_0} \to \mathbb{R}$ and $b, c \in aq^{\mathbb{N}_0}, b < c$, we have the integration by parts formulas*

$$\int_b^c u(t)D_qv(t)D_qt = u(t)v(t)\Big|_b^c - \int_b^c v(\sigma(t))D_qu(t)D_qt$$

*and*

$$\int_b^c u(\sigma(t))D_qv(t)D_qt = u(t)v(t)\Big|_b^c - \int_b^c v(t)D_qu(t)D_qt.$$

*Proof.* The second integration by parts formula follows from the first by interchanging $u$ and $v$. Hence, it suffices to prove the first integration by parts formula. Assume $u, v : aq^{\mathbb{N}_0} \to \mathbb{R}$ and $b, c \in aq^{\mathbb{N}_0}$, $b < c$. Using the product rule we have that

$$D_q(u(t)v(t)) = D_q u(t)v(\sigma(t)) + u(t)D_q v(t),$$

which implies

$$u(t)D_q v(t) = D_q(u(t)v(t)) - D_q u(t)v(\sigma(t)).$$

Integrating both sides we obtain the following

$$\int_b^c u(t)D_q v(t)D_q t = \int_b^c D_q(u(t)v(t))D_q t - \int_b^c D_q u(t)v(\sigma(t))D_q t,$$

which implies the following

$$\int_b^c u(t)D_q v(t)D_q t = u(t)v(t)\Big|_b^c - \int_b^c v(\sigma(t))D_q u(t)D_q t.$$

This completes the proof. □

## 4.6 Quantum Variation of Constants Formula

In this section we will derive a quantum variation of constants formula and give an example of this result. First we will need to prove a Leibniz rule for the $q$-calculus.

**Theorem 4.35 (Leibniz Rule).** *Assume $f : aq^{\mathbb{N}_0} \times aq^{\mathbb{N}_0} \to \mathbb{R}$. Then*

$$D_q\left(\int_a^t f(t,s)D_q s\right) = \int_a^t D_q f(t,s)D_q s + f(\sigma(t),t), \quad t \in aq^{\mathbb{N}_0}.$$

*Here it is understood that the expression $D_q f(t,s)$ in the last integral means the (partial) q-difference of $f$ with respect to $t$.*

*Proof.* Note that

$$\begin{aligned} D_q\left(\int_a^t f(t,s)D_q s\right) &= \frac{\int_a^{\sigma(t)} f(\sigma(t),s)D_q s - \int_a^t f(t,s)D_q s}{\mu(t)} \\ &= \int_a^t \frac{f(\sigma(t),s) - f(t,s)}{\mu(t)} D_q s + \frac{\int_t^{\sigma(t)} f(\sigma(t),s)D_q s}{\mu(t)} \\ &= \int_a^t D_q f(t,s)D_q s + f(\sigma(t),t), \end{aligned}$$

for $t \in aq^{\mathbb{N}_0}$. □

**Theorem 4.36 (Variations of Constants Formula).** *Assume* $f : aq^{\mathbb{N}_0} \to \mathbb{R}$. *Then the solution of the IVP*

$$D_q^n y = f(t), \quad t \in aq^{\mathbb{N}_0}$$

$$D_q^i y(a) = 0, \quad 0 \leq i \leq n-1$$

*is given by*

$$y(t) = \int_a^t h_{n-1}(t, \sigma(s)) f(s) D_q s, \tag{4.2}$$

*for* $t \in aq^{\mathbb{N}_0}$.

*Proof.* We will use repeated applications of the Leibniz rule to prove this theorem. First we see that if $y(t)$ is given by (4.2), then

$$\begin{aligned} D_q y(t) &= D_q \left( \int_a^t h_{n-1}(t, \sigma(s)) f(s) D_q s \right) \\ &= \int_a^t h_{n-2}(t, \sigma(s)) f(s) D_q s + h_{n-1}(\sigma(t), \sigma(t)) f(t) \\ &= \int_a^t h_{n-2}(t, \sigma(s)) f(s) D_q s. \end{aligned}$$

It then follows that

$$\begin{aligned} D_q^2 y(t) &= D_q \left( \int_a^t h_{n-2}(t, \sigma(s)) f(s) D_q s \right) \\ &= \int_a^t h_{n-3}(t, \sigma(s)) f(s) D_q s + h_{n-2}(\sigma(t), \sigma(t)) f(t) \\ &= \int_a^t h_{n-3}(t, \sigma(s)) f(s) D_q s. \end{aligned}$$

Proceeding by induction we get that

$$\begin{aligned} D_q^{n-1} y(t) &= D_q \left( \int_a^t h_1(t, \sigma(s)) f(s) D_q s \right) \\ &= \int_a^t h_0(t, \sigma(s)) f(s) D_q s + h_1(\sigma(t), \sigma(t)) f(t) \\ &= \int_a^t h_0(t, \sigma(s)) f(s) D_q s \\ &= \int_a^t f(s) D_q s. \end{aligned}$$

Finally, taking one more $q$-difference, we get

$$D_q^n y(t) = f(t), \quad t \in aq^{\mathbb{N}_0}.$$

Also, $y(t)$ satisfies the initial conditions

$$D_q^i y(a) = 0, \quad 0 \leq i \leq n-1.$$

This completes the proof. □

*Example 4.37.* Assume $p$ is a nonzero real constant with $\pm p \in \mathcal{R}_q$. Use the variation of constants formula (4.2) to solve the IVP

$$D_q^2 y(t) = \sin_p(t,a), \quad t \in aq^{\mathbb{N}_0}$$
$$y(a) = 0 = D_q y(a).$$

By the variation of constants formula (4.2), we have the solution of our IVP is given by

$$\begin{aligned} y(t) &= \int_a^t h_1(t,\sigma(s)) \sin_p(s,a) D_q s \\ &= \int_a^t (t-\sigma(s)) \sin_p(s,a) D_q s. \end{aligned}$$

Using the second integration by parts formula in Theorem 4.34 with

$$u(\sigma(s)) = h_1(t,\sigma(s)), \quad \text{and} \quad D_q v(s) = \sin_p(s,a)$$

we get

$$\begin{aligned} y(t) &= -\frac{1}{p}(t-s)\cos_p(s,a)\Big|_{s=a}^{t} + \frac{1}{p}\int_a^t \cos_p(s,a) D_q s \\ &= -\frac{1}{p}(t-a)\cos_p(t,a) + \frac{1}{p^2}\sin_p(t,a)\Big|_{s=a}^{t} \\ &= \frac{1}{p^2}\sin_p(t,a) - \frac{1}{p}(t-a)\cos_p(t,a). \end{aligned}$$

*Example 4.38.* Use the variation of constants formula to solve the IVP

$$D_q^2 y(t) = e_p(t,1) \quad t \in aq^{\mathbb{N}_0}$$
$$y(1) = D_q y(1) = 0,$$

where $p$ is a regressive constant. From the variation of constants formula, the solution of this IVP is given by

$$y(t) = \int_1^t h_1(t, \sigma(s)) e_p(s, 1) D_q s.$$

Integrating by parts, we have

$$\begin{aligned}
y(t) &= \int_1^t h_1(t, \sigma(s)) e_p(s, 1) D_q s \\
&= \frac{1}{p} h_1(t, s) e_p(s, 1)\Big|_{s=1}^{t} - \frac{1}{p}\int_1^t e_p(s, 1) D_q h_1(t, s) D_q s \\
&= -\frac{1}{p} h_1(t, 1) + \frac{1}{p^2}\Big[e_p(s, 1)\Big]_{s=1}^{t} \\
&= -\frac{1}{p} h_1(t, 1) + \frac{1}{p^2} e_p(t, 1) - \frac{1}{p^2} e_p(1, 1) \\
&= -\frac{1}{p} h_1(t, 1) + \frac{1}{p^2} e_p(t, 1) - \frac{1}{p^2} \\
&= -\frac{1}{p}(t-1) + \frac{1}{p^2}\prod_{\tau=1}^{\frac{t}{q}}[1 + p(\tau)\mu(\tau)] - \frac{1}{p^2}.
\end{aligned}$$

## 4.7 The $q$-Laplace Transform

Motivated by the real case where the Laplace transform is given by

$$\mathcal{L}\{f\}(s) = \int_0^\infty e^{-st} f(t) dt,$$

we define the $q$-Laplace transform by replacing the real exponential $e^{-st}$ by the quantum exponential $e_{\ominus s}(\sigma(t), a)$.

**Definition 4.39.** Assume $f : aq^{\mathbb{N}_0} \to \mathbb{R}$. Then we define the $q$-Laplace transform of $f$ by

$$\mathcal{L}_a\{f\}(s) := \int_a^\infty e_{\ominus s}(\sigma(t), a) f(t) D_q t$$

for those $s \in \mathcal{R}_q(\mathbb{C})$ such that the above improper integral converges.

In the following theorem we give a formula for $\mathcal{L}_a\{f\}(s)$ as an infinite sum.

**Theorem 4.40.** *Assume $f : aq^{\mathbb{N}_0} \to \mathbb{R}$. Then the q-Laplace transform of $f$ is given by*

$$\mathcal{L}_a\{f\}(s) := \sum_{n=0}^{\infty} \frac{f(aq^n)\mu(aq^n)}{\prod_{k=0}^{n}[1 + s\mu(aq^k)]},$$

*for those values of $s \in \mathcal{R}_q(\mathbb{C})$ such that the above infinite sum converges.*

*Proof.* We have that for $t = aq^n$,

$$\begin{aligned}
\mathcal{L}_a\{f\}(s) &= \int_a^{\infty} e_{\ominus s}(\sigma(t), a)f(t)D_q t \\
&= \sum_{t=a}^{\infty} e_{\ominus s}(\sigma(t), a)f(t)\mu(t) \\
&= \sum_{t=a}^{\infty} \frac{f(t)\mu(t)}{e_s(\sigma(t), a)} \\
&= \sum_{t=a}^{\infty} \frac{f(t)\mu(t)}{[1 + s\mu(t)]e_s(t, a)} \\
&= \sum_{t=a}^{\infty} \frac{f(t)\mu(t)}{[1 + s\mu(t)]\prod_{\tau=a}^{t/q}[1 + s\mu(\tau)]} \\
&= \sum_{t=a}^{\infty} \frac{f(t)\mu(t)}{\prod_{\tau=a}^{t}[1 + s\mu(\tau)]} \\
&= \sum_{t=a}^{\infty} \frac{f(t)\mu(t)}{\prod_{k=0}^{n}[1 + s\mu(aq^k)]} \\
&= \sum_{n=0}^{\infty} \frac{f(aq^n)\mu(aq^n)}{\prod_{k=0}^{n}[1 + s\mu(aq^k)]},
\end{aligned}$$

which is what we wanted to prove. □

To begin the study of finding functions whose Laplace transforms exist, we introduce the next definition.

**Definition 4.41.** We say $f : aq^{\mathbb{N}_0} \to \mathbb{R}$ is of **exponential order** $r \geq 0$ if there is a constant $A > 0$ such that

$$|f(aq^n)| \leq A\left[\mu(a)q^{\frac{n-1}{2}}\right]^n r^n$$

for all sufficiently large $n \in \mathbb{N}_0$.

**Theorem 4.42 (Existence of $q$-Laplace Transform).** *If $f : aq^{\mathbb{N}_0} \to \mathbb{R}$ is of exponential order $r > 0$, then $\mathcal{L}_a\{f\}(s)$ exists for $|s| > r$.*

*Proof.* Assume $f(t)$ is of exponential order $r$. Then there is a constant $A > 0$ and a $t_0 = aq^N \in aq^{\mathbb{N}_0}$ such that $|f(t)| \leq A\left[\mu(a)q^{\frac{n-1}{2}}\right]^n r^n$ for all $t \in aq^{\mathbb{N}_N}$. We now show that

$$\mathcal{L}_a\{f\}(s) = \int_a^\infty e_{\ominus s}(\sigma(t), a)f(t)D_q t = \sum_{n=0}^\infty \frac{f(aq^n)\mu(aq^n)}{\prod_{k=0}^n[1 + s\mu(aq^k)]}$$

converges for $|s| > r$. First note that

$$\begin{aligned}\sum_{n=N}^\infty \left|\frac{f(aq^n)\mu(aq^n)}{\prod_{k=0}^n(1 + s\mu(aq^k))}\right| &\leq \sum_{n=N}^\infty \frac{|f(aq^n)|\mu(aq^n)}{\prod_{k=0}^n|1 + s\mu(aq^k)|}\\ &\leq \sum_{n=N}^\infty \frac{A\left(\mu(a)q^{\frac{n-1}{2}}\right)^n r^n\mu(aq^n)}{\prod_{k=0}^n|1 + s\mu(aq^k)|}.\end{aligned}$$

Consider

$$\begin{aligned}\lim_{n\to\infty}\frac{\dfrac{A(\mu(a)q^{\frac{n}{2}})^{n+1}r^{n+1}\mu(aq^{n+1})}{\prod_{k=0}^{n+1}|1 + s\mu(aq^k)|}}{\dfrac{A(\mu(a)q^{\frac{n-1}{2}})^n r^n\mu(aq^n)}{\prod_{k=0}^n|1 + s\mu(aq^k)|}} &= \mu(a)r\lim_{n\to\infty}\frac{q^{n+1}}{|1 + s\mu(a)q^{n+1}|}\\ &= \mu(a)r\lim_{n\to\infty}\frac{1}{|\frac{1}{q^{n+1}} + s\mu(a)|}\\ &= \frac{\mu(a)r}{\mu(a)|s|} = \frac{r}{|s|} < 1,\end{aligned}$$

for $|s| > r$. Therefore, by the ratio test, for $|s| > r$,

$$\mathcal{L}_a\{f\}(s) = \sum_{n=0}^\infty \frac{f(aq^n)\mu(aq^n)}{\prod_{k=0}^n[1 + s\mu(aq^k)]}$$

converges absolutely. □

**Corollary 4.43.** *If $f : aq^{\mathbb{N}_0} \to \mathbb{R}$ is of exponential order $r > 0$, then for each $|s| > r$ we have that*

$$\lim_{t\to\infty} e_{\ominus s}(\sigma(t), a)f(t) = 0.$$

*Proof.* From Theorem 4.42 we have that for $|s| > r$

$$\mathcal{L}_a\{f\}(s) = \int_a^\infty e_{\ominus s}(\sigma(t), a)f(t)D_qt$$
$$= \sum_{n=0}^\infty e_{\ominus s}(\sigma(aq^n), a)f(aq^n)\mu(aq^n)$$

converges. Therefore by the $n$-th term test we have that

$$\lim_{n\to\infty} e_{\ominus s}(\sigma(aq^n), a)f(aq^n)\mu(aq^n) = 0.$$

Letting $t = aq^n$, we get that

$$\lim_{t\to\infty} e_{\ominus s}(\sigma(t), a)f(t)\mu(t) = 0.$$

Since $\mu(t) = (q-1)t$, this implies that

$$\lim_{t\to\infty} e_{\ominus s}(\sigma(t), a)f(t) = 0.$$

□

**Theorem 4.44 (Linearity).** *Assume $f, g : aq^{\mathbb{N}_0} \to \mathbb{R}$ are of exponential order $r > 0$. Then for $|s| > r$ and $\alpha, \beta \in \mathbb{C}$,*

$$\mathcal{L}_a\{\alpha f + \beta g\}(s) = \alpha\mathcal{L}_a\{f\}(s) + \beta\mathcal{L}_a\{g\}(s).$$

*Proof.* Let $f, g : aq^{\mathbb{N}_0} \to \mathbb{R}$ be of exponential order $r > 0$. Then for $|s| > r$ and $\alpha, \beta \in \mathbb{C}$ we have

$$\mathcal{L}_a\{\alpha f + \beta g\}(s) = \int_a^\infty e_{\ominus s}(\sigma(t), a)[\alpha f(t) + \beta g(t)]D_qt$$
$$= \alpha\int_a^\infty e_{\ominus s}(\sigma(t), a)f(t)D_q + \beta\int_a^\infty e_{\ominus s}(\sigma(t), a)g(t)D_qt$$
$$= \alpha\mathcal{L}_a\{f\}(s) + \beta\mathcal{L}_a\{g\}(s).$$

□

In the next theorem we are concerned with the $q$-Laplace transform of $f : aq^{\mathbb{N}_0} \to \mathbb{R}$ based at a point $aq^m \in aq^{\mathbb{N}_0}$, which is defined by

$$\mathcal{L}_{aq^m}\{f\}(s) = \int_{aq^m}^\infty e_{\ominus s}(\sigma(t), a)f(t)D_qt$$
$$= \sum_{n=m}^\infty \frac{\mu(aq^n)f(aq^n)}{\prod_{k=0}^n[1 + s\mu(aq^k)]}.$$

**Theorem 4.45.** *Let $m \in \mathbb{N}_0$ be given and suppose $f : aq^{\mathbb{N}_0} \to \mathbb{R}$ is of exponential order $r > 0$. Then for $|s| > r$*

$$\mathcal{L}_{aq^m}\{f\}(s) = \mathcal{L}_a\{f\}(s) - \sum_{n=0}^{m-1} \frac{\mu(aq^n)f(aq^n)}{\prod_{k=0}^{n}[1 + s\mu(aq^k)]}.$$

*Proof.* Assume $m \in \mathbb{N}_0$ and $f : aq^{\mathbb{N}_0} \to \mathbb{R}$ is of exponential order $r > 0$. Then for $|s| > r$,

$$\begin{aligned}
\mathcal{L}_{aq^m}\{f\}(s) &= \int_{aq^m}^{\infty} e_{\ominus s}(\sigma(t), a)f(t)D_q t \\
&= \sum_{n=m}^{\infty} \frac{\mu(aq^n)f(aq^n)}{\prod_{k=0}^{n}[1 + s\mu(aq^k)]} \\
&= \sum_{n=0}^{\infty} \frac{\mu(aq^n)f(aq^n)}{\prod_{k=0}^{n}[1 + s\mu(aq^k)]} - \sum_{n=0}^{m-1} \frac{\mu(aq^n)f(aq^n)}{\prod_{k=0}^{n}[1 + s\mu(aq^k)]} \\
&= \mathcal{L}_a\{f\}(s) - \sum_{n=0}^{m-1} \frac{\mu(aq^n)f(aq^n)}{\prod_{k=0}^{n}[1 + s\mu(aq^k)]}.
\end{aligned}$$

□

**Theorem 4.46.** *The function $f(t) = e_p(t, a)$, for $p \in \mathcal{R}_q(\mathbb{C})$ a constant, is of exponential order $r = |p| + \epsilon$, for each $\epsilon > 0$.*

*Proof.* For $t = aq^n, n \in \mathbb{N}_0$, we have

$$\left|e_p(t, a)\right| = |e_p(aq^n, a)| = \left|\prod_{k=0}^{n-1}[1 + \mu(aq^k)p]\right| = \prod_{k=0}^{n-1} |1 + \mu(aq^k)p|.$$

By applying the triangle inequality to each term in the product we obtain

$$|e_p(t, a)| \le \prod_{k=0}^{n-1}[1 + \mu(aq^k)|p|].$$

Let $\epsilon > 0$ be given. Then there exists an $N$ such that for all $n \ge N$ we have $\mu(aq^n)\epsilon > 1$, which implies that

$$\mu(aq^k)(|p| + \epsilon) \ge \mu(aq^k)|p| + 1 \quad \text{for all} \quad k \in \mathbb{N}_N.$$

But it is also true that

$$\begin{aligned}\mu(aq^N)(|p|+\epsilon) &\geq \mu(aq^k)|p| + \mu(aq^N)\epsilon \\ &\geq \mu(aq^k)|p| + 1 \quad \text{for all} \quad k \leq N,\end{aligned}$$

So, for $n$ sufficiently large, we have

$$\begin{aligned}|e_p(t,a)| &\leq \prod_{k=0}^{n-1}[1+\mu(aq^k)|p|] \\ &= \prod_{k=0}^{N}[1+\mu(aq^k)|p|] \prod_{s=N+1}^{n-1}[1+\mu(aq^s)|p|] \\ &\leq \left[\mu(aq^N)(|p|+\epsilon))\right]^{N+1} \prod_{s=N+1}^{n-1} \mu(aq^s)(|p|+\epsilon) \\ &\leq \left[a(q-1)q^N(|p|+\epsilon)\right]^{N+1} [a(q-1)(|p|+\epsilon)]^{n-N-1} q^{\frac{n(n-1)}{2}} q^{-\frac{N(N+1)}{2}} \\ &= a^n(q-1)^n(|p|+\epsilon)^n q^{\frac{N(N+1)}{2}} q^{\frac{n(n-1)}{2}} \\ &= q^{\frac{N(N+1)}{2}} \left[\mu(a)q^{\frac{n-1}{2}}\right]^n (|p|+\epsilon)^n.\end{aligned}$$

Hence, $e_p(t,a)$ is of exponential order $r = |p| + \epsilon$ for each $\epsilon > 0$. □

**Theorem 4.47.** *Assume p is a regressive constant. Then*

$$\mathcal{L}_a\left\{e_p(t,a)\right\}(s) = \frac{1}{s-p}, \quad \textit{for} \quad |s| > |p|.$$

*Proof.* Recall that the quantum exponential function $e_p(t,a)$ is given by

$$e_p(t,a) = \prod_{\tau=a}^{t/q}[1+\mu(\tau)p(\tau)].$$

Since, for any $\epsilon > 0$, $e_p(t,a)$ is of exponential order $|p|+\epsilon$, we have for $|s| > |p|+\epsilon$

$$\begin{aligned}\mathcal{L}_a\{e_p(t,a)\}(s) &= \int_a^\infty e_{\ominus s}(\sigma(t),a)e_p(t,a)D_qt \\ &= \int_a^\infty \frac{e_p(t,a)}{\prod_{\tau=a}^{t}[1+s\mu(r)]}D_qt \\ &= \int_a^\infty \frac{e_p(t,a)}{[1+s\mu(t)]\prod_{\tau=a}^{t/q}[1+s\mu(\tau)]}D_qt\end{aligned}$$

$$\begin{aligned}
&= \int_a^\infty \frac{e_{\ominus s}(t,a)e_p(t,a)}{1+s\mu(t)} D_q t \\
&= \int_a^\infty \frac{e_{p\ominus s}(t,a)}{1+s\mu(t)} D_q t \\
&= \frac{1}{p-s}\int_a^\infty (p\ominus s)\, e_{p\ominus s}(t,a) D_q t \\
&= \frac{1}{p-s} e_{p\ominus s}(t,a)\Big|_a^\infty \\
&= \frac{1}{s-p}.
\end{aligned}$$

Hence,

$$\mathcal{L}_a\{e_p(t,a)\}(s) = \frac{1}{s-p}, \quad |s| > |p| + \epsilon.$$

Since $\epsilon > 0$ is arbitrary, this holds for all $|s| > |p|$. □

*Remark 4.48.* Since $e_p(t,a) \equiv 1$ when $p = 0$, the $q$-Laplace transform of a constant function follows from the above theorem and the linearity of the $q$-Laplace transform:

$$\mathcal{L}_a\{c\}(s) = c\mathcal{L}_a\{e_0(t,a)\} = \frac{c}{s}, \quad |s| > 0. \tag{4.3}$$

**Theorem 4.49.** *The Taylor monomial $h_n(t,a)$, $n \in \mathbb{N}_0$, is of exponential order $r = \epsilon$, for each $\epsilon > 0$.*

*Proof.* Fix $n \in \mathbb{N}_0$. For $t = aq^m \in aq^{\mathbb{N}_0}$ we have that

$$|h_n(t,a)| = \frac{(t-a)_q^{\underline{n}}}{[n]_q!} \leq (t-a)^n \leq t^n = (aq^m)^n = a^n q^{mn}. \tag{4.4}$$

Note that for any fixed $\delta > 0$ and any fixed constant $\alpha > 1$

$$\lim_{m\to\infty} \alpha^{\frac{m^2}{2}} \delta^m = \infty.$$

Therefore, taking $\delta = \mu(a)q^{-\frac{1}{2}}\epsilon q^{-n}$, where $\epsilon > 0$ is fixed but arbitrary and $\alpha = q$, we have there is a positive integer $M$ such that

$$q^{\frac{m^2}{2}}[\mu(a)q^{-\frac{1}{2}}\epsilon q^{-n}]^m \geq 1, \quad \text{for } m \geq M.$$

That is

$$\frac{[q^{\frac{m-1}{2}}\mu(a)\epsilon]^m}{q^{mn}} \geq 1, \quad \text{for } m \geq M.$$

Hence,

$$q^{mn} \leq [q^{\frac{m-1}{m}}\mu(a)\epsilon]^m, \quad \text{for } m \geq M. \tag{4.5}$$

Then using (4.4) and (4.5) we get

$$|h_n(t,a)| \leq a^n q^{mn} \leq a^n[\mu(a)q^{\frac{m-1}{2}}]^m \epsilon^m, \quad \text{for } m \geq M.$$

Hence, $h_n(t,a)$ is of exponential order $\epsilon$. □

**Theorem 4.50.** *For $n \in \mathbb{N}_0$,*

$$\mathcal{L}_a\{h_n(t,a)\}(s) = \frac{1}{s^{n+1}}, \quad \textit{for } |s| > 0. \tag{4.6}$$

*Proof.* From Theorems 4.42 and 4.47, we have that $\mathcal{L}_a\{h_n(t,a)\}(s)$ exists for $|s| > 0$. We prove that the formula (4.6) holds by mathematical induction. From Remark 4.48, we have that (4.6) holds for $n = 0$. Now assume that (4.6) holds for some $n \geq 0$.

Integrating by parts, we obtain

$$\begin{aligned}
\mathcal{L}_a\{h_{n+1}(t,a)\}(s) &= \int_a^\infty h_{n+1}(t,a)e_{\ominus s}(\sigma(t),a)D_qt \\
&= \frac{1}{\ominus s}h_{n+1}(t,a)e_{\ominus s}(\sigma(t),a)\Big|_{t=a}^{\infty} - \int_a^\infty \frac{h_n(t,a)e_{\ominus s}(\sigma(\sigma(t)),a)}{\ominus s}D_qt \\
&= -\int_a^\infty h_n(t,a)\left(\frac{1}{-s\prod_{r=a}^{t}(1+\mu(r)s)}\right) \quad \text{(using Corollary 4.43)} \\
&= \frac{1}{s}\int_a^\infty h_n(t,a)e_{\ominus s}(\sigma(t),a)D_qt \\
&= \frac{1}{s}\mathcal{L}_a\{h_n(t,a)\}(s) \\
&= \frac{1}{s}\frac{1}{s^{n+1}} \\
&= \frac{1}{s^{n+2}}
\end{aligned}$$

for $|s| > 0$. □

**Theorem 4.51.** *Assume $p$ is a constant. If $\pm p \in \mathcal{R}_q$, then* $\cosh_p(t,a)$ *and* $\sinh_p(t,a)$ *are of exponential order $|p|+\epsilon$ for any $\epsilon > 0$. Also if $\pm ip \in \mathcal{R}_q$, then* $\cos_p(t,a)$ *and* $\sin_p(t,a)$ *are of exponential order $|p|+\epsilon$ for any $\epsilon > 0$.*

*Proof.* By Exercise 4.5, $|e_{\pm p}(t,a)| \le e_{|p|}(t,a)$ for $t \in q^{\mathbb{N}_0}$. It follows that

$$|\cosh_p(t,a)| \le e_{|p|}(t,a), \quad \text{for} \quad t \in q^{\mathbb{N}_0},$$

and so by Theorem 4.46 $\cosh_p(t,a)$ is of exponential order $|p|+\epsilon$ for any $\epsilon > 0$. Similarly, $\sinh_p(t,a)$ is of exponential order $|p|+\epsilon$ for any $\epsilon > 0$. Likewise, using $|e_{\pm ip}(t,a)| \le e_{|p|}(t,a)$ for $t \in aq^{\mathbb{N}_0}$ we get that $\cos_p(t,a)$, and $\sin_p(t,a)$ are of exponential order $|p|+\epsilon$ for any $\epsilon > 0$. □

**Theorem 4.52.** *Assume $p$ is a constant. If $\pm p \in \mathcal{R}_q(\mathbb{C})$, then*

(i) $\mathcal{L}_a\{\sinh_p(t,a)\}(s) = \frac{p}{s^2-p^2}$;
(ii) $\mathcal{L}_a\{\cosh_p(t,a)\}(s) = \frac{s}{s^2-p^2}$,

*for $|s| > |p|$. If $\pm ip \in \mathcal{R}_q(\mathbb{C})$, then*

(i) $\mathcal{L}_a\{\cos_p(t,a)\}(s) = \frac{s}{s^2+p^2}$;
(ii) $\mathcal{L}_a\{\sin_p(t,a)\}(s) = \frac{p}{s^2+p^2}$,

*for $|s| > |p|$.*

*Proof.* From Theorem 4.51, the exponential order of each of the functions $\cosh_p(t,a)$, $\sinh_p(t,a)$, $\cos_p(t,a)$, and $\sin_p(t,a)$ is $|p|+\epsilon$ for any $\epsilon > 0$. It then follows from Theorem 4.42 that the Laplace transform of each of these functions exists for $|s| > |p|$.

For $|s| > |p|$, we have

$$\begin{aligned}\mathcal{L}_a\{\sinh_p(t,a)\}(s) &= \mathcal{L}_a\left\{\frac{e_p(t,a)-e_{-p}(t,a)}{2}\right\}(s)\\ &= \frac{1}{2}\left(\mathcal{L}_a\{e_p(t,a)\}(s) - \mathcal{L}_a\{e_{-p}(t,a)\}(s)\right)\\ &= \frac{1}{2}\left(\frac{1}{s-p} - \frac{1}{s-(-p)}\right)\\ &= \frac{2p}{2(s-p)(s+p)}\\ &= \frac{p}{s^2-p^2}.\end{aligned}$$

On the other hand, in case $|s| > |p|$ we calculate

$$\begin{aligned}\mathcal{L}_a\{\cosh_p(t,a)\}(s) &= \mathcal{L}_a\left\{\frac{e_p(t,a)+e_{-p}(t,a)}{2}\right\}(s)\\ &= \frac{1}{2}\left(\mathcal{L}_a\{e_p(t,a)\}(s) + \mathcal{L}_a\{e_{-p}(t,a)\}(s)\right)\end{aligned}$$

$$
\begin{aligned}
&= \frac{1}{2}\left(\frac{1}{s-p} + \frac{1}{s-(-p)}\right) \\
&= \frac{2s}{2(s-p)(s+p)} \\
&= \frac{s}{s^2-p^2}.
\end{aligned}
$$

If it holds instead that $|s| > |p|$, then we find that

$$
\begin{aligned}
\mathcal{L}_a\{\cos_p(t,a)\}(s) &= \mathcal{L}_a\left\{\frac{e_{ip}(t,a)+e_{-ip}(t,a)}{2}\right\}(s) \\
&= \frac{1}{2}\left(\mathcal{L}_a\{e_{ip}(t,a)\}(s) + \mathcal{L}_a\{e_{-ip}(t,a)\}(s)\right) \\
&= \frac{1}{2}\left(\frac{1}{s-ip} + \frac{1}{s+ip}\right) \\
&= \frac{2s}{2(s^2-i^2p^2)} \\
&= \frac{s}{s^2+p^2}.
\end{aligned}
$$

Finally, in the remaining case where $|s| > |p|$, we obtain

$$
\begin{aligned}
\mathcal{L}_a\{\sin_p(t,a)\}(s) &= \mathcal{L}_a\left\{\frac{e_{ip}(t,a)-e_{-ip}(t,a)}{2i}\right\}(s) \\
&= \frac{1}{2i}\left(\mathcal{L}_a\{e_{ip}(t,a)\}(s) - \mathcal{L}_a\{e_{-ip}(t,a)\}(s)\right) \\
&= \frac{1}{2i}\left(\frac{1}{s-ip} - \frac{1}{s+ip}\right) \\
&= \frac{1}{2i}\left(\frac{2ip}{s^2+p^2}\right) \\
&= \frac{p}{s^2+p^2}.
\end{aligned}
$$

And this completes the proof. □

**Theorem 4.53.** *If $f : aq^{\mathbb{N}_0} \to \mathbb{R}$ and $f(t)$ is of exponential order $r > 0$, then*

$$
\mathcal{L}_a\{D_q^n f\}(s) = s^n F_a(s) - \sum_{k=0}^{n-1} s^{n-1-k} D_q^k f(a), \quad \text{for} \quad |s| > r, \tag{4.7}
$$

*where $n \in \mathbb{N}_1$.*

*Proof.* We prove (4.7) by induction for $n \in \mathbb{N}_1$. First if $n = 1$, we obtain (using integration by parts) the following for $|s| > r$.

$$\begin{aligned}\mathcal{L}_a\{D_q^1 f\}(s) &= \int_a^\infty e_{\ominus s}(\sigma(t), a) D_q f(t) D_q t \\ &= e_{\ominus s}(t,a) f(t)\Big|_a^\infty - \int_a^\infty \ominus s e_{\ominus s}(t,a) f(t) D_q t \\ &= -f(a) + s\int_a^\infty e_{\ominus s}(\sigma(t),a) f(t) D_q t \\ &= sF_a(s) - f(a)\end{aligned}$$

Now assume that the theorem is true for some $k \in \mathbb{N}$. Consider

$$\begin{aligned}\mathcal{L}_a\{D_q^{k+1} f\}(s) &= \mathcal{L}_a\{D_q D_q^k f\}(s) \\ &= s\mathcal{L}_a\{D_q^k f\}(s) - D_q^k f(a) \\ &= s\left(s^k F_a(s) - \sum_{r=0}^{k-1} s^{k-1-r} D_q^r f(a)\right) - D_q^k f(a) \\ &= s^{k+1} F_a(s) - \sum_{r=0}^{k-1} s^{k-r} D_q^r f(a) - D_q^k f(a) \\ &= s^{k+1} F_a(s) - \sum_{r=0}^{k} s^{k-r} D_q^r f(a).\end{aligned}$$

Therefore, (4.7) holds for any $n \in \mathbb{N}$. □

**Theorem 4.54.** *Assume* $\alpha, \pm\frac{\beta}{1+\alpha\mu(t)} \in \mathcal{R}_q$. *Then*

(i) $\mathcal{L}_a\left\{e_\alpha(t,a)\cosh_{\frac{\beta}{1+\alpha\mu(t)}}(t,a)\right\}(s) = \frac{s-\alpha}{(s-\alpha)^2-\beta^2}$;

(ii) $\mathcal{L}_a\left\{e_\alpha(t,a)\sinh_{\frac{\beta}{1+\alpha\mu(t)}}(t,a)\right\}(s) = \frac{\beta}{(s-\alpha)^2-\beta^2}$,

*for* $|s| > \max\{|\alpha+\beta|, |\alpha-\beta|\}$.

*Proof.* Let $\alpha$ and $\beta$ be as above, and assume $|s| > \max\{|\alpha+\beta|, |\alpha-\beta|\}$. First consider

$$\begin{aligned}&\mathcal{L}_a\left\{e_\alpha(t,a)\cosh_{\frac{\beta}{1+\alpha\mu(t)}}(t,a)\right\}(s) \\ &= \mathcal{L}_a\left\{e_\alpha(t,a)\left(\frac{1}{2}\left[e_{\frac{\beta}{1+\alpha\mu(t)}}(t,a) + e_{\frac{-\beta}{1+\alpha\mu(t)}}(t,a)\right]\right)\right\}(s)\end{aligned}$$

$$= \frac{1}{2}\mathcal{L}_a\left\{e_\alpha(t,a)e_{\frac{\beta}{1+\alpha\mu(t)}}(t,a)\right\}(s) + \frac{1}{2}\mathcal{L}_a\left\{e_\alpha(t,a)e_{\frac{-\beta}{1+\alpha\mu(t)}}(t,a)\right\}(s)$$

$$= \frac{1}{2}\mathcal{L}_a\left\{e_{\alpha\oplus\frac{\beta}{1+\alpha\mu(t)}}(t,a)\right\}(s) + \frac{1}{2}\mathcal{L}_a\left\{e_{\alpha\oplus\frac{-\beta}{1+\alpha\mu(t)}}(t,a)\right\}(s)$$

$$= \frac{1}{2}\mathcal{L}_a\left\{e_{\alpha+\beta}(t,a)\right\} + \frac{1}{2}\mathcal{L}_a\left\{e_{\alpha-\beta}(t,a)\right\}(s).$$

Therefore,

$$\begin{aligned}\mathcal{L}_a\left\{e_\alpha(t,a)\cosh_{\frac{\beta}{1+\alpha\mu(t)}}(t,a)\right\}(s) &= \frac{1}{2}\left[\frac{1}{s-(\alpha+\beta)} + \frac{1}{s-(\alpha-\beta)}\right]\\ &= \frac{1}{2}\left[\frac{1}{(s-\alpha)-\beta} + \frac{1}{(s-\alpha)+\beta}\right]\\ &= \frac{1}{2}\frac{2(s-\alpha)}{(s-\alpha)^2-\beta^2}\\ &= \frac{s-\alpha}{(s-\alpha)^2-\beta^2}.\end{aligned}$$

Next consider

$$\mathcal{L}_a\left\{e_\alpha(t,a)\sinh_{\frac{\beta}{1+\alpha\mu(t)}}(t,a)\right\}(s)$$

$$= \mathcal{L}_a\left\{e_\alpha(t,a)\left(\frac{1}{2}\left[e_{\frac{\beta}{1+\alpha\mu(t)}}(t,a) - e_{\frac{-\beta}{1+\alpha\mu(t)}}(t,a)\right]\right)\right\}(s)$$

$$= \frac{1}{2}\mathcal{L}_a\left\{e_\alpha(t,a)e_{\frac{\beta}{1+\alpha\mu(t)}}(t,a)\right\} - \frac{1}{2}\mathcal{L}_a\left\{e_{\alpha)}(t,a)e_{\frac{-\beta}{1+\alpha\mu(t)}}(t,a)\right\}$$

$$= \frac{1}{2}\mathcal{L}_a\left\{e_{\alpha\oplus\frac{\beta}{1+\alpha\mu(t)}}(t,a)\right\} - \frac{1}{2}\mathcal{L}_a\left\{e_{\alpha\oplus\frac{-\beta}{1+\alpha\mu(t)}}(t,a)\right\}$$

$$= \frac{1}{2}\mathcal{L}_a\left\{e_{\alpha+\beta}(t,a)\right\} - \frac{1}{2}\mathcal{L}_a\left\{e_{\alpha-\beta}(t,a)\right\}.$$

Hence,

$$\begin{aligned}\mathcal{L}_a\left\{e_\alpha(t,a)\sinh_{\frac{\beta}{1+\alpha\mu(t)}}(t,a)\right\}(s) &= \frac{1}{2}\left[\frac{1}{s-(\alpha+\beta)} - \frac{1}{s-(\alpha-\beta)}\right]\\ &= \frac{1}{2}\left[\frac{1}{(s-\alpha)-\beta} - \frac{1}{(s-\alpha)+\beta}\right]\\ &= \frac{1}{2}\frac{2\beta}{(s-\alpha)^2-\beta^2}\\ &= \frac{\beta}{(s-\alpha)^2-\beta^2}.\end{aligned}$$

□

Similar to Theorem 4.54, one can prove the following theorem.

**Theorem 4.55.** *Assume* $\alpha$ *and* $\beta$ *are constants with* $\alpha, \pm i\frac{\beta}{1+\alpha\mu(t)} \in \mathcal{R}_q$. *Then*

(i) $\mathcal{L}_a\left\{e_\alpha(t,a)\cos_{\frac{\beta}{1+\alpha\mu(t)}}(t,a)\right\}(s) = \frac{s-\alpha}{(s-\alpha)^2+\beta^2}$;

(ii) $\mathcal{L}_a\left\{e_\alpha(t,a)\sin_{\frac{\beta}{1+\alpha\mu(t)}}(t,a)\right\}(s) = \frac{\beta}{(s-\alpha)^2+\beta^2}$,

*for* $|s| > \max\{|\alpha+i\beta|, |\alpha-i\beta|\}$.

*Proof.* Let $\alpha$ and $\beta$ be as above, and assume $|s| > \max\{|\alpha+i\beta|, |\alpha-i\beta|\}$. First consider

$$\begin{aligned}
&\mathcal{L}_a\left\{e_\alpha(t,a)\cos_{\frac{\beta}{1+\alpha\mu(t)}}(t,a)\right\}(s)\\
&= \mathcal{L}_a\left\{e_\alpha(t,a)\left(\frac{1}{2}\left[e_{\frac{i\beta}{1+\alpha\mu(t)}}(t,a) + e_{\frac{-i\beta}{1+\alpha\mu(t)}}(t,a)\right]\right)\right\}(s)\\
&= \frac{1}{2}\mathcal{L}_a\left\{e_\alpha(t,a)e_{\frac{i\beta}{1+\alpha\mu(t)}}(t,a)\right\}(s) + \frac{1}{2}\mathcal{L}_a\left\{e_\alpha(t,a)e_{\frac{-i\beta}{1+\alpha\mu(t)}}(t,a)\right\}(s)\\
&= \frac{1}{2}\mathcal{L}_a\left\{e_{\alpha\oplus\frac{i\beta}{1+\alpha\mu(t)}}(t,a)\right\}(s) + \frac{1}{2}\mathcal{L}_a\left\{e_{\alpha\oplus\frac{-i\beta}{1+\alpha\mu(t)}}(t,a)\right\}(s)\\
&= \frac{1}{2}\mathcal{L}_a\left\{e_{\alpha+i\beta}(t,a)\right\} + \frac{1}{2}\mathcal{L}_a\left\{e_{\alpha-i\beta}(t,a)\right\}(s).
\end{aligned}$$

Therefore,

$$\begin{aligned}
\mathcal{L}_a\left\{e_\alpha(t,a)\cos_{\frac{\beta}{1+\alpha\mu(t)}}(t,a)\right\}(s) &= \frac{1}{2}\left[\frac{1}{s-(\alpha+i\beta)} + \frac{1}{s-(\alpha-i\beta)}\right]\\
&= \frac{1}{2}\left[\frac{1}{(s-\alpha)-i\beta} + \frac{1}{(s-\alpha)+i\beta}\right]\\
&= \frac{1}{2}\frac{2(s-\alpha)}{(s-\alpha)^2+\beta^2}\\
&= \frac{s-\alpha}{(s-\alpha)^2+\beta^2}.
\end{aligned}$$

Next consider

$$\begin{aligned}
&\mathcal{L}_a\left\{e_\alpha(t,a)\sin_{\frac{\beta}{1+\alpha\mu(t)}}(t,a)\right\}(s)\\
&= \mathcal{L}_a\left\{e_\alpha(t,a)\left(\frac{1}{2i}\left[e_{\frac{i\beta}{1+\alpha\mu(t)}}(t,a) - e_{\frac{-i\beta}{1+\alpha\mu(t)}}(t,a)\right]\right)\right\}(s)
\end{aligned}$$

$$= \frac{1}{2i}\mathcal{L}_a\left\{e_\alpha(t,a)e_{\frac{i\beta}{1+\alpha\mu(t)}}(t,a)\right\}(s) - \frac{1}{2i}\mathcal{L}_a\left\{e_\alpha(t,a)e_{\frac{-i\beta}{1+\alpha\mu(t)}}(t,a)\right\}(s)$$

$$= \frac{1}{2i}\mathcal{L}_a\left\{e_{\alpha\oplus\frac{i\beta}{1+\alpha\mu(t)}}(t,a)\right\}(s) - \frac{1}{2i}\mathcal{L}_a\left\{e_{\alpha\oplus\frac{-i\beta}{1+\alpha\mu(t)}}(t,a)\right\}(s)$$

$$= \frac{1}{2i}\mathcal{L}_a\left\{e_{\alpha+i\beta}(t,a)\right\}(s) - \frac{1}{2i}\mathcal{L}_a\left\{e_{\alpha-i\beta}(t,a)\right\}(s).$$

Hence,

$$\begin{aligned}\mathcal{L}_a\left\{e_\alpha(t,a)\sin_{\frac{\beta}{1+\alpha\mu(t)}}(t,a)\right\}(s) &= \frac{1}{2i}\left[\frac{1}{s-(\alpha+i\beta)} - \frac{1}{s-(\alpha-i\beta)}\right]\\ &= \frac{1}{2i}\left[\frac{1}{(s-\alpha)-i\beta} - \frac{1}{(s-\alpha)+i\beta}\right]\\ &= \frac{1}{2i}\frac{2i\beta}{(s-\alpha)^2+\beta^2}\\ &= \frac{\beta}{(s-\alpha)^2+\beta^2}.\end{aligned}$$

□

*Example 4.56.* Use the $q$-Laplace transform to solve the following IVP:

$$D_q^2 y(t) - 2D_q y(t) - 8y(t) = 0, \quad t \in aq^{\mathbb{N}_0}$$
$$y(a) = -3/2, \quad D_q y(a) = 0.$$

Taking the Laplace transform of both sides, we have

$$\left(s^2 Y_a(s) - sy(a) - sD_q y(a)\right) - 2\left(sY_a(s) - y(a)\right) - 8Y_a(s) = 0.$$

Using the initial conditions, we have

$$s^2 Y_a(s) + \frac{3s}{2} - 2sY_a(s) - 3 - 8Y_a(s) = 0.$$

Solving for $Y_a(s)$ we get

$$Y_a(s) = \frac{3 - \frac{3s}{2}}{s^2 - 2s - 8} = -\frac{3}{2}\frac{s-1}{(s-1)^2-9} + \frac{1}{2}\frac{3}{(s-1)^2-9}.$$

Taking the inverse Laplace transform we obtain

$$y(t) = -\frac{3}{2}e_1(t,a)\cosh_{\frac{3}{1+\mu(t)}}(t,a) + \frac{1}{2}e_1(t,a)\sinh_{\frac{3}{1+\mu(t)}}(t,a).$$

## 4.8 Matrix Exponential

In this section we will study matrix exponentials defined on the quantum set $q^{\mathbb{N}_0}$. We say that a square matrix function $A$ defined on $q^{\mathbb{N}_0}$ is regressive provided

$$\det[I + \mu(t)A(t)] \neq 0, \quad t \in aq^{\mathbb{N}_0}.$$

**Definition 4.57.** Assume $A$ is a regressive matrix function on $aq^{\mathbb{N}_0}$. Then we define the matrix exponential $e_A(t,s)$ for each fixed $s \in aq^{\mathbb{N}_0}$ to be the unique solution of the matrix IVP

$$D_q X(t) = A(t)X(t), \quad t \in aq^{\mathbb{N}_0}$$
$$X(s) = I,$$

where $I$ is the identity matrix.

The following theorem gives a formula for the matrix exponential function $e_A(t,a)$. First we introduce some notation. Assume $B(t)$ is a matrix function defined on $aq^{\mathbb{N}_0}$ and $s \leq t$ are in $aq^{\mathbb{N}_0}$. Then

$$* \prod_{\tau=s}^{\frac{t}{q}} B(\tau) := B\left(\frac{t}{q}\right) B\left(\frac{t}{q^2}\right) \cdots B(qs)B(s),$$

with the convention that

$$* \prod_{\tau=s}^{\frac{s}{q}} B(\tau) = I.$$

**Theorem 4.58.** *Assume $A(t)$ is a regressive matrix function on $aq^{\mathbb{N}_0}$ and $s \in aq^{\mathbb{N}_0}$. Then*

$$e_A(t,s) = \begin{cases} * \prod_{\tau=s}^{\frac{t}{q}} [I + \mu(\tau)A(\tau)], & t \in \mathbb{N}_s \\ \prod_{\tau=t}^{\frac{s}{q}} [I + \mu(\tau)A(\tau)]^{-1}, & t \in \mathbb{N}_a^{\frac{s}{q}}. \end{cases}$$

*Also, if $t = aq^n$ and $s = aq^m$, then*

$$e_A(t,s) = \begin{cases} * \prod_{k=m}^{n-1} \left[I + \mu(aq^k)A(aq^k)\right], & n \in \mathbb{N}_m \\ \prod_{k=n}^{m-1} \left[I + \mu(aq^k)A(aq^k)\right]^{-1}, & 0 \leq n < m. \end{cases}$$

*Proof.* Assume $X(t)$ solves the IVP

$$D_qX(t) = A(t)X(t), \quad t \in aq^{\mathbb{N}_0}$$
$$X(s) = I.$$

Since $D_qX(t) = A(t)X(t)$, we have that

$$X(qt) = [I + \mu(t)A(t)]\,X(t). \tag{4.8}$$

Now assume that $t \in \mathbb{N}_{s+1}$. Letting $t = s$ in (4.8) we have that

$$X(qs) = [I + \mu(s)A(s)]\,X(s) = [I + \mu(s)A(s)]\,.$$

Letting $t = qs$ in (4.8) we have that

$$\begin{aligned} X(q^2s) &= [I + \mu(qs)A(qs)]\,X(qs) \\ &= [I + \mu(qs)A(qs)]\,[I + \mu(s)A(s)]\,. \end{aligned}$$

Letting $t = q^2s$ in (4.8) we have that

$$\begin{aligned} X(q^3s) &= \left[I + \mu(q^2s)A(q^2s)\right]X(q^2s) \\ &= \left[I + \mu(q^2s)A(q^2s)\right][I + \mu(qs)A(qs)]\,[I + \mu(s)A(s)]\,. \end{aligned}$$

Proceeding in this manner we have by mathematical induction that for $t \in \mathbb{N}_{s+1}$

$$e_A(t,s) = X(t) = \;^*\prod_{\tau=s}^{\frac{t}{q}} [I + \mu(\tau)A(\tau)]\,.$$

If $t = s$,

$$e_A(t,s) = I = \;^*\prod_{\tau=s}^{\frac{s}{q}} [I + \mu(\tau)A(\tau)]\,,$$

by our convention on products. Finally, consider the case when $t \in \mathbb{N}_u^{\frac{s}{q}}$. Solving (4.8) for $X(t)$, we get

$$X(t) = [I + \mu(t)A(t)]^{-1}\,X(qt). \tag{4.9}$$

Letting $t = \frac{s}{q}$ in (4.9) we get

$$X\left(\frac{s}{q}\right) = \left[I + \mu\left(\frac{s}{q}\right)A\left(\frac{s}{q}\right)\right]^{-1} X(s) = \left[I + \mu\left(\frac{s}{q}\right)A\left(\frac{s}{q}\right)\right]^{-1}.$$

Letting $t = \frac{s}{q^2}$ in (4.9) we get

$$X\left(\frac{s}{q^2}\right) = \left[I + \mu\left(\frac{s}{q^2}\right)A\left(\frac{s}{q^2}\right)\right]^{-1} X\left(\frac{s}{q}\right)$$
$$= \left[I + \mu\left(\frac{s}{q^2}\right)A\left(\frac{s}{q^2}\right)\right]^{-1} \left[I + \mu\left(\frac{s}{q}\right)A\left(\frac{s}{q}\right)\right]^{-1}.$$

Proceeding by mathematical induction we get that for $t \in \mathbb{N}_a^{\frac{s}{q}}$

$$e_A(t,s) = X(t) = \prod_{\tau=t}^{\frac{s}{q}} [I + \mu(\tau)A(\tau)]^{-1}.$$

The second formula for $e_A(t,s)$ in this theorem follows from the first formula in this theorem for $e_A(t,s)$. □

*Example 4.59.* Find $e_A(t,1)$ for the quantum set $q^{\mathbb{N}_0}$, where $A(t) := \frac{1}{t}B$, $t \in q^{\mathbb{N}_0}$, where $B$ is a constant matrix satisfying $\det[I + (q-1)B] \neq 0$. By Theorem 4.58 for $t = q^n$

$$e_A(t,a) = {}^*\prod_{k=1}^{n-1}\left[I + \frac{\mu(q^k)}{q^k}B\right]$$
$$= [I + (q-1)B]^{n-1}$$
$$= [I + (q-1)B]^{\log_q(t)-1}.$$

In the next theorem we give several properties of the matrix exponential function $e_A(t,s)$, based at $s \in \mathbb{N}_a$.

**Theorem 4.60.** *Assume $A(t), B(t)$ are regressive matrix functions on $aq^{\mathbb{N}_0}$ and $t, s, r \in \mathbb{N}_a$. Then*

(i) *if 0 denotes the zero matrix function, then $e_0(t,s) = I$ and $e_A(t,t) = I$;*
(ii) $\det e_A(t,s) \neq 0, \quad t \in \mathbb{N}_a$;
(iii) $\Delta e_A(t,s) = A(t)e_A(t,s)$;
(iv) $e_A(qt,s) = [I + (q-1)tA(t)]e_A(t,s)$;
(v) $e_A(t,s)e_A(s,r) = e_A(t,r)$;
(vi) $e_A(t,s)e_B(t,s) = e_{A\oplus B}(t,s)$;
(vii) $e_{\ominus A}(t,s) = \frac{1}{e_A(t,s)}$;
(viii) $\frac{e_A(t,s)}{e_B(t,s)} = e_{A\ominus B}(t,s), \quad if \quad A(t)B(s) = B(s)A(t) \quad t,s \in aq^{\mathbb{N}_0}$;
(ix) $e_A(t,s) = \{e_A(s,t)\}^{-1}$.

*Proof.* We prove many of these properties when $s = a$ and leave it to the reader to show that the same results hold for any $s \in \mathbb{N}_a$. By the definition of the matrix exponential we have that (i) and (iv) hold. To see that (ii) holds when $s = a$ note that since $A$ is a regressive matrix function on $aq^{\mathbb{N}_0}$, $\det[I + (q-1)tA(t)] \neq 0$ for $t \in \mathbb{N}_a$ and hence we have that

$$\det e_A(t,a) = \prod_{\tau=a}^{\frac{t}{q}} \det[I + \mu(\tau)A(\tau)] \neq 0,$$

for $t \in \mathbb{N}_a$. The proof of (iii) is similar to the proof of (ii).

Since

$$\begin{aligned} e_A(\sigma(t),a) &= {}^*\prod_{\tau=a}^{t}[I + \mu(\tau)A(\tau)] \\ &= [I + \mu(t)A(t)]\left({}^*\prod_{\tau=a}^{\frac{t}{q}}[I + \mu(\tau)A(\tau)]\right) \\ &= [I + \mu(t)A(t)]e_A(t,a), \end{aligned}$$

we have that (v) holds when $s = a$.

We only show (vi) holds when $t \geq s \geq r$ and leave the other cases to the reader. In particular, we merely observe that

$$\begin{aligned} e_A(t,s)e_A(s,r) &= {}^*\prod_{\tau=s}^{\frac{t}{q}}[I + \mu(\tau)A(\tau)]\left({}^*\prod_{\tau=r}^{s-1}[I + \mu(\tau)A(\tau)]\right) \\ &= {}^*\prod_{\tau=r}^{\frac{t}{q}}[I + \mu(\tau)A(\tau)] \\ &= e_A(t,r). \end{aligned}$$

We proved (vii) holds when with $s = a$, earlier to motivate the definition of the circle plus addition. To see that (viii) holds with $s = a$ note that

$$\begin{aligned} e_{\ominus A}(t,a) &= {}^*\prod_{\tau=a}^{\frac{t}{q}}[I + \mu(\tau)(\ominus A)(\tau)] \\ &= {}^*\prod_{\tau=a}^{\frac{t}{q}}\frac{1}{I + \mu(\tau)A(\tau)} \\ &= {}^*\frac{1}{\prod_{\tau=a}^{\frac{t}{q}}[I + \mu(\tau)A(\tau)]} \\ &= e_A(t,a)^{-1}. \end{aligned}$$

Since

$$\frac{e_p(t,a)}{e_q(t,a)} = e_p(t,a)e_{\ominus q}(t,a) = e_{p\oplus[\ominus q]}(t,a) = e_{p\ominus q}(t,a),$$

we have (ix) holds when $s = a$. Since

$$e_p(t,a) = \prod_{s=a}^{\frac{t}{q}}[1+p(s)] = \frac{1}{\prod_{s=a}^{\frac{t}{q}} \frac{1}{[1+p(s)]}} = \frac{1}{e_p(a,t)},$$

we have that (x) holds. □

**Theorem 4.61 (Variation of Constants Formula).** *Assume $A(t)$ is an $n \times n$ regressive matrix function and $b(t)$ is an $n \times 1$ vector function on $aq_0^{\mathbb{N}}$. Then the unique solution of the IVP*

$$D_q y(t) = A(t)y(t) + b(t), \quad t \in aq^{\mathbb{N}_0}$$
$$y(a) = y_0,$$

*where $y_0$ is a given $n \times 1$ constant vector, is given by*

$$y(t) = e_A(t,a)y_0 + \int_a^t e_A(t,\sigma(s))b(s)D_q s, \quad t \in aq^{\mathbb{N}_0}.$$

*Proof.* The proof of the uniqueness is left to the reader. Let

$$y(t) = e_A(t,a)y_0 + \int_a^t e_A(t,\sigma(s))b(s)D_q s, \quad t \in aq^{\mathbb{N}_0}.$$

Then using the Leibniz rule we get

$$\begin{aligned} D_q y(t) &= A(t)e_A(t,a)y_0 + \int_a^t A(t)e_A(t,\sigma(s))b(s)D_q s + e_A(\sigma(t),\sigma(t))b(t) \\ &= A(t)\left[e_A(t,a)y_0 + \int_a^t e_A(t,\sigma(s))b(s)D_q s\right] + b(t) \\ &= A(t)y(t) + b(t), \end{aligned}$$

for $t \in aq^{\mathbb{N}_0}$. Also, $y(a) = y_0$. □

## 4.9 Floquet Theory

Most of the results in this section appear in a paper by Bohner and Chieochan [61]. We now define what Bohner and Chieochan call an $\omega$-periodic function on the quantum set $q^{\mathbb{N}_0}$.

**Definition 4.62.** Assume $\omega \in \mathbb{N}$ and $f$ is a function defined on $q^{\mathbb{N}_0}$. Then we say $f$ is an $\omega$-periodic function on $q^{\mathbb{N}_0}$ provided

$$q^{\omega} f(q^{\omega} t) = f(t), \quad t \in q^{\mathbb{N}_0}.$$

*Remark 4.63.* By Exercise 4.10 if have that if $f$ is an $\omega$-periodic function on $q^{\mathbb{N}_0}$, then

$$q^{n\omega+k} f(q^{n\omega+k} t) = q^k f(q^k t) \tag{4.10}$$

for all $n \in \mathbb{N}_0$, $0 \leq k \leq \omega - 1$, $t \in q^{\mathbb{N}_0}$.

Next we give examples (see [61]) of $\omega$-periodic functions on $q^{\mathbb{N}_0}$.

*Example 4.64.* Let $r_0, r_1, \cdots, r_{\omega-1}$ be $\omega$ real numbers and define $f$ on $q^{\mathbb{N}_0}$ by

$$f(t) = f(q^{n\omega+k}) = \frac{r_k}{q^{n\omega+k}} = \frac{r_k}{t},$$

where $n \in \mathbb{N}_0$ and $0 \leq k \leq \omega - 1$. To see that $f$ is an $\omega$-periodic function on $q^{\mathbb{N}_0}$, note that, since $t \in q^{\mathbb{N}_0}$, it thus follows that $t = q^{n\omega+k}$ for some $n \in \mathbb{N}$ and some $k \in \{0, 1, 2, \ldots, \omega - 1\}$, and hence

$$\begin{aligned}
q^{\omega} f(q^{\omega} t) &= q^{\omega} f(q^{\omega} q^{n\omega+k}) \\
&= q^{\omega} f(q^{(n+1)\omega+k}) \\
&= q^{\omega} \frac{r_k}{q^{(n+1)\omega+k}} \\
&= \frac{r_k}{q^{n\omega+k}} \\
&= f(t).
\end{aligned}$$

Similarly, one gets the following corresponding result for examples of $\omega$-periodic matrix functions on $q^{\mathbb{N}_0}$.

*Example 4.65.* Let $R_0, R_1, \cdots, R_{\omega-1}$ be $\omega$ constant square matrices of the same dimension and define the matrix function $A$ on $q^{\mathbb{N}_0}$ by

$$A(t) = A(q^{n\omega+k}) = \frac{1}{q^{n\omega+k}} R_k = \frac{1}{t} A(t),$$

where $n \in \mathbb{N}_0$ and $0 \leq k \leq \omega - 1$. Then $A$ is an $\omega$-periodic matrix function on $q^{\mathbb{N}_0}$.

Bohner et al. [61] motivated the definition of an $\omega$-periodic function by giving the following theorem.

**Theorem 4.66.** *If $f : q^{\mathbb{N}_0} \to \mathbb{R}$ is $\omega$-periodic, then*

$$\int_{q^{n\omega}}^{q^{(n+1)\omega}} f(t)D_qt = \int_1^{q^\omega} f(t)D_qt \tag{4.11}$$

*for $n \geq 0$.*

*Proof.* Clearly, (4.11) holds for $n = 0$. Now assume $n \geq 1$ is fixed but arbitrary. Then

$$\begin{aligned}\int_{q^{n\omega}}^{q^{(n+1)\omega}} f(t)D_qt &= \sum_{k=n\omega}^{(n+1)\omega-1} f(q^k)\mu(q^k)\\ &= \sum_{k=n\omega}^{(n+1)\omega-1} f(q^k)(q-1)q^k\\ &= \sum_{k=0}^{\omega-1}(q-1)q^{n\omega+k}f(q^{n\omega+k})\\ &= \sum_{k=0}^{\omega-1} f(q^k)(q-1)q^k, \quad \text{(using (4.10))}\\ &= \sum_{k=0}^{\omega-1} f(q^k)\mu(q^k)\\ &= \int_1^{q^\omega} f(t)D_qt.\end{aligned}$$

This completes the proof. □

Floquet theory in the quantum case is the study of the Floquet system

$$D_qx(t) = A(t)x(t), \quad t \in q^{\mathbb{N}_0}, \tag{4.12}$$

where the matrix function $A$ is assumed to be regressive with smallest positive integer period $\omega$. In this case we say $\omega$ is the prime period of $A$.

**Definition 4.67.** The boundary value problem

$$D_qy(t) = A(t)y(t) \tag{4.13}$$

$$x(t_0) = q^\omega x(q^\omega t_0), \tag{4.14}$$

where $t_0 \in q^{\mathbb{N}_0}$, is called a **periodic boundary value problem** and the boundary condition (4.14) is called a **periodic boundary condition**.

The motivation for Definition 4.67 is the following theorem.

**Theorem 4.68.** *Assume $A(t)$ is a regressive matrix function which is $\omega$-periodic on $q^{\mathbb{N}_0}$. If $x(t)$ is a solution of the periodic BVP (4.13), (4.14), then $x$ is $\omega$-periodic on $q^{\mathbb{N}_0}$.*

*Proof.* Assume $x(t)$ is a solution of the periodic BVP (4.13), (4.14). If we let

$$y(t) := q^\omega x(q^\omega t), \quad t \in q^{\mathbb{N}_0},$$

we have, using the chain rule formula (4.1), that

$$\begin{aligned} D_q y(t) &:= q^\omega D_q x(q^\omega t) \\ &= q^\omega q^\omega D_q x(q^\omega t) \\ &= q^\omega q^\omega A(q^\omega t) x(q^\omega t) \\ &= q^\omega A(t) x(q^\omega t) \\ &= A(t) q^\omega x(q^\omega t) \\ &= A(t) y(t). \end{aligned}$$

Hence, $y(t)$ is a solution of the Floquet system (4.12). Since

$$y(t_0) = q^\omega x(q^\omega t_0) = x(t_0),$$

we have that $x(t)$ and $y(t)$ satisfy the same initial conditions. Thus by the uniqueness of solutions of IVPs we have that

$$x(t) = y(t) = q^\omega x(q^\omega t), \quad t \in q^{\mathbb{N}_0}.$$

That is, $x(t)$ is $\omega$-periodic. □

**Theorem 4.69.** *If $B(t)$ is an $\omega$-periodic and regressive matrix function on $q^{\mathbb{N}_0}$, then*

$$e_B(q^\omega t, q^\omega s) = e_B(t, s) \quad t, s \in q^{\mathbb{N}_0}. \tag{4.15}$$

*Proof.* We just prove the case where $s < t$ are in $q^{\mathbb{N}_0}$. Let $s = q^m$, $t = q^n$, where $m, n \in \mathbb{N}_0$ with $m < n$. Then

$$\begin{aligned} e_B(q^\omega t, q^\omega s) &= e_B(q^{\omega+n}, q^{\omega+m}) \\ &= \prod_{k=\omega+m}^{\omega+n-1} [I + (q-1)q^k B(q^k)] \end{aligned}$$

$$= \prod_{k=m}^{n-1} [I + (q-1)q^{k+\omega} B(q^{k+\omega})]$$

$$= \prod_{k=m}^{n-1} [I + (q-1)q^k q^\omega B(q^\omega q^k)]$$

$$= \prod_{k=m}^{n-1} [I + (q-1)q^k B(q^k)]$$

$$= e_B(t, s),$$

which is what we wanted to prove. □

**Theorem 4.70 (Quantum Floquet Theorem).** *Assume the $n \times n$ matrix function $A(t)$ is regressive on $q^{\mathbb{N}_0}$ and is periodic with prime period $\omega$. Assume that $\Phi(t)$ is a fundamental matrix of* (4.12). *Then $\Psi(t) := q^\omega \Phi(q^\omega t)$ is also a fundamental matrix of* (4.12) *and $\Psi(t) = \Phi(t)D$, where $D = q^\omega \Phi^{-1}(1)\Phi(q^\omega)$. Moreover there exist an $\omega$-periodic regressive nonsingular matrix function $B(t)$ and an $\omega$-periodic nonsingular matrix function $B(t)$ such that*

$$\Phi(t) = P(t) e_B(t, 0), \quad t \in q^{\mathbb{N}_0}.$$

*Proof.* The proof of the first part of this theorem is similar to the proof of Theorem 4.70. Assume $\Phi(t)$ is a fundamental matrix of the Floquet system (4.12) and let

$$\Psi(t) := q^\omega \Phi(q^\omega t), \quad t \in \mathbb{N}_a.$$

Then

$$\begin{aligned} D_q \Psi(t) &= q^\omega D_q \Phi(q^\omega t) \\ &= q^\omega q^\omega D_q \Phi(q^\omega t) \\ &= q^\omega q^\omega A(q^\omega t) \Phi(q^\omega t) \\ &= q^\omega A(t) \Phi(q^\omega t) \\ &= A(t) q^\omega \Phi(q^\omega t) \\ &= A(t) \Psi(t). \end{aligned}$$

Hence, $\Psi(t)$ is a solution of the Floquet system (4.12). Since

$$\det \Psi(t) = q^{\mathbb{N}_0} \det \Phi(q^\omega t) \neq 0, \quad t \in q^{\mathbb{N}_0}$$

we have that $\Psi(t)$ is a fundamental matrix of the Floquet system (4.12). But, since $\Phi(t)$ and $\Psi(t)$ are fundamental matrices of (4.12), we have that there is a nonsingular constant matrix $D$ so that $\Phi(t) = \Psi(t)D$, for $t \in q^{\mathbb{N}_0}$. Letting $t = 1$ and solving for $D$ we get $D = q^{\omega}\Phi^{-1}(1)\Phi(q^{\omega})$. □

**Definition 4.71.** Assume $\Phi$ is a fundamental matrix of the Floquet system (4.12). Then the eigenvalues of the matrix

$$D := q^{\omega}\Phi^{-1}(1)\Phi(q^{\omega})$$

are called the Floquet multipliers of the Floquet system (4.12).

Since the Floquet system (4.12) has infinitely many fundamental matrices we need to prove that Floquet multipliers are well defined. To see this assume $\Phi_1(t)$ and $\Phi_2(t)$ are two fundamental matrices of the Floquet system (4.12) and let

$$D_1 := q^{\omega}\Phi_1^{-1}(1)\Phi_1(q^{\omega}), \quad D_2 := q^{\omega}\Phi_2^{-1}(1)\Phi_2(q^{\omega}).$$

It remains to show that $D_1$ and $D_2$ have the same eigenvalues. Since $\Phi_1(t)$ and $\Phi_2(t)$ are fundamental matrices of the Floquet system (4.12), we have that there is a nonsingular constant matrix $M$ so that

$$\Phi_2(t) = \Phi_1(t)M, \quad t \in q^{\mathbb{N}_0}.$$

It follows that

$$D_1 = q^{\omega}\Phi_1^{-1}(1)\Phi_2(q^{\omega}) = q^{\omega}M^{-1}\Phi_2^{-1}(1)\Phi_2(q^{\omega})M = M^{-1}D_2M,$$

from which we conclude that $D_1$ and $D_2$ have the same eigenvalues.

**Theorem 4.72.** *The number* $\mu_0$ *is a Floquet multiplier of the Floquet system* (4.12) *if and only if there is a nontrivial solution* $x(t)$ *of the Floquet system* (4.12) *satisfying*

$$q^{\omega}x(q^{\omega}t) = \mu_0 x(t), \quad t \in q^{\mathbb{N}_0}.$$

*Proof.* Assume the number $\mu_0$ is a Floquet multiplier of the Floquet system (4.12). Let $\Phi(t)$ be a fundamental matrix of the Floquet system (4.12). Then by Floquet's Theorem $q^{\omega}\Phi(q^{\omega}t) = \Phi(t)D$, where $D = q^{\omega}\Phi^{-1}\Phi(q^{\omega})$. Since $\mu_0$ is an eigenvalue of $D$ there is a corresponding eigenvector $x_0$ of the constant matrix $D$. Let $x(t) := \Phi(t)x_0$. Then $x(t)$ is a nontrivial solution of the Floquet system (4.12). Furthermore

$$\begin{aligned} q^{\omega}x(q^{\omega}t) &= q^{\omega}\Phi(q^{\omega}t)x_0 = \Phi(t)Dx_0 \\ &= \Phi(t)\mu_0 x_0 = \mu_0\Phi(t)x_0 = \mu_0 x(t). \end{aligned}$$

The proof of the converse statement in this proof is Exercise 4.11. □

**Corollary 4.73.** *If $p : q^{\mathbb{N}_0} \to \mathbb{R}$ is an $\omega$-periodic function on $q^{\mathbb{N}_0}$, then $\frac{1}{t}e_p(q^\omega t, t)$, $\frac{1}{t}\cosh_p(q^\omega t, t)$, $\frac{1}{t}\sinh_p(q^\omega t, t)$, $\frac{1}{t}\cos_p(q^\omega t, t)$, and $\frac{1}{t}\sin_p(q^\omega t, t)$ are $\omega$-periodic functions on $q^{\mathbb{N}_0}$.*

*Proof.* We will just show that $f(t) := \frac{1}{t}e_p(q^\omega t, t)$ is an $\omega$-periodic function on $q^{\mathbb{N}_0}$. To see this note that

$$\begin{aligned} q^\omega f(q^\omega t) &= \frac{q^\omega}{q^\omega t} e_p(q^{2\omega} t, q^\omega t) \\ &= \frac{1}{t} e_p(q^\omega t, t), \quad \text{by} \quad (4.15) \\ &= f(t) \end{aligned}$$

for $t \in q^{\mathbb{N}_0}$. The rest of the proof is left to the reader (see Exercise 4.12). □

## 4.10 Nabla Fractional $q$-Calculus

Most of the material in this section can be found in Baoguo et al. [50]. In Chap. 2 we considered the delta fractional calculus and in Chap. 3 we considered the nabla fractional calculus. One might have noticed that the nabla fractional calculus was easier to work with than the delta fractional calculus. The same applies to the quantum fractional calculus. Hence, in the remainder of this chapter we will be concerned with what we will call the nabla quantum fractional calculus. In particular, we will be interested in the nabla quantum operator, denoted $\nabla_q$, which is defined by

$$\nabla_q x(t) := \frac{x(t) - x(\rho(t))}{(1-p)t} = \frac{x(t) - x(pt)}{(1-p)t}, \quad t \in q^{\mathbb{N}_0},$$

where we assume that $q > 1$, $p := \frac{1}{q}$, and $x : q^{\mathbb{N}_{-1}} \to \mathbb{R}$. Note that since $q > 1$, it follows that $0 < p < 1$. From Exercise 4.13 we have the relationship

$$\nabla_q x(qt) = D_q x(t).$$

Also, we will use the quantum integral of $x : q^{\mathbb{N}_0} \to \mathbb{R}$, which is defined by

$$\begin{aligned} \int_p^t x(\tau)\nabla_q \tau &= \sum_{\tau=1}^{t} x(\tau)\nu(\tau) \\ &= \sum_{i=0}^{k} x(q^i)[q^i(1-p)], \end{aligned} \tag{4.16}$$

where $t = q^k$, $p = \frac{1}{q} = \rho(1)$, $\nu(t) := q^k - q^{k-1} = q^k(1-p)$, and by convention $\int_p^p x(\tau)\nabla_q\tau = 0$. We would like to point out for the reader that the papers [6] and [26] present an introduction to fractional $q$-differences and $q$-sums. In this section, we consider the asymptotic behavior of solutions of the nabla fractional $q$-difference equation

$$\nabla_{q,p}^{\alpha} x(t) = c(t)x(t), \qquad t \in q^{\mathbb{N}_1}, \tag{4.17}$$

where $q > 1$, $\mathbb{N}_1 = \{1, 2, \ldots\}$, $p = q^{-1}$, and the fractional operator, $\nabla_{q,p}^{\alpha}$, is defined later in Definition 4.78.

**Definition 4.74.** Assume $0 < p < 1$. Then the quantum gamma function is defined by

$$\Gamma_p(t) := \frac{(p,p)_\infty (1-p)^{1-t}}{(p^t, p)_\infty},$$

where $(a,p)_\infty := \prod_{j=0}^{\infty}(1 - ap^j)$ and $t \in \mathbb{R} \setminus \{0, -1, -2, \ldots\}$.

Note that since $0 < p < 1$, the series $\sum_{j=0}^{\infty} ap^j$ converges and so $(a,p)_\infty$, $a \in \mathbb{R}$ is well defined. The next theorem gives an important formula for this quantum gamma function.

**Theorem 4.75.** *Assume* $0 < p < 1$. *Then*

$$\Gamma_p(t+1) = [t]_p \Gamma_p(t), \quad t \in \mathbb{R}. \tag{4.18}$$

*In particular,* $\Gamma_p(1) = 1$ *and* $\Gamma_p(n) = [n]_p!$ *for* $n \in \mathbb{N}_1$.

*Proof.* We have that

$$\begin{aligned}
\Gamma_p(t+1) &= \frac{(p,p)_\infty (1-p)^{-t}}{\prod_{j=0}^{\infty}(1 - p^{t+1+j})} \\
&= \frac{1-p^t}{1-p} \frac{(p,p)_\infty (1-p)^{1-t}}{\prod_{j=0}^{\infty}(1 - p^{t+j})} \\
&= \frac{1-p^t}{1-p} \frac{(p,p)_\infty (1-p)^{1-t}}{(p^t,p)_\infty} \\
&= [t]_p \Gamma_p(t)
\end{aligned}$$

for $t \in \mathbb{R}$. By the definition of $\Gamma_p(t)$ we get $\Gamma_p(1) = 1$. Then using the formula (4.18) we get that $\Gamma_p(n) = [n]_p!$ for $n \in \mathbb{N}_1$. □

The $p = q^{-1}$-power function (quantum power function), $q > 1$, is given by

$$(t-s)_{q^{-1}}^{(\alpha)} = (t-s)_p^{(\alpha)} := t^\alpha \frac{(\frac{s}{t}, p)_\infty}{(p^\alpha \frac{s}{t}, p)_\infty}, \quad t \neq 0, \quad 0 < p := \frac{1}{q} < 1, \quad \alpha \in \mathbb{R}. \tag{4.19}$$

Note that we used a different notation for $(t-s)_{q^{-1}}^{(\alpha)}$ in Definition 4.9. Furthermore, one should observe that the power rule formulas for $D_q(t-s)_{\overline{p}}^{\underline{\alpha}}$ in Theorem 4.12 are not as nice as the nabla quantum power rule formulas that we obtain momentarily in Theorem 4.76. For $\alpha = n$, a positive integer, this expression reduces to (see Exercise 4.15)

$$(t-s)_{q^{-1}}^{(n)} = (t-s)_p^{(n)} = t^n \prod_{j=0}^{n-1} \left(1 - p^j \frac{s}{t}\right). \tag{4.20}$$

We next give nabla power rule formulas.

**Theorem 4.76.** *Assume $q > 1$, $p = q^{-1}$ and $\alpha \in \mathbb{R}$.*

(i) *The nabla q-difference of the p-factorial function $(t-s)_p^{(\alpha)} = (t-s)_p^{(\alpha)}$ with respect to t is given by*

$$_t\nabla_q (t-s)_p^{(\alpha)} = [\alpha]_p (t-s)_p^{(\alpha-1)}.$$

(ii) *The nabla q-difference of the p-factorial function $(t-s)_p^{(\alpha)} = (t-s)_p^{(\alpha)}$ with respect to s is given by*

$$_s\nabla_q (t-s)_p^{(\alpha)} = -[\alpha]_p (t-ps)_p^{(\alpha-1)}.$$

*Proof.* To see that (i) holds consider

$$\begin{aligned}
_t\nabla_q (t-s)_p^{(\alpha)} &= \frac{(t-s)_p^{(\alpha)} - (pt-s)_p^{(\alpha)}}{(1-p)t} \\
&= \frac{t^\alpha \frac{(s/t,p)_\infty}{(p^\alpha s/t,p)_\infty} - p^\alpha t^\alpha \frac{(s/pt,p)_\infty}{(p^{\alpha-1}s/t,p)_\infty}}{(1-p)t} \\
&= \frac{t^\alpha \prod_{n=0}^\infty \frac{1-p^n s/t}{1-p^{\alpha+n}s/t} - p^\alpha t^\alpha \prod_{n=0}^\infty \frac{1-p^{n-1}s/t}{1-p^{\alpha+n-1}s/t}}{(1-p)t} \\
&= \frac{t^\alpha \prod_{n=0}^\infty \frac{1-p^n s/t}{1-p^{\alpha+n}s/t} \left[1 - \frac{p^\alpha(1-s/pt)}{1-p^{\alpha-1}s/t}\right]}{(1-p)t} \\
&= \frac{t^{\alpha-1} \prod_{n=0}^\infty \frac{1-p^n s/t}{1-p^{\alpha+n}s/t} \left[\frac{1-p^\alpha}{1-p^{\alpha-1}s/t}\right]}{1-p}
\end{aligned}$$

$$= \frac{1-p^\alpha}{1-p} t^{\alpha-1} \prod_{n=0}^{\infty} \frac{1-p^n s/t}{1-p^{\alpha-1+n}s/t}$$

$$= \frac{1-p^\alpha}{1-p} \cdot \frac{t^{\alpha-1}(s/t,p)_\infty}{(p^{\alpha+1}s/t,p)_\infty}$$

$$= \frac{1-p^\alpha}{1-p}(t-s)_p^{(\alpha-1)}$$

$$= [\alpha]_p (t-s)_p^{(\alpha-1)}.$$

Hence, (i) holds. To see that (ii) holds consider

$$_s\nabla_q (t-s)_p^{(\alpha)} = \frac{(t-s)_p^{(\alpha)} - (t-ps)_p^{(\alpha)}}{s-ps}$$

$$= \frac{t^\alpha \frac{(s/t,p)_\infty}{(p^\alpha s/t,p)_\infty} - t^\alpha \frac{(ps/t,p)_\infty}{(p^{\alpha+1}s/t,p)_\infty}}{s-ps}$$

$$= \frac{t^\alpha \prod_{n=0}^{\infty} \frac{1-p^n s/t}{1-p^{\alpha+n}s/t} - t^\alpha \prod_{n=0}^{\infty} \frac{1-p^{1+n}s/t}{1-p^{\alpha+1+n}s/t}}{s-ps}$$

$$= \frac{t^\alpha \prod_{n=1}^{\infty} \frac{1-p^n s/t}{1-p^{\alpha+n}s/t}\left[\frac{1-s/t}{1-p^\alpha s/t} - 1\right]}{s-ps}$$

$$= \frac{t^{\alpha-1} \prod_{j=0}^{\infty} \frac{1-p^{j+1} s/t}{1-p^{\alpha+j+1}s/t}\left[\frac{-1+p^\alpha}{1-p^\alpha s/t}\right]}{1-p}$$

$$= -\frac{1-p^\alpha}{1-p} t^{\alpha-1} \prod_{j=0}^{\infty} \frac{1-p^{j+1}s/t}{1-p^{\alpha-1+j+1}s/t}$$

$$= -\frac{1-p^\alpha}{1-p} \cdot \frac{t^{\alpha-1}(ps/t,p)_\infty}{(p^{\alpha+1}ps/t,p)_\infty}$$

$$= -[\alpha]_p (t-ps)_p^{(\alpha-1)}.$$

And this completes the proof. □

For $q > 1, p = q^{-1}$ we define the $\alpha$-th order nabla $q$-fractional Taylor monomial by

$$K_\alpha(t,s) := \frac{(t-s)_p^{(\alpha)}}{\Gamma_p(\alpha+1)}.$$

Then by Theorem 4.76 we get the important formulas

$$
\begin{aligned}
{}_t\nabla_q K_\alpha(t,s) &= K_{\alpha-1}(t,s),\\
{}_t\nabla_q K_\alpha(t,\rho(s)) &= K_{\alpha-1}(t,\rho(s)) = K_{\alpha-1}(t,ps), \text{ and}\\
{}_s\nabla_q K_\alpha(t,s) &= -K_{\alpha-1}(t,ps),
\end{aligned}
$$

which we will use frequently in the remainder of this section. Next we define the $\alpha$-th nabla quantum fractional sum as in J. Čermák and L. Nechvátal [65] in terms of the $K_{\alpha-1}(t,s)$

**Definition 4.77 (Nabla Fractional Sum).** Assume $\alpha > 0$, $q > 1$, $p = q^{-1}$, and $f : q^{\mathbb{N}_0} \to \mathbb{R}$. Then we define the nabla $q$-fractional sum of $f$ at $t = q^k$ by

$$
\begin{aligned}
(\nabla_{q,p}^{-\alpha} f)(t) &:= \int_p^t K_{\alpha-1}(t,p\tau) f(\tau)\nabla_q\tau\\
&= \sum_{\tau=1}^{t} K_{\alpha-1}(t,p\tau) f(\tau)\nu(\tau)\\
&= \sum_{i=0}^{k} K_{\alpha-1}(t,q^{i-1}) f(q^i)\nu(q^i),
\end{aligned}
$$

where by convention $\nabla_{q,p}^{-\alpha} f(p) = 0$. Note that the second subscript gives the lower limit of integration in the above integral definition.

Next we define the $\alpha$-order nabla $q$-difference for $\alpha > 0$ in terms of a nabla $q$-sum.

**Definition 4.78 (Nabla Fractional Difference).** Assume $\alpha > 0$, $q > 1$, $p = q^{-1}$, $m - 1 < \alpha < m$, where $m \in \mathbb{N}_1$, and $f : q^{\mathbb{N}_0} \to \mathbb{R}$. Then we define the nabla $q$-fractional difference of $f$ at $t$ by

$$(\nabla_{q,p}^{\alpha} f)(t) := (\nabla_q^m \nabla_{q,p}^{-(m-\alpha)} f)(t),$$

for $t \in q^{\mathbb{N}_0}$. Also

$$(\nabla_{q,1}^{\alpha} f)(t) := (\nabla_q^m \nabla_{q,1}^{-(m-\alpha)} f)(t).$$

Similar to the proof of the $q$-difference Leibniz rule in Theorem 4.35 one can prove the following nabla quantum difference Leibniz rule (see Exercise 4.14).

**Lemma 4.79 (Leibniz Rule).** *Assume $f : q^{\mathbb{N}_1} \times q^{\mathbb{N}_1} \to \mathbb{R}$ and $q > 1$. Then*

$$\nabla_q\left[\int_1^t f(t,s)\nabla_q s\right] = \int_1^t {}_t\nabla_q f(t,s)\nabla_q s + f(pt,t)$$

*for $t \in q^{\mathbb{N}_1}$.*

The following theorem shows that we can get the formula for the nabla alpha fractional difference, for $\alpha > 0$, from the formula for the nabla alpha fractional sum by replacing $\alpha$ by $-\alpha$.

**Theorem 4.80.** *Assume* $\alpha > 0$, $q > 1$, $p = q^{-1}$, $m - 1 < \alpha < m$, *where* $m \in \mathbb{N}_1$, *and* $f : q^{\mathbb{N}_0} \to \mathbb{R}$. *Then*

$$\nabla_{q,p}^{\alpha} f(t) = \int_p^t K_{-\alpha-1}(t, ps) f(s) \nabla_q s, \quad t \in q^{\mathbb{N}_0}. \tag{4.21}$$

*Also*

$$\nabla_{q,1}^{\alpha} f(t) = \int_1^t K_{-\alpha-1}(t, ps) f(s) \nabla_q s. \tag{4.22}$$

*Proof.* Using the definition of the nabla fractional difference (Definition 4.78) and the Leibniz rule in Lemma 4.79 we have that

$$\begin{aligned}
\nabla_{q,p}^{\alpha} f(t) &= \nabla_q^m \nabla_{p,q}^{-(m-\alpha)} f(t) \\
&= \nabla_q^{m-1} \nabla_q \nabla_{p,q}^{-(m-\alpha)} f(t) \\
&= \nabla_q^{m-1} \left[ \nabla_q \int_p^t K_{m-\alpha-1}(t, ps) f(s) \nabla_q \right] \\
&= \nabla_q^{m-1} \left[ \int_p^t K_{m-\alpha-2}(t, ps) f(s) \nabla_q s + K_{m-\alpha-1}(pt, pt) f(t) \right] \\
&= \nabla_q^{m-1} \left[ \int_p^t K_{m-\alpha-2}(t, ps) f(s) \nabla_q s \right].
\end{aligned}$$

Repeating this argument $m-1$ more times we obtain by mathematical induction that

$$\nabla_{q,p}^{\alpha} f(t) = \int_p^t K_{-\alpha-1}(t, ps) f(s) \nabla_q s.$$

The proof of the last sentence in the statement of this theorem is similar and hence is omitted. □

**Theorem 4.81.** *Assume* $0 < \alpha < 1$, $c(t) \leq 0$, $t \in q^{\mathbb{N}_1}$. *Then any solution of the equation*

$$\nabla_{q,p}^{\alpha} x(t) = c(t) x(t), \qquad t \in q^{\mathbb{N}_1} \tag{4.23}$$

*satisfying* $x(1) > 0$ *is positive on* $q^{\mathbb{N}_0}$.

*Proof.* Equation 4.22 gives us that

$$\nabla_{q,p}^{\alpha}x(t) = \int_{p}^{t} K_{-\alpha-1}(t,ps)x(s)\nabla_q s,$$

and hence if $t = q^k$, we see that

$$\nabla_{q,p}^{\alpha}x(t) = \sum_{i=0}^{k} K_{-\alpha-1}(t,p^{1-i})x(q^i)\nu(q^i). \tag{4.24}$$

Since $x(t)$ is a solution of (4.23), we obtain, using (4.24), that

$$\Big[K_{-\alpha-1}(t,p^{1-k})\nu(t) - c(t)\Big]x(q^k) \tag{4.25}$$
$$= -\sum_{i=0}^{k-1} K_{-\alpha-1}(t,p^{1-i})\nu(q^i)x(q^i).$$

We now show that since $0 < \alpha < 1$, it follows that $K_{-\alpha-1}(t,p^{1-i}) < 0$ for $0 \le i \le k-1$. To see this first note that

$$\begin{aligned}\Gamma_p(-\alpha) &= \frac{(p,p)_\infty(1-p)^{1+\alpha}}{(p^{-\alpha},p)_\infty}\\ &= \frac{(1-p)^{1+\alpha}\prod_{j=0}^{\infty}[1-p^{j+1}]}{\prod_{j=0}^{\infty}[1-p^{j-\alpha}]}\\ &= \frac{(1-p)^{1+\alpha}\prod_{j=0}^{\infty}[1-p^{j+1}]}{(1-p^{-\alpha})\prod_{j=1}^{\infty}[1-p^{j-\alpha}]} < 0.\end{aligned}$$

Then note that for $0 \le i \le k-1$, $t = q^k$,

$$\begin{aligned}(t-q^{i-1})_p^{(-\alpha-1)} &= t^{-\alpha-1}\frac{(p^{k-i+1},p)_\infty}{(p^{k-i-\alpha},p)_\infty}\\ &= t^{-\alpha-1}\frac{\prod_{j=0}^{\infty}[1-p^{k-i+1+j}]}{\prod_{j=0}^{\infty}[1-p^{k-i-\alpha+j}]} > 0\end{aligned}$$

It follows from the last two inequalities that

$$K_{-\alpha-1}(t,q^{i-1}) = \frac{(t-q^{i-1})_p^{-\alpha-1}}{\Gamma_p(-\alpha)} < 0$$

for $0 \le i \le k-1$.

Since $c(t) \leq 0$ and $K_{-\alpha-1}(t, p^{1-k}) > 0$ the coefficient of $x(t) = x(q^k)$ on the right-hand side of (4.25) is positive. Since $K_{-\alpha-1}(t, p^{1-i}) < 0$ for $0 \leq i \leq k-1$, $v(t) > 0$, and $x(1) > 0$ it follows from (4.25) and the strong induction principle and $x(1) > 0$ that $x(t) = x(q^k) > 0$, for $k \in \mathbb{N}_0$. This completes the proof. □

Next we consider the case where the coefficient in (4.17) satisfies $c(t) \leq 0$, $t \in q^{\mathbb{N}_0}$. The following results will be useful in proving our Theorem 4.88.

**Lemma 4.82.** *Assume $\alpha \in \mathbb{R}$ and $0 < p < 1$. Then for $t \neq 0$*

$$(1-p)K_{-\alpha-1}(t,p) + K_{-\alpha}(t,1) = K_{-\alpha}(t,p).$$

*Proof.* For $t \neq 0$ we have that

$$\begin{aligned}
&(1-p^{-\alpha})(t-p)_p^{(-\alpha-1)} + (t-1)_p^{(-\alpha)}\\
&= (1-p^{-\alpha})t^{-\alpha-1}\frac{(pt^{-1},p)_\infty}{(p^{-\alpha}t^{-1},p)_\infty} + t^{-\alpha}\frac{(t^{-1},p)\infty}{(p^{-\alpha}t^{-1},p)\infty}\\
&= \frac{t^{-\alpha}}{(p^{-\alpha}t^{-1},p)_\infty}\left[t^{-1}(1-p^{-\alpha})(pt^{-1},p)_\infty + (t^{-1},p)_\infty\right]\\
&= \frac{t^{-\alpha}}{(p^{-\alpha}t^{-1},p)_\infty}\left[(1-p^{-\alpha})t^{-1}\prod_{j=0}^{\infty}\left(1-\frac{p^{j+1}}{t}\right) + \prod_{j=0}^{\infty}\left(1-\frac{p^j}{t}\right)\right]\\
&= \frac{t^{-\alpha}\prod_{j=0}^{\infty}\left(1-\frac{p^{j+1}}{t}\right)}{(p^{-\alpha}t^{-1},p)_\infty}\left[(1-p^{-\alpha})t^{-1} + (1-t^{-1})\right]\\
&= \frac{t^{-\alpha}\prod_{j=0}^{\infty}\left(1-\frac{p^{j+1}}{t}\right)}{(p^{-\alpha}t^{-1},p)_\infty}[1-p^{-\alpha}t^{-1}]\\
&= \frac{t^{-\alpha}\prod_{j=0}^{\infty}\left(1-\frac{p^{j+1}}{t}\right)}{(p^{1-\alpha}t^{-1},p)_\infty}\\
&= \frac{t^{-\alpha}\prod_{j=0}^{\infty}\left(1-\frac{p^{j+1}}{t}\right)}{\prod_{j=0}^{\infty}(1-t^{-1}p^{j+1-\alpha})}\\
&= (t-p)_p^{(-\alpha)}.
\end{aligned}$$

We now divide both sides of this last equation by $\Gamma_p(-\alpha+1) = [-\alpha]_p\Gamma_p(-\alpha)$ to obtain

$$\frac{(1-p^{-\alpha})(t-p)_p^{(-\alpha-1)}}{[-\alpha]_p\Gamma_p(-\alpha)} + K_{-\alpha}(t,1) = K_{-\alpha}(t,p).$$

This yields

$$(1-p)K_{-\alpha-1}(t,p) + K_{-\alpha}(t,1) = K_{-\alpha}(t,p),$$

which is the desired result. □

**Lemma 4.83.** *Assume $f : q^{\mathbb{N}_0} \to \mathbb{R}$, $0 < \alpha < 1$, $q > 1$, and $p = q^{-1}$. Then*

$$\begin{aligned}\nabla_{q,1}^{-(1-\alpha)} \nabla_q f(t) &= \nabla_q \nabla_{q,1}^{-(1-\alpha)} f(t) - f(1)K_{-\alpha}(t,1) && (4.26)\\ &= \nabla_{q,1}^{\alpha} f(t) - f(1)K_{-\alpha}(t,1), && (4.27)\end{aligned}$$

*and*

$$\begin{aligned}\nabla_{q,1}^{-(1-\alpha)} \nabla_q f(t) &= \nabla_q \nabla_{q,p}^{-(1-\alpha)} f(t) - f(1)K_{-\alpha}(t,p) && (4.28)\\ &= \nabla_{q,p}^{\alpha} f(t) - f(1)K_{-\alpha}(t,p). && (4.29)\end{aligned}$$

*Proof.* Integrating by parts and using (4.22), we have

$$\begin{aligned}\nabla_{q,1}^{-(1-\alpha)} \nabla_q f(t) &= \int_1^t K_{-\alpha}(t,ps) \nabla_q f(s) \nabla_q s\\ &= \Big[K_{-\alpha}(t,s)f(s)\Big]_{s=1}^t + \int_1^t K_{-\alpha-1}(t,ps) f(s) \nabla_q s\\ &= \nabla_{q,1}^{\alpha} f(t) - f(1)K_{-\alpha}(t,1)\end{aligned}$$

and hence (4.27) holds. By the Leibniz rule (Lemma 4.79) we get that

$$\nabla_q \nabla_{q,1}^{-(1-\alpha)} f(t) = \nabla_{q,1}^{\alpha} f(t). \tag{4.30}$$

Using (4.30) and (4.27) we get that (4.26) holds.

From (4.30) and (4.21) it follows that

$$\begin{aligned}\nabla_q \nabla_{q,p}^{-(1-\alpha)} f(t) = \nabla_{q,1}^{\alpha} f(t) &= \int_p^t K_{-\alpha-1}(t,ps) f(s) \nabla_q s\\ &= \int_p^1 K_{-\alpha-1}(t,ps) f(s) \nabla_q s + \int_1^t K_{-\alpha-1}(t,ps) f(s) \nabla_q s\\ &= K_{-\alpha-1}(t,p) f(1) \nu(1) + \int_1^t K_{-\alpha-1}(t,ps) f(s) \nabla_q s\\ &= K_{-\alpha-1}(t,p) f(1)(1-p) + \nabla_{q,1}^{\alpha} f(t),\end{aligned}$$

where in the last step we used (4.22). Subtracting $f(1)K_{-\alpha}(t,p)$ from both sides of this last equation we get

$$\begin{aligned}\nabla_q \nabla_{q,p}^{-(1-\alpha)} f(t) - f(1)K_{-\alpha}(t,p) \\ &= K_{-\alpha-1}(t,p)f(1)(p-1) + \nabla_{q,1}^{\alpha} f(t) - f(1)K_{-\alpha}(t,p) \\ &= \nabla_{q,1}^{\alpha} f(t) - f(1)\left[(1-p)K_{-\alpha-1}(t,p) + K_{-\alpha}(t,p)\right] \\ &= \nabla_{q,1}^{\alpha} f(t) - f(1)K_{-\alpha}(t,1) \\ &= \nabla_{q,1}^{-(1-\alpha)} \nabla_q f(t),\end{aligned}$$

where in the second to the last step we used Lemma 4.82 and in the last step we used (4.27). Hence (4.28) holds. Since by the Leibniz rule $\nabla_q \nabla_{q,p}^{-(1-\alpha)} f(t) = \nabla_{q,p}^{\alpha} f(t)$ we have that (4.28) implies (4.29) holds. □

Replacing $\alpha$ by $1-\alpha$ in Lemma 4.83, we get the following corollary will be used later.

**Corollary 4.84.** *Assume* $f : q^{\mathbb{N}_0} \to \mathbb{R}$, $0 < \alpha < 1$. *Then*

$$\nabla_{q,1}^{-\alpha} \nabla_q f(t) = \nabla_q \nabla_{q,1}^{-\alpha} f(t) - f(1)K_{\alpha-1}(t,1), \tag{4.31}$$

$$\nabla_{q,1}^{-\alpha} \nabla_q f(t) = \nabla_q \nabla_{q,p}^{-\alpha} f(t) - f(1)K_{\alpha-1}(t,p). \tag{4.32}$$

**Lemma 4.85.** *Assume that* $f : q^{\mathbb{N}_0} \times q^{\mathbb{N}_0} \to \mathbb{R}$. *Then we have that*

$$\int_a^t \int_{\rho(\xi)}^t f(\eta,\xi)\nabla_q \eta \nabla_q \xi = \int_a^t \int_a^{\eta} f(\eta,\xi)\nabla_q \xi \nabla_q \eta.$$

*Proof.* Let

$$\phi(t) := \int_a^t \int_{\rho(\xi)}^t f(\eta,\xi)\nabla_q \eta \nabla_q \xi - \int_a^t \int_a^{\eta} f(\eta,\xi)\nabla_q \xi \nabla_q \eta;$$

From Lemma 4.79, we have

$$\nabla_q \phi(t) = \int_a^t f(t,\xi)\nabla_q \xi - \int_a^t f(t,\xi)\nabla_q \xi = 0$$

and $\phi(a) = 0$. So $\phi(t) \equiv 0$. This completes the proof. □

The following lemma appears in [66].

**Lemma 4.86.** *Let* $\alpha \in \mathbb{R}^+, \beta \in \mathbb{R}$. *Then*

$$\nabla_{q,a}^{-\alpha} (t-a)_p^{(\beta)} = \frac{\Gamma_p(\beta+1)}{\Gamma_p(\alpha+\beta+1)} (t-a)_p^{(\alpha+\beta)}.$$

We next state and prove the following composition rule.

**Lemma 4.87.** *Let f be a real valued function, and $\alpha, \beta > 0$. Then for $t \in q^{\mathbb{N}_1}$, we have*

$$\nabla_{q,1}^{-\alpha}[\nabla_{q,1}^{-\beta} f(t)] = \nabla_{q,1}^{-(\beta+\alpha)} f(t).$$

*Proof.* To see this note that

$$\begin{aligned}
\nabla_{q,1}^{-\alpha}[\nabla_{q,1}^{-\beta} f(t)] &= \int_1^t K_{\alpha-1}(t, ps) \left[\nabla_{q,1}^{-\beta} f(s)\right] \nabla_q s \\
&= \int_1^t K_{\alpha-1}(t, ps) \left[\int_1^s K_{\beta-1}(t, p\tau) f(\tau) \nabla_q \tau\right] \nabla_q s \\
&\overset{(4.85)}{=} \int_1^t f(\tau) \left[\int_{p\tau}^t K_{\alpha-1}(t, ps) K_{\beta-1}(s, p\tau) \nabla_q s\right] \nabla_q \tau \\
&= \int_1^t {}_t\nabla_{q,p\tau}^{-\alpha} K_{\beta-1}(t, p\tau) f(\tau) \nabla_q \tau \\
&= \int_1^t K_{\alpha+\beta-1}(t, p\tau) f(\tau) \nabla_q \tau \\
&= \nabla_{q,1}^{-(\alpha+\beta)} f(t).
\end{aligned}$$

□

**Theorem 4.88.** *Assume that $q = p^{-1} > 1$, $0 < \alpha < 1$, $c(t) \leq 0$, for $t \in q^{\mathbb{N}_1}$, and $x(t)$ is a solution of the fractional q-difference equation*

$$\nabla_{q,p}^{\alpha} x(t) = c(t) x(t), \qquad t \in q^{\mathbb{N}_1} \tag{4.33}$$

*satisfying $x(1) > 0$. Then*

$$\lim_{t \to \infty} x(t) = 0.$$

*Proof.* Applying the operator $\nabla_{q,1}^{-\alpha}$ to each side of Eq. (4.33) we obtain

$$\nabla_{q,1}^{-\alpha} \nabla_{q,p}^{\alpha} x(t) = \nabla_{q,1}^{-\alpha} c(t) x(t),$$

which can be written in the form

$$\nabla_{q,1}^{-\alpha} \nabla_q \nabla_{q,p}^{-(1-\alpha)} x(t) = \nabla_{q,1}^{-\alpha} c(t) x(t).$$

Using (4.32), we obtain

$$\nabla_q \nabla_{q,p}^{-\alpha} \nabla_{q,p}^{-(1-\alpha)} x(t) - K_{\alpha-1}(t, p) \nabla_{q,p}^{-(1-\alpha)} x(t)|_{t=1} = \nabla_{q,1}^{-p} c(t) x(t),$$

whereupon combining this with the identity

$$\begin{aligned}\nabla_{q,p}^{-(1-\alpha)}x(t)|_{t=1} &= \int_p^1 K_{-\alpha}(1,ps)x(s)\nabla_q s\\ &= K_{-\alpha}(1,p)x(1)\nu(1)\\ &= (1-p)^{1-\alpha}x(1),\end{aligned}$$

we arrive at

$$\nabla_q \nabla_{q,p}^{-\alpha}\nabla_{q,p}^{-(1-\alpha)}x(t) = K_{\alpha-1}(t,p)(1-p)^{1-\alpha}x(1) + \nabla_{q,1}^{-\alpha}c(t)x(t).$$

Using the composition rule, Lemma 4.87, it follows that $\nabla_{q,p}^{-\alpha}\nabla_{q,p}^{-(1-\alpha)}x(t) = \nabla_{q,p}^{-1}x(t)$ and that

$$\begin{aligned}\nabla_q\nabla_{q,p}^{-1}x(t) &= \nabla_q\left(\int_p^t K_0(t,ps)x(s)\nabla_q s\right)\\ &= \nabla_q\left(\int_p^t x(s)\nabla_q s\right)\\ &= x(t),\end{aligned}$$

where we used $K_0(t,ps) = 1$. Hence, we see that

$$x(t) = K_{\alpha-1}(t,p)(1-p)^{1-\alpha}x(1) + \nabla_{q,1}^{-\alpha}c(t)x(t).$$

That is,

$$x(t) = K_{\alpha-1}(t,p)(1-p)^{1-\alpha}x(1) + \int_1^t K_{\alpha-1}(t,ps)c(s)x(s)\nabla_q s. \tag{4.34}$$

Since $x(1) > 0$, $0 < \alpha < 1$, $c(t) \le 0$ and Lemma 4.81, we have $x(t) > 0$ for $t \in q^{\mathbb{N}_0}$. It is easy to see for $t \ge s$

$$(t-ps)_p^{(\alpha-1)} = t^{\alpha-1}\frac{\prod_{j=1}^{\infty}(1-\frac{s}{qt}q^{-j})}{\prod_{j=0}^{\infty}(1-\frac{s}{q^{\alpha}t}q^{-j})} \ge 0.$$

Since $c(s) \le 0$ and (4.34) we get that (taking $t = q^k$)

$$\begin{aligned}0 < x(q^k) &\le K_{\alpha-1}(q^k,p)(1-p)^{1-\alpha}x(1)\\ &= \frac{(q^k-p)_p^{(\alpha-1)}}{\Gamma_p(\alpha)}(1-p)^{1-\alpha}x(1).\end{aligned} \tag{4.35}$$

Note that

$$
\begin{aligned}
&(q^k - p)_p^{(\alpha-1)} \qquad (4.36)\\
&= q^{k(\alpha-1)} \frac{(q^{-k-1}, p)_\infty}{(p^{\alpha-1}p^{k+1}, p)_\infty}\\
&= q^{k(\alpha-1)} \frac{\prod_{j=0}^{\infty}(1 - p^{k+1}q^{-j})}{\prod_{j=0}^{\infty}(1 - q^{-\alpha-k}q^{-j})}\\
&= q^{k(\alpha-1)} \frac{(1-p^\alpha)\cdots(1-q^{-\alpha-(k-1)})}{(1-p)\cdots(1-q^{-k})} \cdot \frac{\prod_{j=0}^{\infty}(1 - pq^{-j})}{\prod_{j=0}^{\infty}(1 - p^\alpha q^{-j})}\\
&= q^{k(\alpha-1)} \frac{(1-p^\alpha)\cdots(1-q^{-\alpha-(k-1)})}{(1-p)\cdots(1-q^{-k})} \cdot \frac{\Gamma_p(\alpha)}{(1-p)^{1-\alpha}}\\
&\to 0,
\end{aligned}
$$

where we used both that $\lim_{k\to 0} q^{k(\alpha-1)} = 0$ and that

$$
\begin{aligned}
\lim_{k\to\infty} &\frac{(1-p^\alpha)\cdots(1-q^{-\alpha-(k-1)})}{(1-p)\cdots(1-q^{-k})}\\
&= \frac{\prod_{j=0}^{\infty}(1 - q^{-\alpha-j})}{\prod_{j=0}^{\infty}(1 - pq^{-j})}\\
&= \frac{(p^\alpha, p)_\infty}{(p,p)_\infty}\\
&= \frac{(1-p)^{1-\alpha}}{\Gamma_p(\alpha)};
\end{aligned}
$$

note that in this second calculation we have used (4.19). From (4.35) and (4.36), we have the desired result

$$\lim_{k\to\infty} x(q^k) = 0,$$

which completes the proof. □

We conclude this section by considering solutions $x(t)$ of the $\alpha$-th order nabla fractional $q$-difference equation

$$\nabla_{q,p}^{\alpha} x(t) = c(t)x(t), \qquad t \in q^{\mathbb{N}_1}, \tag{4.37}$$

satisfying $x(1) < 0$. By making the transformation $x(t) = -y(t)$ and using Theorem 4.88 we get the following theorem.

**Theorem 4.89.** *Assume that* $0 < \alpha < 1$, $q > 1$, *and* $c(t) \leq 0$ *for* $t \in q^{\mathbb{N}_1}$. *Then any solution of equation* (4.37) *with* $x(1) < 0$ *satisfies*

$$\lim_{t \to \infty} x(t) = 0.$$

## 4.11 Exercises

**4.1.** Prove parts (i)–(iv) and (vii) of Theorem 4.2.

**4.2.** Fix $s \in q^{\mathbb{N}_0}$ and let $f(t) = (t-s)^2$, $t \in q^{\mathbb{N}_0}$. Express $f(t)$ in terms of Taylor monomials based at $s$. Use your answer to find $D_q f(t)$. Then check your answer by finding $D_q f(t)$ using the product rule.

**4.3.** Prove that if $p, r \in \mathcal{R}_q$, then

$$(p \ominus r)(t) = \frac{p(t) - r(t)}{1 + \mu(t) r(t)}, \quad t \in aq^{\mathbb{N}_0}. \tag{4.38}$$

**4.4.** Show that if $\alpha, \pm \frac{\beta}{1+\alpha\mu(t)} \in \mathcal{R}_q$, then $\alpha \pm \beta \in \mathcal{R}_q$.

**4.5.** Show that if $p : aq^{\mathbb{N}_0} \to \mathbb{C}$, then $|p| \in \mathcal{R}_q^+$. Then show that if $p \in \mathcal{R}_q$, it follows that

$$|e_p(t,a)| \leq e_{|p|}(t,a) \quad \text{for} \quad t \in aq^{\mathbb{N}_0}.$$

**4.6.** Assume $a > 0$, $f : aq^{\mathbb{N}_0} \to \mathbb{R}$ and $b = aq^m \leq c = aq^n$. Show that

$$\int_b^c f(t) D_q t = a(q-1) \sum_{k=m}^{n-1} q^k f(aq^k).$$

**4.7.** Assume $p$ is a nonzero real number. Use the variation of constants formula (4.2) to solve each of the following IVPs:

(i)

$$D_q^2 y = \cos_p(t, 0), \quad t \in aq^{\mathbb{N}_0};$$

$$y(a) = 0 = D_q y(a);$$

(ii)

$$D_2^2 y = t - a, \quad t \in a2^{\mathbb{N}_0};$$

$$y(a) = 0 = D_2 y(a);$$

(iii)

$$D_q^2 y = \sinh_p(t, 0), \quad t \in q^{\mathbb{N}_0},$$

$$y(1) = 0, \ D_2 y(1) = 2.$$

**4.8.** Use the $q$-Laplace transform to solve each of the following:

(i)

$$D_q^2 y(t) - 3D_q y(t) + 2y(t) = 2e_3(t, a), \quad t \in aq^{\mathbb{N}_0},$$

$$y(a) = -1, \quad D_q y(a) = -1;$$

(ii)

$$D_q^2 y(t)) + 4y(t) = 8e_2(t, a), \quad t \in aq^{\mathbb{N}_0},$$

$$y(a) = 1, \quad D_q y(a) = 5;$$

(iii)

$$D_q^2 y(t) + 16y(t) = 0, \quad t \in aq^{\mathbb{N}_0},$$

$$y(a) = 0, \quad D_q y(a) = 3;$$

(iv)

$$D_q^2 y(t)) - 6D_q y(t) + 25y(t) = 0, \quad t \in aq^{\mathbb{N}_0},$$

$$y(a) = 1, \quad D_q y(a) = 2.$$

**4.9.** Show that any linear combination of $\omega$-periodic functions on $q^{\mathbb{N}_0}$ is also $\omega$-periodic.

**4.10.** Prove Remark 4.63 that if $f$ is an $\omega$-periodic function on $q^{\mathbb{N}_0}$, then

$$q^{n\omega+k} f(q^{n\omega+k} t) = q^k f(q^k t) \tag{4.39}$$

for $n \in \mathbb{N}_0$, $0 \leq k \leq \omega - 1$, $t \in q^{\mathbb{N}_0}$.

**4.11.** Prove that if there is a nontrivial solution of the Floquet system (4.12) and a number $\mu_0$ such that

$$q^{\omega}x(q^{\omega}t) = \mu_0 x(t), \quad t \in q^{\mathbb{N}_0},$$

then $\mu_0$ is a Floquet multiplier.

**4.12.** Prove that if $p : q^{\mathbb{N}_0} \to \mathbb{R}$ is periodic with period $\omega$, then $\frac{1}{t}\cos_p(q^{\omega}t, t)$ is an $\omega$-periodic function on $q^{\mathbb{N}_0}$.

**4.13.** Show that the nabla quantum operator and the quantum operator are related by the formula

$$\nabla_q x(qt) = D_q x(t).$$

**4.14.** Prove the nabla $q$-difference Leibniz rule in Lemma 4.79.

**4.15.** Prove that (4.19) reduces to (4.20) when $\alpha = n$ is a positive integer.

**4.16.** Show that if $K_{\alpha}(t,s) := \frac{(t-s)_p^{(\alpha)}}{\Gamma_p(\alpha+1)}$, then $K_{-\alpha}(1,p) = (1-p)^{1-\alpha}$.

# Chapter 5
# Calculus on Mixed Time Scales

## 5.1 Introduction

This chapter focuses on what we call the **calculus on a mixed time scale** whose elements we will define in terms of a point $\alpha$ and two linear functions. There has been recent interest in mixed time scales by Auch [37, 38], Auch et al. [39], Estes [34, 78], Erbe et al. [76], and Mert [145].

## 5.2 Basic Mixed Time Scale Calculus

In this section, we introduce some fundamental concepts and properties concerning what we will call a mixed time scale. Throughout this chapter we assume $a, b$ are constants satisfying

$$a \geq 1, \quad b \geq 0, \quad a + b > 1.$$

We will use two linear functions to define our so-called mixed time scale. First we let $\sigma : \mathbb{R} \to \mathbb{R}$ be defined by

$$\sigma(t) = at + b, \quad t \in \mathbb{R}.$$

Then we define the linear function $\rho$ to be the inverse function of $\sigma$, that is

$$\rho(t) = \frac{t-b}{a}, \quad t \in \mathbb{R}.$$



C. Goodrich, A.C. Peterson, *Discrete Fractional Calculus*,
DOI 10.1007/978-3-319-25562-0_5

We call $\sigma$ the **forward jump operator** and $\rho$ the **backward jump operator**. Only for these two functions we use the following notation. For $n \geq 1$ we define the function $\sigma^n$ recursively by

$$\sigma^n(t) = \sigma(\sigma^{n-1}(t)), \quad t \in \mathbb{R},$$

where $\sigma^0(t) := t$, and

$$\rho^n(t) = \rho(\rho^{n-1}(t)), \quad t \in \mathbb{R},$$

where $\rho^0(t) := t$. We now define our mixed time scale $\mathbb{T}_\alpha$, where for simplicity we always assume $\alpha \geq 0$:

$$\mathbb{T}_\alpha := \{\cdots, \rho^2(\alpha), \rho(\alpha), \alpha, \sigma(\alpha), \sigma^2(\alpha), \cdots\}.$$

By Exercise 5.1, we have that

$$\cdots < \rho^2(\alpha) < \rho(\alpha) < \alpha < \sigma(\alpha) < \sigma^2(\alpha) < \cdots.$$

Usually the domains of $\sigma$ and $\rho$ will be either $\mathbb{R}$ or $\mathbb{T}_\alpha$.

**Theorem 5.1.** *If $a > 1$, then*

$$t > \frac{b}{1-a}, \quad t \in \mathbb{T}_\alpha.$$

*Proof.* Since $\sigma^n(\alpha) \geq 0 > \frac{b}{1-a}$ for all $n \geq 0$, it remains to show that $\rho^n(\alpha) > \frac{b}{1-a}$ for all $n \geq 1$. We prove this by induction. First, for the base case note that

$$\rho(\alpha) = \frac{\alpha - b}{a} > \frac{b}{1-a}.$$

Now assume $n \geq 1$ and $\rho^n(\alpha) > \frac{b}{1-a}$. Then it follows that

$$\rho^{n+1}(\alpha) = \rho\left(\rho^n(t)\right) = \frac{\rho^n(\alpha) - b}{a} > \frac{\frac{b}{1-a} - b}{a} = \frac{b}{1-a}.$$

$\square$

Note that the above theorem does not hold if $a = 1$. Also note that when $a > 1$, $\mathbb{T}_\alpha$ is not a closed set (see Theorem 5.6 (iii)).

**Definition 5.2.** For $c, d \in \mathbb{T}_\alpha$ such that $d \geq c$, we define

$$\mathbb{T}_{[c,d]} := \mathbb{T}_\alpha \cap [c, d] = \{c, \sigma(c), \sigma^2(c), \ldots, \rho(d), d\}.$$

We define $\mathbb{T}_{(c,d)}$, $\mathbb{T}_{(c,d]}$, and $\mathbb{T}_{[c,d)}$ similarly. Additionally, we may use the notation $\mathbb{T}_c^d$, where $\mathbb{T}_c^d := \mathbb{T}_{[c,d]}$.

**Definition 5.3.** We define a forward graininess function, $\mu$, by

$$\mu(t) := \sigma(t) - t = (at + b) - t = (a-1)t + b.$$

In the following theorem we give some properties of the graininess function $\mu$.

**Theorem 5.4.** *For $t \in \mathbb{T}_\alpha$ and $n \in \mathbb{N}_0$, the following hold:*

(i) $\mu(t) > 0$;
(ii) $\mu(\sigma^n(t)) = a^n \mu(t)$;
(iii) $\mu(\rho^n(t)) = a^{-n}\mu(t)$.

*Proof.* We just prove (iii) and leave the rest of the proof (see Exercise 5.2) to the reader. To see that (iii) holds consider for $t \in \mathbb{T}_\alpha$ the base case

$$\mu(\rho(t)) = (a-1)\rho(t) + b = (a-1)\frac{t-b}{a} + b = \frac{(a-1)t+b}{a} = a^{-1}\mu(t).$$

Now assume $n \geq 1$ and $\mu(\rho^n(t)) = a^{-n}\mu(t)$ for $t \in \mathbb{T}_\alpha$. Then using the induction assumption we get for $t \in \mathbb{T}_\alpha$

$$\mu(\rho^{n+1}(t)) = \mu(\rho^n(\rho(t))) = a^{-n}\mu(\rho(t)) = a^{-(n+1)}\mu(t).$$

Hence, (iii) holds. □

**Theorem 5.5 (Properties of Forward Jump Operator).** *Given $m, n \in \mathbb{N}_0$ and $t \in \mathbb{T}_\alpha$*

(i) *for* $n \geq 1$, $\sigma^n(t) = a^n t + b\sum_{j=0}^{n-1} a^j$;
(ii) *if* $m > n$, $\sigma^m(t) > \sigma^{(n)}(t)$;
(iii) *if* $t > 0$, $\lim_{n\to\infty} \sigma^n(t) = \infty$.

*Proof.* We will only prove (i) and (iii) here. First we will prove property (i) by an induction argument. The base case clearly holds. Assume that $n \geq 1$ and $\sigma^n(t) = a^n t + \sum_{j=0}^{n-1} a^j b$. It follows that

$$\begin{aligned}\sigma^{n+1}(t) = \sigma(\sigma^n(t)) &= \sigma\left(a^n t + b\sum_{j=0}^{n-1} a^j\right)\\ &= a\left(a^n t + b\sum_{j=0}^{n-1} a^j\right) + b = a^{n+1}t + \left(b\sum_{j=0}^{n} a^j\right).\end{aligned}$$

This completes the proof of (i).

Next we prove that (iii) holds. First, consider the subcase in which $a > 1$. Then

$$\sigma^n(t) = a^n t + \sum_{j=0}^{n-1} a^j b \geq a^n t.$$

Since $t > 0$ and $a > 1$, we have that $\lim_{n\to\infty} \sigma^n(t) = \infty$. Next, consider the case in which $a = 1$. Then $b > 0$, and

$$\sigma^n(t) = a^n t + \sum_{j=0}^{n-1} a^j b = t + \sum_{j=0}^{n-1} b = t + nb \geq nb.$$

Since $b > 0$, we have that $\lim_{n\to\infty} \sigma^n(t) = \infty$. This completes the proof of (iii). □

**Theorem 5.6 (Properties of Backward Jump Operator).** *Given positive integers $m$, $n$, and $t \in \mathbb{T}_\alpha$, the following properties hold:*

(i) $\rho^n(t) = a^{-n}\left(t - \sum_{j=0}^{n-1} a^j b\right)$;

(ii) *if* $m > n$, *then* $\rho^m(t) < \rho^n(t)$;

(iii) $\lim_{n\to\infty} \rho^n(t) = -\infty$ *if* $a = 1$ *and* $= \frac{b}{1-a}$ *if* $a > 1$.

*Proof.* We will just prove (i) holds (see Exercise 5.3 for parts (ii) and (iii)). So, first we note that

$$\rho^1(t) = \frac{t-b}{a} = a^{-1}\left(t - \sum_{j=0}^{0} a^j b\right).$$

Assume that $n \geq 1$ and $\rho^n(t) = a^{-n}\left(t - \sum_{j=0}^{n-1} a^j b\right)$ holds. Then

$$\begin{aligned}
\rho^{n+1}(t) &= \rho(\rho^n(t)) \\
&= \rho\left(a^{-n}\left[t - \sum_{j=0}^{n-1} a^j b\right]\right) \\
&= \frac{a^{-n}\left[t - \sum_{j=0}^{n-1} a^j b\right] - b}{a}
\end{aligned}$$

$$= a^{-n-1}\left(t - \sum_{j=0}^{n-1} a^j b - a^n b\right)$$

$$= a^{-(n+1)}\left(t - \sum_{j=0}^{n} a^j b\right).$$

This completes the proof of (i) by induction. □

We now define a function $N(t,s)$, whose value (we will see in Theorem 5.8), when $s,t \in \mathbb{T}_\alpha$ with $s \leq t$, gives the cardinality, $card(\mathbb{T}_{[s,t)})$, of the set $\mathbb{T}_{[s,t)}$.

**Definition 5.7.** For $a > 1$, we define the function $N : \mathbb{T}_\alpha \times \mathbb{T}_\alpha \to \mathbb{Z}$ by

$$N(t,s) := \log_a\left(\frac{\mu(t)}{\mu(s)}\right).$$

For simplicity, we will use the notation $N(t) := N(t,\alpha)$, for $t \in \mathbb{T}_\alpha$.

As presented in Estes [34, 78], some properties of the function $N$ are given in the following theorem.

**Theorem 5.8.** *Assume $a > 1$ and $t,s,r \in \mathbb{T}_\alpha$. Then the following hold:*

(i) $N(t,t) = 0$;
(ii) $N(t,s) = card(\mathbb{T}_{[s,t)})$, *if* $s \leq t$;
(iii) $N(s,t) = -N(t,s)$;
(iv) $N(t,s) = N(t,r) + N(r,s)$.

*Proof.* Since

$$N(t,t) = \log_a\left(\frac{\mu(t)}{\mu(t)}\right) = \log_a 1 = 0,$$

we have that (i) holds. To see that (ii) holds, let $s,t \in \mathbb{T}_\alpha$ with $s \leq t$. If $k = card(\mathbb{T}_{[s,t)})$, then $t = \sigma^k(s)$, and so we have that

$$N(t,s) = \log_a\left[\frac{\mu(t)}{\mu(s)}\right] = \log_a\left[\frac{\mu(\sigma^k(s))}{\mu(s)}\right] = \log_a\left[\frac{a^k\mu(s)}{\mu(s)}\right] = \log_a a^k = k.$$

To see that (iii) holds, consider

$$N(t,s) = \log_a\left[\frac{\mu(t)}{\mu(s)}\right] = -\log_a\left[\frac{\mu(s)}{\mu(t)}\right] = -N(s,t).$$

The proof of (iv) is Exercise 5.4. □

## 5.3 Discrete Difference Calculus

In this section, we define a difference operator on our mixed time scale $\mathbb{T}_\alpha$ and study its properties. Note that if $a = q > 1$ and $b = 0$, then $D$ is the $q$-difference operator (see Chap. 4), and if $a = b = 1$, then $D$ is the forward difference operator.

**Definition 5.9.** Given $f : \mathbb{T}_\alpha \to \mathbb{R}$, the **mixed time scale difference operator** is defined by

$$Df(t) := \frac{f(\sigma(t)) - f(t)}{\mu(t)}, \quad t \in \mathbb{T}_\alpha.$$

**Theorem 5.10 (Properties of Difference Operator).** *Let $f, g : \mathbb{T}_\alpha \to \mathbb{R}$ and $\alpha \in [0, \infty)$ be given. Then for $t \in \mathbb{T}_\alpha$ the following hold:*

(i) $D\alpha = 0$;
(ii) $D\alpha f(t) = \alpha Df(t)$;
(iii) $D(f(t) + g(t)) = Df(t) + Dg(t)$;
(iv) $D(f(t)g(t)) = f(\sigma(t))Dg(t) + (Df(t))g(t)$;
(v) $D(f(t)g(t)) = f(t)Dg(t) + (Df(t))g(\sigma(t))$;
(vi) $D\left(\frac{f(t)}{g(t)}\right) = \frac{g(t)Df(t) - (Dg(t))f(t)}{g(t)g(\sigma(t))}$ *if* $g(t)g(\sigma(t)) \neq 0$.

*Proof.* Since $D\alpha = \frac{\alpha - \alpha}{\mu(t)} = 0$ we have that (i) holds. Also

$$D\alpha f(t) = \frac{\alpha f(\sigma(t)) - \alpha f(t)}{\mu(t)} = \alpha \left(\frac{f(\sigma(t)) - f(t)}{\mu(t)}\right) = \alpha Df(t),$$

so (ii) holds. To see that (iii) holds note that

$$\begin{aligned} D(f(t) + g(t)) &= \frac{[f(\sigma(t)) + g(\sigma(t))] - [f(t) + g(t)]}{\mu(t)} \\ &= \frac{f(\sigma(t)) - f(t)}{\mu(t)} + \frac{g(\sigma(t)) - g(t)}{\mu(t)} = Df(t) + Dg(t). \end{aligned}$$

The proof of property (iv) is left to the reader. Property (v) follows from (iv) by interchanging $f(t)$ and $g(t)$. Finally, property (vi) follows from the following:

$$\begin{aligned} D\left(\frac{f(t)}{g(t)}\right) &= \frac{\left(\frac{f(\sigma(t))}{g(\sigma(t))}\right) - \left(\frac{f(t)}{g(t)}\right)}{\mu(t)} \\ &= \frac{f(\sigma(t))g(t) - g(\sigma(t))f(t)}{g(t)g(\sigma(t))\mu(t)} \end{aligned}$$

$$= \frac{f(\sigma(t))g(t) - f(t)g(t) + f(t)g(t) - g(\sigma(t))f(t)}{g(\sigma(t))g(t)\mu(t)}$$

$$= \frac{g(t)\left(\frac{f(\sigma(t)) - f(t)}{\mu(t)}\right) - f(t)\left(\frac{g(\sigma(t)) - g(t)}{\mu(t)}\right)}{g(t)g(\sigma(t))}$$

$$= \frac{g(t)Df(t) - (Dg(t))f(t)}{g(t)g(\sigma(t))}.$$

And this completes the proof. □

## 5.4 Discrete Delta Integral

In this section, we will define the integral of a function defined on the mixed time scale $\mathbb{T}_\alpha$. We will develop several properties of this integral, including the two fundamental theorems for the calculus on mixed time scales.

**Definition 5.11.** Let $f : \mathbb{T}_\alpha \to \mathbb{R}$ and $c, d \in \mathbb{T}_\alpha$ be given. Then

$$\int_c^d f(t)Dt := \begin{cases} \sum_{j=0}^{N(d,c)-1} f(\sigma^j(c))\mu(\sigma^j(c)) & \text{if } c < d \\ 0 & \text{if } c = d \\ -\sum_{j=0}^{N(c,d)-1} f(\sigma^j(d))\mu(\sigma^j(d)) & \text{if } c > d. \end{cases}$$

**Theorem 5.12 (Properties of Integral).** *Given $f, g : \mathbb{T}_\alpha \to \mathbb{R}$ and $c, d, l \in \mathbb{T}_\alpha$, the following properties hold:*

(i) $\int_c^d f(t)Dt = -\int_d^c f(t)Dt$;
(ii) $\int_c^d \alpha f(t)Dt = \alpha \int_c^d f(t)Dt$;
(iii) $\int_c^d (f(t) + g(t))Dt = \int_c^d f(t)Dt + \int_c^d g(t)Dt$;
(iv) $\int_c^c f(t)Dt = 0$;
(v) $\int_c^d f(t)Dt = \int_c^l f(t)Dt + \int_l^d f(t)Dt$;
(vi) *if $d \geq c$, then* $\left|\int_c^d f(t)Dt\right| \leq \int_c^d |f(t)|Dt$;
(vii) *if $f(t) \geq g(t)$ for $t \in \mathbb{T}_{[c,d)}$, then* $\int_c^d f(t)Dt \geq \int_c^d g(t)Dt$, *if $d \geq c$.*

*Proof.* These properties follow from properties of summations. As an example, we will just prove property (vi). To this end, we note that

$$\left| \int_c^d f(t)Dt \right| = \left| \sum_{j=0}^{K(d,c)-1} f(\sigma^j(c))\mu(\sigma^j(c)) \right|$$
$$\leq \sum_{j=0}^{K(d,c)-1} |f(\sigma^j(c))\mu(\sigma^j(c))|$$
$$= \int_c^d |f(t)|Dt.$$

□

**Definition 5.13.** Assume $c, d \in \mathbb{T}_\alpha$ and $c < d$. Given $f : \mathbb{T}_{[c,d]} \to \mathbb{R}$. We say $F$ is an antidifference of $f$ on $\mathbb{T}_{[c,d]}$ provided $DF(t) = f(t)$ for all $t \in \mathbb{T}_{[c,\rho(d)]}$.

The following theorem shows that every function $f : \mathbb{T}_\alpha \to \mathbb{R}$ has an antidifference on $\mathbb{T}_\alpha$.

**Theorem 5.14 (Fundamental Theorem of Difference Calculus: Part II).** *Assume $f : \mathbb{T}_\alpha \to \mathbb{R}$ and $c \in \mathbb{T}_\alpha$. If we define $F : \mathbb{T}_\alpha \to \mathbb{R}$ by $F(t) = \int_c^t f(s)Ds$, then $F$ is an antidifference of $f$ on $\mathbb{T}_\alpha$.*

*Proof.* Let $F$ be as defined as in the statement of this theorem. Then for $t \in \mathbb{T}_\alpha$,

$$DF(t) = \frac{\int_c^{\sigma(t)} f(s)Ds - \int_c^t f(s)Ds}{\mu(t)} = \frac{\int_t^{\sigma(t)} f(s)Ds}{\mu(t)} = \frac{f(t)\mu(t)}{\mu(t)} = f(t),$$

which is what we wanted to show. □

**Theorem 5.15.** *Assume $f : \mathbb{T}_\alpha \to \mathbb{R}$ and $F$ is an antidifference of $f$ on $\mathbb{T}_\alpha$. Then a general antidifference of $f$ on $\mathbb{T}_\alpha$ is given by*

$$G(t) = F(t) + C, \quad t \in \mathbb{T}_\alpha,$$

*where $C$ is an arbitrary constant.*

*Proof.* Let $F$ be an antidifference of $f$ on $\mathbb{T}_\alpha$. Set $G(t) = F(t) + C$ for $t \in \mathbb{T}_\alpha$, where $C$ is a constant. Then

$$DG(t) = D[F(t) + C] = f(t) + 0 = f(t), \quad \text{for} \quad t \in \mathbb{T}_\alpha.$$

Conversely, assume $G(t)$ is any antidifference of $f$ on $\mathbb{T}_\alpha$. Then

$$D[G(t) - F(t)] = DG(t) - DF(t) = f(t) - f(t) = 0, \quad t \in \mathbb{T}_\alpha.$$

From Exercise 5.6, there is a constant $C$ so that

$$G(t) - F(t) = C, \quad t \in \mathbb{T}_\alpha.$$

Hence,

$$F(t) = G(t) + C, \quad t \in \mathbb{T}_\alpha,$$

as desired. □

**Definition 5.16.** We define the indefinite integral as follows:

$$\int f(t)Dt = F(t) + C,$$

where $F(t)$ is any antidifference of $f(t)$.

**Theorem 5.17 (Fundamental Theorem of Difference Calculus: Part I).** *Assume $f : \mathbb{T}_\alpha \to \mathbb{R}$ and $c, d \in \mathbb{T}_\alpha$. Then, if $F$ is any antidifference of $f$ on $\mathbb{T}_\alpha$, it follows that*

$$\int_c^d f(t)Dt = \int_c^d DF(t)Dt = F(d) - F(c).$$

*Proof.* Put

$$G(t) := \int_c^t f(s)Ds, \quad t \in \mathbb{T}_\alpha.$$

By Theorem 5.14 $G(t)$ is an antidifference of $f(t)$ on $\mathbb{T}_\alpha$. Let $F(t)$ be any fixed antidifference of $f(t)$ on $\mathbb{T}_\alpha$. Then by Theorem 5.15 we have that

$$F(t) = G(t) + A, \quad \text{where } A \text{ is a constant.}$$

It follows that

$$F(d) - F(a) = [G(d) + A] - [G(c) + A] = G(d) = \int_c^d f(s)Ds.$$

□

*Remark 5.18.* Note that the Fundamental Theorem of Calculus tells us that given $f : \mathbb{T}_\alpha \to \mathbb{R}$, a point $t_0 \in \mathbb{T}_\alpha$, and a real number $C$, the unique solution of the IVP

$$Dy(t) = f(t)$$
$$y(t_0) = C$$

is given by $y(t) = \int_{t_0}^t f(s)Ds + C$, for $t \in \mathbb{T}_\alpha$.

The integration by parts formulas in the next theorem are very useful.

**Theorem 5.19 (Integration by Parts).** *Given two functions* $u, v : \mathbb{T}_\alpha \to \mathbb{R}$, *if* $c, d \in \mathbb{T}_\alpha$, $c < d$, *then*

$$\int_c^d u(t)Dv(t)Dt = u(t)v(t)\Big|_c^d - \int_c^d v(\sigma(t))Du(t)Dt$$

*and*

$$\int_c^d u(\sigma(t))Dv(t)Dt = u(t)v(t)\Big|_c^d - \int_c^d v(t)Du(t)Dt.$$

*Proof.* By the product rule

$$D[u(t)v(t)] = v(\sigma(t))Du(t) + (Dv(t))u(t).$$

Using the fundamental theorem of calculus, we get

$$\int_c^d u(t)Dv(t)Dt + v(\sigma(t))Du(t)Dt = u(d)v(d) - u(c)v(c).$$

It follows that

$$\int_c^d u(t)Dv(t)Dt = u(t)v(t)|_c^d - \int_c^d v(\sigma(t))Du(t)Dt.$$

This proves the first integration by parts formula. Interchanging $u(t)$ and $v(t)$ leads to the second integration by parts formula. □

## 5.5 Falling and Rising Functions

In this section, we define the falling and rising functions for the mixed time scale $\mathbb{T}_\alpha$, which are analogous to the falling and rising functions for the delta calculus in Chap. 1. Several properties of these functions will be given, including the appropriate power rule.

First we define the appropriate rising and falling functions for the mixed time scale calculus.

**Definition 5.20.** Assume $n \in \mathbb{N}$. We define the **rising function**, $t^{\overline{n}}$, read "$t$ to the $n$ rising," by

$$t^{\overline{n}} := \prod_{j=0}^{n-1} \sigma^j(t), \quad t^{\overline{0}} := 1,$$

for $t \in \mathbb{R}$. We also define the **falling function**, $t^{\underline{n}}$, read "$t$ to the $n$ falling," by

$$t^{\underline{n}} := \prod_{j=0}^{n-1} \rho^j(t), \quad t^{\underline{0}} := 1.$$

for $t \in \mathbb{R}$.

**Definition 5.21.** For $n \in \mathbb{Z}$, we define **$a$-square bracket of $n$** by

$$[n]_a := \begin{cases} \dfrac{a^n - 1}{a - 1} & \text{for } a > 1 \\ n & \text{for } a = 1 \end{cases}.$$

**Theorem 5.22 (Properties of $a$-Square Bracket of $n$).** *For $n \in \mathbb{Z}$, and $a \geq 1$,*

(i) $[0]_a = 0$;
(ii) $[1]_a = 1$;
(iii) $[n]_a + a^n = [n+1]_a$;
(iv) $a[n]_a + 1 = [n+1]_a$;
(v) $[-n]_a = -\dfrac{[n]_a}{a^n}$.

*Proof.* To see that (iii) holds for $a > 1$, note that

$$[n]_a + a^n = \frac{a^n - 1}{a - 1} + a^n = \frac{a^{n+1} - 1}{a - 1} = [n+1]_a.$$

Also (iii) trivially holds for $a = 1$.

To see that (iv) holds for $a > 1$ note that

$$\begin{aligned} a[n]_a + 1 &= a\frac{a^n - 1}{a - 1} + 1 = \frac{a^{n+1} - a}{a - 1} + \frac{a - 1}{a - 1} \\ &= \frac{a^{n+1} - 1}{a - 1} = [n+1]_a. \end{aligned}$$

Also for $a = 1$ we have that

$$a[n]_a + 1 = n + 1 = [n+1]_a.$$

Property (v) holds for $a > 1$, since

$$[-n]_a = \frac{a^{-n} - 1}{a - 1} = -a^{-n}\frac{a^n - 1}{a - 1} = -\frac{[n]_a}{a^n}.$$

Furthermore, (v) is trivially true for $a = 1$. □

We may use the $a$-square bracket function to simplify the expressions that we found for the forward and backward jump operators.

**Theorem 5.23.** *For $n \in \mathbb{N}$ the following hold:*

(i) $\sigma^n(t) = a^n t + [n]_a b$;
(ii) $\rho^n(t) = a^{-n} t + [-n]_a b$;
(iii) $\sigma^n(t) - t = [n]_a \mu(t)$.

*Proof.* In order to prove property (i), we have by part (i) of Theorem 5.5 that

$$\sigma^n(t) = a^n t + b \sum_{j=0}^{n-1} a^j = a^n t + b \left( \frac{a^n - 1}{a - 1} \right) = a^n t + [n]_a b.$$

Similarly part (i) of Theorem 5.6 gives us that (ii) holds. Finally, using property (i) we have that

$$\begin{aligned} \sigma^n(t) - t = a^n t + [n]_a b - t &= (a^n - 1)t + \left( \frac{a^n - 1}{a - 1} \right) b \\ &= \left( \frac{a^n - 1}{a - 1} \right) [(a-1)t + b] = [n]_a \mu(t), \end{aligned}$$

and hence (iii) holds. □

Next we prove a power rule.

**Theorem 5.24 (Power Rule).** *For $n \in \mathbb{N}$ the following holds:*

$$Dt^{\overline{n}} = [n]_a (\sigma(t))^{\overline{n-1}}, \quad \textit{for} \quad t \in \mathbb{R}.$$

*Proof.* For $t \in \mathbb{R}$ we have that

$$\begin{aligned} Dt^{\overline{n}} = \frac{\prod_{j=0}^{n-1} \sigma^j(\sigma(t)) - \prod_{j=0}^{n-1} \sigma^j(t)}{\mu(t)} &= \frac{[\sigma^{(n)}(t) - t]}{\mu(t)} \prod_{j=1}^{n-1} \sigma^j(t) \\ &= \frac{[\sigma^{(n)}(t) - t]}{\mu(t)} \prod_{j=0}^{n-2} \sigma^j(\sigma(t)) \\ &= [n]_a \, [\sigma(t)]^{\overline{n-1}}, \end{aligned}$$

where in the last step we used part (iii) of Theorem 5.23. □

**Definition 5.25.** For $n \in \mathbb{Z}$ and $a \geq 1$, we define the ***a*-bracket of** $n$ by

$$\{n\}_a := \begin{cases} \dfrac{a^n - 1}{(a-1)a^{n-1}} & \text{for } a > 1 \\ n & \text{for } a = 1. \end{cases}$$

The following theorem gives us several properties of the $a$-bracket function.

**Theorem 5.26.** *The following hold:*

(i) $\{0\}_a = 0$;
(ii) $\{1\}_a = 1$;
(iii) $\{n\}_a = \dfrac{[n]_a}{a^{n-1}}$;
(iv) $\{n\}_a = -a[-n]_a$;
(v) $\sigma(t) - \rho^{n-1}(t) = \{n\}_a \mu(t)$.

*Proof.* We will just prove part (iv) holds when $a > 1$. This follows from

$$\{n\}_a = \frac{[n]_a}{a^{n-1}} = -a[-n]_a,$$

where the first equality is by part (iii) of this theorem and the second equality is by part (v) of Theorem 5.22. □

**Theorem 5.27 (Power Rule).** *For $n \in \mathbb{N}$ the following holds:*

$$Dt^{\underline{n}} = \{n\}_a t^{\underline{n-1}}, \quad \text{for} \quad t \in \mathbb{R}.$$

*Proof.* To establish the result, we calculate

$$\begin{aligned} Dt^{\underline{n}} = \frac{[\sigma(t)]^{\underline{n}} - t^{\underline{n}}}{\mu(t)} &= \frac{\prod_{j=0}^{n-1} \rho^j(\sigma(t)) - \prod_{j=0}^{n-1} \rho^j(t)}{\mu(t)} \\ &= \frac{\sigma(t) \prod_{j=1}^{n-1} \rho^{j-1}(t) - \rho^{n-1}(t) \prod_{j=1}^{n-1} \rho^{j-1}(t)}{\mu(t)} \\ &= \frac{\sigma(t) - \rho^{n-1}(t)}{\mu(t)} \prod_{j=0}^{n-2} \rho^j(t) \\ &= \{n\}_a t^{\underline{n-1}}, \end{aligned}$$

with the last equality by part (v) of Theorem 5.26. □

## 5.6 Discrete Taylor's Theorem

In this section we will develop Taylor's Theorem for functions on $\mathbb{T}_\alpha$. First we define the Taylor monomials for the mixed time scale $\mathbb{T}_\alpha$ as follows.

**Definition 5.28.** We define the **Taylor monomials for the mixed time scale** $\mathbb{T}_\alpha$ as follows. First put $h_0(t,s) = 1$ for $t, s \in \mathbb{T}_\alpha$. Then for each $n \in \mathbb{N}_1$ we recursively define $h_n(t,s)$, for each fixed $s \in \mathbb{T}_\alpha$, to be the unique solution (see Remark 5.18) of the IVP

$$\begin{aligned} Dy(t) &= h_{n-1}(t,s), \quad t \in \mathbb{T}_\alpha \\ y(s) &= 0. \end{aligned}$$

In the next theorem we derive a formula for $h_n(t,s)$ (see Erbe et al. [76]).

**Theorem 5.29.** *The Taylor monomials, $h_n(t,s)$, $n \in \mathbb{N}_0$, for the mixed time scale $\mathbb{T}_\alpha$ are given by*

$$h_n(t,s) = \prod_{k=1}^{n} \frac{t - \sigma^{k-1}(s)}{[k]_a}$$

*for $t, s \in \mathbb{T}_\alpha$.*

*Proof.* For $n \in \mathbb{N}_0$, let

$$f_n(t,s) := \prod_{k=1}^{n} \frac{t - \sigma^{k-1}(s)}{[k]_a}, \quad \text{for} \quad t, s \in \mathbb{T}_\alpha.$$

We prove by induction on $n$, for $n \in \mathbb{N}_0$, that $f_n(t,s) = h_n(t,s)$ for $t, s \in \mathbb{T}_\alpha$. By our convention on products $f_0(t,s) = 1 = h_0(t,s)$, and it is easy to see that $f_1(t,s) = t - s = h_1(t,s)$ for $t, s \in \mathbb{T}_\alpha$. Assume $n \geq 1$ and $f_k(t,s) = h_k(t,s)$ for $t, s \in \mathbb{T}_\alpha$, $0 \leq k \leq n$. It remains to show that $f_{n+1}(t,s) = h_{n+1}(t,s)$ for $t, s \in \mathbb{T}_\alpha$. First, note that

$$\begin{aligned} f_{n+1}(t,s) = \prod_{k=1}^{n+1} \frac{t - \sigma^{k-1}(s)}{[k]_a} &= \frac{t - \sigma^n(s)}{[n+1]_a} \prod_{k=1}^{n} \frac{t - \sigma^{k-1}(s)}{[k]_a} \\ &= \frac{t - \sigma^n(s)}{[n+1]_a} f_n(t,s) \\ &= \frac{t - \sigma^n(s)}{[n+1]_a} h_n(t,s) \end{aligned}$$

by the induction hypothesis. Fix $s \in \mathbb{T}_\alpha$, then using the product rule

$$\begin{aligned}
Df_{n+1}(t,s) &= D\left(\frac{t-\sigma^n(s)}{[n+1]_a}\, h_n(t,s)\right)\\
&= \frac{\sigma(t)-\sigma^n(s)}{[n+1]_a}\, h_{n-1}(t,s) + \frac{h_n(t,s)}{[n+1]_a}\\
&= \frac{(at+b-a^n s-[n]_a b)}{[n+1]_a} f_{n-1}(t,s) + \frac{f_n(t,s)}{[n+1]_a} \qquad \text{using Theorem 5.23, (i)}\\
&= \frac{a(t-a^{n-1}s-b[n-1]_a)}{[n+1]_a} f_{n-1}(t,s) + \frac{f_n(t,s)}{[n+1]_a} \quad \text{using Theorem 5.22, (iv)}\\
&= \frac{a(t-\sigma^{n-1}(s))}{[n+1]_a} f_{n-1}(t,s) + \frac{f_n(t,s)}{[n+1]_a} \qquad \text{using Theorem 5.23, (i)}\\
&= \frac{a[n]_a f_n(t,s)}{[n+1]_a} + \frac{f_n(t,s)}{[n+1]_a}\\
&= \frac{a[n]_a+1}{[n+1]_a} f_n(t,s)\\
&= f_n(t,s) \qquad \text{using Theorem 5.22, (iv)}\\
&= h_n(t,s).
\end{aligned}$$

Since, for each fixed $s$, $y(t) = f_{n+1}(t,s)$ solves the IVP

$$\begin{aligned}
Dy(t) &= h_n(t,s), \quad t \in \mathbb{T}_\alpha\\
y(s) &= 0,
\end{aligned}$$

we have by Remark 5.18 that $h_{n+1}(t,s) = f_{n+1}(t,s)$ for $t \in \mathbb{T}_\alpha$. Finally, notice that since $s \in \mathbb{T}_\alpha$ is arbitrary we conclude that $h_{n+1}(t,s) = f_{n+1}(t,s)$ for all $t, s \in \mathbb{T}_\alpha$. □

**Definition 5.30.** For $n \in \mathbb{N}_0$, we define the **a-falling-bracket (of n) factorial**, denoted by $\{n\}_a!$, recursively by $\{0\}_a! = 1$ and for $n \in \mathbb{N}_1$

$$\{n\}_a! = \{n\}_a\left(\{n-1\}_a!\right).$$

**Definition 5.31.** For $n \in \mathbb{N}_0$, we define the **a-rising-bracket (of n) factorial**, denoted by $[n]_a!$, recursively by $[0]_a! = 1$ and for $n \in \mathbb{N}_1$

$$[n]_a! = [n]_a\left([n-1]_a!\right).$$

The following theorem is a generalization of the binomial expansion of $(t-t)^n = 0$, $n \in \mathbb{N}_1$.

**Theorem 5.32 (Estes [34, 78]).** *Assume $n \geq 1$ and $t \in \mathbb{T}_\alpha$. Then*

$$\sum_{i=0}^{n} \frac{(-1)^i \left(t^{\overline{i}}\right)\left(t^{\underline{n-i}}\right)}{[i]_a!\{n-i\}_a!} = 0.$$

*Proof.* For $n \in \mathbb{N}_1$, consider

$$f_n(t) := \sum_{i=0}^{n} \frac{(-1)^i \left(t^{\overline{i}}\right)\left(t^{\underline{n-i}}\right)}{[i]_a!\{n-i\}_a!}.$$

We will prove by induction on $n$ that $f_n(t) = 0$. For the base case $n = 1$ we have

$$f_1(t) = t - t = 0.$$

Assume $n \in \mathbb{N}_1$ and $f_n(t) = 0$. It remains to show $f_{n+1}(t) = 0$. Using the product rule

$$\begin{aligned}
Df_{n+1}(t) &= D\left(\sum_{i=0}^{n+1} \frac{(-1)^i \left(t^{\overline{i}}\right)\left(t^{\underline{n+1-i}}\right)}{[i]_a!\{n+1-i\}_a!}\right) \\
&= D\left(\frac{(-1)^{n+1} t^{\overline{n+1}}}{[n+1]_a!} + \sum_{i=1}^{n} \frac{(-1)^i \left(t^{\overline{i}}\right)\left(t^{\underline{n+1-i}}\right)}{[i]_a!\{n+1-i\}_a!} + \frac{t^{\underline{n+1}}}{\{n+1\}_a!}\right) \\
&= \sum_{i=1}^{n} \frac{(-1)^i (\sigma(t))^{\overline{i-1}} (\sigma(t))^{\underline{n+1-i}}}{[i-1]_a!\{n+1-i\}_a!} \\
&\quad + \sum_{i=1}^{n} \frac{(-1)^i \left(t^{\overline{i}}\right)\left(t^{\underline{n-i}}\right)}{[i]_a!\{n-i\}_a!} + \frac{(-1)^{n+1} (\sigma(t))^{\overline{n}}}{[n]_a!} + \frac{t^{\underline{n}}}{\{n\}_a!} \\
&= \sum_{i=1}^{n+1} \frac{(-1)^i (\sigma(t))^{\overline{i-1}} (\sigma(t))^{\underline{n+1-i}}}{[i-1]_a!\{n+1-i\}_a!} + \sum_{i=0}^{n} \frac{(-1)^i \left(t^{\overline{i}}\right)\left(t^{\underline{n-i}}\right)}{[i-1]_a!\{n-i\}_a!} \\
&= -\sum_{i=0}^{n} \frac{(-1)^i (\sigma(t))^{\overline{i}} (\sigma(t))^{\underline{n-i}}}{[i]_a!\{n-i\}_a!} + \sum_{i=0}^{n} \frac{(-1)^i \left(t^{\overline{i}}\right)\left(t^{\underline{n-i}}\right)}{[i]_a!\{n-i\}_a!} \\
&= -f_n(\sigma(t)) + f_n(t) \\
&= 0.
\end{aligned}$$

Since $Df_{n+1}(t) = 0$, we have that $f_{n+1}(t) = C$, with $t \in \mathbb{T}_\alpha$, for some constant $C$. Note that $f_{n+1}(t)$ can be expanded to a polynomial in $t$, and that each term of the sum

$$f_{n+1}(t) = \sum_{i=0}^{n+1} \frac{(-1)^i \binom{t^{\bar{i}}}{} \binom{t^{\underline{n+1-i}}}{}}{[i]_a!\{n+1-i\}_a!}$$

is divisible by $t$. Thus, the polynomial expansion of $f_{n+1}(t)$ has no constant term and by the polynomial principle, $C = 0$. We have shown that $f_{n+1}(t) = 0$. This completes the proof by induction. □

Next we prove an alternate formula for the Taylor monomials due to Estes [34, 78].

**Theorem 5.33.** *Assume $n \in \mathbb{N}_0$ and $t, s \in \mathbb{T}_\alpha$. Then*

$$h_n(t,s) = \sum_{i=0}^{n} \frac{(-1)^i \left(s^{\bar{i}}\right) \left(t^{\underline{n-i}}\right)}{[i]_a!\{n-i\}_a!}.$$

*Proof.* Fix $s$ and let

$$f_n(t,s) := \sum_{i=0}^{n} \frac{(-1)^i \left(s^{\bar{i}}\right) \left(t^{\underline{n-i}}\right)}{[i]_a!\{n-i\}_a!}.$$

We will show by induction that $f_n(t,s) = h_n(t,s)$. The base case $f_0(t,s) = h_0(t,s)$ follows from the definitions. Assume that $n \in \mathbb{N}_1$ and $f_n(t,s) = h_n(t,s)$. From Theorem 5.32, we know that $f_n(s,s) = 0$. Also

$$\begin{aligned} Df_{n+1}(t,s) &= D\left(\sum_{i=0}^{n} \frac{(-1)^i \left(s^{\bar{i}}\right) \left(t^{\underline{n+1-i}}\right)}{[i]_a!\{n+1-i\}_a!}\right) \\ &= \sum_{i=0}^{n-1} \frac{(-1)^i \left(s^{\bar{i}}\right) \left(t^{\underline{n-i}}\right)}{[i]_a!\{n-i\}_a!} \\ &= f_n(t,s) = h_n(t,s). \end{aligned}$$

Hence, for each fixed $s \in \mathbb{T}_\alpha$, $y(t) = f_{n+1}(t,s)$ solves the IVP

$$\begin{aligned} Dy(t) &= h_n(t,s), \quad t \in \mathbb{T}_\alpha \\ y(s) &= 0. \end{aligned}$$

So, by the uniqueness of solutions to IVPs (see Remark 5.18), we have that $f_{n+1}(t,s) = h_{n+1}(t,s)$ for each fixed $s \in \mathbb{T}_\alpha$. This completes the proof by induction. □

Later we are going to see that we need to take the mixed time scale difference of $h_n(t,s)$ with respect to its second variable. To do this we now introduce the type-two Taylor monomials, $g_n(t,s)$, $n \in \mathbb{N}_0$.

**Definition 5.34.** We define the **type-two Taylor monomials** $g_n : \mathbb{T}_\alpha \times \mathbb{T}_\alpha \to \mathbb{R}$, for $n \in \mathbb{N}_0$, recursively as follows:

$$g_0(t,s) = 1 \quad \text{for} \quad t, s \in \mathbb{T}_\alpha,$$

and for each $n \in \mathbb{N}_1$, $g_n(t,s)$, for each fixed $s \in \mathbb{T}_\alpha$, is the unique solution (see Remark 5.18) of the IVP

$$\begin{aligned} Dy(t) &= -g_{n-1}(\sigma(t), s), \quad t \in \mathbb{T}_\alpha \\ y(s) &= 0. \end{aligned}$$

In the next theorem we give two different formulas for the type-two Taylor monomials.

**Theorem 5.35.** *The type-two Taylor monomials are given by*

$$g_n(t,s) = h_n(s,t) = \sum_{i=0}^{n} \frac{(-1)^i \left(t^{\bar{i}}\right)\left(s^{\underline{n-i}}\right)}{[i]_a!\{n-i\}_a!} = \prod_{k=1}^{n} \frac{s - \sigma^{k-1}(t)}{[k]_a},$$

*for* $t, s \in \mathbb{T}_\alpha$.

*Proof.* We prove by induction on $n$ that $h_n(s,t) = g_n(t,s)$, $t, s \in \mathbb{T}_\alpha$, $n \in \mathbb{N}_0$. Obviously this holds for $n = 0$. Assume $n \geq 0$ and $h_n(s,t) = g_n(t,s)$ for $t, s \in \mathbb{T}_\alpha$. Fix $s \in \mathbb{T}_\alpha$ and consider

$$\begin{aligned} Dh_{n+1}(s,t) &= D\left(\sum_{i=0}^{n+1} \frac{(-1)^i \left(t^{\bar{i}}\right)\left(s^{\underline{n+1-i}}\right)}{[i]_a!\{n+1-i\}_a!}\right) \\ &= \sum_{i=1}^{n+1} \frac{(-1)^i (\sigma(t))^{\overline{i-1}} \left(s^{\underline{n+1-i}}\right)}{[i-1]_a!\{n+1-i\}_a!} \qquad \text{by Theorem 5.24} \\ &= -\sum_{i=0}^{n} \frac{(-1)^i (\sigma(t))^{\bar{i}} \left(s^{\underline{n-i}}\right)}{[i]_a!\{n-i\}_a!} \\ &= -h_n(s,\sigma(t)) = -g_n(\sigma(t), s) \quad \text{by the induction hypothesis} \end{aligned}$$

for $t \in \mathbb{T}_\alpha$. Also, by Theorem 5.32, $h_{n+1}(s,s) = 0$. So, $y(t) = h_{n+1}(s,t)$ satisfies for each fixed $s$ the same IVP

$$\begin{aligned} Dy(t) &= -g_{n-1}(\sigma(t), s) \\ y(s) &= 0 \end{aligned}$$

as $g_{n+1}(t,s)$. Hence, by the uniqueness (see Remark 5.18) of solutions to IVPs

$$f_{n+1}(t,s) = h_{n+1}(s,t) = \sum_{i=0}^{n} \frac{(-1)^i \left(t^{\overline{i}}\right)\left(s^{\underline{n-i}}\right)}{[i]_a!\{n-i\}_a!},$$

for $t \in \mathbb{T}_\alpha$. □

We now can state our power rule as follows.

**Theorem 5.36.** *Assume $n \in \mathbb{N}_0$. Then for each fixed $s \in \mathbb{T}_\alpha$*

$$Dh_{n+1}(t,s) = h_n(t,s)$$

*and*

$$Dh_{n+1}(s,t) = -h_n(\sigma(t),s)$$

*for $t \in \mathbb{T}_\alpha$.*

*Proof.* By the definition (Definition 5.28) of $h_n(t,s)$ we have for each fixed $s \in \mathbb{T}_\alpha$ that $Dh_{n+1}(t,s) = h_n(t,s)$ for all $t \in \mathbb{T}_\alpha$. To see that $Dh_{n+1}(s,t) = -h_n(\sigma(t),s)$ for $t \in \mathbb{T}_\alpha$, note that by Theorem 5.33

$$\begin{aligned} Dh_{n+1}(s,t) &= D\sum_{i=0}^{n+1} \frac{(-1)^i \left(t^{\overline{i}}\right)\left(s^{\underline{n+1-i}}\right)}{[i]_a!\{n+1-i\}_a!} \\ &= \sum_{i=1}^{n+1} \frac{(-1)^i \left([\sigma(t)]^{\overline{i-1}}\right)\left(s^{\underline{n+1-i}}\right)}{[i-1]_a!\{n+1-i\}_a!} \\ &= -\sum_{i=0}^{n} \frac{(-1)^i \left([\sigma(t)]^{\overline{i}}\right)\left(s^{\underline{n-i}}\right)}{[i]_a!\{n-i\}_a!} \\ &= -h_n(\sigma(t),s), \end{aligned}$$

which completes the proof. □

**Theorem 5.37 (Taylor's Formula).** *Assume $f : \mathbb{T}_\alpha \to \mathbb{R}$, $n \in \mathbb{N}_0$ and $s \in \mathbb{T}_\alpha$. Then*

$$f(t) = p_n(t,s) + R_n(t,s), \quad t \in \mathbb{T}_\alpha,$$

*where the n-th degree Taylor polynomial, $p_n(t,s)$, based at s, is given by*

$$p_n(t,s) = \sum_{k=0}^{n} D^k f(s) h_k(t,s), \quad t \in \mathbb{T}_\alpha,$$

*and the remainder term, $R_n(t,s)$, based at s, is given by*

$$R_n(t,s) = \int_s^t h_n(t,\sigma(\tau))D^{n+1}f(\tau)D\tau, \quad t \in \mathbb{T}_\alpha.$$

*Proof.* We prove Taylor's Formula by induction. For $n = 0$ we have

$$R_0(t,s) = \int_s^t h_0(t,\sigma(\tau))Df(\tau)D\tau = \int_s^t Df(\tau)D\tau = f(t) - f(s).$$

Solving for $f(t)$ we get the desired result

$$f(t) = f(s) + R_0(t,s) = p_0(t,s) + R_0(t,s), \quad t \in \mathbb{T}_\alpha.$$

Now assume that $n \geq 0$ and $f(t) = p_n(t,s) + R_n(t,s)$, for $t \in \mathbb{T}_\alpha$. Then integrating by parts we obtain

$$\begin{aligned}
R_{n+1}(t,s) &= \int_s^t h_{n+1}(t,\sigma(\tau))D^{n+2}f(\tau)D\tau \\
&= h_{n+1}(t,\tau)D^{n+1}f(\tau)\Big|_{\tau=s}^{t} + \int_s^t h_n(t,\sigma(\tau))D^{n+1}f(\tau)D\tau \\
&= -h_{n+1}(t,s)D^{n+1}f(s) + \int_s^t h_n(t,\sigma(\tau))D^{n+1}f(\tau)D\tau \\
&= -h_{n+1}(t,s)D^{n+1}f(s) + R_n(t,s) \\
&= -h_{n+1}(t,s)D^{n+1}f(s) + f(t) - p_n(t,s) \\
&= -p_{n+1}(t,s) + f(t).
\end{aligned}$$

Solving for $f(t)$ we obtain the desired result

$$f(t) = p_{n+1}(t,s) + R_{n+1}(t,s), \quad t \in \mathbb{T}_\alpha.$$

This completes the proof by induction. □

We can now use Taylor's Theorem to prove the following variation of constants formula.

**Theorem 5.38 (Variation of Constants Formula).** *Assume $f : \mathbb{T}_\alpha \to \mathbb{R}$ and $s \in \mathbb{T}_\alpha$. Then the unique solution of the IVP*

$$\begin{aligned}
D^n y(t) &= f(t), \quad t \in \mathbb{T}_\alpha \\
D^i y(s) &= C_k, \quad 0 \leq k \leq n-1,
\end{aligned}$$

*where $C_k$, $0 \le k \le n-1$, are given constants, is given by*

$$y(t) = \sum_{k=0}^{n-1} C_k h_k(t,s) + \int_s^t h_{n-1}(t,\sigma(\tau))f(\tau)D\tau,$$

*for $t \in \mathbb{T}_\alpha$.*

*Proof.* It is easy to see that the given IVP has a unique solution $y(t)$. By Taylor's Formula (see Theorem 5.37) applied to $y(t)$ with $n$ replaced by $n-1$, we get

$$\begin{aligned} y(t) &= \sum_{k=0}^{n-1} D^k y(s) h_k(t,s) + \int_s^t h_{n-1}(t,\sigma(\tau))D^n y(\tau)D\tau \\ &= \sum_{k=0}^{n-1} C_k h_k(t,s) + \int_s^t h_{n-1}(t,\sigma(\tau))f(\tau)D\tau, \end{aligned}$$

for $t \in \mathbb{T}_\alpha$. □

*Example 5.39.* Consider the mixed time scale where $\alpha = 1$ and $\sigma(t) = 2t + 1$ (so $a = 2$ and $b = 1$). Use the variation of constants formula in Theorem 5.38 to solve the IVP

$$\begin{aligned} D^2 y(t) &= t - 1, \quad t \in \mathbb{T}_1 \\ y(1) &= 2, \quad Dy(1) = 0. \end{aligned}$$

By the variation of constants formula in Theorem 5.38, we have that

$$\begin{aligned} y(t) &= 2h_0(t,1) + 0h_1(t,1) + \int_1^t h_1(t,\sigma(s))(s-1)Ds \\ &= 2 + \int_1^t h_1(t,\sigma(s))h_1(s,1)Ds. \end{aligned}$$

Integrating by parts we calculate

$$\begin{aligned} y(t) &= 2 + h_1(t,s)h_2(s,1)\Big|_{s=1}^{t} + \int_1^t h_2(s,1)Ds \\ &= 2 + h_3(s,1)\Big|_{s=1}^{t} = 2 + h_3(t,1) \\ &= 2 + \frac{1}{21}(t-1)(t-3)(t-7), \end{aligned}$$

for $t \in \mathbb{T}_1$. The reader could check this result by just twice integrating both sides of the equation $D^2 y(t) = t - 1$ from 1 to $t$.

## 5.7 Exponential Function

In this section we define the exponential function on a mixed time scale and give several of its properties. First we define the set of regressive functions by

$$\mathcal{R} = \{p : \mathbb{T}_\alpha \to \mathbb{C} \text{ such that } 1 + p(t)\mu(t) \neq 0 \text{ for } t \in \mathbb{T}_\alpha\}.$$

**Definition 5.40.** The mixed time scale exponential function based at $s \in \mathbb{T}_\alpha$, denoted by $e_p(t, s)$, where $p \in \mathcal{R}$ is defined to be the unique solution, $y(t)$, of the initial value problem

$$Dy(t) = p(t)y(t), \tag{5.1}$$

$$y(s) = 1. \tag{5.2}$$

In the next theorem we give a formula for $e_p(t, s)$.

**Theorem 5.41.** *Assume $p \in \mathcal{R}$ and $s \in \mathbb{T}_\alpha$. Then*

$$e_p(t,s) = \begin{cases} \prod_{j=0}^{K(t,s)-1} \left[1 + p(\sigma^j(s))\mu(\sigma^j(s))\right], & if\, t > s \\ 1, & if\, t = s \\ \prod_{j=1}^{-K(t,s)} \frac{1}{[1 + p(\rho^j(s))\mu(\rho^j(s))]}, & if\, t < s. \end{cases}$$

*Proof.* It suffices to show (see Remark 5.18) that $y(t) = e_p(t, s)$ as defined in the statement of this theorem satisfies the IVP

$$Dy(t) = p(t)y(t), \quad t \in \mathbb{T}_\alpha$$

$$y(s) = 1.$$

It is clear that $e_p(s, s) = 1$. It remains to show that $De_p(t, s) = p(t)e_p(t, s)$. Consider the case that $t = \sigma^k(s)$ for $k \geq 1$. Then

$$De_p(t,s) = \frac{1}{\mu(t)} \prod_{j=0}^{N(\sigma(t),s)-1} \left[1 + p\left(\sigma^j(s)\right)\mu\left(\sigma^j(s)\right)\right]$$

$$- \frac{1}{\mu(t)} \prod_{j=0}^{K(t,s)-1} \left[1 + p\left(\sigma^j(s)\right)\mu\left(\sigma^j(s)\right)\right]$$

$$= \frac{p(\sigma^{N(t,s)}(s))\mu\left(\sigma^{N(t,s)}(s)\right)\prod_{j=0}^{N(t,s)-1}\left[1+p\left(\sigma^j(s)\right)\mu\left(\sigma^j(s)\right)\right]}{\mu(t)}$$

$$= p(t)e_p(t,s).$$

Consider the case when $t = s$. Then,

$$De_p(s,s) = \frac{[1+p(s)\mu(s)]-1}{\mu(s)} = p(s) = p(s)e_p(s,s).$$

Consider the case when $t = \rho(s)$. In this case it follows that

$$\begin{aligned} De_p(\rho(s),s) &= \frac{1 - \dfrac{1}{1+p\left(\rho(s)\right)\mu\left(\rho(s)\right)}}{\mu\left(\rho(s)\right)} \\ &= \frac{p\left(\rho(s)\right)}{1+p\left(\rho(s)\right)\mu\left(\rho(s)\right)} \\ &= p\left(\rho(s)\right)e_p(\rho(s),s). \end{aligned}$$

Finally, consider the case when $t = \rho^k(s)$ for $k \geq 2$. In this final case it then holds that

$$\begin{aligned} De_p(t,s) &= \frac{1}{\mu(t)}\prod_{j=1}^{-N(\sigma(t),s)}\frac{1}{1+p\left(\rho^{(j)}(s)\right)\mu\left(\rho^{(j)}(s)\right)} \\ &\quad - \frac{1}{\mu(t)}\prod_{j=1}^{-N(t,s)}\frac{1}{1+p\left(\rho^{(j)}(s)\right)\mu\left(\rho^{(j)}(s)\right)} \\ &= \frac{1}{\mu(t)}\left[1-\frac{1}{1+p\left(\rho^{(-N(t,s))}(s)\right)\mu\left(\rho^{(-N(t,s))}(s)\right)}\right] \\ &\quad \times \prod_{j=1}^{-N(\sigma(t),s)}\frac{1}{1+p\left(\rho^{(k)}(s)\right)\mu\left(\rho^{(k)}(s)\right)} \\ &= \frac{p(t)}{1+p\mu(t)}\prod_{j=1}^{-N(\sigma(t),s)}\frac{1}{1+p\left(\rho^{(k)}(s)\right)\mu\left(\rho^{(k)}(s)\right)} \\ &= p(t)e_p(t,s), \end{aligned}$$

which completes the proof. □

Next we define an addition on $\mathcal{R}$.

**Definition 5.42.** We define the circle plus addition, $\oplus$, on the set of regressive functions $\mathcal{R}$ on the mixed time scale $\mathbb{T}_\alpha$ by

$$(p \oplus r)(t) := p(t) + r(t) + \mu(t)p(t)r(t), \quad t \in \mathbb{T}_\alpha.$$

Similar to the proof of Theorem 4.16, we can prove the following theorem.

**Theorem 5.43.** *The set of regressive functions $\mathcal{R}$ with the addition $\oplus$ is an Abelian group.*

Like in Chap. 4, the additive inverse of a function $p \in \mathcal{R}$ is given by

$$\ominus p := \frac{-p}{1+\mu p}.$$

We then define the circle minus subtraction, $\ominus$, on $\mathcal{R}$ by

$$p \ominus r := p \oplus (\ominus r).$$

It follows that

$$p \ominus r = \frac{p-r}{1+\mu r}.$$

In the following theorem we give several properties of the exponential function $e_p(t,s)$.

**Theorem 5.44.** *Let $t, s, r \in \mathbb{T}_\alpha$ and $p, l \in \mathcal{R}$. Then the following properties hold:*

(i) $e_0(t,s) = 1$;
(ii) $e_p(s,s) = 1$;
(iii) $De_p(t,s) = p(t)e_p(t,s)$;
(iv) $e_p(\sigma(t),s) = [1 + p(t)\mu(t)]\, e_p(t,s)$;
(v) $e_p(\rho(t),s) = \dfrac{e_p(t,s)}{[1 + p(\rho(t))\, \mu(\rho(t))]}$;
(vi) *if* $1 + p(t)\mu(t) > 0$ *for all* $t \in \mathbb{T}_\alpha$ *then* $e_p(t,s) > 0$;
(vii) $e_p(t,s)e_p(s,r) = e_p(t,r)$;
(viii) $e_p(s,t) = \dfrac{1}{e_p(t,s)} = e_{\ominus p}(t,s)$;
(ix) $e_p(t,s)e_l(t,s) = e_{p\oplus l}(t,s)$;
(x) $\dfrac{e_l(t,s)}{e_p(t,s)} = e_{l\ominus p}(t,s)$.

*Proof.* The proof of this theorem is very similar to the proof of Theorem 4.18. Here we will just prove first half of part (viii). Consider the case $t > s$. Then

$$e_p(s,t) = \prod_{j=1}^{N(t,s)} \frac{1}{1+p\left(\rho^j(t)\right)\mu\left(\rho^j(t)\right)}$$

$$= \prod_{j=1}^{N(t,s)} \frac{1}{1+p\left(\rho^j(\sigma^{N(t,s)}(s)\right)\mu\left(\rho^j(\sigma^{N(t,s)}(s)\right)}$$

$$= \prod_{j=1}^{N(t,s)} \frac{1}{1+p\left(\sigma^{N(t,s)-j}(s)\right)\mu\left(\sigma^{N(t,s)-j}(s)\right)}$$

$$= \prod_{j=0}^{N(t,s)-1} \frac{1}{1+p\left(\sigma^j(s)\right)\mu\left(\sigma^j(s)\right)} = \frac{1}{e_p(t,s)}.$$

When $t = s$, it follows that $e_p(s,s) = \frac{1}{e_p(s,s)} = 1$. Finally, consider the case when $t < s$. Then

$$e_p(s,t) = \prod_{j=0}^{N(s,t)-1} \left[1+p\left(\sigma^j(t)\right)\mu\left(\sigma^j(t)\right)\right]$$

$$= \prod_{j=0}^{N(s,t)-1} \left[1+p\left(\rho^{N(s,t)-j}(s)\right)\mu(\rho^{N(s,t)-j}(s))\right]$$

$$= \prod_{j=1}^{N(s,t)} \left[1+p\left(\rho^j(s)\right)\mu\left(\rho^j(s)\right)\right] = \frac{1}{e_p(t,s)},$$

which was to be shown. □

Next we define the scalar dot multiplication, $\odot$, on the set of positively regressive functions $\mathcal{R}^+ := \{p \in \mathcal{R} : 1 + \mu(t)p(t) > 0, t \in \mathbb{T}_\alpha\}$.

**Definition 5.45.** We define the scalar dot multiplication, $\odot$, on $\mathcal{R}+$ by

$$(\alpha \odot p)(t) := \frac{[1+\mu(t)p(t)]^\alpha - 1}{\mu(t)}, \quad t \in \mathbb{T}_\alpha.$$

Similar to the proof of Theorem 4.21 we can prove the following theorem.

**Theorem 5.46.** *If $\alpha \in \mathbb{R}$ and $p \in \mathcal{R}^+$, then*

$$e_p^\alpha(t,a) = e_{\alpha \odot p}(t,a)$$

*for $t \in \mathbb{T}_\alpha$.*

Then similar to the proof of Theorem 4.22, we get the following result.

**Theorem 5.47.** *The set of positively regressive functions $\mathcal{R}^+$ on a mixed time scale with the addition $\oplus$ and the scalar multiplication $\odot$ is a vector space.*

## 5.8 Trigonometric Functions

In this section, we use the exponential function defined in the previous section to define the hyperbolic and trigonometric functions for the mixed time scale.

**Definition 5.48.** For $\pm p \in \mathcal{R}$, we define the mixed time scale hyperbolic cosine function $\cosh_p(\cdot, s)$ based at $s \in \mathbb{T}_\alpha$ by

$$\cosh_p(t,s) := \frac{e_p(t,s) + e_{-p}(t,s)}{2}, \quad t \in \mathbb{T}_\alpha.$$

**Definition 5.49.** Likewise we define the mixed time scale hyperbolic sine function $\sinh_p(\cdot, s)$ based at $s \in \mathbb{T}_\alpha$ by

$$\sinh_p(t,s) := \frac{e_p(t,s) - e_{-p}(t,s)}{2}, \quad t \in \mathbb{T}_\alpha.$$

Similar to the proof of Theorem 4.24 one can prove the following theorem.

**Theorem 5.50.** *Assume $\pm p \in \mathcal{R}$ and $t, s \in \mathbb{T}_\alpha$. Then the following properties hold:*

(i) $\cosh_p(s,s) = 1$;
(ii) $\sinh_p(s,s) = 0$;
(iii) $\cosh_{-p}(t,s) = \cosh_p(t;s)$;
(iv) $\sinh_{-p}(t,s) = -\sinh_p(t,s)$;
(v) $D\cosh_p(t,s) = p(t)\sinh_p(t,s)$;
(vi) $D\sinh_p(t,s) = p(t)\cosh_p(t,s)$;
(vii) $\cosh_p^2(t,s) - \sinh_p^2(t,s) = e_{-\mu p^2}(t,s)$.

**Definition 5.51.** Assume $\pm ip \in \mathcal{R}$. Then we define the mixed time scale cosine function $\cos_p(\cdot, s)$ based at $s \in \mathbb{T}_\alpha$ by

$$\cos_p(t,s) := \frac{e_{ip}(t,s) + e_{-ip}(t,s)}{2}, \quad t \in \mathbb{T}_\alpha.$$

**Definition 5.52.** We define the mixed time scale sine function $\sin_p(\cdot, s)$ based at $s \in \mathbb{T}_\alpha$ by

$$\sin_p(t,s) := \frac{e_{ip}(t,s) - e_{-ip}(t,s)}{2i}, \quad t \in \mathbb{T}_\alpha.$$

Similar to the proof of Theorem 4.27 one can prove the following theorem.

**Theorem 5.53.** *Assume* $\pm ip \in \mathcal{R}$ *and* $t, s \in \mathbb{T}_\alpha$. *Then the following properties hold:*

(i) $\cos_p(s, s) = 1$;
(ii) $\sin_p(s, s) = 0$;
(iii) $\cos_{-p}(t, s) = \cos_p(t, s)$;
(iv) $\sin_{-p}(t, s) = -\sin_p(t, s)$;
(v) $D\cos_p(t, s) = -p(t)\sin_p(t, s)$;
(vi) $D\sin_p(t, s) = p(t)\cos_p(t, s)$;
(vii) $\cos_p^2(t, s) + \sin_p^2(t, s) = e_{\mu p^2}(t, s)$.

Similar to the proof of Theorem 4.26 one can prove the following theorem.

**Theorem 5.54.** *Assume* $\pm ip \in \mathcal{R}$ *and* $t, s \in \mathbb{T}_\alpha$. *Then the following properties hold:*

(i) $\sin_{ip}(t, s) = i\sinh_p(t, s)$;
(ii) $\cos_{ip}(t, s) = \cosh_p(t, s)$;
(iii) $\sinh_{ip}(t, s) = i\sin_p(t, s)$;
(iv) $\cosh_{ip}(t, s) = \cos_p(t, s)$,

*for* $t \in \mathbb{T}_\alpha$.

It is easy to prove the following theorem.

**Theorem 5.55.** *If* $p \in \mathcal{R}$, *then a general solution of*

$$Dy(t) = p(t)y(t), \quad t \in \mathbb{T}_\alpha$$

*is given by*

$$y(t) = ce_p(t, a), \quad t \in \mathbb{T}_\alpha.$$

**Theorem 5.56.** *Assume* $t, s \in \mathbb{T}_\alpha$ *and* $p$ *is a constant. Then the following Taylor series converge on* $\mathbb{T}_{[s,\infty)}$.

(i) $e_p(t, s) = \sum_{n=0}^{\infty} p^n h_n(t, s)$, *if* $p \in \mathcal{R}$;

(ii) $\sin_p(t, s) = \sum_{n=0}^{\infty} (-1)^n p^{2n+1} h_{2n+1}(t, s)$, *if* $\pm ip \in \mathcal{R}$;

(iii) $\cos_p(t, s) = \sum_{n=0}^{\infty} (-1)^n p^{2n} h_{2n}(t, s)$, *if* $\pm ip \in \mathcal{R}$;

(iv) $\sinh_p(t, s) = \sum_{n=0}^{\infty} p^{2n+1} h_{2n+1}(t, s)$, *if* $\pm p \in \mathcal{R}$;

(v) $\cosh_p(t, s) = \sum_{n=0}^{\infty} p^{2n} h_{2n}(t, s)$, *if* $\pm p \in \mathcal{R}$.

*Proof.* Fix $t \in \mathbb{T}_{[s,\infty)}$. Then for some $k \geq 0$, it holds that $t = \sigma^k(s) \geq s$. Let $M = \max\{|e_p(\tau, s)| : \tau \in \mathbb{T}_{[s,t]}\}$. Then for $n \geq 1$ we have

$$\begin{aligned}|R_n(t,s)| &= \left|\int_s^t h_n(t,\sigma(\tau))D^{n+1}e_p(\tau,s)D\tau\right| \\ &= \left|\int_s^t h_n(t,\sigma(\tau))p^{n+1}e_p(\tau,s)D\tau\right| \\ &\leq M|p|^{n+1}\left|\int_s^t h_n(t,\sigma(\tau))D\tau\right| \\ &= M|p|^{n+1}\left|h_{n+1}(t,s)\right|. \end{aligned} \tag{5.3}$$

Now if $m \geq k$, we have that

$$|R_m(t,s)| \leq M|p|^{m+1}\left|h_{m+1}(t,s)\right| = M|p|^{m+1}\prod_{i=1}^{m+1}\frac{t-\sigma^{i-1}(s)}{[i]_a}.$$

Note that since $m \geq k$, the product in the above expression contains the factor

$$\frac{t-\sigma^k(s)}{[k+1]_a} = \frac{t-t}{[k+1]_a} = 0.$$

Thus, for all $m \geq k$,

$$R_m(t,s) = 0.$$

Hence, by Taylor's Formula (Theorem 5.37) the Taylor series for $e_p(t,s)$ converges for any $t \in \mathbb{T}_{[s,\infty)}$. The remainder of this theorem follows from the fact that the functions $\cos_p(t,s)$ $\sin_p(t,s)$, $\cosh_p(t,s)$, and $\sinh_p(t,s)$ are defined in terms of appropriate exponential functions. □

**Theorem 5.57.** *Fix $s \in \mathbb{T}_\alpha$. Then the Taylor series for each of the functions in Theorem 5.56 converges on $\mathbb{T}_{(-\infty,s)}$ when $|p| < \frac{a}{\mu(s)}$.*

*Proof.* Let $s \in \mathbb{T}_\alpha$ be fixed. We will first prove that if $|p| < \frac{\mu(s)}{a}$, then the Taylor series for $e_p(t,s)$ converges for each $t \in \mathbb{T}_{(-\infty,s)}$. Fix $t \in \mathbb{T}_{(-\infty,s)}$. We claim that for each $a \geq 1$

$$\lim_{n\to\infty}\frac{\sigma^{n-1}(s)-t}{[n]_a} = \frac{\mu(s)}{a}. \tag{5.4}$$

First we prove (5.4) for $a = 1$. This follows from the following calculations:

$$\lim_{n\to\infty}\frac{\sigma^{n-1}(s)-t}{[n]_a} = \lim_{n\to\infty}\frac{a^{n-1}s+[n-1]_a b-t}{[n]_a}$$

$$= \lim_{n\to\infty} \frac{s + [n-1]_1 b - t}{[n]_1}$$
$$= \lim_{n\to\infty} \frac{s - (n-1)b - t}{n} = \frac{b}{1} = \frac{\mu(s)}{a}.$$

Next we prove (5.4) for $a > 1$. To this end consider

$$\lim_{n\to\infty} \frac{\sigma^{n-1}(s) - t}{[n]_a} = \lim_{n\to\infty} \frac{a^{n-1}s + [n-1]_a b - t}{[n]_a} \quad \text{by Theorem 5.5, (i)}$$
$$= \lim_{n\to\infty} \frac{a^{n-1}s + \frac{a^{n-1}-1}{a-1}b - t}{\frac{a^n-1}{a-1}} \quad \text{by the definition of } [n]_a$$
$$= \lim_{n\to\infty} \frac{(a-1)a^{n-1}s + (a^{n-1}-1)b - (a-1)t}{a^n - 1}$$
$$= \frac{(a-1)s + b}{a}$$
$$= \frac{\mu(s)}{a}.$$

Now consider the remainder term

$$R_n(t,s) = \int_s^t h_n(t,\sigma(\tau))D^{n+1}e_p(\tau,s)D\tau = \int_s^t h_n(t,\sigma(\tau))p^{n+1}e_p(\tau,s)D\tau.$$

It follows that

$$|R_n(t,s)| \le |p|^{n+1} \int_t^s |h_n(t,\sigma(\tau))||e_p(\tau,s)|D\tau.$$

If we let

$$M := \max\{|e_p(t,s)| : t \le \tau \le s-1\},$$

then

$$|R_n(t,s)| \le M|p|^{n+1} \int_t^s |h_n(t,\sigma(\tau))|D\tau$$
$$= M|p|^{n+1} \int_t^s \left| \prod_{k=1}^n \frac{t - \sigma^{k-1}(\tau)}{[k]_a} \right| D\tau$$
$$= M|p|^{n+1} \int_t^s (-1)^n h_n(t,\sigma(\tau))D\tau$$
$$= M|p|^{n+1} \left[(-1)^{n+1} h_{n+1}(t,\tau)\right]_{\tau=t}^{\tau=s}$$

$$= M|p|^{n+1}(-1)^{n+1}h_{n+1}(t,s)$$
$$= M|p|^{n+1}\prod_{k=1}^{n+1}\frac{\sigma^{k-1}(s)-t}{[k]_a}.$$

Using (5.4) and $|p| < \frac{a}{\mu(s)}$, there is a number $r$ and a positive integer $N$ so that

$$0 \le |p|\frac{\sigma^{n-1}(s)-t}{[n]_a} \le r < 1, \quad \text{for} \quad n \ge N.$$

It follows that

$$\lim_{n\to\infty}|R_n(t,s)| = \lim_{n\to\infty} M\prod_{k=1}^{\infty}|p|\frac{\sigma^{k-1}(s)-t}{[k]_a} = 0.$$

Therefore, by Taylor's Formula,

$$e_p(t,s) = \sum_{n=0}^{\infty} p^n h_n(t,s)$$

for $t \in \mathbb{T}_{(-\infty,s)}$.

The remainder of this theorem follows from the fact that the functions $\cos_p(t,s)$ $\sin_p(t,s)$, $\cosh_p(t,s)$, and $\sinh_p(t,s)$ are defined in terms of appropriate exponential functions. □

**Theorem 5.58.** *For fixed $t, s \in \mathbb{T}_\alpha$, the power series*

$$f(x) = \sum_{n=0}^{\infty} h_n(t,s)x^n$$

*converges for* $|x| < \dfrac{a}{\mu(s)}$.

*Proof.* First, consider the power series

$$A(x) = \sum_{n=0}^{\infty}\frac{t^{\underline{n}}}{\{n\}_a!}x^n.$$

We will perform the ratio test with this series:

$$\lim_{n\to\infty}\left|\frac{a_{n+1}}{a_n}\right| = \lim_{n\to\infty}\left|\frac{t^{\underline{n+1}}}{\{n+1\}_a t^{\underline{n}}}\right|$$

$$= \lim_{n\to\infty}\left|\frac{\rho^n(t)}{\{n+1\}_a}\right|$$

$$= \lim_{n\to\infty}\left|\frac{\dfrac{a^{-n}t+[-n]_a b}{[n+1]_a}}{a^n}\right| = \lim_{n\to\infty}\left|\frac{t}{[n+1]_a} - \frac{[n]_a b}{[n+1]_a}\right|.$$

Since $\lim_{n\to\infty}\frac{t}{[n+1]_a} = 0$, we have that

$$\lim_{n\to\infty}\left|\frac{a_{n+1}}{a_n}\right| = \lim_{n\to\infty}\left|\frac{s^{\underline{n+1}}}{\{n+1\}_a t^{\underline{n}}}\right| = \lim_{n\to\infty}\left|\frac{[n]_a b}{[n+1]_a}\right|$$

$$= \lim_{n\to\infty}\left|\frac{(a^n-1)b}{(a^{n+1}-1)}\right| = \frac{b}{a}.$$

So, $A(x)$ converges when $|x| < \frac{a}{b}$. Next, consider the power series

$$B(x) = \sum_{n=0}^{\infty}\frac{(-1)^n s^{\overline{n}}}{[n]_a!}x^n.$$

Again, we perform the ratio test. Then

$$\lim_{n\to\infty}\left|\frac{a_{n+1}}{a_n}\right| = \lim_{n\to\infty}\left|\frac{\sigma^n(s)}{[n+1]_a}\right| = \lim_{n\to\infty}\frac{a^n s+[n]_a b}{[n+1]_a}$$

$$= \lim_{n\to\infty}\left(\frac{a^n s}{[n+1]_a} + \frac{[n]_a b}{[n+1]_a}\right)$$

$$= \lim_{n\to\infty}\left(\frac{a^n(a-1)s}{a^n-1} + \frac{[n]_a b}{[n+1]_a}\right)$$

$$= \lim_{n\to\infty}\left(\frac{a^n(a-1)s}{a^{n+1}-1} + \frac{[n]_a b}{[n+1]_a}\right)$$

$$= \frac{(a-1)s}{a} + \frac{b}{a} = \frac{\mu(s)}{a}.$$

So $B(x)$ converges when $|x| < \frac{a}{\mu(s)}$. Note that $\mu(s) > b$, so $\frac{a}{\mu(s)} < \frac{a}{b}$ for all $s$.

Now, $f(x) = A(x)B(x)$. So, $f(x)$ converges when $|x| < \frac{a}{\mu(s)}$. □

## 5.9 The Laplace Transform

Most of the results in this section are due to Auch et al. [39]. In this chapter when discussing Laplace transforms we assume that $r \in \mathbb{T}_\alpha$ satisfies $r \geq \alpha \geq 0$, and we let $\mathbb{T}_r = \{t \geq r : t \in \mathbb{T}_\alpha\}$. Also we let $\mathcal{R}^c$ denote the set of regressive complex constants.

**Definition 5.59.** If $f : \mathbb{T}_r \to \mathbb{R}$, then we define the **discrete Laplace transform** of $f$ based at $r \in \mathbb{T}$ by

$$\mathcal{L}_r\{f\}(s) := \int_r^\infty e_{\ominus s}(\sigma(t), r)f(t)\, Dt,$$

where $\mathcal{L}_r\{f\} : \mathcal{R}^c \to \mathbb{C}$.

**Definition 5.60.** We say that a function $f : \mathbb{T}_\alpha \to \mathbb{R}$ is of exponential order $k > 0$ if for every fixed $r \in \mathbb{T}_\alpha$, there exists a constant $M > 0$ such that

$$|f(t)| \leq Me_k(t, r),$$

for all sufficiently large $t \in \mathbb{T}_r$.

**Theorem 5.61.** *Suppose $f : \mathbb{T}_r \to \mathbb{R}$ is of exponential order $k > 0$. Then $\mathcal{L}_r\{f\}(s)$ exists for $|s| > k$.*

*Proof.* Since $f$ is of exponential order $k > 0$, there is a constant $M > 0$ and a $T \in \mathbb{T}_r$ such that $|f(t)| \leq Me_k(t, r)$ for $t \geq T$. Pick $N \in \mathbb{N}_0$ such that $t = \sigma^N(r)$. Then we have

$$\begin{aligned}
\left| \int_T^\infty e_{\ominus s}(\sigma(t), r)f(t)\, Dt \right| &\leq \int_T^\infty |e_{\ominus s}(\sigma(t), r)f(t)\,|\, Dt \\
&\leq M \int_T^\infty |e_{\ominus s}(\sigma(t), r)e_k(t, r)\,|\, Dt \\
&= M \int_T^\infty \left| \frac{1}{1 + s\mu(t)} e_{k\ominus s}(t, r) \right| Dt \\
&= M \sum_{i=N}^\infty \left| \frac{\mu(\sigma^i(r))}{1 + s\mu(\sigma^i(r))} e_{k\ominus s}(\sigma^i(r), r) \right| \\
&= M \sum_{i=N}^\infty \left| \frac{a^i\mu(r)}{1 + sa^i\mu(r)} e_{k\ominus s}(\sigma^i(r), r) \right|.
\end{aligned}$$

We will show that this sum converges absolutely for $|s| > k$ by the ratio test. We have

$$\begin{aligned}
&\lim_{i\to\infty}\left|\left(\frac{a^{i+1}\mu(r)e_{k\ominus s}(\sigma^{i+1}(r),r)}{1+sa^{i+1}\mu(r)}\right)\left(\frac{1+sa^{i}\mu(r)}{a^{i}\mu(r)e_{k\ominus s}(\sigma^{i}(r),r)}\right)\right|\\
&=\lim_{i\to\infty}\left|\left(\frac{ae_{k\ominus s}(\sigma^{i+1}(r),r)}{1+a^{i+1}s\mu(r)}\right)\left(\frac{1+sa^{i}\mu(r)}{e_{k\ominus s}(\sigma^{i}(r),r)}\right)\right|\\
&=\lim_{i\to\infty}\left|\left(\frac{ae_{k\ominus s}(\sigma^{i}(r),r)(1+ka^{i}\mu(r))}{(1+a^{i+1}s\mu(r))(1+sa^{i}\mu(r))}\right)\left(\frac{1+sa^{i}\mu(r)}{e_{k\ominus s}(\sigma^{i}(r),r)}\right)\right|\\
&=\lim_{i\to\infty}\left|\frac{a+ka^{i+1}\mu(r)}{1+sa^{i+1}\mu(r)}\right|\\
&=\lim_{i\to\infty}\left|\frac{\frac{1}{a^{i}}+k\mu(r)}{\frac{1}{a^{i+1}}+s\mu(r)}\right|=\frac{k}{|s|}.
\end{aligned}$$

Hence the sum converges absolutely when $|s| > k$, and therefore $\mathcal{L}_r\{f\}(s)$ converges if $|s| > k$. □

**Theorem 5.62 (Linearity).** *Suppose $f, g : \mathbb{T}_r \to \mathbb{R}$ are of exponential order $k > 0$, and $c, d \in \mathbb{R}$. Then*

$$\mathcal{L}_r\{cf + dg\}(s) = c\mathcal{L}_r\{f\}(s) + d\mathcal{L}_r\{g\}(s),$$

*for $|s| > k$.*

*Proof.* The result follows easily from the linearity of the delta integral. We have, for $|s| > k$, that

$$\begin{aligned}
\mathcal{L}_r\{cf + dg\}(s) &= \int_r^\infty (cf(t) + dg(t))e_{\ominus s}(\sigma(t), r)\,Dt\\
&= c\int_r^\infty f(t)e_{\ominus s}(\sigma(t), r)\,Dt + d\int_r^\infty g(t)e_{\ominus s}(\sigma(t), r)\,Dt\\
&= c\mathcal{L}_r\{f\}(s) + d\mathcal{L}_r\{g\}(s),
\end{aligned}$$

which completes the proof. □

The following lemma will be useful for computing Laplace transforms of various functions.

**Lemma 5.63.** *If $p, q \in \mathcal{R}^c$ and $|p| < |q|$, then for $t \in \mathbb{T}_r$ we have*

$$\lim_{t\to\infty} e_{p\ominus q}(t, r) = 0.$$

*Proof.* Let $p, q \in \mathcal{R}^c$ with $|p| < |q|$. First, note that

$$\left|\lim_{t\to\infty} \frac{1+p\mu(t)}{1+q\mu(t)}\right| = \left|\lim_{t\to\infty} \frac{\frac{1}{\mu(t)}+p}{\frac{1}{\mu(t)}+q}\right|$$
$$= \left|\frac{p}{q}\right| < 1, \quad \text{since} \quad |p| < |q|.$$

This implies that

$$\lim_{t\to\infty} \left|e_{p\ominus q}(t,r)\right| = \lim_{t\to\infty} \left|\frac{e_p(t,r)}{e_q(t,r)}\right| = \prod_{i=0}^{\infty} \left|\frac{1+p\mu(\sigma^i(r))}{1+q\mu(\sigma^i(r))}\right| = 0.$$

Thus, $\lim_{t\to\infty} e_{p\ominus q}(t,r) = 0$. □

*Remark 5.64.* In particular, note that if $s > 0$, then

$$\lim_{t\to\infty} e_{\ominus s}(t,r) = 0.$$

**Theorem 5.65.** *Let $p \in \mathcal{R}^c$. Then for $|s| > |p|$, we have*

$$\mathcal{L}_r\{e_p(t,r)\}(s) = \frac{1}{s-p}.$$

*Proof.* First, note that $e_p(t,r)$ is of exponential order $|p|$ since

$$|e_p(t,r)| = \prod_{i=0}^{K(t,r)-1} |1+p\mu(\sigma^i(r))|$$
$$\leq \prod_{i=0}^{K(t,r)-1} 1+|p|\,\mu(\sigma^i(r)) = e_{|p|}(t,r).$$

Thus, if $|s| > |p|$, we have

$$\mathcal{L}_r\{e_p(t,r)\}(s) = \int_r^\infty e_{\ominus s}(\sigma(t),r)e_p(t,r)Dt$$
$$= \int_r^\infty e_{p\ominus s}(t,r)\frac{1}{1+s\mu(t)}Dt$$
$$= \frac{1}{p-s}\int_r^\infty e_{p\ominus s}(t,r)\frac{p-s}{1+s\mu(t)}Dt.$$

Then note both that $De_{p\ominus s}(t,r) = (p\ominus s)(t)\,e_{p\ominus s}(t,r)$ and that $(p\ominus s)(t) = \frac{p-s}{1+s\mu(t)}$. This gives us

$$\begin{aligned}\mathcal{L}_r\{e_p(t,r)\}(s) &= \frac{1}{p-s}\int_r^\infty De_{p\ominus s}(t,r)Dt\\ &= \frac{1}{p-s}\left[\lim_{t\to\infty} e_{p\ominus s}(t,r) - e_{p\ominus s}(r,r)\right]\\ &= \frac{1}{s-p},\end{aligned}$$

since $\lim_{t\to\infty} e_{p\ominus s}(t,r) = 0$ by Lemma 5.63. □

The following results describe the relationship between the Laplace transform and the delta difference.

**Lemma 5.66.** *If $f : \mathbb{T}_r \to \mathbb{R}$ is of exponential order $k > 0$, then $Df$ is also of exponential order $k > 0$.*

*Proof.* Let $|f(t)| \leq Me_k(t,r)$ for sufficiently large $t$. We will prove this lemma by showing that

$$\lim_{t\to\infty}\left|\frac{Df(t)}{e_k(t,r)}\right| \leq M < \infty.$$

First, consider

$$\begin{aligned}\left|\frac{Df(t)}{e_k(t,r)}\right| &= \left|\frac{f(\sigma(t)) - f(t)}{\mu(t)e_k(t,r)}\right|\\ &\leq \frac{|f(\sigma(t))| + |f(t)|}{|\mu(t)e_k(t,r)|}\\ &\leq \frac{Me_k(\sigma(t),r) + Me_k(t,r)}{|\mu(t)e_k(t,r)|}\\ &\leq \frac{Me_k(t,r)(1+k\mu(t)) + Me_k(t,r)}{\mu(t)e_k(t,r)}\\ &\leq \frac{Me_k(t,r)(2+k\mu(t))}{\mu(t)e_k(t,r)}\\ &\leq \frac{2M}{\mu(t)} + Mk.\end{aligned}$$

Thus, we have

$$\lim_{t\to\infty}\left|\frac{Df(t)}{e_k(t,r)}\right| \leq Mk.$$

Then, for any $\epsilon > 0$ and $t$ sufficiently large,

$$|Df(t)| \leq (Mk + \epsilon)e_k(t,r).$$

Therefore, $Df$ is of exponential order $k > 0$. □

**Corollary 5.67.** *If $f : \mathbb{T}_r \to \mathbb{R}$ is of exponential order $k > 0$, then $D^n f$ is also of exponential order $k > 0$ for every $n \in \mathbb{N}$.*

To see that the corollary holds, use the previous result.

**Theorem 5.68.** *If $f : \mathbb{T}_r \to \mathbb{R}$ is of exponential order $k > 0$ and $n \in \mathbb{N}$, then*

$$\mathcal{L}_r\{D^n f(t)\}(s) = s^n \mathcal{L}_r\{f\}(s) - \sum_{i=0}^{n-1} s^{n-1-i} D^i f(r),$$

*for $|s| > k$.*

*Proof.* We will proceed by induction on $n$. First consider the base case $n = 1$. Using integration by parts we have

$$\begin{aligned}
\mathcal{L}_r\{Df(t)\}(s) &= \int_r^\infty e_{\ominus s}(\sigma(t), r) Df(t)\, Dt \\
&= e_{\ominus s}(t, r) f(t)\Big|_{t=r}^{t\to\infty} + \int_r^\infty \frac{s}{1 + s\mu(t)} e_{\ominus s}(t, r) f(t)\, Dt \\
&= s\int_r^\infty e_{\ominus s}(\sigma(t), r) f(t)\, Dt - \frac{f(r)}{e_s(r, r)} \\
&= s\mathcal{L}_r\{f(t)\}(s) - f(r).
\end{aligned}$$

Now assume the statement is true for some $n > 1$. Then by the base case we have

$$\begin{aligned}
\mathcal{L}_r\{D^{n+1} f(t)\}(s) &= s\mathcal{L}_r\{D^n f(t)\}(s) - D^n f(r) \\
&= s\left[s^n \mathcal{L}_r\{f\}(s) - \sum_{i=0}^{n-1} s^{n-1-i} D^i f(r)\right] - D^n f(r) \\
&= s^{n+1} \mathcal{L}_r\{f\}(s) - \sum_{i=0}^{n-1} s^{n-i} D^i f(r) - D^n f(r) \\
&= s^{n+1} \mathcal{L}_r\{f\}(s) - \sum_{i=0}^{n} s^{n-i} D^i f(r),
\end{aligned}$$

as desired. □

To show that the Laplace transform is injective (Theorem 5.70) and therefore invertible we will use the following lemma.

**Lemma 5.69.** *Let $f : \mathbb{T}_r \to \mathbb{R}$ be of exponential order $k > 0$. Then*

$$\lim_{s\to\infty} \mathcal{L}_r\{f\}(s) = 0.$$

*Proof.* Let $t_n = \sigma^n(r)$. Since $f$ is of exponential order $k$, there is a constant $M > 0$ and a positive integer $N$ such that $|f(t_n)| \leq M e_k(t_n, r)$ for all $n \geq N$. Then we have

$$\lim_{s\to\infty} |\mathcal{L}_r\{f\}(s)| \leq \lim_{s\to\infty} \sum_{n=0}^{\infty} |f(t_n)\, e_{\ominus s}(t_{n+1}, r)|\; \mu(t_n)$$

$$= \lim_{s\to\infty} \sum_{n=0}^{N-1} |f(t_n)||e_{\ominus s}(t_{n+1}, r)|\mu(t_n)$$

$$+ \lim_{s\to\infty} \sum_{n=N}^{\infty} |e_{\ominus s}(t_{n+1}, r)||f(t_n)|\mu(t_n)$$

$$\leq M \lim_{s\to\infty} \sum_{n=N}^{\infty} e_k(t_n, r)|e_{\ominus s}(t_{n+1}, r)|\mu(t_n)$$

$$\leq M \lim_{s\to\infty} \sum_{n=N}^{\infty} e_k(t_n, r)|[1 + \ominus s(t_n)\mu(t_n)]e_{\ominus s}(t_n, r)|\mu(t_n)$$

$$= M \lim_{s\to\infty} \sum_{n=N}^{\infty} \left|\frac{e_k(t_n, r)}{[1 + s\mu(t_n)]e_s(t_n, r)}\right|\mu(t_n)$$

$$\leq M \lim_{s\to\infty} \sum_{n=N}^{\infty} \left|\frac{e_k(t_n, r)}{e_s(t_n, r)}\right|\mu(t_n).$$

We now show that

$$\left|\frac{1 + k\mu(r)}{1 + s\mu(r)}\right| \geq \left|\frac{1 + k\mu(t_m)}{1 + s\mu(t_m)}\right|$$

for $\Re(s), \Im(s) > k$ and for all integers $m \geq 0$. To this end, we prove that

$$|(1 + k\mu(r))(1 + s\mu(t_m))| \geq |(1 + k\mu(t_m))(1 + s\mu(r))|$$

holds by writing $s = \Re(s) + \mathrm{i}\Im(s)$ and showing

$$(1 + k\mu(r))(1 + \Re(s)\mu(t_m)) \geq (1 + k\mu(t_m))(1 + \Re(s)\mu(r))$$

and

$$\Im(s)\mu(t_m) + k\Im(s)\mu(r)\mu(t_m) \geq \Im(s)\mu(r) + k\Im(s)\mu(t_m)\mu(r).$$

Rewriting these inequalities yields

$$(\Re(s) - k)\mu(t_m) \geq (\Re(s) - k)\mu(r)$$

and

$$\Im(s)\mu(t_m) \geq \Im(s)\mu(r),$$

respectively, which are true by assumption. Thus for sufficiently large $|s|$, we have

$$\begin{aligned}
|\mathcal{L}_r\{f\}(s)| &\leq \sum_{n=N}^{\infty} \left| \frac{e_k(t_n, r)}{e_s(t_n, r)} \mu(t_n) \right| \\
&= \sum_{n=N}^{\infty} \mu(t_n) \prod_{m=0}^{n-1} \left| \frac{1 + k\mu(t_m)}{1 + s\mu(t_m)} \right| \\
&\leq \sum_{n=N}^{\infty} \mu(r)\, a^n \left| \frac{1 + k\mu(r)}{1 + s\mu(r)} \right|^n \\
&= \mu(r) \frac{\left| \frac{a + ak\mu(r)}{1 + s\mu(r)} \right|^N}{1 - \left| \frac{a + ak\mu(r)}{1 + s\mu(r)} \right|} \to 0
\end{aligned}$$

as $|s| \to \infty$. □

**Theorem 5.70 (Injectivity).** *If $f, g : \mathbb{T}_r \to \mathbb{R}$ and $\mathcal{L}_r\{f\}(s) = \mathcal{L}_r\{g\}(s)$, then $f(t) = g(t)$ for all $t \geq r$.*

*Proof.* We will first prove that $\mathcal{L}_r\{f\}(s) = 0$ implies $f(t) = 0$ for all $t \geq r$. First, note that by Lemma 5.69, we have

$$\lim_{s \to \infty} \mathcal{L}_r\{f\}(s) = 0,$$

for any $r \in \mathbb{T}_\alpha$. We will show that $f(\sigma^n(r)) = 0$ for all $n \geq 0$ by induction. We first prove the case $n = 0$. To this end, we observe that if

$$\mathcal{L}_r\{f\}(s) = 0,$$

then it follows that

$$\int_r^\infty f(t)\, e_{\ominus s}(\sigma(t), r)\, Dt = 0.$$

And from this it follows that

$$e_s(\sigma(r), r) \int_r^\infty f(t)\, e_{\ominus s}(\sigma(t), r)\, Dt = 0,$$

whence

$$\int_r^\infty f(t)\, e_{\ominus s}(\sigma(t), \sigma(r))\, Dt = 0.$$

Consequently, it holds that

$$f(r)\mu(r) + \int_{\sigma(r)}^\infty f(t)\, e_{\ominus s}(\sigma(t), \sigma(r))\, Dt = 0,$$

from which it follows that

$$f(r)\mu(r) + \mathcal{L}_{\sigma(r)}\{f\}(s) = 0.$$

Taking the limit as $s \to \infty$ yields

$$f(r)\mu(r) + \lim_{s\to\infty} \mathcal{L}_{\sigma(r)}\{f\}(s) = 0,$$

and so, it holds that

$$f(r)\mu(r) = 0,$$

hence

$$f(r) = 0.$$

For the inductive step, assume $f(\sigma^i(r)) = 0$ for all $i < n$. Then it follows that

$$\mathcal{L}_r\{f\}(s) = 0,$$

from which we obtain

$$\int_{\sigma^n(r)}^\infty f(t)\, e_{\ominus s}(\sigma(t), r)\, Dt = 0.$$

Thus,

$$e_s(\sigma^{n+1}(r), r) \int_{\sigma^n(r)}^\infty f(t)\, e_{\ominus s}(\sigma(t), r)\, Dt = 0.$$

So, we deduce that

$$\int_{\sigma^n(r)}^\infty f(t)\, e_{\ominus s}(\sigma(t), \sigma^{n+1}(r))\, Dt = 0.$$

All in all, we conclude that

$$f(\sigma^n(r))\mu(\sigma^n(r)) + \int_{\sigma^{n+1}(r)}^{\infty} f(t)\, e_{\ominus s}(\sigma(t), \sigma^{n+1}(r))\, Dt = 0,$$

and so,

$$f(\sigma^n(r))\mu(\sigma^n(r)) + \mathcal{L}_{\sigma^{n+1}(r)}\{f\}(s) = 0.$$

Taking the limit as $s \to \infty$ yields

$$f(\sigma^n(r))\mu(\sigma^n(r)) + \lim_{s\to\infty} \mathcal{L}_{\sigma^{n+1}(r)}\{f\}(s) = 0.$$

Therefore,

$$f(\sigma^n(r))\mu(\sigma^n(r)) = 0.$$

Hence, we deduce that

$$f(\sigma^n(r)) = 0.$$

So, $f(t) = 0$ for all $t \geq r$.

Thus, $\mathcal{L}_r\{f\}(s) = 0$ if and only if $f = 0$. Now let $g$ be an arbitrary function such that $\mathcal{L}_r\{f\}(s) = \mathcal{L}_r\{g\}(s)$. Then by linearity, we have $\mathcal{L}_r\{f - g\}(s) = 0$, which implies $f - g = 0$. Hence, $f(t) = g(t)$ for all $t \geq r$. □

**Theorem 5.71 (Shifting).** *Suppose $f : \mathbb{T}_r \to \mathbb{R}$ is of exponential order $k > 0$. Then for $|s| > k$, we have:*

$$i)\ \mathcal{L}_{\sigma^n(r)}\{f\}(s) = e_s(\sigma^n(r), r)\mathcal{L}_r\{f\}(s) - \int_r^{\sigma^n(r)} e_s(\sigma^n(r), \sigma(t))f(t)\, Dt;$$

$$ii)\ \mathcal{L}_{\rho^n(r)}\{f\}(s) = e_{\ominus s}(r, \rho^n(r))\mathcal{L}_r\{f\}(s) + \int_{\rho^n(r)}^{r} e_{\ominus s}(\sigma(t), \rho^n(r))f(t)\, Dt.$$

*Proof.* For part (i), we will proceed by induction on $n$. For the base case, consider $n = 1$:

$$\mathcal{L}_{\sigma(r)}\{f\}(s) = \int_{\sigma(r)}^{\infty} e_{\ominus s}(\sigma(t), \sigma(r))f(t)\, Dt.$$

Note that

$$e_{\ominus s}(\sigma(t),\sigma(r)) = \frac{(1+s\mu(r))}{(1+s\mu(r))e_s(\sigma(t),\sigma(r))}$$
$$= \frac{(1+s\mu(r))}{e_s(\sigma(t),r)} = (1+s\mu(r))e_{\ominus s}(\sigma(t),r).$$

Therefore, we have

$$\begin{aligned}
&\mathcal{L}_{\sigma(r)}\{f\}(s)\\
&= \int_{\sigma(r)}^{\infty} (1+s\mu(r))e_{\ominus s}(\sigma(t),r)f(t)\,Dt\\
&= \sum_{j=0}^{\infty}\left[(1+s\mu(r))e_{\ominus s}(\sigma^{j+1}(\sigma(r)),r)f(\sigma^j(\sigma(r)))\mu(\sigma^j(\sigma(r)))\right]\\
&= \sum_{j=0}^{\infty}\left[(1+s\mu(r))e_{\ominus s}(\sigma^{j+2}(r),r)f(\sigma^{j+1}(r))\mu(\sigma^{j+1}(r))\right]\\
&= \sum_{j=0}^{\infty}\left[(1+s\mu(r))e_{\ominus s}(\sigma^{j+1}(r),r)f(\sigma^j(r))\mu(\sigma^j(r))\right]\\
&\qquad - (1+s\mu(r))e_{\ominus s}(\sigma(r),r)f(r)\mu(r).
\end{aligned}$$

Note that

$$(1+s\mu(r))e_{\ominus s}(\sigma(r),r) = \frac{(1+s\mu(r))}{e_s(\sigma(r),r)} = \frac{(1+s\mu(r))}{(1+s\mu(r))e_s(r,r)} = 1.$$

We can thus further simplify the expression to

$$\begin{aligned}
\mathcal{L}_{\sigma(r)}\{f\}(s) &= \int_r^{\infty} (1+s\mu(r))e_{\ominus s}(\sigma(t),r)f(t)\,Dt - f(r)\mu(r)\\
&= (1+s\mu(r))\mathcal{L}_r\{f\}(s) - f(r)\mu(r)\\
&= e_s(\sigma(r),r))\mathcal{L}_r\{f\}(s) - \int_r^{\sigma(r)} e_s(\sigma(r),\sigma(t))f(t)\,Dt,
\end{aligned}$$

which proves the base case.

For the inductive step, assume the hypothesis is true for some $n \geq 1$. Then by the base case, we have

$$
\begin{aligned}
&\mathcal{L}_{\sigma^{n+1}(r)}\{f\}(s)\\
&= e_s(\sigma^{n+1}(r), \sigma^n(r))\mathcal{L}_{\sigma^n(r)}\{f\}(s)\\
&\quad - \int_{\sigma^n(r)}^{\sigma^{n+1}(r)} e_s(\sigma^{n+1}(r), \sigma(t))f(t)\,Dt\\
&= e_s(\sigma^{n+1}(r), \sigma^n(r))\Bigg[e_s(\sigma^n(r), r)\mathcal{L}_r\{f\}(s)\\
&\quad - \int_r^{\sigma^n(r)} e_s(\sigma^n(r), \sigma(t))f(t)\,Dt\Bigg] - \int_{\sigma^n(r)}^{\sigma^{n+1}(r)} e_s(\sigma^{n+1}(r), \sigma(t))f(t)\,Dt\\
&= e_s(\sigma^{n+1}(r), r)\mathcal{L}_r\{f\}(s) - \int_r^{\sigma^n(r)} e_s(\sigma^{n+1}(r), \sigma(t))f(t)\,Dt\\
&\quad - \int_{\sigma^n(r)}^{\sigma^{n+1}(r)} e_s(\sigma^{n+1}(r), \sigma(t))f(t)\,Dt\\
&= e_s(\sigma^{n+1}(r), r)\mathcal{L}_r\{f\}(s) - \int_r^{\sigma^{n+1}(r)} e_s(\sigma^{n+1}(r), \sigma(t))f(t)\,Dt.
\end{aligned}
$$

We now prove part (ii) similarly. For the base case, consider $n = 1$. We obtain

$$
\begin{aligned}
\mathcal{L}_{\rho(r)}\{f\}(s) &= \int_{\rho(r)}^{\infty} e_{\ominus s}(\sigma(t), \rho(r))f(t)\,Dt\\
&= \int_{\rho(r)}^{\infty} \frac{e_{\ominus s}(\sigma(t), r)}{1 + s\mu(\rho(r))} f(t)\,Dt\\
&= \frac{1}{1 + s\mu(\rho(r))}\mathcal{L}_r\{f\}(s) + \frac{f(\rho(r))\mu(\rho(r))}{1 + s\mu(\rho(r))}\\
&= e_{\ominus s}(r, \rho(r))\mathcal{L}_r\{f\}(s) + \int_{\rho(r)}^{r} e_{\ominus s}(\sigma(t), \rho(r))f(t)\,Dt,
\end{aligned}
$$

proving the base case. For the inductive step, assume the hypothesis is true for some $n \geq 1$. Then by the base case, we have

$$
\begin{aligned}
&\mathcal{L}_{\rho^{n+1}(r)}\{f\}(s)\\
&= e_{\ominus s}(\rho^n(r), \rho^{n+1}(r))\mathcal{L}_{\rho^n(r)}\{f\}(s) + \int_{\rho^{n+1}(r)}^{\rho^n(r)} e_{\ominus s}(\sigma(t), \rho^{n+1}(r))f(t)\,Dt
\end{aligned}
$$

$$= e_{\ominus s}(\rho^n(r), \rho^{n+1}(r)) \left[ e_{\ominus s}(r, \rho^n(r)) \mathcal{L}_r\{f\}(s) + \int_{\rho^n(r)}^{r} e_{\ominus s}(\sigma(t), \rho^n(t)) f(t)\, Dt \right] + \int_{\rho^{n+1}(r)}^{\rho^n(r)} e_{\ominus s}(\sigma(t), \rho^{n+1}(t)) f(t)\, Dt$$

$$= e_{\ominus s}(r, \rho^{n+1}(r)) \mathcal{L}_r\{f\}(s) + \int_{\rho^n(r)}^{r} e_{\ominus s}(\sigma(t), \rho^{n+1}(t)) f(t)\, Dt + \int_{\rho^{n+1}(r)}^{\rho^n(r)} e_{\ominus s}(\sigma(t), \rho^{n+1}(t)) f(t)\, Dt$$

$$= e_{\ominus s}(r, \rho^{n+1}(r)) \mathcal{L}_r\{f\}(s) + \int_{\rho^{n+1}(r)}^{r} e_{\ominus s}(\sigma(t), \rho^{n+1}(t)) f(t)\, Dt.$$

And this completes the proof. □

## 5.10 Laplace Transform Formulas

**Theorem 5.72.** *If $c \in \mathbb{R}$, then $\mathcal{L}_r\{c\}(s) = \frac{c}{s}$ for $|s| > 0$.*

*Proof.* Note $c = c\, e_0(t, r)$ and apply Theorems 5.65 and 5.62. □

**Definition 5.73.** If $f : \mathbb{T}_r \to \mathbb{R}$, then for any $n \in \mathbb{N}$, we define the $n$-th antidifference of $f$ based at $r$ by

$$D_r^{-n} f(t) = \int_r^t \int_r^{x_n} \cdots \int_r^{x_3} \int_r^{x_2} f(x_1)\, Dx_1 \cdots Dx_n.$$

**Theorem 5.74.** *Let $f : \mathbb{T}_r \to \mathbb{R}$. Then for any $n \in \mathbb{N}$,*

$$D_r^{-n} f(t) = \int_r^t h_{n-1}(t, \sigma(s)) f(s)\, Ds.$$

*Proof.* Consider the initial value problem

$$\begin{cases} D^n y(t) = f(t), \\ D^k y(r) = 0 \text{ for } 0 \le k \le n-1. \end{cases}$$

It is easy to see that the unique solution to this system is given by $D_r^{-n} f(t)$.

However, by Taylor's Theorem, $y(t) = P_{n-1}(t, r) + R_{n-1}(t, r)$ where

$$P_{n-1}(t, r) = \sum_{k=0}^{n-1} D^k f(r) h_k(t, r),$$

and

$$R_{n-1}(t,r) = \int_r^t h_{n-1}(t,\sigma(s))D^n f(s)\,Ds.$$

Then we have

$$f(t) = D^n y(t) = D^n P_{n-1}(t,r) + D^n R_{n-1}(t,r) = D^n R_{n-1}(t,r),$$

which implies that $R_{n-1}(t,r)$ is also a solution. Thus we have

$$\begin{aligned} D_r^{-n} f(t) &= R_{n-1}(t,r) \\ &= \int_r^t h_{n-1}(t,\sigma(s))D^n y(s)\,Ds \\ &= \int_r^t h_{n-1}(t,\sigma(s))D^n D_r^{-n} f(s)\,Ds \\ &= \int_r^t h_{n-1}(t,\sigma(s))f(s)\,Ds, \end{aligned}$$

since the solution is unique. □

The following results are used to obtain the exponential order of the Taylor monomial, $h_n(t,r)$, and give its Laplace transform.

**Lemma 5.75.** *Let $f(t)$ be of exponential order $k > 0$. Then $D_r^{-1} f(t)$ is also of exponential order $k > 0$.*

*Proof.* Let $|f(t)| \le M e_k(t,r)$ for all $t \ge x$, and let $C = \left|\int_r^x f(u)\,Du\right|$. Then for $t \ge x$, we have

$$\begin{aligned} \left|D_r^{-1} f(t)\right| &= \left|\int_r^t f(u)\,Du\right| \\ &\le \left|\int_r^x f(u)\,Du\right| + \left|\int_x^t f(u)\,Du\right| \\ &\le C + \int_r^t |f(u)|\,Du \\ &\le C + M\int_r^t e_k(u,r)\,Du \\ &= C + \frac{M}{k}\left(e_k(u,r)\right)\Big|_{u=r}^{u=t} \\ &= C + \frac{M}{k}(e_k(t,r) - 1) \end{aligned}$$

$$\le C + \frac{M}{k} e_k(t,r)$$
$$\le \left(C + \frac{M}{k}\right) e_k(t,r).$$

As this demonstrates that $D_r^{-1} f$ is of exponential order, the proof of the lemma is complete. □

**Corollary 5.76.** *Let $f(t)$ be of exponential order $k > 0$. Then $D_r^{-n} f(t)$ is also of exponential order $k > 0$ for all $n \in \mathbb{N}$.*

To see that the corollary holds, use the previous result.

**Theorem 5.77.** *Let $f$ be of exponential order $k > 0$. Then for $|s| > k$,*

$$\mathcal{L}_r\{D_r^{-n} f(t)\}(s) = \frac{1}{s^n} \mathcal{L}_r\{f\}(s).$$

*Proof.* We will proceed by induction on $n$. For the base case, consider $n = 1$. Then using integration by parts, we have

$$\begin{aligned}
\mathcal{L}_r\{f\}(s) &= \int_r^\infty e_{\ominus s}(\sigma(t), r) f(t)\, Dt \\
&= e_{\ominus s}(t,r) D_r^{-1} f(t)\Big|_{t=r}^{t\to\infty} + s \int_r^\infty e_{\ominus s}(\sigma(t), r) D_r^{-1} f(t)\, Dt \\
&= s\mathcal{L}_r\{D_r^{-1} f(t)\}(s).
\end{aligned}$$

Thus, $\mathcal{L}_r\{D_r^{-1} f(t)\}(s) = \frac{1}{s}\mathcal{L}_r\{f\}(s)$.

For the inductive step, assume the statement is true for some integer $n > 0$. Then we have

$$\begin{aligned}
\mathcal{L}_r\{D_r^{-n-1} f(t)\}(s) &= \frac{1}{s} \mathcal{L}_r\{D_r^{-n} f(t)\} \\
&= \frac{1}{s}\left[\frac{1}{s^n}\mathcal{L}_r\{f\}(s)\right] \\
&= \frac{1}{s^{n+1}} \mathcal{L}_r\{f\}(s),
\end{aligned}$$

as was to be shown. □

**Lemma 5.78.** *The n-th Taylor monomial $h_n(t,r)$ is of exponential order $k = 1$.*

*Proof.* We will prove this result for $a > 1$ (leaving the case $a = 1$ to the reader) by induction on $n$. Consider the base case $n = 1$. We have

$$\begin{aligned}
\lim_{t\to\infty}\left|\frac{h_1(t,r)}{e_1(t,r)}\right| &= \lim_{t\to\infty}\left|\frac{t-r}{e_1(t,r)}\right| \\
&\le \lim_{t\to\infty}\left|\frac{t}{e_1(t,r)}\right| + \lim_{t\to\infty}\left|\frac{-r}{e_1(t,r)}\right| \\
&\le \lim_{t\to\infty}\left|\frac{t}{1+\mu(\rho(t))}\right| \\
&= \lim_{t\to\infty}\left|\frac{t}{1+a^{-1}\mu(t)}\right| \\
&= \lim_{t\to\infty}\left|\frac{at}{a+(a-1)t+b}\right| \\
&= \lim_{t\to\infty}\left|\frac{a}{(a-1)+\frac{a+b}{t}}\right| \\
&= \frac{a}{a-1}.
\end{aligned}$$

Thus, for sufficiently large $t$ and any $\epsilon > 0$,

$$|h_1(t,r)| \le \left(\frac{a}{a-1}+\epsilon\right) e_1(t,r).$$

For the inductive step, assume $h_n(t,r)$ is of exponential order 1 for some $n$. Then, since $D_r^{-1}h_n(t,r) = h_{n+1}(t,r)$, applying Lemma 5.75 implies $h_{n+1}$ is of exponential order 1. □

**Theorem 5.79.** *Let* $|s| > 1$. *Then*

$$\mathcal{L}_r\{h_n(t,r)\}(s) = \frac{1}{s^{n+1}}.$$

*Proof.* The base case, $n = 0$, is trivial since $\mathcal{L}_r\{1\}(s) = \frac{1}{s}$. Note that

$$\begin{aligned}
&\mathcal{L}_r\{h_n(t,r)\}(s) \\
&\quad = \int_r^\infty e_{\ominus s}(\sigma(t),r)h_n(t,r)Dt \\
&\quad = e_{\ominus s}(\sigma(t),r)h_{n+1}(t,r)\Big|_{t=r}^{t\to\infty} + \int_r^\infty \frac{s}{1+s\mu(t)}e_{\ominus s}(t,r)h_{n+1}(t,r) \\
&\quad = \int_r^\infty \frac{s}{1+s\mu(t)}e_{\ominus s}(t,r)h_{n+1}(t,r) \\
&\quad = s\mathcal{L}_r\{h_{n+1}(t,r)\}(s).
\end{aligned}$$

Thus, $\mathcal{L}_r\{h_{n+1}(t,r)\}(s) = \frac{1}{s}\mathcal{L}_r\{h_n(t,r)\}(s)$. Suppose that the theorem holds for some $n$. Then it follows that

$$\mathcal{L}_r\{h_{n+1}(t,r)\}(s) = \frac{1}{s}\mathcal{L}_r\{h_n(t,r)\}(s) = \frac{1}{s}\frac{1}{(s^{n+1})} = \frac{1}{s^{n+2}},$$

which completes the induction step and thus proves the result. □

**Lemma 5.80.** *The discrete trigonometric functions,* $\sin_p$ *and* $\cos_p$*, and the hyperbolic trigonometric functions,* $\sinh_p$ *and* $\cosh_p$*, are all of exponential order* $|p|$*.*

*Proof.* Let $p$ be such that $\pm p \in \mathcal{R}^c$. Then for sufficiently large $t$, we have

$$\begin{aligned}|\sinh_p(t,r)| &= \frac{1}{2}\left|e_p(t,r) - e_{-p}(t,r)\right| \\ &\le \frac{1}{2}|e_p(t,r)| + \frac{1}{2}|e_{-p}(t,r)| \\ &\le e_{|p|}(t,r).\end{aligned}$$

The proof for $\cosh_p(t,r)$ is analogous.

For $\cos_p$, we can use the identity $\cos_p(t,r) = \cosh_{ip}(t,r)$ to obtain

$$\begin{aligned}|\cos_p(t,r)| &= |\cosh_{ip}(t,r)| \\ &\le e_{|ip|}(t,r) \\ &= e_{|p|}(t,r).\end{aligned}$$

The proof for $\sin_p(t,r)$ is analogous. □

**Theorem 5.81.** *For* $|s| > |p|$ *and* $\pm p \in \mathcal{R}^c$*, we have*

(i) $\mathcal{L}_r\{\cosh_p(t,r)\}(s) = \frac{s}{s^2-p^2}$;
(ii) $\mathcal{L}_r\{\sinh_p(t,r)\}(s) = \frac{p}{s^2-p^2}$.

*Proof.* To see that (i) holds, note that

$$\begin{aligned}\mathcal{L}_r\{\cosh_p(t,r)\}(s) &= \frac{1}{2}\left[\mathcal{L}_r\{e_p(t,r)\}(s) + \mathcal{L}_r\{e_{-p}(t,r)\}(s)\right] \\ &= \frac{1}{2}\frac{1}{(s-p)} + \frac{1}{2}\frac{1}{(s+p)} \\ &= \frac{s}{s^2-p^2}.\end{aligned}$$

The proof of (ii) is similar. □

**Theorem 5.82.** *For* $|s| > |p|$ *and* $\pm ip \in \mathcal{R}^c$, *we have*

(i) $\mathcal{L}_r\{\cos_p(t,r)\}(s) = \frac{s}{s^2+p^2}$;
(ii) $\mathcal{L}_r\{\sin_p(t,r)\}(s) = \frac{p}{s^2+p^2}$.

*Proof.* To see that (i) holds, recall that $\cos_p(t,r) = \cosh_{ip}(t,r)$ and thus,

$$\begin{aligned}\mathcal{L}_r\{\cos_p(t,r)\}(s) &= \mathcal{L}_r\{\cosh_{ip}(t,r)\}(s)\\ &= \frac{1}{2}\frac{1}{(s-ip)} + \frac{1}{2}\frac{1}{(s+ip)}\\ &= \frac{s}{s^2+p^2}.\end{aligned}$$

The proof of (ii) is analogous. □

**Lemma 5.83.** *For* $p, q \in \mathcal{R}^c$ *and* $t, r \in \mathbb{T}_\alpha$, *let* $k(t) = \frac{q}{1+p\mu(t)}$. *Then the following functions are of exponential order* $|p| + |q|$*:*

(i) $e_p(t,r)\cosh_k(t,r)$;
(ii) $e_p(t,r)\sinh_k(t,r)$;
(iii) $e_p(t,r)\cos_k(t,r)$;
(iv) $e_p(t,r)\sin_k(t,r)$.

*Proof.* We will prove the result for (i). First, note that

$$\begin{aligned}p \oplus \frac{q}{1+p\mu(t)} &= p + \frac{q}{1+p\mu(t)} + \frac{pq\mu(t)}{1+p\mu(t)}\\ &= p + \frac{q(1+p\mu(t))}{1+p\mu(t)}\\ &= p+q.\end{aligned}$$

Therefore,

$$\begin{aligned}|e_p(t,r)\cosh_k(t,r)| &= \left|e_p(t,r)\frac{e_k(t,r)+e_{-k}(t,r)}{2}\right|\\ &= \left|\frac{e_{p\oplus k}(t,r)+e_{p\oplus -k}(t,r)}{2}\right|\\ &\le \left|\frac{e_{p\oplus k}(t,r)| + |e_{p\oplus -k}(t,r)}{2}\right|\\ &\le |e_s(t,r)|,\end{aligned}$$

where

$$\begin{aligned} s &= max\{|p \oplus k|, |p \oplus -k|\} \\ &= max\{|p+q|, |p-q|\} \\ &\leq |p| + |q|. \end{aligned}$$

Thus,

$$|e_p(t,r)\cosh_k(t,r)| \leq |e_s(t,r)| \leq Me_s(t,r),$$

for some $M > 0$, and so, $e_p(t,r)\cosh_k(t,r)$ is of exponential order $|p| + |q|$. The proofs of (ii)–(iv) are analogous. □

**Theorem 5.84.** *Let* $k(t) = \frac{q}{1+p\mu(t)}$ *for* $p, q \in \mathcal{R}^c$. *Then for* $|s| > |p| + |q|$, *we have*

(i) $\mathcal{L}_r\{e_p(t,r)\cosh_k(t,r)\}(s) = \frac{s-p}{(s-p)^2-q^2}$;
(ii) $\mathcal{L}_r\{e_p(t,r)\sinh_k(t,r)\}(s) = \frac{q}{(s-p)^2-q^2}$;
(iii) $\mathcal{L}_r\{e_p(t,r)\cos_k(t,r)\}(s) = \frac{s-p}{(s-p)^2+q^2}$;
(iv) $\mathcal{L}_r\{e_p(t,r)\sin_k(t,r)\}(s) = \frac{q}{(s-p)^2+q^2}$.

*Proof.* To prove (i), first note that

$$p \oplus k = p + q,$$

as stated above. Therefore,

$$\begin{aligned} &\mathcal{L}_r\{e_p(t,r)\cosh_k(t,r)\}(s) \\ &\quad = \frac{1}{2}\left[\mathcal{L}_r\left\{e_{p\oplus\frac{q}{1+p\mu(t)}}(t,r)\right\}(s) + \mathcal{L}_r\left\{e_{p\oplus\frac{-q}{1+p\mu(t)}}(t,r)\right\}(s)\right] \\ &\quad = \frac{1}{2}\left[\mathcal{L}_r\left\{e_{p+q}(t,r)\right\}(s) + \mathcal{L}_r\left\{e_{p-q}(t,r)\right\}(s)\right] \\ &\quad = \frac{1}{2}\left[\frac{1}{s-(p+q)} + \frac{1}{s-(p-q)}\right] \\ &\quad = \frac{s-p}{(s-p)^2-q^2}. \end{aligned}$$

The proof of (ii) is similar.

To see that (iii) holds, consider that $\cos_q(t,r) = \cosh_{iq}(t,r)$ and let $q = iq$ in the result of the proof of (i) to get

$$\mathcal{L}_r\{e_p(t,r)\cos_k(t,r)\}(s) = \frac{s-p}{(s-p)^2-(iq)^2} = \frac{s-p}{(s-p)^2+q^2}.$$

The proof of (iv) is analogous. □

## 5.11 Solving IVPs Using the Laplace Transform

In this section we will demonstrate how the discrete Laplace transform can be applied to solve difference equations on $\mathbb{T}_r$.

*Example 5.85.* Solve:

$$\begin{cases} D^2f(t) - 2Df(t) - 8f(t) = 0, \\ Df(r) = 0;\ f(r) = -\frac{3}{2}. \end{cases}$$

We will take the Laplace transform of both sides of the equation and use the initial conditions to solve this problem. We begin with

$$\begin{aligned} 0 &= \mathcal{L}_r\{D^2f(t) - 2Df(t) - 8f(t)\}(s) \\ &= \mathcal{L}_r\{D^2f(t)\}(s) - 2\mathcal{L}_r\{Df(t)\}(s) - 8\mathcal{L}_r\{f\}(s) \\ &= \left[s^2\mathcal{L}\{f\}(s) - sf(r) - Df(r)\right] - 2\left[s\mathcal{L}\{f\}(s) - f(r)\right] - 8\mathcal{L}\{f\}(s) \\ &= \left[s^2\mathcal{L}\{f\}(s) - s\left(-\frac{3}{2}\right)\right] - 2\left[s\mathcal{L}\{f\}(s) - \left(-\frac{3}{2}\right)\right] - 8\mathcal{L}\{f\}(s) \\ &= (s^2 - 2s - 8)\mathcal{L}\{f\}(s) + \frac{3}{2}s - 3, \end{aligned}$$

from which it follows that

$$\mathcal{L}\{f\}(s) = \frac{3 - \frac{3}{2}s}{s^2 - 2s - 8}.$$

Using partial fractions, we obtain

$$\mathcal{L}\{f\}(s) = \frac{3 - \frac{3}{2}s}{s^2 - 2s - 8} = \frac{-\frac{1}{2}}{s - 4} + \frac{-1}{s + 2}.$$

Therefore, by the injectivity of the Laplace transform,

$$f(t) = -\frac{1}{2}e_4(t, r) - e_{-2}(t, r).$$

*Example 5.86.* Solve the following IVP:

$$\begin{aligned} D^2y(t) + 4y(t) &= 0 \\ y(0) &= 1 \\ Dy(0) &= 1. \end{aligned}$$

To solve the above problem, we first take the Laplace transform of both sides. This yields

$$\mathcal{L}_0\{D^2y(t) + 4y(t)\} = \mathcal{L}_0\{0\},$$

from which it follows that

$$s^2\mathcal{L}_0\{y\} - (s+1) + 4\mathcal{L}_0\{y\} = 0.$$

We then solve for $\mathcal{L}_0\{y\}$ and invert by writing

$$(s^2 + 4)\mathcal{L}_0\{y\} = s + 1,$$

from which it follows that

$$\mathcal{L}_0\{y\} = \frac{s}{(s^2+4)} + \frac{1}{2}\frac{2}{(s^2+4)}.$$

Thus,

$$y(t) = \cos_2(t, 0) + \frac{1}{2}\sin_2(t, 0).$$

## 5.12 Green's Functions

In this section we will consider boundary value problems on a mixed time scale with Sturm–Liouville type boundary value conditions for $a > 1$. We will find a Green's function for a boundary value problem on a mixed time scale with Dirichlet boundary conditions, and investigate some of its properties. Many of the results in this section can be viewed as analogues to results for the continuous case given in Kelley and Peterson [137].

**Theorem 5.87.** *Let $\beta \in \mathbb{T}_{\sigma^2(\alpha)}$ and $A, B, E, F \in \mathbb{R}$ be given. Then the homogeneous boundary value problem (BVP)*

$$\begin{cases} -D^2y(t) = 0, & t \in \mathbb{T}_\alpha \\ Ay(\alpha) - BDy(\alpha) = 0 \\ Ey(\beta) + FDy(\beta) = 0 \end{cases}$$

*has only the trivial solution if and only if*

$$\gamma := AE(\beta - \alpha) + BE + AF \neq 0.$$

*Proof.* A general solution of $-D^2y(t) = 0$ is given by

$$y(t) = c_0 + c_1 h_1(t, \alpha).$$

Using the boundary conditions, we have

$$Ay(\alpha) - BDy(\alpha) = Ac_0 - Bc_1 = 0$$

and

$$Ey(\beta) + FDy(\beta) = E[c_0 + c_1(\beta - \alpha)] + Fc_0 = 0.$$

Thus, we have the following system of equations

$$\begin{aligned} c_0 A - c_1 B &= 0 \\ c_0 E + c_1[E(\beta - \alpha) + F] &= 0, \end{aligned}$$

which has only the trivial solution if and only if

$$\gamma := \begin{vmatrix} A & -B \\ E & E\beta - E\alpha + F \end{vmatrix} \neq 0.$$

It follows that

$$\begin{aligned} \gamma &:= A[E(\beta - \alpha) + F)] + BE \\ &= AE(\beta - \alpha) + BE + AF, \end{aligned}$$

as claimed. □

**Lemma 5.88.** *Assume $\beta \in \mathbb{T}_{\sigma^2(\alpha)}$ and $A_1, A_2 \in \mathbb{R}$. Then the boundary value problem*

$$\begin{aligned} -D^2y(t) = 0, &\quad t \in \mathbb{T}_\alpha^{\rho^2(\beta)} \\ y(\alpha) = A_1, &\quad y(\beta) = A_2 \end{aligned}$$

*has the solution*

$$y(t) = A_1 + \frac{A_2 - A_1}{\beta - \alpha}(t - \alpha).$$

*Proof.* A general solution to the mixed difference equation $D^2y(t) = 0$ is given by

$$y(t) = c_0 + c_1 h_1(t, \alpha) = c_0 + c_1(t - \alpha).$$

Using the first boundary condition, we get

$$y(\alpha) = c_0 = A_1.$$

Using the second boundary condition, we have that

$$y(\beta) = A_1 + c_1(\beta - \alpha) = A_2.$$

Solving for $c_1$ we get

$$c_1 = \frac{A_2 - A_1}{\beta - \alpha}.$$

Hence,

$$y(t) = A_1 + \frac{A_2 - A_1}{\beta - \alpha}(t - \alpha).$$

□

**Theorem 5.89.** *Assume* $f : \mathbb{T}_\alpha^{\rho^2(\beta)} \to \mathbb{R}$ *and* $\beta \in \mathbb{T}_{\sigma^2(\alpha)}$. *Then the unique solution of the BVP*

$$-D^2y(t) = f(t), \quad t \in \mathbb{T}_\alpha^{\rho^2(\beta)} \tag{5.5}$$

$$y(\alpha) = 0 = y(\beta), \tag{5.6}$$

*is given by*

$$y(t) = \int_\alpha^\beta G(t,s)f(s)Ds = \sum_{j=0}^{K(\beta)-1} G(t, \sigma^j(\alpha))f(\sigma^j(\alpha))\mu(\sigma^j(\alpha)),$$

*for* $t \in \mathbb{T}_\alpha^\beta$, *where* $G : \mathbb{T}_\alpha^\beta \times \mathbb{T}_\alpha^{\rho(\beta)} \to \mathbb{R}$ *is called the Green's function for the homogeneous BVP*

$$-D^2y(t) = 0, \quad t \in \mathbb{T}_\alpha^{\rho^2(\beta)} \tag{5.7}$$

$$y(\alpha) = 0 = y(\beta), \tag{5.8}$$

*and is defined by*

$$G(t,s) := \begin{cases} u(t,s), & 0 \le K(s) \le K(t) - 1 \le K(\beta) - 1 \\ v(t,s), & 0 \le K(t) \le K(s) \le K(\beta) - 1, \end{cases}$$

*where for* $(t,s) \in \mathbb{T}_\alpha^\beta \times \mathbb{T}_\alpha^{\rho(\beta)}$

$$u(t,s) := \frac{h_1(\beta,\sigma(s))}{h_1(\beta,\alpha)} h_1(t,\alpha) - h_1(t,\sigma(s))$$

*and*

$$v(t,s) := \frac{h_1(\beta,\sigma(s))}{h_1(\beta,\alpha)} h_1(t,\alpha).$$

*Proof.* Note for the $\gamma$ defined in Theorem 5.87 we have that for $A = E = 1$, $B = F = 0$,

$$\gamma = AC(\beta-\alpha) + BC + AD = (\beta - \alpha) \neq 0.$$

Hence, by Exercise 5.13, the BVP (5.5), (5.6) has a unique solution $y(t)$. Using the variation of constants formula (Theorem 5.38 with $n = 2$) we have that

$$\begin{aligned} y(t) &= c_0 h_0(t,\alpha) + c_1 h_1(t,\alpha) - \int_\alpha^t h_1(t,\sigma(s)f(s)Ds \\ &= c_0 + c_1 h_1(t,\alpha) - \sum_{j=0}^{K(t)-1} h_1(t,\sigma(\sigma^j(\alpha)))f(\sigma^j(\alpha))\mu(\sigma^j(\alpha)) \\ &= c_0 + c_1 h_1(t,\alpha) - \sum_{j=0}^{K(t)-1} h_2(t,\sigma^{j+1}(\alpha))f(\sigma^j(\alpha))a^j\mu(\alpha). \end{aligned}$$

Using the first boundary condition, we get

$$\begin{aligned} y(\alpha) &= c_0 + c_1 h_1(\alpha,\alpha) - \int_\alpha^\alpha h_1(t,\sigma(s))f(s)Ds \\ &= c_0 \\ &= 0, \end{aligned}$$

and using the second boundary condition, we have that

$$y(\beta) = c_1 h_1(\beta,\alpha) - \sum_{j=0}^{K(\beta)-1} h_1(\beta,\sigma^{j+1}(\alpha))f(\sigma^j(\alpha))\mu(\sigma^j(\alpha)) = 0.$$

Solving for $c_1$ yields

$$c_1 = \frac{\sum_{j=0}^{K(\beta)-1} h_2(\beta,\sigma^{j+1}(\alpha))f(\sigma^j(\alpha))\mu(\sigma^j(\alpha))}{h_1(\beta,\alpha)}.$$

Thus,

$$y(t) = \frac{\sum_{j=0}^{K(\beta)-1} h_1(\beta, \sigma^{j+1}(\alpha))f(\sigma^j(\alpha))\mu(\sigma^j(\alpha))}{h_1(\beta, \alpha)} h_1(t, \alpha)$$
$$- \sum_{j=0}^{K(t)-1} h_1(t, \sigma^{j+1}(\alpha))f(\sigma^j(\alpha))\mu(\sigma^j(\alpha))$$
$$= \sum_{j=0}^{K(t)-1} \left[ \frac{h_1(\beta, \sigma^{j+1}(\alpha))}{h_1(\beta, \alpha)} h_1(t, \alpha) - h_1(t, \sigma^{j+1}(\alpha)) \right] f(\sigma^j(\alpha))\mu(\sigma^j(\alpha))$$
$$+ \sum_{j=K(t)}^{K(\beta)-1} \frac{h_1(\beta, \sigma^{j+1}(\alpha))}{h_1(\beta, \alpha)} h_1(t, \alpha) f(\sigma^j(\alpha))\mu(\sigma^j(\alpha))$$
$$= \sum_{j=0}^{K(\beta)-1} G(t, \sigma^j(\alpha))f(\sigma^j(\alpha))\mu(\sigma^j(\alpha))$$
$$= \int_\alpha^\beta G(t, s)f(s)Ds,$$

for $G(t, s)$ defined as in the statement of this theorem. □

**Theorem 5.90.** *The Green's function for the BVP* (5.7), (5.8), *satisfies*

$$G(t, s) \geq 0, \quad (t, s) \in \mathbb{T}_\alpha^\beta \times \mathbb{T}_\alpha^{\rho(\beta)}$$

*and*

$$\max_{t \in \mathbb{T}_\alpha^\beta} G(t, s) = G(\sigma(s), s), \quad s \in \mathbb{T}_\alpha^{\rho(\beta)}.$$

*Proof.* First, note both that

$$G(\alpha, s) = \frac{h_1(\beta, \sigma(s))}{h_1(\beta, \alpha)} h_1(\alpha, \alpha) = \frac{h_1(\beta, \sigma(s))}{h_1(\beta, \alpha)} (\alpha - \alpha) = 0$$

and that

$$G(\beta, s) = \frac{h_1(\beta, \sigma(s))}{h_1(\beta, \alpha)} h_1(\beta, \alpha) - h_1(\beta, \sigma(s)) = 0.$$

Now we will show that $DG(t,s) \geq 0$ for $t \leq s$, $DG(t,s) \leq 0$ for $s < t$, and $G(\sigma(s),s) \geq G(s,s)$. So first consider the domain $0 \leq K(t) \leq K(s) \leq K(\beta) - 1$:

$$\begin{aligned} DG(t,s) &= \frac{h_1(\beta,\sigma(s))}{h_1(\beta,\alpha)} Dh_1(t,\alpha) \\ &= \frac{\beta-\sigma(s)}{\beta-\alpha} h_0(t,\alpha) \\ &= \frac{\beta-\sigma(s)}{\beta-\alpha} \\ &\geq 0. \end{aligned}$$

Now consider the domain $0 \leq K(s) \leq K(t) - 1 \leq K(\beta) - 1$:

$$\begin{aligned} DG(t,s) &= \frac{h_1(\beta,\sigma(s))}{h_1(\beta,\alpha)} Dh_1(t,\alpha) - D\overline{h_1}(t,\sigma(s)) \\ &= \frac{\beta-\sigma(s)}{\beta-\alpha} h_0(t,\alpha) - h_0(t,\sigma(s)) \\ &= \frac{\beta-\sigma(s)}{\beta-\alpha} - 1 \\ &\leq 0, \end{aligned}$$

since $\beta - \sigma(s) \leq \beta - \alpha$. Now, since $G$ is increasing for $t \leq s$ and decreasing for $s < t$, we need to see which is larger: $G(\sigma(s),s)$ or $G(s,s)$. So consider

$$\begin{aligned} &G(\sigma(s),s) - G(s,s) \\ &= \frac{\beta-\sigma(s)}{\beta-\alpha}(\sigma(s)-\alpha) - (\sigma(s)-\sigma(s)) - \frac{\beta-\sigma(s)}{\beta-\alpha}(s-\alpha) \\ &= \frac{\beta-\sigma(s)}{\beta-\alpha}[\sigma(s)-\alpha-s+\alpha] \\ &= \frac{\beta-\sigma(s)}{\beta-\alpha}(\sigma(s)-s) \\ &\geq 0, \end{aligned}$$

which implies that $\max_{t\in\mathbb{T}_\alpha^\beta} G(t,s) = G(\sigma(s),s)$. Also, since $DG(t,s) \geq 0$ for $t \in \mathbb{T}_{[\alpha,s]}$, $DG(t,s) \leq 0$ for $t \in \mathbb{T}_{(s,\beta)}$, and $G(\alpha,s) = 0 = G(\beta,s)$, we have $G(t,s) \geq 0$ on its domain. □

*Remark 5.91.* Note that in the above proof, we have $DG(t,s) > 0$ for $t \leq s < \rho(\beta)$, $DG(t,s) < 0$ for $\alpha < s < t$.

In the next theorem we give some more properties of the Green's function for the BVP (5.7), (5.8).

**Theorem 5.92.** *Let $G(t,s)$, $u(t,s)$, and $v(t,s)$ be as defined in Theorem 5.89. Then the following hold:*

(i) $G(\alpha, s) = 0 = G(\beta, s), \quad s \in \mathbb{T}_\alpha^{\rho(\beta)}$;
(ii) *for each fixed* $s \in \mathbb{T}_\alpha^{\rho(\beta)}$, $-D^2u(t,s) = 0 = -D^2v(t,s)$ *for* $t \in \mathbb{T}_\alpha^{\rho^2(\beta)}$;
(iii) $v(t,s) = u(t,s) + h_1(t, \sigma(s)), \quad (t,s) \in \mathbb{T}_\alpha^\beta \times \mathbb{T}_\alpha^{\rho(\beta)}$;
(iv) $u(\sigma(s), s) = v(\sigma(s), s), \quad s \in \mathbb{T}_\alpha^{\rho(\beta)}$;
(v) $-D^2G(t,s) = \frac{\delta_{ts}}{\mu(s)}, \quad (t,s) \in \mathbb{T}_\alpha^{\rho^2(\beta)} \times \mathbb{T}_\alpha^{\rho(\beta)}$, *where $\delta_{ts}$ is the Kronecker delta, i.e., $\delta_{ts} = 1$ for $t = s$ and $\delta_{ts} = 0$ for $t \neq s$.*

*Proof.* In the proof of Theorem 5.90 we proved (i). The proofs of the properties (ii)–(iv) are straightforward and left to the reader (see Exercise 5.15). We now use these properties to prove (v). It follows that for $t < s$,

$$-D^2G(t,s) = -D^2u(t,s) = 0 = \frac{\delta_{ts}}{\mu(s)}$$

and for $t > s$,

$$-D^2G(t,s) = -D^2v(t,s) = 0 = \frac{\delta_{ts}}{\mu(s)}.$$

Finally, when $t = s$, we have using Exercise 5.5

$$\begin{aligned}
&D^2G(t,s)\\
&= \frac{G(\sigma^2(t), s)\mu(t) - G(\sigma(t), s)[\mu(t) + \mu(\sigma(t))] + G(t,s)\mu(\sigma(t))}{[\mu(t)]^2\mu(\sigma(t))}\\
&= \frac{v(\sigma^2(t), s)\mu(t) - u(\sigma(t), s)[\mu(t) + \mu(\sigma(t))] + u(t,s)\mu(\sigma(t))}{[\mu(t)]^2\mu(\sigma(t))}\\
&= \frac{v(\sigma^2(s), s)\mu(s) - u(\sigma(s), s)[\mu(s) + \mu(\sigma(s))] + u(s,s)\mu(\sigma(s))}{[\mu(s)]^2\mu(\sigma(s))}\\
&= \frac{[u(\sigma^2(s), s) + h_1(\sigma^2(s), \sigma(s))]\mu(s)}{[\mu(s)]^2\mu(\sigma(s))}\\
&\quad + \frac{-u(\sigma(s), s)[\mu(s) + \mu(\sigma(s))] + u(s,s)\mu(\sigma(s))}{[\mu(s)]^2\mu(\sigma(s))}\\
&= \frac{h_1(\sigma^2(s), \sigma(s))\mu(s)}{[\mu(s)]^2\mu(\sigma(s))} + D^2u(s,s)
\end{aligned}$$

$$= \frac{h_1(\sigma^2(s), \sigma(s))}{\mu(s)\mu(\sigma(s))}$$
$$= \frac{\sigma^2(s) - \sigma(s)}{\mu(s)\mu(\sigma(s))}$$
$$= \frac{1}{\mu(s)}.$$

Therefore,

$$-D^2G(t,s) = \frac{\delta_{ts}}{\mu(s)},$$

for $(t,s) \in \mathbb{T}_\alpha^{\rho^2(\beta)} \times \mathbb{T}_\alpha^{\rho(\beta)}$. □

The following theorem along with Exercise 5.13 is a uniqueness result for the Green's function for the BVP (5.7), (5.8).

**Theorem 5.93.** *There is a unique function* $G : \mathbb{T}_\alpha^\beta \times \mathbb{T}_\alpha^{\rho(\beta)} \to \mathbb{R}$ *such that* $G(\alpha, s) = 0 = G(\beta, s)$, *for each* $s \in \mathbb{T}_\alpha^{\rho(\beta)}$, *and that* $-D^2G(t,s) = \frac{\delta_{ts}}{\mu(s)}$, *for each fixed* $s \in \mathbb{T}_\alpha^{\rho(\beta)}$.

*Proof.* Fix $s \in \mathbb{T}_\alpha^{\rho(\beta)}$. Then by Theorem 5.89 with $f(t) = \frac{\delta_{ts}}{\mu(s)}$, $t \in \mathbb{T}_\alpha^{\rho^2(\beta)}$, the BVP

$$-D^2y(t) = \frac{\delta_{ts}}{\mu(s)}, \quad t \in \mathbb{T}_\alpha^{\rho^2(\beta)}$$
$$y(\alpha) = 0 = y(\beta),$$

has a unique solution on $\mathbb{T}_\alpha^\beta$. Hence for each fixed $s \in \mathbb{T}_\alpha^{\rho(\beta)}$, $G(t,s)$ is uniquely determined for $t \in \mathbb{T}_\alpha^\beta$. Since $s \in \mathbb{T}_\alpha^{\rho(\beta)}$ is arbitrary, $G(t,s)$ is uniquely determined on $\mathbb{T}_\alpha^\beta \times \mathbb{T}_\alpha^{\rho(\beta)}$. □

**Theorem 5.94.** *Assume* $f : \mathbb{T}_\alpha^{\rho^2(\beta)} \to \mathbb{R}$. *Then the unique solution of the BVP*

$$-D^2y(t) = f(t), \quad t \in \mathbb{T}_\alpha^{\rho^2(\beta)}$$
$$y(\alpha) = A_1, \quad y(\beta) = A_2$$

*is given by*

$$y(t) = u(t) + \int_\alpha^\beta G(t,s)f(s)Ds = u(t)$$
$$+ \sum_{j=0}^{K(\beta)-1} G(t, \sigma^j(\alpha))f(\sigma^j(\alpha))\mu(\sigma^j(\alpha)),$$

*where $u(t)$ solves the BVP*

$$\begin{cases} -D^2y(t) = 0, & t \in \mathbb{T}_\alpha^{\rho^2(\beta)} \\ y(\alpha) = A_1, & y(\beta) = A_2 \end{cases}$$

*and $G(t, s)$ is the Green's function for the BVP* (5.7), (5.8).

*Proof.* By Exercise 5.13 the given BVP has a unique solution $y(t)$. By Theorem 5.89

$$\begin{aligned} y(t) &= u(t) + \int_\alpha^\beta G(t, s)f(s)Ds \\ &= u(t) + z(t), \end{aligned}$$

where $z(t) := \int_\alpha^\beta G(t, s)f(s)Ds$ is by Theorem 5.89 the solution of the BVP

$$-D^2z(t) = f(t), \quad z(\alpha) = 0 = z(\beta).$$

It follows that

$$y(\alpha) = u(\alpha) + z(\alpha) = A_1$$

and

$$y(\beta) = u(\beta) + z(\beta) = A_2.$$

Furthermore,

$$-D^2y(t) = -D^2u(t) - D^2z(t) = 0 + f(t) = f(t)$$

for $t \in \mathbb{T}_\alpha^{\rho^2(\beta)}$. □

We now prove a comparison theorem for solutions of boundary value problems of the type treated by Theorem 5.94.

**Theorem 5.95 (Comparison Theorem).** *If $u, v : \mathbb{T}_\alpha^\beta \to \mathbb{R}$ satisfy*

$$\begin{aligned} &D^2u(t) \le D^2v(t), \quad t \in \mathbb{T}_\alpha^{\rho^2(\beta)} \\ &u(\alpha) \ge v(\alpha), \\ &u(\beta) \ge v(\beta). \end{aligned}$$

*Then $u(t) \ge v(t)$ on $\mathbb{T}_\alpha^\beta$.*

*Proof.* Let $w(t) := u(t) - v(t)$, for $t \in \mathbb{T}_\alpha^\beta$. Then for $t \in \mathbb{T}_\alpha^{\rho^2(\beta)}$

$$f(t) := -D^2 w(t) = -D^2 u(t) + D^2 v(t) \geq 0.$$

If $A_1 := u(\alpha) - v(\alpha) \geq 0$ and $A_2 := u(\beta) - v(\beta) \geq 0$, then $w(t)$ solves the boundary value problem

$$\begin{cases} -D^2 w(t) = f(t), & t \in \mathbb{T}_\alpha^{\rho^2(\beta)} \\ w(\alpha) = A_1, & w(\beta) = A_2. \end{cases}$$

Thus, by Theorem 5.94

$$w(t) = y(t) + \int_\alpha^\beta G(t,s) f(s) Ds \quad t \in \mathbb{T}_\alpha^\beta,$$

where $G(t,s)$ is the Green's function defined earlier and $y(t)$ is the solution of

$$\begin{cases} -D^2 y(t) = 0, & t \in \mathbb{T}_\alpha^{\rho^2(\beta)} \\ y(\alpha) = A_1, & y(\beta) = A_2. \end{cases}$$

Since $-D^2 y(t) = 0$ has the general solution

$$y(t) = c_0 + c_1 h_1(t,\alpha) = c_0 + c_1(t - \alpha),$$

and both $y(\alpha), y(\beta) \geq 0$, we have $y(t) \geq 0$. By Theorem 5.90, $G(t,s) \geq 0$, and, thus, we have

$$w(t) = y(t) + \int_\alpha^\beta G(t,s) f(s) Ds \geq 0,$$

for $t \in \mathbb{T}_\alpha^\beta$. □

## 5.13 Exercises

**5.1.** Show that the points in $\mathbb{T}_\alpha$ satisfy

$$\cdots < \rho^2(\alpha) < \rho(\alpha) < \alpha < \sigma(\alpha) < \sigma^2(\alpha) < \cdots.$$

**5.2.** Prove part (ii) of Theorem 5.4.

**5.3.** Prove parts (ii) and (iii) of Theorem 5.6.

**5.4.** Prove part (iv) of Theorem 5.8.

**5.5.** Assume $f : \mathbb{T}_\alpha \to \mathbb{R}$. Show that

$$D^2 f(t) = \frac{f(\sigma^2(t))\mu(t) - f(\sigma(t))[\mu(t) + \mu(\sigma(t))] + f(t)\mu(\sigma(t))}{[\mu(t)]^2 \mu(\sigma(t))}.$$

**5.6.** Assume $c, d \in \mathbb{T}_\alpha$ with $c < d$. Prove that if $f : \mathbb{T}_{[c,d]} \to \mathbb{R}$ and $Df(t) = 0$ for $t \in \mathbb{T}_{[c,\rho(d)]}$, then $f(t) = C$ for all $t \in \mathbb{T}_{[c,d]}$, where $C$ is a constant.

**5.7.** Show that if $n \in \mathbb{N}_1$ and $a \geq 1$, then

$$[n]_a = \sum_{k=0}^{n-1} a^k.$$

Then use this formula to prove parts (iii)–(v) of Theorem 5.22.

**5.8.** Prove part (ii) of Theorem 5.23.

**5.9.** Assume $f : \mathbb{T}_\alpha \times \mathbb{T}_\alpha \to \mathbb{R}$. Derive the Leibniz formula

$$D \int_a^t f(t,s)Ds = \int_a^t Df(t,s)Ds + f(\sigma(t), t)$$

for $t \in \mathbb{T}_\alpha$.

**5.10.** Consider the mixed time scale where $\alpha = 2$ and $\sigma(t) = 3t + 2$ (so $a = 3$ and $b = 2$). Use the variation of constants formula in Theorem 5.38 to solve the IVP

$$D^2 y(t) = 2t - 4, \quad t \in \mathbb{T}_2$$
$$y(2) = 0, \quad Dy(2) = 0.$$

**5.11.** Use the Leibniz formula in Exercise 5.9 to prove the Variation of Constants Theorem (Theorem 5.37).

**5.12.** Prove Theorem 5.43.

**5.13.** Assume $\beta \in \mathbb{T}_{\sigma^2(\alpha)}$, $A, B, E, F \in \mathbb{R}$ and $f : \mathbb{T}_\alpha^{\rho^2(\beta)} \to \mathbb{R}$. Then the nonhomogeneous BVP

$$-D^2 y(t) = f(t), \quad t \in \mathbb{R}_\alpha^\beta$$
$$Ay(\alpha) - BDy(\alpha) = C_1$$
$$Ey(\beta) + FDy(\beta) = C_2,$$

where the constants $C_1$, $C_2$ are given, has a unique solution if and only if the corresponding homogeneous BVP

$$\begin{aligned} -D^2y(t) &= 0, \quad t \in \mathbb{R}_\alpha^\beta \\ Ay(\alpha) - BDy(\alpha) &= 0 \\ Ey(\beta) + FDy(\beta) &= 0 \end{aligned}$$

has only the trivial solution.

**5.14.** Show that for the BVP

$$\begin{aligned} -D^2y(t) &= 0, \quad t \in \mathbb{T}_\alpha^{\rho^2(\alpha)} \\ Dy(\alpha) &= 0 = Dy(\beta), \end{aligned}$$

the $\gamma$ in Theorem 5.87 satisfies $\gamma = 0$. Then show that the given BVP has infinitely many solutions.

**5.15.** Prove parts (ii)–(iv) of Theorem 5.92.

**5.16.** Use Theorem 5.92 to prove directly that the function

$$y(t) := \int_\alpha^\beta G(t,s)f(s)Ds,$$

for $t \in \mathbb{T}_\alpha^\beta$, where $G(t,s)$ is the Green's function for the BVP (5.7), (5.8), solves the BVP (5.5), (5.6).

# Chapter 6
# Fractional Boundary Value Problems

## 6.1 Introduction

In this chapter we derive the Green's function for the fractional boundary value problem (FBVP)

$$\begin{cases} -\Delta_{\nu-2}^{\nu} y(t) = 0, & t \in \mathbb{N}_0^{b+2} \\ y(\nu-2) = 0 = y(\nu+b+1), \end{cases} \tag{6.1}$$

where $\nu \in (1, 2]$ and $b \in \mathbb{N}_0$. Next we derive several important properties of this Green's function which will be useful for proving some interesting results regarding solutions of the nonlinear FBVP

$$\begin{cases} -\Delta_{\nu-2}^{\nu} y(t) = f(t, y(t+\nu-1)), & t \in \mathbb{N}_0^{b+2} \\ y(\nu-2) = A, \quad y(\nu+b+1) = B, \end{cases} \tag{6.2}$$

where $\nu \in (1, 2]$, $f : \mathbb{N}_0^{b+1} \times \mathbb{R} \to \mathbb{R}$ and $b \in \mathbb{N}_0$. In particular, we will prove some results regarding the existence and uniqueness of solutions to the conjugate FBVP (6.2) via various fixed point theorems. We also consider boundary value problems for the so-called linear self-adjoint fractional equation. Most of the results in this chapter are due to Awasthi [40–43].

## 6.2 Two Point Green's Function

In this section we derive the Green's function for the two point FBVP (6.1) and prove several of its properties. In the next theorem we consider the following nonhomogeneous FBVP with homogeneous boundary conditions:



C. Goodrich, A.C. Peterson, *Discrete Fractional Calculus*,
DOI 10.1007/978-3-319-25562-0_6

$$\begin{cases} -\Delta_{\nu-2}^{\nu} y(t) = f(t), & t \in \mathbb{N}_0^{b+2} \\ y(\nu-2) = 0, & y(\nu+b+1) = 0, \end{cases} \tag{6.3}$$

where $f : \mathbb{N}_0^{b+2} \to \mathbb{R}$.

**Theorem 6.1 (Atici and Eloe [31]).** *The solution y of the FBVP* (6.3) *is given by*

$$y(t) = \int_0^{b+3} G(t,s)f(s)\Delta s = \sum_{s=0}^{b+2} G(t,s)f(s), \quad t \in \mathbb{N}_{\nu-2}^{b+\nu+1},$$

*where the so-called Green's function G(t, s) for the FBVP* (6.1) *is given by*

$$G(t,s) := \begin{cases} u(t,s), & 0 \le t-\nu+1 \le s \le b+2, \\ v(t,s), & 0 \le s \le t-\nu+1 \le b+2, \\ 0, & (t,s) \in \{\nu-2\} \times [0,b+2]_{\mathbb{N}_0}, \end{cases}$$

*where*

$$u(t,s) = \frac{t^{\underline{\nu-1}} h_{\nu-1}(\nu+b+1, \sigma(s))}{(\nu+b+1)^{\underline{\nu-1}}},$$

*and*

$$v(t,s) = u(t,s) - h_{\nu-1}(t, \sigma(s)),$$

*where* $h_{\nu-1}(t,s) = \frac{(t-s)^{\underline{\nu-1}}}{\Gamma(\nu)}$ *is the ν-th fractional Taylor monomial based at the point s (see Definition 2.24).*

*Proof.* Let $y$ be a solution of the fractional difference equation

$$-\Delta_{\nu-2}^{\nu} y(t) = f(t), \quad t \in \mathbb{N}_0^{b+2},$$

on $\mathbb{N}_{\nu-2}^{b+\nu+1}$. It follows from Theorems 2.43 and 2.52 that

$$y(t) = -\sum_{s=0}^{t-\nu} \frac{(t-\sigma(s))^{\underline{\nu-1}}}{\Gamma(\nu)} f(s) + C_1 t^{\underline{\nu-1}} + C_2 t^{\underline{\nu-2}}. \tag{6.4}$$

We want $y(t)$ to satisfy the boundary conditions

$$y(\nu-2) = 0, \quad y(\nu+b+1) = 0.$$

Using the first BC, $y(\nu-2)=0$, gives

$$0 = C_1(\nu-2)^{\underline{\nu-1}} + C_2(\nu-2)^{\underline{\nu-2}} = C_2\Gamma(\nu-1),$$

which implies that $C_2 = 0$. Using the second BC, $y(\nu+b+1)=0$, we get that

$$0 = -\sum_{s=0}^{b+1} h_{\nu-1}(\nu+b+1,\sigma(s))h(s) + C_1(\nu+b+1)^{\underline{\nu-1}}.$$

Solving for $C_1$, we have that

$$C_1 = \sum_{s=0}^{b+1} \frac{h_{\nu-1}(\nu+b+1,\sigma(s))}{(\nu+b+1)^{\underline{\nu-1}}} f(s).$$

Thus, from (6.4),

$$y(t) = -\sum_{s=0}^{t-\nu} h_{\nu-1}(t,\sigma(s))f(s) + \sum_{s=0}^{b+1} \frac{t^{\underline{\nu-1}} h_{\nu-1}(\nu+b+1,\sigma(s))}{(\nu+b+1)^{\underline{\nu-1}}} f(s).$$

Since $(\nu+b+1-\sigma(s))^{\underline{\nu-1}} = 0$ for $s = b+2$, the above expression can be rewritten as

$$y(t) = -\sum_{s=0}^{t-\nu} h_{\nu-1}(t,\sigma(s))f(s) + \sum_{s=0}^{b+2} \frac{t^{\underline{\nu-1}} h_{\nu-1}(\nu+b+1,\sigma(s))}{(\nu+b+1)^{\underline{\nu-1}}} f(s).$$

It follows that

$$\begin{aligned} y(t) &= \sum_{s=0}^{b+2} u(t,s)f(s) - \sum_{s=0}^{t-\nu} h_{\nu-1}(t,\sigma(s))f(s) \\ &= \sum_{s=0}^{t-\nu} [u(t,s) - h_{\nu-1}(t,\sigma(s))]f(s) + \sum_{s=t-\nu+1}^{b+2} u(t,s)f(s) \\ &= \sum_{s=0}^{t-\nu} v(t,s)f(s) + \sum_{s=t-\nu+1}^{b+2} u(t,s)f(s) \\ &= \sum_{s=0}^{b+2} G(t,s)f(s), \end{aligned}$$

where $G(t,s)$ is the Green's function for the FBVP (6.1). $\square$

*Remark 6.2.* By Theorem 6.1 the Green's function for the FBVP (6.1) is given by

$$G(t,s) := \begin{cases} \frac{t^{\underline{\nu-1}}\,(\nu+b+1-\sigma(s))^{\underline{\nu-1}}}{\Gamma(\nu)\,(\nu+b+1)^{\underline{\nu-1}}} - \frac{(t-\sigma(s))^{\underline{\nu-1}}}{\Gamma(\nu)}, & s \le t-\nu \\ \frac{t^{\underline{\nu-1}}\,(\nu+b+1-\sigma(s))^{\underline{\nu-1}}}{\Gamma(\nu)\,(\nu+b+1)^{\underline{\nu-1}}}, & t-\nu+1 \le s \end{cases},$$

for $\nu-2 \le t \le \nu+b-1$, $0 \le s \le b+2$.

**Theorem 6.3.** *The Green's function for the FBVP* (6.1) *satisfies*

$$G(t,s) \ge 0,$$

*for* $\nu-2 \le t \le \nu+b-1$, $0 \le s \le b+2$.

*Proof.* From Theorem 6.1 we have that for $0 \le t-\nu+1 \le s \le b+2$

$$u(t,s) = \frac{t^{\underline{\nu-1}}\,(\nu+b+1-\sigma(s))^{\underline{\nu-1}}}{\Gamma(\nu)\,(\nu+b+1)^{\underline{\nu-1}}} \ge 0.$$

To show that $v(t,s) \ge 0$ it suffices to prove that

$$\frac{(t-\sigma(s))^{\underline{\nu-1}}}{\Gamma(\nu)} \le \frac{t^{\underline{\nu-1}}\,(\nu+b+1-\sigma(s))^{\underline{\nu-1}}}{\Gamma(\nu)\,(\nu+b+1)^{\underline{\nu-1}}}, \quad t-\nu+1 \ge s.$$

Equivalently, it suffices to show

$$\frac{(t-\sigma(s))^{\underline{\nu-1}}\,(\nu+b+1)^{\underline{\nu-1}}}{(\nu+b+1-\sigma(s))^{\underline{\nu-1}}\,t^{\underline{\nu-1}}} \le 1.$$

Thus, consider

$$\frac{(t-\sigma(s))^{\underline{\nu-1}}\,(\nu+b+1)^{\underline{\nu-1}}}{(\nu+b+1-\sigma(s))^{\underline{\nu-1}}\,t^{\underline{\nu-1}}}$$
$$= \frac{\Gamma(t-s)}{\Gamma(t-s-\nu+1)}\,\frac{\Gamma(\nu+b+2)}{\Gamma(b+3)}\,\frac{\Gamma(t-\nu+2)}{\Gamma(t+1)}\,\frac{\Gamma(b+2-s)}{\Gamma(\nu+b-s+1)}.$$

But $t = \nu+s+k$ for some $k$, so

$$\frac{(t-\sigma(s))^{\underline{\nu-1}}\,(\nu+b+1)^{\underline{\nu-1}}}{(\nu+b+1-\sigma(s))^{\underline{\nu-1}}\,t^{\underline{\nu-1}}}$$
$$= \left[\frac{\Gamma(\nu+k)}{\Gamma(k+1)}\right]\left[\frac{\Gamma(k+s+2)}{\Gamma(\nu+k+s+1)}\right]\left[\frac{\Gamma(b+2-s)}{\Gamma(b+3)}\right]\left[\frac{\Gamma(\nu+b+2)}{\Gamma(\nu+b+1-s)}\right]$$
$$= \left[\frac{\Gamma(\nu+k)}{\Gamma(\nu+k+s+1)}\right]\left[\frac{\Gamma(k+s+2)}{\Gamma(k+1)}\right]\left[\frac{\Gamma(b+2-s)}{\Gamma(b+3)}\right]\left[\frac{\Gamma(\nu+b+2)}{\Gamma(\nu+b+1-s)}\right].$$

Using a property of the Gamma function, we have that

$$\begin{aligned}
&\frac{(t-\sigma(s))^{\underline{\nu-1}}(\nu+b+1)^{\underline{\nu-1}}}{(\nu+b+1-\sigma(s))^{\underline{\nu-1}}t^{\underline{\nu-1}}}\\
&=\left[\frac{1}{(\nu+k+s)\cdots(\nu+k)}\right]\left[\frac{(k+s+1)\cdots(k+1)}{1}\right]\\
&\cdot\left[\frac{1}{(b+2)\cdots(b+2-s)}\right]\left[\frac{(\nu+b+1)\cdots(\nu+b+1-s)}{1}\right]\\
&=\left[\frac{(k+s+1)(\nu+b+1)}{(\nu+k+s)(b+2)}\right]\cdots\left[\frac{(k+1)(\nu+b+1-s)}{(\nu+k)(b+2-s)}\right].
\end{aligned}$$

It suffices to show that each factor in the last expression is less than or equal to one. We will just show that the first factor is less than or equal to one, since the proof of the other factors being less than or equal to one is similar. Therefore, consider the first factor

$$\frac{(k+s+1)(\nu+b+1)}{(\nu+k+s)(b+2)}=\frac{(k+s)(b+1)+(k+s)\nu+\nu+(b+1)}{(k+s)(b+1)+(k+s)+\nu+\nu(b+1)}.$$

Considering the numerator and the denominator of this last expression, to show that this fraction is less than or equal to one, it suffices to show that

$$(k+s)\nu+(b+1)\leq(k+s)+\nu(b+1).$$

To see this first note that since $(k+s)\leq(b+1)$ and $1<\nu\leq 2$, we get

$$(k+s)(\nu-1)\leq(b+1)(\nu-1).$$

Adding $(k+1)+(b+1)$ to both sides we get the desired result, namely

$$(k+s)\nu+(b+1)\leq(k+s)+\nu(b+1).$$

Therefore, we have that $G(t,s)\geq 0$. □

**Theorem 6.4 (Awasthi [40, 41]).** *Let $G(t,s)$ be the Green's function for the FBVP* (6.1). *Then*

$$\sum_{s=0}^{b+2}G(t,s)=\frac{t^{\underline{\nu-1}}(\nu+b+1-t)}{\Gamma(\nu+1)}$$

*for $t\in\mathbb{N}_{\nu-2}^{\nu+b+1}$.*

*Proof.* By the definition of the Green's function and using the fact that for all $t \in [\nu-2, \nu+b+1]_{\mathbb{N}_{\nu-2}}$, $G(t, b+2) = 0$, we have that

$$\begin{aligned}
\sum_{s=0}^{b+2} G(t,s) &= \sum_{s=0}^{b+1} G(t,s) \\
&= \sum_{s=0}^{t-\nu} v(t,s) + \sum_{s=t+\nu+1}^{b+1} u(t,s) \\
&= \sum_{s=0}^{t-\nu} [u(t,s) - h_{\nu-1}(t,\sigma(s))] + \sum_{s=t+\nu+1}^{b+1} u(t,s) \\
&= \sum_{s=0}^{b+1} u(t,s) - \sum_{s=0}^{t-\nu} h_{\nu-1}(t,\sigma(s)) \\
&= \int_0^{b+2} u(t,s)\Delta s - \int_0^{t-\nu+1} h_{\nu-1}(t,\sigma(s))\Delta s.
\end{aligned}$$

Using integration by parts and the formula for $u(t,s)$, we get that

$$\begin{aligned}
&\sum_{s=0}^{b+2} G(t,s) \\
&= \frac{t^{\underline{\nu-1}}}{(\nu+b+1)^{\underline{\nu-1}}} \int_0^{b+2} h_{\nu-1}(\nu+b+1,\sigma(s))\Delta s + [h_\nu(t,s)]_{s=0}^{t-\nu+1} \\
&= -\frac{t^{\underline{\nu-1}}}{(\nu+b+1)^{\underline{\nu-1}}} [h_\nu(\nu+b+1,s)]_{s=0}^{b+2} - h_\nu(t,0) \\
&= \frac{t^{\underline{\nu-1}}}{(\nu+b+1)^{\underline{\nu-1}}} h_\nu(\nu+b+1,0) - h_\nu(t,0) \\
&= \frac{t^{\underline{\nu-1}}(\nu+b+1)^{\underline{\nu}}}{(\nu+b+1)^{\underline{\nu-1}}\Gamma(\nu+1)} - \frac{t^{\underline{\nu}}}{\Gamma(\nu+1)} \\
&= \frac{t^{\underline{\nu-1}}(\nu+b+1-t)}{\Gamma(\nu+1)}
\end{aligned}$$

for $t \in \mathbb{N}_{\nu-2}^{\nu+b+1}$. □

In the following theorem we find an upper bound for $\sum_{s=0}^{b+2} G(t,s)$, where $t \in \mathbb{N}_{\nu-2}^{\nu+b+1}$.

**Theorem 6.5 (Awasthi [40, 41]).** *The Green's function for the FBVP* (6.1) *satisfies*

$$\max_{t\in\mathbb{N}_{\nu-2}^{\nu+b+1}} \sum_{s=0}^{b+2} G(t,s)$$
$$= \frac{1}{\Gamma(\nu+1)}\left[\nu+\left\lceil b-\frac{b+2}{\nu}\right\rceil\right]^{\underline{\nu-1}}\left[b+1-\left\lceil b-\frac{b+2}{\nu}\right\rceil\right],$$

*where* $\lceil\cdot\rceil$ *is the ceiling function.*

*Proof.* By Theorem 6.4, we have

$$\sum_{s=0}^{b+2} G(t,s) = \frac{(\nu+b+1-t)\,t^{\underline{\nu-1}}}{\Gamma(\nu+1)} = \frac{1}{\Gamma(\nu+1)}F(t),$$

where $F(t) := t^{\underline{\nu-1}}\,(\nu+b+1-t)$. We observe that $F(t)\geq 0$ on $\mathbb{N}_{\nu-2}^{\nu+b+1}$, with $F(\nu-2)=0$ and $F(\nu+b+1)=0$. So $F$ has a nonnegative maximum and to find this maximum we consider

$$\begin{aligned}\Delta F(t) &= (-1)\,t^{\underline{\nu-1}} + (\nu-1)(\nu+b-t)\,t^{\underline{\nu-2}}\\ &= t^{\underline{\nu-2}}\,[-(t-\nu+2)+(\nu-1)(\nu+b-t)]\\ &= t^{\underline{\nu-2}}\,\left[\nu^2+b\nu-t\nu-b-2\right].\end{aligned}$$

It follows from this preceding expression that $F(t)$ has its maximum on $\mathbb{N}_{\nu-2}^{\nu+b+1}$ at $t=\nu+\left\lceil b-\frac{b+2}{\nu}\right\rceil$. Hence,

$$\max_{t\in\mathbb{N}_{\nu-2}^{\nu+b+1}} \sum_{s=0}^{b+2} G(t,s)$$
$$= \frac{1}{\Gamma(\nu+1)}\left[\nu+\left\lceil b-\frac{b+2}{\nu}\right\rceil\right]^{\underline{\nu-1}}\left[b+1-\left\lceil b-\frac{b+2}{\nu}\right\rceil\right].$$

This completes the proof. □

*Remark 6.6.* It is very important to mention that we are able to calculate the actual maximum of the summation of the Green's function. By putting $\nu = 2$ and choosing $b$ to be any nonnegative integer, one can compare the above result with the classical cases both in the theory of ordinary differential equations and of difference equations.

**Theorem 6.7.** *Assume* $\nu \in (1,2]$ *and* $h : [0, b+2]_{\mathbb{N}_{\nu-2}} \to \mathbb{R}$. *Then the solution to the nonhomogeneous FBVP*

$$\begin{cases} -\Delta^{\nu}_{\nu-2}\, y(t) = h(t), & t \in \mathbb{N}_0^{b+2} \\ y(\nu-2) = A, \quad y(\nu+b+1) = B \end{cases} \tag{6.5}$$

*is given by*

$$y(t) = z(t) + \sum_{s=0}^{b+2} G(t,s)h(s), \quad t \in \mathbb{N}_{\nu-2}^{b+\nu+1},$$

*where* $G(t,s)$ *is the Green's function for the FBVP* (6.1) *and* $z(t)$ *is the unique solution to the FBVP*

$$\begin{cases} \Delta^{\nu}_{\nu-2}\, z(t) = 0, & t \in \mathbb{N}_0^{b+2} \\ z(\nu-2) = A, \quad z(\nu+b+1) = B. \end{cases} \tag{6.6}$$

*Proof.* Let

$$w(t) := \sum_{s=0}^{b+2} G(t,s)h(s), \quad t \in \mathbb{N}_{\nu-2}^{b+\nu+1}.$$

By Theorem 6.1 $w(t)$ is the solution of the FBVP (6.3) on $\mathbb{N}_{\nu-2}^{b+\nu+1}$. Let $y(t)$ and $z(t)$ be as in the statement of this theorem. Then

$$y(\nu-2) = z(\nu-2) + w(\nu-2) = A + 0 = A,$$

and

$$y(b+\nu+1) = z(b+\nu+1) + w(b+\nu+1) = B + 0 = B.$$

Finally,

$$-\Delta^{\nu}_{\nu-2}y(t) = -\Delta^{\nu}_{\nu-2}z(t) - \Delta^{\nu}_{\nu-2}w(t) = h(t)$$

for $t \in \mathbb{N}_0^{b+2}$. □

## 6.3 Various Fixed Point Theorems

Fixed point theorems are useful tools for guaranteeing the existence and uniqueness of solutions of nonlinear equations in ordinary differential equations, partial differential equations, and many other areas of pure and applied mathematics. In this section, we will discuss the application of several fixed point theorems to the solution of nonlinear fractional boundary value problems. In particular, conjugate discrete fractional boundary value problems will be our main interest. We start with the following well-known result [167].

**Theorem 6.8 (Contraction Mapping Theorem).** *Let $(X, \|\cdot\|)$ be a Banach space and $T : X \to X$ be a contraction mapping. Then $T$ has a unique fixed point in $X$.*

The following theorem is an application of the above theorem.

**Theorem 6.9 (Awasthi [40, 41]).** *Assume that $f : [0, b+2]_{\mathbb{N}_0} \times \mathbb{R} \to \mathbb{R}$ satisfies a uniform Lipschitz condition with respect to its second variable with Lipschitz constant $k > 0$. If*

$$\left(\nu + \left\lceil b - \frac{b+2}{\nu} \right\rceil\right)^{\underline{\nu-1}} \left(b + 1 - \left\lceil b - \frac{b+2}{\nu} \right\rceil\right) < \frac{\Gamma(\nu+1)}{k},$$

*then the nonlinear fractional boundary value problem*

$$\begin{cases} -\Delta^{\nu}_{\nu-2}\, y(t) = f(t, y(t+\nu-1)), & t \in [0, b+2] \\ y(\nu-2) = A, \quad y(\nu+b+1) = B \end{cases} \tag{6.7}$$

*has a unique solution.*

*Proof.* Let $Z$ be the space of real-valued functions defined on $\mathbb{N}^{\nu+b+1}_{\nu-2}$. Then we define a norm $\|\cdot\|$ on $Z$ by $\|x\| = \max\{|x(t)| : t \in \mathbb{N}^{\nu+b+1}_{\nu-2}\}$ so that the pair $(Z, \|\cdot\|)$ is a Banach space. Now we define the map $T : Z \to Z$ by

$$Tx(t) = z(t) + \sum_{s=0}^{b+2} G(t,s) f(s, x(s+\nu-1)), \quad t \in \mathbb{N}^{\nu+b+1}_{\nu-2},$$

where $z$ is the unique solution to the FBVP (6.6) and $G$ is the Green's function for the FBVP (6.1). Next we will show that $T$ defined as above is a contraction map. Observe for all $t \in \mathbb{N}^{\nu+b+1}_{\nu-2}$ and for all $x, y \in Z$ that

$$\begin{aligned}
&\|Tx(t) - Ty(t)\| \\
&= \max_{t \in \mathbb{N}^{\nu+b+1}_{\nu-2}} \left| \sum_{s=0}^{b+2} G(t,s)[f(s, x(s+\nu-1)) - f(s, y(s+\nu-1))] \right| \\
&\le \max_{t \in \mathbb{N}^{\nu+b+1}_{\nu-2}} \sum_{s=0}^{b+2} G(t,s) |f(s, x(s+\nu-1)) - f(s, y(s+\nu-1))| \\
&\le \max_{t \in \mathbb{N}^{\nu+b+1}_{\nu-2}} \sum_{s=0}^{b+2} G(t,s) k |x(s+\nu-1) - y(s+\nu-1)|
\end{aligned}$$

$$
\begin{aligned}
&\leq k\|x-y\| \max_{t\in\mathbb{N}_{\nu-2}^{\nu+b+1}} \sum_{s=0}^{b+2} G(t,s)\\
&= \frac{k\|x-y\|}{\Gamma(\nu+1)}\left(\nu+\left\lceil b-\frac{b+2}{\nu}\right\rceil\right)^{\underline{\nu-1}}\left(b+1-\left\lceil b-\frac{b+2}{\nu}\right\rceil\right)\\
&\leq \alpha\|x-y\|,
\end{aligned}
$$

where

$$\alpha=\frac{k}{\Gamma(\nu+1)}\left(\nu+\left\lceil b-\frac{b+2}{\nu}\right\rceil\right)^{\underline{\nu-1}}\left(b+1-\left\lceil b-\frac{b+2}{\nu}\right\rceil\right)<1$$

by assumption. Therefore $T$ is a contraction mapping on $Z$. Hence $T$ has a unique fixed point in $Z$. Thus there exists a unique $\bar{x}\in Z$ such that $T(\bar{x})=\bar{x}$.

Next we will show that $\bar{x}$ is the unique solution of (6.7). Consider the FBVP

$$\begin{cases}\Delta_{\nu-2}^{\nu}u(t)=f(t,\bar{x}(t+\nu-1))\\ u(\nu-2)=A, \quad u(\nu+b+1)=B.\end{cases}$$

Using Theorem 6.7 we get that the solution is given by

$$u(t)=z(t)+\sum_{s=0}^{b+2}G(t,s)f(s,\bar{x}(s+\nu-1)).$$

But

$$u(t)=z(t)+\sum_{s=0}^{b+2}G(t,s)f(s,\bar{x}(s+\nu-1))=T\bar{x}(t)=\bar{x}(t),$$

which implies that $\bar{x}$ is a solution to (6.5) since $u(t)=\bar{x}(t)$ for all $t\in\mathbb{N}_{\nu-2}^{\nu+b+1}$. The uniqueness of the solution to the FBVP (6.7) follows from the fact that the Contraction Mapping Theorem guarantees the uniqueness of a fixed point of the mapping $T$. □

The following result appears in [167].

**Theorem 6.10 (Schauder's Fixed Point Theorem).** *Every continuous function from a compact, convex subset of a topological vector space to itself has a fixed point.*

The following theorem is an application of Schauder's Fixed Point Theorem.

**Theorem 6.11 (Awasthi [40, 41]).** *Assume that $f:[0,b+2]_{\mathbb{N}_0}\times\mathbb{R}\to\mathbb{R}$ is continuous in its second variable and $M\geq\max_{t\in\mathbb{N}_{\nu-2}^{\nu+b+1}}|z(t)|$, where $z$ is the unique solution to the FBVP*

$$\begin{cases} \Delta_{\nu-2}^{\nu} z(t) = 0, & t \in \mathbb{N}_0^{b+2}, \\ z(\nu-2) = A, & z(\nu+b+1) = B. \end{cases}$$

*Let* $C = \max\{|f(t,u)| : 0 \leq t \leq b+2, u \in \mathbb{R}, |u| \leq 2M\} > 0$. *Then the nonlinear FBVP* (6.7) *has a solution provided*

$$\left(\nu + \left\lceil b - \frac{b+2}{\nu} \right\rceil\right)^{\underline{\nu-1}} \left(b + 1 - \left\lceil b - \frac{b+2}{\nu} \right\rceil\right) \leq \frac{\Gamma(\nu+1)M}{C}.$$

*Proof.* Let Z be the Banach space defined in the proof of Theorem 6.9. Thus $Z$ is a topological vector space. Let $K = \{y \in Z : \|y\| \leq 2M\}$, then $K$ is a compact, convex subset of Z. Next define the map $T : Z \to Z$ by

$$Ty(t) = z(t) + \sum_{s=0}^{b+2} G(t,s) f(s, y(s+\nu-1)).$$

We will first show that $T$ maps $K$ into $K$. Observe that for $t \in \mathbb{N}_{\nu-2}^{\nu+b+1}$ and $y \in K$ we have

$$\begin{aligned} |Ty(t)| &= \left| z(t) + \sum_{s=0}^{b+2} G(t,s) f(s, y(s+\nu-1)) \right| \\ &\leq |z(t)| + \sum_{s=0}^{b+2} G(t,s) |f(s, y(s+\nu-1))| \\ &\leq M + C \sum_{s=0}^{b+2} G(t,s) \\ &\leq M + C \frac{1}{\Gamma(\nu+1)} \left(\nu + \left\lceil b - \frac{b+2}{\nu} \right\rceil\right)^{\underline{\nu-1}} \left(b + 1 - \left\lceil b - \frac{b+2}{\nu} \right\rceil\right) \\ &\leq M + C \frac{1}{\Gamma(\nu+1)} \frac{\Gamma(\nu+1)M}{C} \\ &\leq 2M. \end{aligned}$$

Since $t \in \mathbb{N}_{\nu-2}^{\nu+b+1}$ was arbitrary, we have that $\|Ty\| \leq 2M$. This proves that $T$ maps $K$ into $K$.

Next we will show that $T$ is continuous on $K$. Let $\epsilon > 0$ be given and put $l := \max_{t \in \mathbb{N}_{\nu-2}^{\nu+b+1}} \sum_{s=0}^{b+2} G(t,s)$. Then by Theorem 6.5 we have that

$$l = \frac{1}{\Gamma(\nu+1)}\left(\nu + \left\lceil b - \frac{b+2}{\nu}\right\rceil\right)^{\underline{\nu-1}}\left(b+1-\left\lceil b - \frac{b+2}{\nu}\right\rceil\right).$$

Since $f$ is continuous in its second variable on $\mathbb{R}$, we have that $f$ is, in fact, uniformly continuous in its second variable on $[-2M, 2M]$. Therefore, there exists a $\delta > 0$ such that for all $t \in [0, b+2]_{\mathbb{N}_0}$, and for all $u, v \in [-2M, 2M]$ with $|(t,u) - (t,v)| < \delta$ we have

$$|f(t,u) - f(t,v)| < \frac{\epsilon}{l}.$$

Thus for all $t \in \mathbb{N}_{\nu-2}^{\nu+b+1}$ it follows that

$$\begin{aligned}
&|Ty(t) - Tx(t)| \\
&= \left|\sum_{s=0}^{b+2} G(t,s)f(s, y(s+\nu-1)) - \sum_{s=0}^{b+2} G(t,s)f(s, x(s+\nu-1))\right| \\
&\leq \sum_{s=0}^{b+2} G(t,s)|f(s, y(s+\nu-1)) - f(s, x(s+\nu-1))| \\
&< \sum_{s=0}^{b+2} G(t,s)\frac{\epsilon}{l} \leq l\frac{\epsilon}{l} = \epsilon.
\end{aligned}$$

Now since $t \in \mathbb{N}_{\nu-2}^{\nu+b+1}$ was arbitrary we have that

$$\|T(x) - T(y)\| < \epsilon.$$

This establishes the continuity of $T$ on $K$. Hence, $T$ is a continuous map from $K$ into $K$. Therefore, by Schauder's Fixed Point Theorem (Theorem 6.10), $T$ has a fixed point in $K$. Thus, there exists a $\bar{x} \in K$ such that $T(\bar{x}) = \bar{x}$. This implies the existence of a solution to the FBVP (6.7), and so this completes the proof. □

*Remark 6.12.* The above theorem not only guarantees the existence of a solution $y(t)$ but also shows that the solution satisfies $|y(t)| \leq 2M$ for $t \in [\nu-2, \nu+b+1]_{\mathbb{N}_{\nu-2}}$.

The following fixed point theorem appears in [167] and is a special case of Brouwer's Theorem.

**Theorem 6.13.** *Let $Z$ be a Banach space and $T : Z \to Z$ be a compact operator. If $T(Z)$ is bounded, then $T$ has a fixed point in $Z$.*

As an application of this theorem we will prove the following theorem regarding the existence of a solution of the FBVP (6.5) under a strong assumption on $f$. This result is important in proving results concerning upper and lower solutions.

**Theorem 6.14 (Awasthi [40, 41]).** *Assume that $f : [0, b+2]_{\mathbb{N}_0} \times \mathbb{R} \to \mathbb{R}$ is continuous with respect to its second variable and is bounded. Then the nonlinear FBVP* (6.7) *has a solution.*

*Proof.* Let $(Z, \|\cdot\|)$ be the Banach space as defined earlier. Now we define the operator $T : Z \to Z$ by

$$Ty(t) = z(t) + \sum_{s=0}^{b+2} G(t,s)f(s, y(s+\nu-1)), \quad t \in \mathbb{N}_{\nu-2}^{\nu+b+1},$$

where $z$ is the unique solution to the FBVP

$$\begin{cases} \Delta_{\nu-2}^{\nu} z(t) = 0 \\ z(\nu-2) = A, \quad z(\nu+b+1) = B. \end{cases} \tag{6.8}$$

It is easy to see that the operator $T$ is compact. Next we will show that $T(Z)$ is bounded. Since $f$ is bounded, there exists $m > 0$ such that for all $t \in [0, b+2]_{\mathbb{N}_0}$ and for all $u \in \mathbb{R}$, it holds that $|f(t,u)| \leq m$. Thus for any $y \in Z$ and $t \in \mathbb{N}_{\nu-2}^{\nu+b+1}$ we observe that

$$\begin{aligned} |Ty(t)| &= \left| z(t) + \sum_{s=0}^{b+2} G(t,s)f(s, y(s+\nu-1)) \right| \\ &\leq |z(t)| + m \sum_{s=0}^{b+2} G(t,s) \\ &\leq \max_{t \in \mathbb{N}_{\nu-2}^{\nu+b+1}} |z(t)| + m \max_{t \in \mathbb{N}_{\nu-2}^{\nu+b+1}} \sum_{s=0}^{b+2} G(t,s). \end{aligned}$$

Hence, $T$ is bounded on $Z$, and the conclusion follows as a result of Theorem 6.13. □

**Theorem 6.15 (Awasthi [40, 41]).** *Assume that $f : [0, b+2]_{\mathbb{N}_0} \times \mathbb{R} \to \mathbb{R}$ satisfies a uniform Lipschitz condition with respect to its second variable, with Lipschitz constant $k$, and that the equation $\Delta_{\nu-2}^{\nu} y(t) + ky(t+\nu-1) = 0$ has a positive solution $u$. Then it follows that the nonlinear FBVP* (6.7) *has a unique solution.*

*Proof.* Since the equation

$$\Delta_{\nu-2}^{\nu} y(t) + ky(t+\nu-1) = 0, \quad t \in \mathbb{N}_0^{b+2}$$

has a positive solution $u$ on $\mathbb{N}_{\nu-2}^{\nu+b+1}$, it follows that $u(t)$ is a solution of the FBVP

$$\begin{cases} -\Delta_{\nu-2}^{\nu} y(t) = ku(t+\nu-1), & t \in \mathbb{N}_0^{b+2} \\ y(\nu-2) = C, \quad y(\nu+b+1) = D, \end{cases} \tag{6.9}$$

where $C := u(\nu-2) > 0$ and $D := u(\nu+b+1) > 0$. By using the conclusion of Theorem 6.7 we have that

$$u(t) = z(t) + \sum_{s=0}^{b+2} G(t,s)\, ku(s+\nu-1), \tag{6.10}$$

where $z$ is the unique solution to the FBVP

$$\begin{cases} \Delta_{\nu-2}^{\nu} z(t) = 0, & t \in \mathbb{N}_0^{b+2} \\ z(\nu-2) = C, \quad z(\nu+b+1) = D, \end{cases}$$

and $G(t,s)$ is the Green's function for the FBVP (6.1). Again by using Theorem 6.7, $z$ is given by

$$\begin{aligned} z(t) = {} & \frac{1}{(\nu+b+1)^{\underline{\nu-1}}} \left[ u(\nu+b+1) - u(\nu-2) \frac{(\nu+b+1)^{\underline{\nu-2}}}{\Gamma(\nu-1)} \right] t^{\underline{\nu-1}} \\ & + \frac{u(\nu-2)}{\Gamma(\nu-1)}\, t^{\underline{\nu-2}}. \end{aligned}$$

We now show that $z(t) > 0$ on $\mathbb{N}_{\nu-2}^{\nu+b+1}$. Since $t^{\underline{\nu-1}} = t^{\underline{\nu-2}}\,(t-\nu+2)$, then by replacing $t^{\underline{\nu-1}}$ with $t^{\underline{\nu-2}}\,(t-\nu+2)$ and rearranging the terms yields

$$\begin{aligned} z(t) &= t^{\underline{\nu-2}} \left[ \frac{(t-\nu+2)}{(\nu+b+1)^{\underline{\nu-1}}} \left( u(\nu+b+1) - \frac{u(\nu-2)(\nu+b+1)^{\underline{\nu-2}}}{\Gamma(\nu-1)} \right) \right. \\ &\quad \left. + \frac{u(\nu-2)}{\Gamma(\nu-1)} \right] \\ &= t^{\underline{\nu-2}} \left[ \frac{(t-\nu+2)u(\nu+b+1)}{(\nu+b+1)^{\underline{\nu-1}}} \right. \\ &\quad \left. + \frac{u(\nu-2)}{\Gamma(\nu-1)} \left( 1 - \frac{(t-\nu+2)(\nu+b+1)^{\underline{\nu-2}}}{(\nu+b+1)^{\underline{\nu-1}}} \right) \right] \\ &= t^{\underline{\nu-2}} \left[ \frac{(t-\nu+2)u(\nu+b+1)}{(\nu+b+1)^{\underline{\nu-1}}} + \frac{u(\nu-2)}{\Gamma(\nu-1)} \left( 1 - \frac{(t-\nu+2)}{b+3} \right) \right] \\ &= t^{\underline{\nu-2}} \left[ \frac{(t-\nu+2)u(\nu+b+1)}{(\nu+b+1)^{\underline{\nu-1}}} + \frac{(\nu+b+1-t)u(\nu-2)}{\Gamma(\nu-1)(b+3)} \right] \\ &= t^{\underline{\nu-2}}\, h(t), \end{aligned}$$

where

$$h(t) = \frac{(t-\nu+2)u(\nu+b+1)}{(\nu+b+1)^{\underline{\nu-1}}} + \frac{(\nu+b+1-t)u(\nu-2)}{\Gamma(\nu-1)(b+3)}.$$

Since $t^{\underline{\nu-2}}$ is a decreasing function of $t$ and $(\nu+b+1)^{\underline{\nu-2}} = \frac{\Gamma(\nu+b+2)}{\Gamma(b+4)} > 0$, we get that $t^{\underline{\nu-2}} > 0$, and in order to show that $z(t)$ is positive we just need to show that $h(t)$ is positive.

We note in the definition of $h(t)$ that the first term is zero only at the left end point $t = \nu - 2$ and is positive elsewhere. Also the second term is zero only at the right end point $t = \nu + b + 1$ and positive elsewhere. Therefore, combining these arguments we conclude that $z(t) > 0$ for all $t \in \mathbb{N}_{\nu-2}^{\nu+b+1}$. Thus, by (6.10) for all $t \in \mathbb{N}_{\nu-2}^{\nu+b+1}$ we have that

$$u(t) > \sum_{s=0}^{b+2} kG(t,s)u(s+\nu-1).$$

Hence,

$$\alpha = \max_{t\in\mathbb{N}_{\nu-2}^{\nu+b+1}} \frac{k}{u(t)} \sum_{s=0}^{b+2} G(t,s)u(s+\nu-1) < 1.$$

Now, let Z be the space of real-valued functions defined on $\mathbb{N}_{\nu-2}^{\nu+b+1}$. Consider the (weighted) norm $\|\cdot\|$ defined by

$$\|x\| = \max\left\{\frac{|x(t)|}{u(t)} : t \in \mathbb{N}_{\nu-2}^{\nu+b+1}\right\}.$$

Then the pair $(Z, \|\cdot\|)$ is a complete normed space. Define $T$ on $Z$ by

$$Tx(t) = z(t) + \sum_{s=0}^{b+2} G(t,s)h(s). \quad t \in \mathbb{N}_{\nu-2}^{\nu+b+1}.$$

Then for all $t \in \mathbb{N}_{\nu-2}^{\nu+b+1}$ we have that

$$\frac{|Tx(t)-Ty(t)|}{u(t)}$$

$$= \frac{1}{u(t)}\left|\sum_{s=0}^{b+2} G(t,s)[f(s,x(s+\nu-1)) - f(s,y(s+\nu-1))]\right|$$

$$
\begin{aligned}
&\le \frac{1}{u(t)} \sum_{s=0}^{b+2} G(t,s)|f(s,x(s+\nu-1)) - f(s,y(s+\nu-1))| \\
&\le \frac{1}{u(t)} \sum_{s=0}^{b+2} G(t,s)k|x(s+\nu-1) - y(s+\nu-1)| \\
&= \frac{1}{u(t)} \sum_{s=0}^{b+2} G(t,s)ku(s+\nu-1)\frac{|x(s+\nu-1) - y(s+\nu-1)|}{u(s+\nu-1)} \\
&\le \|x-y\| \frac{k}{u(t)} \sum_{s=0}^{b+2} G(t,s)u(s+\nu-1) \\
&\le \|x-y\| \max_{t\in\mathbb{N}_{\nu-2}^{\nu+b+1}} \frac{k}{u(t)} \sum_{s=0}^{b+2} G(t,s)\, u(s+\nu-1) \\
&= \alpha\, \|x-y\|.
\end{aligned}
$$

Since $\alpha < 1$, it follows that $T$ is a contraction mapping on $Z$. Therefore, $T$ has a unique fixed point in $Z$. This implies the existence of a unique solution to the nonlinear FBVP (6.7). □

## 6.4 Self-Adjoint Linear Fractional Difference Equation

In this section, we prove an existence and uniqueness theorem for a class of boundary value problems known as the self-adjoint linear fractional difference equation—that is,

$$\Delta^{\mu}_{\mu-1}(p\Delta x)(t) + q(t+\mu-1)x(t+\mu-1) = f(t), \quad t \in \mathbb{N}_0^b, \tag{6.11}$$

where $0 < \mu \le 1$, $b \in \mathbb{N}_1$, $p : \mathbb{N}_{\mu-1}^{\mu+b} \to (0,\infty)$, $f : \mathbb{N}_0^b \to \mathbb{R}$, and $q : \mathbb{N}_{\mu-1}^{\mu+b-1} \to \mathbb{R}$. Note that if $\mu = 1$, then we obtain the standard self-adjoint difference equation

$$\Delta(p\Delta x)(t) + q(t)x(t) = f(t), \quad t \in \mathbb{N}_0^b, \tag{6.12}$$

and it is for this reason that we call (6.11) a fractional self-adjoint equation. A slight variation of the difference equation (6.12) is well studied in Kelley–Peterson [135].

We next state and prove an existence and uniqueness theorem for the following fractional initial value problem (FIVP):

$$\begin{cases} \Delta^{\mu}_{\mu-1}(p\Delta x)(t) + q(t+\mu-1)x(t+\mu-1) = h(t), & t \in \mathbb{N}_0^b \\ x(\mu-1) = A, \quad x(\mu) = B. \end{cases} \tag{6.13}$$

**Theorem 6.16 (Awasthi [40, 41]).** *Assume* $h : \mathbb{N}_0^b \to \mathbb{R}$, $q : \mathbb{N}_{\mu-1}^{\mu+b-1} \to \mathbb{R}$, $p : \mathbb{N}_{\mu-1}^{\mu+b} \to \mathbb{R}$ *with* $p(t) > 0$, $A, B \in \mathbb{R}$. *Then the FIVP*

$$\begin{cases} \Delta_{\mu-1}^{\mu}(p\Delta x)(t) + q(t+\mu-1)x(t+\mu-1) = h(t), & t \in \mathbb{N}_0^b \\ x(\mu-1) = A, \quad x(\mu) = B, \end{cases} \tag{6.14}$$

*has a unique solution that exists on* $\mathbb{N}_{\mu-1}^{\mu+b+1}$.

*Proof.* Consider the self-adjoint equation

$$\Delta_{\mu-1}^{\mu}(p\Delta x)(t) + q(t+\mu-1)x(t+\mu-1) = h(t), \qquad t \in \mathbb{N}_0^b.$$

By applying the alternate definition of the fractional difference of a function (Theorem 2.33) we have this fractional difference equation can be written in the form (for $t \in \mathbb{N}_0^b$)

$$\frac{1}{\Gamma(-\mu)} \sum_{s=\mu-1}^{t+\mu} (t-\sigma(s))^{\underline{-\mu-1}} p(s)\Delta x(s) + q(t+\mu-1)x(t+\mu-1) = h(t). \tag{6.15}$$

We now show the existence and uniqueness of our solution. Letting $t = 0$ in Eq. (6.15), we get

$$\begin{aligned} h(0) &= \frac{1}{\Gamma(-\mu)} \sum_{s=\mu-1}^{\mu} (-\sigma(s))^{\underline{-\mu-1}} p(s)\Delta x(s) + q(\mu-1)x(\mu-1) \\ &= \frac{1}{\Gamma(-\mu)} \left[(-\mu)^{\underline{-\mu-1}} p(\mu-1)(x(\mu)-x(\mu-1))\right] \\ &\quad + \frac{1}{\Gamma(-\mu)} \left[(-\mu-1)^{\underline{-\mu-1}} p(\mu)(x(\mu+1)-x(\mu))\right] + Aq(\mu-1). \end{aligned}$$

Hence,

$$h(0) = (-\mu)p(\mu-1)(B-A) + p(\mu)(x(\mu+1)-B) + Aq(\mu-1).$$

Solving for $x(\mu+1)$ we have that

$$x(\mu+1) = \frac{h(0) + \mu p(\mu-1)(B-A) - Aq(\mu-1)}{p(\mu)} + B. \tag{6.16}$$

Thus we see that the value of $x(t)$ at $t = \mu+1$ is uniquely determined by the two initial values of $x(t)$ at $t = \mu-1$ and $t = \mu$. Hence, we get the existence and uniqueness of the solution of the FIVP (6.14) on $\mathbb{N}_{\mu-1}^{\mu+1}$.

We now show the existence and uniqueness of our solution on $\mathbb{N}_{\mu-1}^{\mu+b+1}$ by induction. To this end assume there is a unique solution $x(t)$ on $\mathbb{N}_{\mu-1}^{t_0}$, where $t_0 \in \mathbb{N}_{\mu+1}^{\mu+b}$. We argue that the values of the solution map $t \mapsto x(t)$, for $t \in \mathbb{N}_{\mu-1}^{t_0}$, uniquely determine the value of the solution at $t_0 + 1$. To prove this, we first substitute $t = t_0 - \mu$ in (6.15) to get

$$\begin{aligned}
h(t_0-\mu) &= \frac{1}{\Gamma(-\mu)} \sum_{s=\mu-1}^{t_0} (t_0-\mu-\sigma(s))^{\underline{-\mu-1}} p(s)\Delta x(s) \\
&\quad + q(t_0-1)x(t_0-1) \\
&= \frac{1}{\Gamma(-\mu)} \left[ \sum_{s=\mu-1}^{t_0-1} (t_0-\mu-\sigma(s))^{\underline{-\mu-1}} p(s)\Delta x(s) \right] \\
&\quad + \frac{1}{\Gamma(-\mu)} \left[ (-\mu-1)^{\underline{-\mu-1}} p(t_0)[x(t_0+1)-x(t_0)] \right] \\
&\quad + q(t_0-1)x(t_0-1) \\
&= \frac{1}{\Gamma(-\mu)} \left[ \sum_{s=\mu-1}^{t_0-1} (t_0-\mu-\sigma(s))^{\underline{-\mu-1}} p(s)\Delta x(s) \right] \\
&\quad + p(t_0)[x(t_0+1)-x(t_0)] + q(t_0-1)x(t_0-1).
\end{aligned}$$

We can uniquely solve the above equation for $x(t_0 + 1)$ to get that

$$\begin{aligned}
x(t_0+1) &= x(t_0) + \frac{1}{p(t_0)} [h(t_0-\mu) - q(t_0-1)x(t_0-1)] \\
&\quad - \frac{1}{p(t_0)} \left[ \frac{1}{\Gamma(-\mu)} \sum_{s=\mu-1}^{t_0-1} (t_0-\mu-\sigma(s))^{\underline{-\mu-1}} p(s)\Delta x(s) \right].
\end{aligned}$$

Since by the induction hypothesis, each of the values of $x(t)$ in the expression

$$\frac{1}{\Gamma(-\mu)} \sum_{s=\mu-1}^{t_0-1} (t_0-\mu-\sigma(s))^{\underline{-\mu-1}} p(s)\Delta x(s)$$

is known, it follows that $x(t_0 + 1)$ is uniquely determined, and hence we have that $x(t)$ is the unique solution of (6.14) on $\mathbb{N}_{\mu-1}^{t_0+1}$. Thus, by mathematical induction, the fractional IVP (6.14) has a unique solution that exists on $\mathbb{N}_{\mu-1}^{\mu+b+1}$. □

*Remark 6.17.* If $\mu = 1$ (i.e., an integer case) in Theorem 6.16, it can be shown that for any $t_0 \in \mathbb{N}_{\mu-1}^{\mu+b}$ the initial conditions $x(t_0) = A$ and $x(t_0 + 1) = B$ determine a unique solution of the IVP (6.14) if $q(t) \neq 0$. Note in the fractional case $0 < \mu < 1$, we just get the existence and uniqueness of the solution of (6.14) for the case $t_0 = \mu - 1$. The reason for this is that in the true fractional case (i.e., $0 < \mu < 1$) the fractional difference depends on all of its values back to its initial value at $\mu - 1$.

## 6.5 Variation of Constants Formula

In this section we are interested in establishing the variation of constants formula for the self-adjoint FIVP

$$\begin{cases} \Delta^{\mu}_{\mu-1}(p\Delta x)(t) = h(t), & t \in \mathbb{N}_0^b \\ x(\mu-1) = \Delta x(\mu-1) = 0, \end{cases} \tag{6.17}$$

where $0 < \mu \leq 1$, $b \in \mathbb{N}_1$, and $p : \mathbb{N}_{\mu-1}^{b+\mu} \to \mathbb{R}$ with $p(t) > 0$. Our variation of constants formula will involve the Cauchy function, which we now define.

**Definition 6.18.** We define the Cauchy function $x(\cdot,\cdot)$ for the homogeneous fractional equation

$$\Delta^{\mu}_{\mu-1}(p\Delta x)(t) = 0$$

to be the function $x : \mathbb{N}_{\mu-1}^{\mu+b+1} \times \mathbb{N}_0^b \to \mathbb{R}$ such that for each fixed $s \in \mathbb{N}_0^b$, $x(\cdot, s)$ is the solution of the fractional initial value problem

$$\begin{cases} \Delta^{\mu}_{\mu-1}(p\Delta x)(t) = 0, & t \in \mathbb{N}_0^b \\ x(s+\mu) = 0, \quad \Delta x(s+\mu) = \frac{1}{p(s+\mu)} \end{cases} \tag{6.18}$$

and is given by the formula

$$x(t,s) = \sum_{\tau=s+\mu}^{t-1} \left[ \frac{1}{p(\tau)} \frac{(\tau - \sigma(s))^{\underline{\mu-1}}}{\Gamma(\mu)} \right], \quad t \in \mathbb{N}_{\mu-1}^{\mu+b+1}.$$

Note that by our convention on sums $x(t,s) = 0$ for $t \leq s + \mu$.

**Theorem 6.19.** *Let $h : \mathbb{N}_0^b \to \mathbb{R}$ and $p : \mathbb{N}_{\mu-1}^{\mu+b} \to \mathbb{R}$ with $p(t) > 0$. Then the solution to the initial value problem*

$$\begin{cases} \Delta^{\mu}_{\mu-1}(p\Delta x)(t) = h(t), & t \in \mathbb{N}_0^b \\ x(\mu-1) = \Delta x(\mu-1) = 0 \end{cases} \tag{6.19}$$

*is given by*

$$x(t) = \sum_{s=0}^{t-\mu-1} x(t,s)h(s), \quad t \in \mathbb{N}_{\mu-1}^{\mu+b+1},$$

*where $x(t,s)$ is the Cauchy function for $\Delta^{\mu}_{\mu-1}(p\Delta x)(t) = 0$.*

*Proof.* Let $x(t)$ be a solution of $\Delta^{\mu}_{\mu-1}(p\Delta x)(t) = h(t)$, $t \in \mathbb{N}_0^b$. Then

$$\Delta^{\mu}_{0+\mu-1}(p\Delta x)(t) = h(t), \quad t \in \mathbb{N}_0^b.$$

Let $y(t) = (p\Delta x)(t)$. Then $y(t)$ is a solution of

$$\Delta^{\mu}_{0+\mu-1} y(t) = h(t), \quad t \in \mathbb{N}_0^b,$$

and hence is given by

$$y(t) = \sum_{s=0}^{t-\mu} \frac{(t-\sigma(s))^{\underline{\mu-1}}}{\Gamma(\mu)} h(s) + c_0 t^{\underline{\mu-1}}, \quad t \in \mathbb{N}_{\mu-1}^{\mu+b}.$$

Dividing both sides by $p(t)$ we get that

$$\Delta x(t) = \frac{1}{p(t)} \left[ \sum_{s=0}^{t-\mu} \frac{(t-\sigma(s))^{\underline{\mu-1}}}{\Gamma(\mu)} h(s) + c_0 t^{\underline{\mu-1}} \right], \quad t \in \mathbb{N}_{\mu-1}^{\mu+b}.$$

Using the second initial condition we obtain

$$0 = \Delta x(\mu-1) = \frac{1}{p(\mu-1)} \left[ \sum_{s=0}^{\mu-1-\mu} \frac{(\mu-1-\sigma(s))^{\underline{\mu-1}}}{\Gamma(\mu)} h(s) + c_0(\mu-1)^{\underline{\mu-1}} \right].$$

Note that the first term in the sum on the right-hand side is zero by our convention on sums, and therefore we are left with

$$0 = \Delta x(\mu-1) = \frac{1}{p(\mu-1)} \left[ c_0(\mu-1)^{\underline{\mu-1}} \right],$$

which implies that $c_0 = 0$. Thus,

$$\Delta x(t) = \frac{1}{p(t)} \left[ \sum_{s=0}^{t-\mu} \frac{(t-\sigma(s))^{\underline{\mu-1}}}{\Gamma(\mu)} h(s) \right], \quad t \in \mathbb{N}_{\mu-1}^{\mu+b}.$$

Summing both sides from $\tau = \mu - 1$ to $\tau = t - 1$ we obtain

$$\sum_{\tau=\mu-1}^{t-1} \Delta x(\tau) = \sum_{\tau=\mu-1}^{t-1} \frac{1}{p(\tau)} \left[ \sum_{s=0}^{\tau-\mu} \frac{(\tau - \sigma(s))^{\underline{\mu-1}}}{\Gamma(\mu)} h(s) \right], \quad t \in \mathbb{N}_{\mu-1}^{\mu+b}.$$

Interchanging the order of the summation then yields

$$x(t) - x(\mu - 1) = \frac{1}{\Gamma(\mu)} \sum_{s=0}^{t-1-\mu} \sum_{\tau=s+\mu}^{t-1} \left[ \frac{1}{p(\tau)} (\tau - \sigma(s))^{\underline{\mu-1}} h(s) \right].$$

By using the first initial condition it follows that

$$x(t) = \frac{1}{\Gamma(\mu)} \sum_{s=0}^{t-1-\mu} \sum_{\tau=s+\mu}^{t-1} \left[ \frac{1}{p(\tau)} (\tau - \sigma(s))^{\underline{\mu-1}} h(s) \right],$$

and so,

$$\begin{aligned} x(t) &= \sum_{s=0}^{t-1-\mu} \sum_{\tau=s+\mu}^{t-1} \left[ \frac{1}{p(\tau)} h_{\mu-1}(\tau, \sigma(s)) h(s) \right] \\ &= \sum_{s=0}^{t-\mu-1} x(t,s) h(s), \quad t \in \mathbb{N}_{\mu-1}^{\mu+b+1}. \end{aligned}$$

This completes the proof. □

**Theorem 6.20.** *Let $p : \mathbb{N}_{\mu-1}^{\mu+b} \to \mathbb{R}$ with $p(t) > 0$ and $b \in \mathbb{N}_1$, $0 < \mu \leq 1$, and assume that*

$$\rho = \alpha\gamma \sum_{\tau=\mu-1}^{\mu+b-1} \frac{\tau^{\underline{\mu-1}}}{p(\tau)} + \frac{\beta\gamma\Gamma(\mu)}{p(\mu-1)} + \frac{\alpha\delta(\mu+b)^{\underline{\mu-1}}}{p(\mu+b)}.$$

*Then the homogeneous fractional boundary value problem (FBVP)*

$$\begin{cases} \Delta_{\mu-1}^{\mu}(p\Delta x)(t) = 0, & t \in \mathbb{N}_0^b \\ \alpha x(\mu-1) - \beta\Delta x(\mu-1) = 0 \\ \gamma x(\mu+b) + \delta\Delta x(\mu+b) = 0 \end{cases} \tag{6.20}$$

*has only the trivial solution if and only if $\rho \neq 0$.*

*Proof.* Consider

$$\Delta^{\mu}_{\mu-1}(p\Delta x)(t) = 0, \quad t \in \mathbb{N}_0^b.$$

Then

$$(p\Delta x)(t) = c_0 t^{\underline{\mu-1}}, \quad t \in \mathbb{N}_{\mu-1}^{\mu+b},$$

and so,

$$\Delta x(t) = \frac{c_0 t^{\underline{\mu-1}}}{p(t)}, \quad t \in \mathbb{N}_{\mu-1}^{\mu+b}.$$

Summing both sides from $\tau = \mu - 1$ to $\tau = t - 1$ we obtain

$$x(t) - x(\mu - 1) = \sum_{\tau=\mu-1}^{t-1} \frac{c_0 \tau^{\underline{\mu-1}}}{p(\tau)}, \quad t \in \mathbb{N}_{\mu-1}^{\mu+b+1}.$$

Let $c_1 = x(\mu - 1)$. Then the general solution of $\Delta^{\mu}_{\mu-1}(p\Delta x)(t) = 0$ is given by

$$x(t) = \sum_{\tau=\mu-1}^{t-1} \frac{c_0 \tau^{\underline{\mu-1}}}{p(\tau)} + c_1, \quad t \in \mathbb{N}_{\mu-1}^{\mu+b+1}.$$

Now by using the boundary conditions we obtain the following linear system in $c_0$ and $c_1$.

$$-c_0 \frac{\beta\Gamma(\mu)}{p(\mu-1)} + c_1\alpha = 0$$

$$c_0 \left( \gamma \sum_{\tau=\mu-1}^{\mu+b-1} \frac{\tau^{\underline{\mu-1}}}{p(\tau)} + \delta \frac{(\mu+b)^{\underline{\mu-1}}}{p(\mu+b)} \right) + c_1\gamma = 0.$$

This system of equations has only the trivial solution if and only if

$$\rho = \begin{vmatrix} -\frac{\beta\Gamma(\mu)}{p(\mu-1)} & \alpha \\ \left( \gamma \sum_{\tau=\mu-1}^{\mu+b-1} \frac{\tau^{\underline{\mu-1}}}{p(\tau)} + \delta \frac{(\mu+b)^{\underline{\mu-1}}}{p(\mu+b)} \right) & \gamma \end{vmatrix} \neq 0,$$

which implies that the self-adjoint fractional boundary valued problem has only trivial solution if and only if

$$\rho = \alpha\gamma \sum_{\tau=\mu-1}^{\mu+b-1} \frac{\tau^{\underline{\mu-1}}}{p(\tau)} + \frac{\beta\gamma\Gamma(\mu)}{p(\mu-1)} + \frac{\alpha\delta(\mu+b)^{\underline{\mu-1}}}{p(\mu+b)} \neq 0.$$

And this completes the proof. □

*Remark 6.21.* Letting $\mu = 1$ in the above theorem gives us the known result that the BVP

$$\Delta(p\Delta x)(t) = 0, t \in \mathbb{N}_0^b$$
$$\alpha x(0) - \beta\Delta x(0) = 0$$
$$\gamma x(b+1) + \delta\Delta x(b+1) = 0$$

has only the trivial solution if and only if

$$\rho = \alpha\gamma \sum_{\tau=0}^{b} \frac{1}{p(\tau)} + \frac{\beta\gamma}{p(0)} + \frac{\alpha\delta}{p(b+1)} \neq 0.$$

## 6.6 Green's Function for a Two-Point FBVP

In this section we will derive a formula for the Green's function for the self-adjoint fractional boundary value problem

$$\begin{cases} \Delta_{\mu-1}^{\mu}(p\Delta x)(t) = 0, & t \in \mathbb{N}_0^b \\ x(\mu-1) = 0, & x(\mu+b+1) = 0, \end{cases} \tag{6.21}$$

where $h : \mathbb{N}_0^b \to \mathbb{R}$ and $p : \mathbb{N}_{\mu-1}^{\mu+b} \to \mathbb{R}$ with $p(t) > 0$.

**Theorem 6.22.** *Let $h : \mathbb{N}_0^b \to \mathbb{R}$ and $p : \mathbb{N}_{\mu-1}^{\mu+b} \to \mathbb{R}$ with $p(t) > 0$. Then the Green's function for the FBVP*

$$\begin{cases} \Delta_{\mu-1}^{\mu}(p\Delta x)(t) = 0, & t \in \mathbb{N}_0^b \\ x(\mu-1) = 0, & x(\mu+b+1) = 0 \end{cases} \tag{6.22}$$

*is given by*

$$G(t,s) = \begin{cases} -\frac{1}{\rho}\left(\sum_{\tau=\mu-1}^{t-1} \frac{\tau^{\underline{\mu-1}}}{p(\tau)}\right) x(\mu+b+1,s), & t \leq s+\mu \\ x(t,s) - \frac{1}{\rho}\left(\sum_{\tau=\mu-1}^{t-1} \frac{\tau^{\underline{\mu-1}}}{p(\tau)}\right) x(\mu+b+1,s), & s \leq t-\mu-1. \end{cases} \tag{6.23}$$

*Proof.* In order to find the Green's function for the above homogeneous FBVP, we first consider the following nonhomogeneous FBVP

$$\begin{cases} \Delta^{\mu}_{\mu-1}(p\Delta x)(t) = h(t), & t \in \mathbb{N}_0^b \\ x(\mu-1) = 0, & x(\mu+b+1) = 0. \end{cases}$$

Since we have already derived the expression for $\Delta x$ in the proof of Theorem 6.19, we have

$$\Delta x(t) = \frac{1}{p(t)}\left[\sum_{s=0}^{t-\mu} \frac{(t-\sigma(s))^{\underline{\mu-1}}}{\Gamma(\mu)} h(s) + c_0(t)^{\underline{\mu-1}}\right], \quad t \in \mathbb{N}_{\mu-1}^{\mu+b}.$$

Now summing from $\tau = \mu - 1$ to $\tau = t - 1$ and interchanging the order of summation we have that

$$x(t) = \sum_{s=0}^{t-1-\mu} \sum_{\tau=s+\mu}^{t-1} \left[\frac{1}{p(\tau)} \frac{(\tau-\sigma(s))^{\underline{\mu-1}} h(s)}{\Gamma(\mu)}\right] + \sum_{\tau=\mu-1}^{t-1} \frac{c_0 \tau^{\underline{\mu-1}}}{p(\tau)} + x(\mu-1).$$

By letting $c_1 = x(\mu-1)$, the above expression for the general solution can be rewritten as

$$x(t) = \sum_{s=0}^{t-1-\mu} \sum_{\tau=s+\mu}^{t-1} \left[\frac{1}{p(\tau)} \frac{(\tau-\sigma(s))^{\underline{\mu-1}} h(s)}{\Gamma(\mu)}\right] + \sum_{\tau=\mu-1}^{t-1} \frac{c_0 \tau^{\underline{\mu-1}}}{p(\tau)} + c_1.$$

Now, if we represent the term

$$\sum_{\tau=s+\mu}^{t-1} \left[\frac{1}{p(\tau)} \frac{(\tau-\sigma(s))^{\underline{\mu-1}}}{\Gamma(\mu)}\right]$$

by $x(t,s)$, the Cauchy function, then the above expression for the general solution can be rewritten as

$$x(t) = \sum_{s=0}^{t-1-\mu} x(t,s)h(s) + \sum_{\tau=\mu-1}^{t-1} \frac{c_0 \tau^{\underline{\mu-1}}}{p(\tau)} + c_1.$$

By using the first boundary condition $x(\mu-1)= 0$ we obtain

$$x(\mu-1) = \sum_{s=0}^{\mu-2-\mu} x(\mu-1,s)h(s) + \sum_{\tau=\mu-1}^{\mu-2} \frac{c_0 \tau^{\underline{\mu-1}}}{p(\tau)} + c_1.$$

Notice that the first two sums in the preceding expression are zero by convention as the upper limit of summation is smaller than the lower limit, and therefore we have that

$$c_1 = 0.$$

Thus,

$$x(t) = \sum_{s=0}^{t-\mu-1} x(t,s)h(s) + \sum_{\tau=\mu-1}^{t-1} \frac{c_0 \tau^{\underline{\mu-1}}}{p(\tau)}.$$

By using the second boundary condition $x(\mu + b + 1) = 0$ we have that

$$x(\mu + b + 1) = \sum_{s=0}^{b} x(\mu + b, s)h(s) + \sum_{\tau=\mu-1}^{\mu+b} \frac{c_0 \tau^{\underline{\mu-1}}}{p(\tau)}. \tag{6.24}$$

Solving for $c_0$, we obtain

$$c_0 = -\frac{\sum_{s=0}^{b} x(\mu + b + 1, s)}{\sum_{\tau=\mu-1}^{\mu+b} \frac{\tau^{\underline{\mu-1}}}{p(\tau)}} h(s).$$

Moreover, if we let $\rho = \sum_{\tau=\mu-1}^{\mu+b} \frac{\tau^{\underline{\mu-1}}}{p(\tau)}$ and substitute the values of $c_0$ and $c_1$ into the expression for $x$, we thus obtain that the solution of the FBVP is

$$\begin{aligned}
x(t) &= \sum_{s=0}^{t-1-\mu} x(t,s)h(s) - \frac{1}{\rho}\left(\sum_{\tau=\mu-1}^{t-1} \frac{\tau^{\underline{\mu-1}}}{p(\tau)}\right) \sum_{s=0}^{b} x(\mu + b + 1, s)h(s) \\
&= \sum_{s=t-\mu}^{b} \left(-\frac{1}{\rho}\right)\left(\sum_{\tau=\mu-1}^{t-1} \frac{\tau^{\underline{\mu-1}}}{p(\tau)}\right) x(\mu + b + 1, s)h(s) \\
&\quad + \sum_{s=0}^{t-1-\mu} \left[x(t,s) - \frac{1}{\rho}\left(\sum_{\tau=\mu-1}^{t-1} \frac{\tau^{\underline{\mu-1}}}{p(\tau)}\right) x(\mu + b + 1, s)\right] h(s).
\end{aligned}$$

In particular, then, the Green's function for the FBVP can be written as

$$G(t,s) = \begin{cases} -\frac{1}{\rho}\left(\sum_{\tau=\mu-1}^{t-1} \frac{\tau^{\underline{\mu-1}}}{p(\tau)}\right) x(\mu+b+1,s), & t \le s+\mu \\ x(t,s) - \frac{1}{\rho}\left(\sum_{\tau=\mu-1}^{t-1} \frac{\tau^{\underline{\mu-1}}}{p(\tau)}\right) x(\mu+b+1,s), & s \le t-\mu-1. \end{cases}$$

And this completes the proof. □

## 6.7 Green's Function for the General Two-Point FBVP

In this section we will derive a formula for the Green's function for the general self-adjoint fractional boundary value problem

$$\begin{aligned} \Delta^{\mu}_{\mu-1}(p\Delta x)(t) &= 0, t \in \mathbb{N}_0^b \\ \alpha x(\mu-1) - \beta\Delta x(\mu-1) &= 0 \\ \gamma x(\mu+b) + \delta\Delta x(\mu+b) &= 0, \end{aligned} \tag{6.25}$$

where $\alpha^2 + \beta^2 > 0$ and $\gamma^2 + \delta^2 > 0$.

**Theorem 6.23.** *Assume* $p : \mathbb{N}_{\mu-1}^{\mu+b} \to \mathbb{R}$ *with* $p(t) > 0$ *and*

$$\rho := \alpha\gamma \sum_{\tau=\mu-1}^{\mu+b-1} \frac{\tau^{\underline{\mu-1}}}{p(\tau)} + \frac{\beta\gamma\Gamma(\mu)}{p(\mu-1)} + \frac{\alpha\delta(\mu+b)^{\underline{\mu-1}}}{p(\mu+b)} \neq 0.$$

*Then the Green's function for the FBVP* (6.25) *is given by*

$$G(t,s) = \begin{cases} u(t,s), & t \le s+\mu \\ v(t,s), & s \le t-\mu-1, \end{cases}$$

*where*

$$\begin{aligned} u(t,s) = -\frac{1}{\rho}\Bigg[&\left(\sum_{\tau=\mu-1}^{t-1} \frac{\tau^{\underline{\mu-1}}}{p(\tau)}\right)\Bigg(\alpha\gamma\, x(\mu+b,s) \\ &+ \frac{\alpha\delta}{p(\mu+b)} \frac{(\mu+b-\sigma(s))^{\underline{\mu-1}}}{\Gamma(\mu)}\Bigg) + \frac{\beta\gamma\Gamma(\mu)}{p(\mu-1)} x(\mu+b,s) \\ &+ \frac{\beta\delta}{p(\mu-1)p(\mu+b)} (\mu+b-\sigma(s))^{\underline{\mu-1}}\Bigg], \end{aligned}$$

*for* $t \leq s + \mu$, *and*

$$v(t,s) := u(t,s) + x(t,s),$$

*for* $s \leq t - \mu - 1$, *where* $x(t,s)$ *is the Cauchy function for*

$$\Delta^{\mu}_{\mu-1}(p\Delta x)(t) = 0.$$

*Proof.* In order to find the Green's function for the above homogeneous FBVP (6.25), we consider the following nonhomogeneous FBVP

$$\begin{aligned} \Delta^{\mu}_{\mu-1}(p\Delta x)(t) &= h(t), \quad t \in \mathbb{N}_0^b \\ \alpha x(\mu-1) - \beta \Delta x(\mu-1) &= 0 \\ \gamma x(\mu+b) + \delta \Delta x(\mu+b) &= 0, \end{aligned} \tag{6.26}$$

where $h : \mathbb{N}_0^b \to \mathbb{R}$ is a given function. Since $\rho \neq 0$, we have by Theorem 6.20 that the corresponding homogeneous FBVP (6.25) has only the trivial solution. It is a standard argument that this implies that the nonhomogeneous FBVP (6.26) has a unique solution. Let $x(t)$ be the solution of the nonhomogeneous FBVP (6.26). As in the proof of Theorem 6.20 we get that since $x(t)$ is a solution of the fractional difference equation

$$\Delta^{\mu}_{\mu-1}(p\Delta x)(t) = h(t), \quad t \in \mathbb{N}_0^b,$$

then

$$\Delta x(t) = \frac{1}{p(t)}\left[\sum_{s=0}^{t-\mu} \frac{(t-\sigma(s))^{\underline{\mu-1}}}{\Gamma(\mu)} h(s) + c_0 (t)^{\underline{\mu-1}}\right], \quad t \in \mathbb{N}_{\mu-1}^{\mu+b}.$$

Now summing both sides from $\tau = \mu - 1$ to $\tau = t - 1$ and interchanging the order of summation we have that

$$x(t) = \sum_{s=0}^{t-1-\mu} \sum_{\tau=s+\mu}^{t-1} \left[\frac{1}{p(\tau)} \frac{(\tau-\sigma(s))^{\underline{\mu-1}} h(s)}{\Gamma(\mu)}\right] + \sum_{\tau=\mu-1}^{t-1} \frac{c_0 \tau^{\underline{\mu-1}}}{p(\tau)} + x(\mu-1).$$

By letting $c_1 = x(\mu - 1)$, the above expression for the general solution can be rewritten as

$$x(t) = \sum_{s=0}^{t-1-\mu} \sum_{\tau=s+\mu}^{t-1} \left[ \frac{1}{p(\tau)} \frac{(\tau - \sigma(s))^{\underline{\mu-1}} h(s)}{\Gamma(\mu)} \right] + \sum_{\tau=\mu-1}^{t-1} \frac{c_0 \tau^{\underline{\mu-1}}}{p(\tau)} + c_1$$

$$= \sum_{s=0}^{t-1-\mu} x(t,s)h(s) + \sum_{\tau=\mu-1}^{t-1} \frac{c_0 \tau^{\underline{\mu-1}}}{p(\tau)} + c_1. \tag{6.27}$$

Applying the two boundary conditions we get the following equations:

$$c_1 \alpha - c_0 \frac{\beta \Gamma(\mu)}{p(\mu - 1)} = 0$$

and

$$\gamma \left( \sum_{s=0}^{b-1} x(\mu + b, s)h(s) + \sum_{\tau=\mu-1}^{\mu+b-1} \frac{c_0 \tau^{\underline{\mu-1}}}{p(\tau)} + c_1 \right) +$$

$$\delta \left[ \frac{1}{p(\mu + b)} \left( \sum_{s=0}^{b} \frac{(\mu + b - \sigma(s))^{\underline{\mu-1}}}{\Gamma(\mu)} h(s) + c_0 (\mu + b)^{\underline{\mu-1}} \right) \right] = 0; \tag{6.28}$$

i.e.,

$$c_1 \gamma + c_0 \left( \gamma \sum_{\tau=\mu-1}^{\mu+b-1} \frac{\tau^{\underline{\mu-1}}}{p(\tau)} + \delta \frac{(\mu + b)^{\underline{\mu-1}}}{p(\mu + b)} \right) = -\gamma \sum_{s=0}^{b-1} x(\mu + b, s)h(s)$$

$$- \frac{\delta}{p(\mu + b)} \sum_{s=0}^{b} \frac{(\mu + b - \sigma(s))^{\underline{\mu-1}}}{\Gamma(\mu)} h(s). \tag{6.29}$$

Now, if we let

$$\rho = \alpha \gamma \sum_{\tau=\mu-1}^{\mu+b-1} \frac{\tau^{\underline{\mu-1}}}{p(\tau)} + \frac{\beta \gamma \Gamma(\mu)}{p(\mu - 1)} + \frac{\alpha \delta (\mu + b)^{\underline{\mu-1}}}{p(\mu + b)}$$

and solve the above system for $c_0$ and $c_1$, we have that

$$c_0 = -\frac{1}{\rho} \left( \alpha \gamma \sum_{s=0}^{b-1} x(\mu + b, s) + \frac{\alpha \delta}{p(\mu + b)} \sum_{s=0}^{b} \frac{(\mu + b - \sigma(s))^{\underline{\mu-1}}}{\Gamma(\mu)} \right) h(s)$$

and

$$c_1 = -\frac{1}{\rho}\frac{\beta\Gamma(\mu)}{\alpha p(\mu-1)}\left(\alpha\gamma\sum_{s=0}^{b-1}x(\mu+b,s)\right.$$
$$\left.+\frac{\alpha\delta}{p(\mu+b)}\sum_{s=0}^{b}\frac{(\mu+b-\sigma(s))^{\underline{\mu-1}}}{\Gamma(\mu)}\right)h(s).$$

It follows that

$$c_1 = -\frac{1}{\rho}\left[\frac{\beta\Gamma(\mu)}{p(\mu-1)}\gamma\sum_{s=0}^{b-1}x(\mu+b,s)\right.$$
$$\left.+\frac{\beta\delta}{p(\mu-1)p(\mu+b)}\sum_{s=0}^{b}(\mu+b-\sigma(s))^{\underline{\mu-1}}\right]h(s).$$

Substituting the above values of $c_0$ and $c_1$ in the Eq. (6.27) we get that

$$x(t) = \sum_{s=0}^{t-1-\mu}x(t,s)h(s) - \frac{1}{\rho}\sum_{\tau=\mu-1}^{t-1}\frac{\tau^{\underline{\mu-1}}}{p(\tau)}\left(\alpha\gamma\sum_{s=0}^{b-1}x(\mu+b,s)h(s)\right.$$
$$\left.+\frac{\alpha\delta}{p(\mu+b)}\sum_{s=0}^{b}\frac{(\mu+b-\sigma(s))^{\underline{\mu-1}}}{\Gamma(\mu)}h(s)\right)$$
$$-\frac{1}{\rho}\left(\frac{\beta\gamma\Gamma(\mu)}{p(\mu-1)}\sum_{s=0}^{b-1}x(\mu+b,s)h(s)\right.$$
$$\left.+\frac{\beta\delta}{p(\mu-1)p(\mu+b)}\sum_{s=0}^{b}(\mu+b-\sigma(s))^{\underline{\mu-1}}h(s)\right).$$

Using $x(\mu+b,b)=0$, we obtain

$$\sum_{s=0}^{b-1}x(\mu+b,s)h(s) = \sum_{s=0}^{b}x(\mu+b,s)h(s).$$

As a result of the above we obtain

$$x(t) = \sum_{s=0}^{t-1-\mu}x(t,s)h(s) - \frac{1}{\rho}\left(\sum_{\tau=\mu-1}^{t-1}\frac{\tau^{\underline{\mu-1}}}{p(\tau)}\right)\left[\alpha\gamma\sum_{s=0}^{b}x(\mu+b,s)h(s)\right.$$
$$\left.+\frac{\alpha\delta}{p(\mu+b)}\sum_{s=0}^{b}\frac{(\mu+b-\sigma(s))^{\underline{\mu-1}}}{\Gamma(\mu)}h(s)\right]$$

$$-\frac{1}{\rho}\left[\frac{\beta\gamma\Gamma(\mu)}{p(\mu-1)}\sum_{s=0}^{b}x(\mu+b,s)h(s)\right.$$
$$\left.+\frac{\beta\delta}{p(\mu-1)p(\mu+b)}\sum_{s=0}^{b}(\mu+b-\sigma(s))^{\underline{\mu-1}}h(s)\right].$$

Hence, we have that

$$x(t)=\sum_{s=0}^{t-1-\mu}x(t,s)h(s)-\frac{1}{\rho}\left(\sum_{\tau=\mu-1}^{t-1}\frac{\tau^{\underline{\mu-1}}}{p(\tau)}\right)\left[\alpha\gamma\Big(\sum_{s=0}^{t-\mu-1}x(\mu+b,s)h(s)\right.$$
$$\left.+\sum_{s=t-\mu}^{b}x(\mu+b,s)h(s)\Big)\right]$$
$$+\frac{\alpha\delta}{p(\mu+b)}\left[\sum_{s=0}^{t-\mu-1}\frac{(\mu+b-\sigma(s))^{\underline{\mu-1}}}{\Gamma(\mu)}h(s)\right.$$
$$\left.+\sum_{s=t-\mu}^{b}\frac{(\mu+b-\sigma(s))^{\underline{\mu-1}}}{\Gamma(\mu)}h(s)\right]$$
$$-\frac{1}{\rho}\left(\frac{\beta\gamma\Gamma(\mu)}{p(\mu-1)}\right)\left[\sum_{s=0}^{t-\mu-1}x(\mu+b,s)h(s)+\sum_{s=t-\mu}^{b}x(\mu+b,s)h(s)\right]$$
$$+\frac{\beta\delta}{p(\mu-1)p(\mu+b)}\left[\sum_{s=0}^{t-\mu-1}(\mu+b-\sigma(s))^{\underline{\mu-1}}h(s)\right.$$
$$\left.+\sum_{s=t-\mu}^{b}(\mu+b-\sigma(s))^{\underline{\mu-1}}h(s)\right].$$

Thus the solution to the given FBVP is given by

$$x(t)=\sum_{s=0}^{t-1-\mu}\left[x(t,s)-\frac{1}{\rho}\left(\left(\sum_{\tau=\mu-1}^{t-1}\frac{\tau^{\underline{\mu-1}}}{p(\tau)}\right)\left(\alpha\gamma x(\mu+b,s)\right.\right.\right.$$
$$\left.+\frac{\alpha\delta}{p(\mu+b)}\frac{(\mu+b-\sigma(s))^{\underline{\mu-1}}}{\Gamma(\mu)}\right)+\frac{\beta\gamma\Gamma(\mu)}{p(\mu-1)}x(\mu+b,s)$$
$$\left.\left.+\frac{\beta\delta}{p(\mu-1)p(\mu+b)}(\mu+b-\sigma(s))^{\underline{\mu-1}}\right)\right]h(s)$$

$$+\sum_{s=t-\mu}^{b}\Bigg[-\frac{1}{\rho}\Bigg(\Bigg(\sum_{\tau=\mu-1}^{t-1}\frac{\tau^{\underline{\mu-1}}}{p(\tau)}\Bigg)\Bigg(\alpha\gamma x(\mu+b,s)$$

$$+\frac{\alpha\delta}{p(\mu+b)}\frac{(\mu+b-\sigma(s))^{\underline{\mu-1}}}{\Gamma(\mu)}\Bigg)+\frac{\beta\gamma\Gamma(\mu)}{p(\mu-1)}x(\mu+b,s)$$

$$+\frac{\beta\delta}{p(\mu-1)p(\mu+b)}(\mu+b-\sigma(s))^{\underline{\mu-1}}\Bigg)\Bigg]h(s).$$

Hence the Green's function for the given FBVP is given by

$$G(t,s)=\begin{cases}u(t,s), & t\leq s+\mu\\ v(t,s), & s\leq t-\mu-1,\end{cases}$$

where

$$u(t,s)=-\frac{1}{\rho}\Bigg[\Bigg(\sum_{\tau=\mu-1}^{t-1}\frac{\tau^{\underline{\mu-1}}}{p(\tau)}\Bigg)\Bigg(\alpha\gamma x(\mu+b,s)$$

$$+\frac{\alpha\delta}{p(\mu+b)}\frac{(\mu+b-\sigma(s))^{\underline{\mu-1}}}{\Gamma(\mu)}\Bigg)+\frac{\beta\gamma\Gamma(\mu)}{p(\mu-1)}x(\mu+b,s)$$

$$+\frac{\beta\delta}{p(\mu-1)p(\mu+b)}(\mu+b-\sigma(s))^{\underline{\mu-1}}\Bigg]$$

for $t\leq s+\mu$, and

$$v(t,s)=x(t,s)-\frac{1}{\rho}\Bigg[\Bigg(\sum_{\tau=\mu-1}^{t-1}\frac{\tau^{\underline{\mu-1}}}{p(\tau)}\Bigg)\Bigg(\alpha\gamma x(\mu+b,s)$$

$$+\frac{\alpha\delta}{p(\mu+b)}\frac{(\mu+b-\sigma(s))^{\underline{\mu-1}}}{\Gamma(\mu)}\Bigg)+\frac{\beta\gamma\Gamma(\mu)}{p(\mu-1)}x(\mu+b,s)$$

$$+\frac{\beta\delta}{p(\mu-1)p(\mu+b)}(\mu+b-\sigma(s))^{\underline{\mu-1}}\Bigg]$$

for $s\leq t-\mu-1$. This completes the proof. □

*Remark 6.24.* If $\alpha = \gamma = \delta = 1$ and $\beta = 0$, then we get the known formula for the Green's function for the conjugate case

$$\begin{cases} \Delta^{\mu}_{\mu-1}(p\Delta x)(t) = 0, & t \in \mathbb{N}_0^b \\ x(\mu - 1) = 0, & x(\mu + b + 1) = 0, \end{cases} \tag{6.30}$$

namely

$$G(t,s) = \begin{cases} -\frac{1}{\rho}\left(\sum_{\tau=\mu-1}^{t-1} \frac{\tau^{\underline{\mu-1}}}{p(\tau)}\right) x(\mu+b+1,s), & t \leq s+\mu \\ x(t,s) - \frac{1}{\rho}\left(\sum_{\tau=\mu-1}^{t-1} \frac{\tau^{\underline{\mu-1}}}{p(\tau)}\right) x(\mu+b+1,s), & s \leq t-\mu-1. \end{cases} \tag{6.31}$$

## 6.8 Green's Function When $p \equiv 1$

Our goal in this section is to deduce some important properties of the Green's function for the conjugate case when $p(t) \equiv 1$ for $t \in \mathbb{N}_{\mu-1}^{\mu+b}$, which will be useful later in this chapter. In order to do that, first we will explicitly give a formula for the Green's function when $p(t) \equiv 1$ in the following theorem.

**Theorem 6.25.** *Letting $p(t) \equiv 1$ in Remark 6.24 we get that the Green's function for the conjugate FBVP*

$$\begin{cases} \Delta^{\mu}_{\mu-1}(\Delta x)(t) = 0, & t \in \mathbb{N}_0^b \\ x(\mu - 1) = 0, & x(\mu + b + 1) = 0 \end{cases} \tag{6.32}$$

*is given by*

$$G(t,s) = \begin{cases} u(t,s) := -\frac{1}{\rho}\left(\frac{t^{\underline{\mu}}(\mu+b+1-\sigma(s))^{\underline{\mu}}}{\mu\Gamma(\mu+1)}\right), & t \leq s+\mu \\ v(t,s) := \frac{(t-\sigma(s))^{\underline{\mu}}}{\Gamma(\mu+1)} - \frac{1}{\rho}\left(\frac{t^{\underline{\mu}}(\mu+b+1-\sigma(s))^{\underline{\mu}}}{\mu\Gamma(\mu+1)}\right), & s \leq t-\mu-1, \end{cases} \tag{6.33}$$

*where* $\rho = \frac{1}{\mu}(\mu + b + 1)^{\underline{\mu}}$.

*Proof.* First we observe that with $p(t) \equiv 1$ for $t \in \mathbb{N}_{\mu-1}^{\mu+b}$, the Cauchy function $x(t,s)$ takes the form

$$x(t,s) = \sum_{\tau=s+\mu}^{t-1} \left[\frac{(\tau - \sigma(s))^{\underline{\mu-1}}}{\Gamma(\mu)}\right]$$

$$= \frac{1}{\Gamma(\mu)} \sum_{\tau=s+\mu}^{t-1} (\tau - \sigma(s))^{\underline{\mu-1}}$$

$$= \frac{1}{\Gamma(\mu)} \sum_{\tau=s+\mu}^{t-1} \Delta_\tau \frac{(\tau - \sigma(s))^{\underline{\mu}}}{\mu}$$

$$= \frac{1}{\Gamma(\mu+1)} \sum_{\tau=s+\mu}^{t-1} \Delta_\tau (\tau - \sigma(s))^{\underline{\mu}}$$

$$= \frac{1}{\Gamma(\mu+1)} \left[(t - \sigma(s))^{\underline{\mu}} - (\mu - 1)^{\underline{\mu}}\right] = \frac{(t - \sigma(s))^{\underline{\mu}}}{\Gamma(\mu+1)}.$$

Thus, $x(\mu + b + 1, s) = \frac{(\mu+b+1-\sigma(s))^{\underline{\mu}}}{\Gamma(\mu+1)}$ and

$$\rho = \sum_{\tau=\mu-1}^{\mu+b} \tau^{\underline{\mu-1}} = \sum_{\tau=\mu-1}^{\mu+b} \Delta_\tau \left[\frac{\tau^{\underline{\mu}}}{\mu}\right]$$

$$= \frac{1}{\mu}\left[(\mu + b + 1)^{\underline{\mu}} - (\mu - 1)^{\underline{\mu}}\right] = \frac{1}{\mu}(\mu + b + 1)^{\underline{\mu}}.$$

In the above, we have used the fact that $(\mu - 1)^{\underline{\mu}} = 0$. Moreover, for $p(t) \equiv 1$ we have

$$\sum_{\tau=\mu-1}^{t-1} \frac{t^{\underline{\mu-1}}}{p(\tau)} = \sum_{\tau=\mu-1}^{t-1} \tau^{\underline{\mu-1}}.$$

So we can write

$$\sum_{\tau=\mu-1}^{t-1} \tau^{\underline{\mu-1}} = \sum_{\tau=\mu-1}^{t-1} \Delta_\tau \left(\frac{\tau^{\underline{\mu}}}{\mu}\right) = \frac{t^{\underline{\mu}}}{\mu}.$$

Thus, with $p(t) \equiv 1$ and with these modified forms of the Cauchy function $x(t, s)$ together with preceding sum, the Green's function as derived in Theorem 6.22 can be written in the simplified form

$$G(t,s) = \begin{cases} u(t,s) := -\frac{1}{\rho}\left(\frac{t^{\underline{\mu}}(\mu+b+1-\sigma(s))^{\underline{\mu}}}{\mu\Gamma(\mu+1)}\right), & t \le s + \mu \\ v(t,s) := \frac{(t-\sigma(s))^{\underline{\mu}}}{\Gamma(\mu+1)} - \frac{1}{\rho}\left(\frac{t^{\underline{\mu}}(\mu+b+1-\sigma(s))^{\underline{\mu}}}{\mu\Gamma(\mu+1)}\right), & s \le t - \mu - 1, \end{cases} \tag{6.34}$$

where $\rho = \frac{1}{\mu}(\mu + b + 1)^{\underline{\mu}}$. This completes the proof. □

In the following four theorems we will derive some properties of this Green's function in Theorem 6.25. First we prove that this Green's function is of constant sign.

**Theorem 6.26.** *The Green's function for the FBVP* (6.32) *in Theorem 6.25 satisfies*

$$G(t,s) \leq 0.$$

*Proof.* We will show that each of $u(t,s)$ and $v(t,s)$ is nonpositive. First consider $u(t,s)$. Since $\rho > 0$ by its definition, $\Gamma(\mu+1) > 0$ since $0 < \mu \leq 1$, $t^{\underline{\mu}} \geq 0$ as $\mu - 1 \leq t \leq \mu + b + 1$, and $(\mu + b + 1 - \sigma(s))^{\underline{\mu}} \geq 0$ as $s \in \mathbb{N}_0^b$. This implies that for $t \leq s + \mu$ we have that

$$u(t,s) = -\frac{1}{\rho}\left(\frac{t^{\underline{\mu}}(\mu + b + 1 - \sigma(s))^{\underline{\mu}}}{\mu\Gamma(\mu+1)}\right) \leq 0.$$

Next we will show that $v(t,s)$ is nonpositive, for $0 \leq s \leq t - \mu - 1 \leq b$. That is, we will show that

$$\frac{(t-\sigma(s))^{\underline{\mu}}}{\Gamma(\mu+1)} - \frac{1}{\rho}\left(\frac{t^{\underline{\mu}}(\mu + b + 1 - \sigma(s))^{\underline{\mu}}}{\mu\Gamma(\mu+1)}\right) \leq 0.$$

After substituting the value of $\rho = \frac{1}{\mu}(\mu + b + 1)^{\underline{\mu}}$ and simplifying the above inequality, we get that

$$\frac{1}{(\mu+b+1)^{\underline{\mu}}\Gamma(\mu+1)}\left[(t-\sigma(s))^{\underline{\mu}}(\mu+b+1)^{\underline{\mu}} - t^{\underline{\mu}}(\mu + b + 1 - \sigma(s))^{\underline{\mu}} \leq 0.\right.$$

Since $(\mu + b + 1)^{\underline{\mu}}$ and $\Gamma(\mu+2) > 0$, it is sufficient to show that

$$\frac{(t-\sigma(s))^{\underline{\mu}}(\mu+b+1)^{\underline{\mu}}}{t^{\underline{\mu}}(\mu+b+1-\sigma(s))^{\underline{\mu}}} \leq 1.$$

Thus for $0 \leq s \leq t - \mu - 1 \leq b$, we consider

$$\begin{aligned}
&\frac{(t-\sigma(s))^{\underline{\mu}}(\mu+b+1)^{\underline{\mu}}}{t^{\underline{\mu}}(\mu+b+1-\sigma(s))^{\underline{\mu}}} \\
&= \frac{\Gamma(t-s)\Gamma(t+1-\mu)}{\Gamma(t-s-\mu)\Gamma(t+1)} \frac{\Gamma(\mu+b+2)\Gamma(b+2-\sigma(s))}{\Gamma(b+2)\Gamma(\mu+b+2-\sigma(s))} \\
&= \frac{\Gamma(t-s)}{\Gamma(t+1)} \frac{\Gamma(t+1-\mu)}{\Gamma(t-s-\mu)} \frac{\Gamma(\mu+b+2)}{\Gamma(\mu+b+2-\sigma(s))} \frac{\Gamma(b+2-\sigma(s))}{\Gamma(b+2)}.
\end{aligned}$$

Using the fact that $\Gamma(s+1) = s\Gamma(s)$, for each $s \in \mathbb{R} \setminus \{\ldots, -2, -1, 0\}$, we obtain

$$\frac{\Gamma(t-s)}{t\cdots(t-s)\Gamma(t-s)} \frac{(t-\mu)\cdots(t-s-\mu)\Gamma(t-s-\mu)}{\Gamma(t-s-\mu)}$$
$$= \frac{(\mu+b+1)\cdots(\mu+b+2-\sigma(s))\Gamma(\mu+b+2-\sigma(s))}{\Gamma(\mu+b+2-\sigma(s))}$$
$$= \frac{\Gamma(b+2-\sigma(s))}{(b+1)\cdots(b+2-\sigma(s))\Gamma(b+2-\sigma(s))}.$$

Simplifying and rearranging the factors from the numerator and denominator we see that the above expression is the same as

$$\left[\frac{(t-\mu)(t-\mu-1)\cdots(t-\mu-s)}{t(t-1)\cdots(t-s)}\right]\left[\frac{(\mu+b+1)(\mu+b)\cdots(\mu+b+1-s)}{(b+1)b\cdots(b+1-s)}\right].$$

A further rearrangement of the factors gives us the expression

$$\left[\frac{(t-\mu)(\mu+b+1)}{t(b+1)}\right]\left[\frac{(t-\mu-1)(\mu+b)}{(t-1)b}\right]\cdots\left[\frac{(t-\mu-s)(\mu+b+1-s)}{(t-s)(b+1-s)}\right]. \tag{6.35}$$

Now, in order to show that the above product in (6.35) is less than or equal to one, we will show that each factor within the square brackets is less than or equal to one. Thus for $0 \leq k \leq s$, we consider

$$\frac{(t-\mu-k)(\mu+b+1-k)}{(t-k)(b+1-k)}$$

and show it is less than or equal to one. Notice that the condition $s \leq t-\mu-1 \leq b$ implies that

$$t-\mu-1 \leq b.$$

Multiplying both sides by $\mu > 0$ and rewriting we get

$$t\mu - \mu^2 - \mu b - \mu \leq 0.$$

Adding to both sides the quantity $tb + t - tk - kb - k + k^2$ we get

$$t\mu + tb + t - tk - \mu^2 - \mu b - \mu + \mu k - \mu k - kb - k + k^2 \leq tb + t - tk - kb - k + k^2,$$

which implies that

$$(t-\mu-k)(\mu+b+1-k) \leq (t-k)(b+1-k).$$

This implies that

$$\frac{(t-\mu-k)(\mu+b+1-k)}{(t-k)(b+1-k)} \leq 1.$$

Thus, we have shown that a general factor in the above product given by (6.35) is less than or equal to one. Therefore, all the factors are less than or equal to one. And this implies that the product given by (6.35) is less than or equal to one. So

$$\frac{(t-\sigma(s))^{\underline{\mu}}}{\Gamma(\mu+1)} - \frac{1}{\rho}\left(\frac{t^{\underline{\mu}}(\mu+b+1-\sigma(s))^{\underline{\mu}}}{\mu\Gamma(\mu+1)}\right) \leq 0.$$

Since $u(t,s) \leq 0$ and $v(t,s) \leq 0$, we conclude that $G(t,s) \leq 0$. This completes the proof. □

Next we find a formula for $\sum_{s=0}^{b}\left|G(t,s)\right|$ in the following theorem.

**Theorem 6.27.** *If* $t \in \mathbb{N}_{\mu-1}^{\mu+b+1}$, *then the Green's function for* (6.32) *satisfies*

$$\sum_{s=0}^{b}\left|G(t,s)\right| = \frac{t^{\underline{\mu}}}{\Gamma(\mu+2)}(\mu+b+1-t). \tag{6.36}$$

*Proof.* Consider

$$\sum_{s=0}^{b}\left|G(t,s)\right| = \sum_{s=0}^{t-\mu-1}\left|\frac{(t-\sigma(s))^{\underline{\mu}}}{\Gamma(\mu+1)} - \frac{1}{\rho}\left(\frac{t^{\underline{\mu}}(\mu+b+1-\sigma(s))^{\underline{\mu}}}{\mu\Gamma(\mu+1)}\right)\right|$$
$$+ \sum_{s=t-\mu}^{b}\left|-\frac{1}{\rho}\left(\frac{t^{\underline{\mu}}(\mu+b+1-\sigma(s))^{\underline{\mu}}}{\mu\Gamma(\mu+1)}\right)\right|.$$

In Theorem 6.26 we proved that $G(t,s) \leq 0$, which implies that $u(t,s) \leq 0$ for $t \leq s+\mu$ and $v(t,s) \leq 0$ for $s \leq t-\mu-1$. Thus

$$\sum_{s=0}^{b}|G(t,s)| = \sum_{s=0}^{t-\mu-1}\left[\frac{1}{\rho}\left(\frac{t^{\underline{\mu}}(\mu+b+1-\sigma(s))^{\underline{\mu}}}{\mu\Gamma(\mu+1)}\right) - \frac{(t-\sigma(s))^{\underline{\mu}}}{\Gamma(\mu+1)}\right]$$
$$+ \sum_{s=t-\mu}^{b}\frac{1}{\rho}\left(\frac{t^{\underline{\mu}}(\mu+b+1-\sigma(s))^{\underline{\mu}}}{\mu\Gamma(\mu+1)}\right)$$
$$= \frac{1}{\rho}\left(\sum_{s=0}^{b}\frac{t^{\underline{\mu}}(\mu+b+1-\sigma(s))^{\underline{\mu}}}{\mu\Gamma(\mu+1)}\right) - \sum_{s=0}^{t-\mu-1}\frac{(t-\sigma(s))^{\underline{\mu}}}{\Gamma(\mu+1)}$$
$$= \frac{t^{\underline{\mu}}}{\rho\mu\Gamma(\mu+1)}\sum_{s=0}^{b}\frac{\Delta_s(\mu+b+1-s)^{\underline{\mu+1}}}{-(\mu+1)}$$
$$- \frac{1}{\Gamma(\mu+1)}\sum_{s=0}^{t-\mu-1}\frac{\Delta_s(t-s)^{\underline{\mu+1}}}{-(\mu+1)}.$$

Using the Fundamental Theorem of Discrete Calculus, we get that

$$\sum_{s=0}^{b}|G(t,s)| = \frac{1}{\Gamma(\mu+2)}\left[\left[(t-s)^{\underline{\mu+1}}\right]_{s=0}^{t-\mu} - \frac{t^{\underline{\mu}}}{\rho\mu}\left[(\mu+b+1-s)^{\underline{\mu+1}}\right]_{s=0}^{b+1}\right]$$

$$= \frac{1}{\Gamma(\mu+2)}\left[\left(\mu^{\underline{\mu+1}} - t^{\underline{\mu+1}}\right) - \frac{t^{\underline{\mu}}}{\rho\mu}\left(\mu^{\underline{\mu+1}} - (\mu+b+1)^{\underline{\mu+1}}\right)\right].$$

Since $\rho = \frac{1}{\mu}(\mu+b+1)^{\underline{\mu}}$ and using the fact that $\mu^{\underline{\mu+1}} = 0$, we obtain

$$\sum_{s=0}^{b}|G(t,s)| = \frac{1}{\Gamma(\mu+2)}\left[\frac{t^{\underline{\mu}}(\mu+b+1)^{\underline{\mu+1}}}{(\mu+b+1)^{\underline{\mu}}} - t^{\underline{\mu+1}}\right].$$

In addition, since $t^{\underline{\mu+1}} = t^{\underline{\mu}}(t-\mu)$, we have that

$$\sum_{s=0}^{b}|G(t,s)| = \frac{t^{\underline{\mu}}}{\Gamma(\mu+2)}(\mu+b+1-t), \quad t \in \mathbb{N}_{\mu-1}^{\mu+b+1}.$$

This completes the proof. □

**Theorem 6.28.** *The Green's function for* (6.32) *satisfies*

$$\max_{t\in\mathbb{N}_{\mu-1}^{\mu+b+1}} \sum_{s=0}^{b+2}|G(t,s)|$$

$$= \frac{1}{\Gamma(\mu+2)}\left[\mu+b-\left\lceil\frac{b+1}{\mu+1}\right\rceil\right]^{\underline{\mu-1}}\left[1+\left\lceil\frac{b+1}{\mu+1}\right\rceil\right],$$

*where $\lceil x \rceil$ denotes ceiling of $x$.*

*Proof.* In the previous theorem we proved that

$$\sum_{s=0}^{b}|G(t,s)| = \frac{t^{\underline{\mu}}}{\Gamma(\mu+2)}(\mu+b+1-t), \quad t \in \mathbb{N}_{\mu-1}^{\mu+b+1}.$$

Thus,

$$\max_{t\in\mathbb{N}_{\mu-1}^{\mu+b+1}} \sum_{s=0}^{b}|G(t,s)| = \frac{1}{\Gamma(\mu+2)} \max_{t\in\mathbb{N}_{\mu-1}^{\mu+b+1}} t^{\underline{\mu}}(\mu+b+1-t), \quad t \in \mathbb{N}_{\mu-1}^{\mu+b+1}.$$

Let $F(t) := t^{\underline{\mu}}(\mu+b+1-t)$, where $\mu-1 \le t \le \mu+b+1$ is a real variable. Then observe that $F(t) \ge 0$ for $t \in \mathbb{N}_{\mu-1}^{\mu+b+1}$ with $F(\mu-1) = 0$ and $F(\mu+b+1) = 0$. So $F$ has a nonnegative maximum, and to find this maximum we consider

$$\begin{aligned}F'(t) &= (-1)\, t^{\underline{\mu}} + (\mu)(\mu + b - t)\, t^{\underline{\mu-1}} \\ &= t^{\underline{\mu-1}}\, [(-1)(t + 1 - \mu) + (\mu)(\mu + b - t)] \\ &= t^{\underline{\mu-1}}\, \left[\mu^2 + b\mu + \mu - t\mu - t - 1\right].\end{aligned}$$

Now, by setting $F'(t) = 0$ and using the fact that $t^{\underline{\mu-1}} > 0$ whenever $t \in \mathbb{N}_{\mu-1}^{\mu+b+1}$, we have that

$$\mu^2 + b\mu + \mu - t\mu - t - 1 = 0.$$

Using standard calculus arguments it turns out that $F(t)$ restricted to the domain $\mathbb{N}_{\mu-1}^{\mu+b+1}$ has a maximum at $t = \mu + b - \left\lceil \frac{b+1}{\mu+1} \right\rceil$. Thus,

$$\max_{t\in\mathbb{N}_{\mu-1}^{\mu+b+1}} \sum_{s=0}^{b} |G(t,s)| = \frac{1}{\Gamma(\mu+2)} \left[\mu + b - \left\lceil \frac{b+1}{\mu+1} \right\rceil\right]^{\underline{\mu-1}} \left[1 + \left\lceil \frac{b+1}{\mu+1} \right\rceil\right],$$

as claimed. □

## 6.9 Existence of a Zero Tending Solution

In this section we will prove the existence of a zero tending solution as $t \to +\infty$ of the forced self-adjoint fractional difference equation (6.11); we obtain this result by using the Banach Fixed Point Theorem (Contraction Mapping Theorem). Also an example illustrating this result will be given.

**Theorem 6.29.** *Let $p : \mathbb{N}_{\mu-1} \to \mathbb{R}$, $f : \mathbb{N}_0 \to \mathbb{R}$, $q : \mathbb{N}_{\mu-1} \to \mathbb{R}$ and assume*

*(1)* $p(t) > 0$ *and* $q(t) \geq 0$, *for all* $t \in \mathbb{N}_{\mu-1}$

*(2)* $\sum_{\tau=\mu}^{\infty} \frac{1}{p(\tau)} < \infty$

*(3)* $\sum_{\tau=0}^{\infty} f(\tau) < \infty$

*(4)* $\sum_{\tau=\mu}^{\infty} \frac{1}{p(\tau)} \left( \sum_{s=0}^{\tau-\mu} h_{\mu-1}(t, \sigma(s)) q(s + \mu - 1) \right) < \infty$

*hold. Then the forced self-adjoint fractional difference equation*

$$\Delta_{\mu-1}^{\mu}(p\Delta x)(t) + q(t + \mu - 1)x(t + \mu - 1) = f(t), \quad t \in \mathbb{N}_0$$

*has a solution $x$ which satisfies* $\lim_{t\to\infty} x(t) = 0$.

*Proof.* In order to prove this theorem we will use the Banach Fixed Point Theorem. Since $\sum_{\tau=\mu}^{\infty} \frac{1}{p(\tau)} \left( \sum_{s=0}^{\tau-\mu} h_{\mu-1}(\tau, \sigma(s)) q(s+\mu-1) \right) < \infty$, we can choose $a \in \mathbb{N}_\mu$ such that

$$\alpha := \sum_{\tau=a}^{\infty} \frac{1}{p(\tau)} \left( \sum_{s=0}^{\tau-\mu} h_{\mu-1}(\tau, \sigma(s)) q(s+\mu-1) \right) < 1.$$

Let $Z$ be the space of all real-valued functions $x : \mathbb{N}_a \to \mathbb{R}$ such that $\lim_{t\to\infty} x(t) = 0$ with norm $\|\cdot\|$ on $Z$ defined by

$$\|x(t)\| = \max_{t\in\mathbb{N}_a} |x(t)|.$$

Define the operator $T$ on $Z$ by

$$Tx(t) = \sum_{\tau=t}^{\infty} \frac{1}{p(\tau)} \left[ \sum_{s=0}^{\tau-\mu} h_{\mu-1}(\tau, \sigma(s))[q(s+\mu-1)x(s+\mu-1) - f(s)] \right],$$

for $t \in \mathbb{N}_a$.

Now we will show that $T : Z \to Z$. To do this we will first show that $Tx$ is a real-valued function. Let $x \in Z$ be arbitrary but fixed. Then this implies that $x$ is a real-valued function satisfying $\lim_{t\to\infty} x(t) = 0$, which implies that for some real number $M > 0$, it holds that $|x(t)| \leq M$ for all $t \in \mathbb{N}_a$. Since $\sum_{\tau=0}^{\infty} f(\tau)$ converges, it follows that for some real number $N > 0$, we have $|f(\tau)| \leq N$ for all $t \in \mathbb{N}_{a-\mu}$. Thus,

$$\begin{aligned}
Tx(t) &= \sum_{\tau=t}^{\infty} \frac{1}{p(\tau)} \left[ \sum_{s=0}^{\tau-\mu} h_{\mu-1}(\tau, \sigma(s))[x(s+\mu-1)q(s+\mu-1) - f(s)] \right] \\
&\leq \sum_{\tau=t}^{\infty} \frac{1}{p(\tau)} \left[ \sum_{s=0}^{\tau-\mu} h_{\mu-1}(\tau, \sigma(s))[Mq(s+\mu-1) + N] \right] \\
&= M \sum_{\tau=t}^{\infty} \frac{1}{p(\tau)} \left[ \sum_{s=0}^{\tau-\mu} h_{\mu-1}(\tau, \sigma(s)) q(s+\mu-1) \right] \\
&\quad + N \sum_{\tau=t}^{\infty} \frac{1}{p(\tau)} \left[ \sum_{s=0}^{\tau-\mu} h_{\mu-1}(\tau, \sigma(s)) \right] \\
&< \infty.
\end{aligned}$$

And this demonstrates that $T$ is well defined on $Z$.

Next we will show that $T(Z) \subset Z$. We will use an argument similar to the previous paragraph. Let $x \in Z$ be arbitrary. Notice that

$$\begin{aligned}
|Tx(t)| &= \left|\sum_{\tau=t}^{\infty} \frac{1}{p(\tau)} \left[\sum_{s=0}^{\tau-\mu} h_{\mu-1}(\tau,\sigma(s))[x(s+\mu-1)q(s+\mu-1)-f(s)]\right]\right| \\
&\leq \sum_{\tau=t}^{\infty} \frac{1}{p(\tau)} \left[\sum_{s=0}^{\tau-\mu} h_{\mu-1}(\tau,\sigma(s))[Mq(s+\mu-1)+N]\right] \\
&= M\sum_{\tau=t}^{\infty} \frac{1}{p(\tau)} \left[\sum_{s=0}^{\tau-\mu} h_{\mu-1}(\tau,\sigma(s))q(s+\mu-1)\right] \\
&\quad + N\sum_{\tau=t}^{\infty} \frac{1}{p(\tau)} \left[\sum_{s=0}^{\tau-\mu} h_{\mu-1}(\tau,\sigma(s))\right].
\end{aligned} \tag{6.37}$$

Again, since

$$\sum_{\tau=t}^{\infty} \frac{1}{p(\tau)} \left(\sum_{s=0}^{\tau-\mu} h_{\mu-1}(\tau,\sigma(s))q(s+\mu-1)\right) < \infty,$$

we have that

$$\lim_{t\to\infty} \sum_{\tau=t}^{\infty} \frac{1}{p(\tau)} \left(\sum_{s=0}^{\tau-\mu} h_{\mu-1}(\tau,\sigma(s))q(s+\mu-1)\right) = 0,$$

for any real-valued nonnegative function $q$ defined on $\mathbb{N}_a$. Thus, by applying the limit as $t \to \infty$ on both sides of the (6.37) we get that $\lim_{t\to\infty} Tx(t) = 0$. Thus, $T : Z \to Z$.

Next we will show that $T$ is a contraction mapping on $Z$. Let $x, y \in Z$ and $t \in \mathbb{N}_a$ be arbitrary but fixed. Consider that

$$\begin{aligned}
&|Tx(t) - Ty(t)| \\
&= \left|\sum_{\tau=t}^{\infty} \frac{1}{p(\tau)} \left(\sum_{s=0}^{\tau-\mu} h_{\mu-1}(\tau,\sigma(s))[(x-y)q(s+\mu-1)-f(s)+f(s)]\right)\right| \\
&= \left|\sum_{\tau=t}^{\infty} \frac{1}{p(\tau)} \left(\sum_{s=0}^{\tau-\mu} h_{\mu-1}(\tau,\sigma(s))(x-y)q(s+\mu-1)\right)\right| \\
&\leq \sum_{\tau=t}^{\infty} \frac{1}{p(\tau)} \left(\sum_{s=0}^{\tau-\mu} h_{\mu-1}(\tau,\sigma(s))|x-y|q(s+\mu-1)\right)
\end{aligned}$$

$$\le \sum_{\tau=t}^{\infty} \frac{1}{p(\tau)} \left( \sum_{s=0}^{\tau-\mu} h_{\mu-1}(\tau, \sigma(s)) q(s+\mu-1) \right) \|x-y\|$$

$$\le \sum_{\tau=a}^{\infty} \frac{1}{p(\tau)} \left( \sum_{s=0}^{\tau-\mu} h_{\mu-1}(\tau, \sigma(s)) q(s+\mu-1) \right) \|x-y\|$$

$$= \alpha \|x-y\|,$$

for $t \in \mathbb{N}_a$. Since $t \in \mathbb{N}_a$ is arbitrary and $\alpha < 1$, we conclude that $T$ is a contraction mapping on Z. Therefore, by the Banach fixed point theorem there exists a unique fixed point $z \in Z$ such that $Tz = z$, which implies that for all $t \in \mathbb{N}_a$

$$z(t) = \sum_{\tau=t}^{\infty} \frac{1}{p(\tau)} \left[ \sum_{s=0}^{\tau-\mu} h_{\mu-1}(\tau, \sigma(s)) [q(s+\mu-1) z(s+\mu-1) - f(s)] \right].$$

We will now show that the unique fixed point $z$ is a solution to the forced self-adjoint fractional difference equation (6.11). To this end, applying the difference operator $\Delta$ to the map $z$ we obtain

$$\Delta z(t) = -\frac{1}{p(t)} \left[ \sum_{s=0}^{t-\mu} \frac{(t-\sigma(s))^{\underline{\mu-1}}}{\Gamma(\mu)} [q(s+\mu-1) z(s-\mu+1) - f(s)] \right].$$

Thus,

$$p(t)\Delta z(t) = -\left[ \sum_{s=0}^{t-\mu} \frac{(t-\sigma(s))^{\underline{\mu-1}}}{\Gamma(\mu)} [q(s+\mu-1) z(s-\mu+1) - f(s)] \right],$$

from which it follows that

$$(p\Delta z)(t) = \left[ \sum_{s=0}^{t-\mu} \frac{(t-\sigma(s))^{\underline{\mu-1}}}{\Gamma(\mu)} [f(s) - q(s+\mu-1) z(s-\mu+1)] \right]$$

$$= \Delta_0^{-\mu} [f(\cdot) - q(\cdot+\mu-1) z(\cdot+\mu-1)](t), \quad t \in \mathbb{N}_\mu.$$

Consequently,

$$\Delta_{\mu-1}^{\mu} (p\Delta z))(t) = \Delta_{\mu-1}^{\mu} (\Delta_0^{-\mu} (f(.) - q(\cdot+\mu-1) z(\cdot+\mu-1)))(t), \quad t \in \mathbb{N}_0.$$

Now we use the fractional composition rule on the right-hand side of the above equation to get

$$\begin{aligned}\Delta^{\mu}_{\mu-1}(p\Delta z)(t) &= \Delta^{0}_{-1}[f(t) - q(t+\mu-1)z(t+\mu-1)], \quad t \in \mathbb{N}_0\\ &= f(t) - q(t+\mu-1)z(t+\mu-1), \quad t \in \mathbb{N}_0\end{aligned}$$

so that

$$\Delta^{\mu}_{\mu-1}(p\Delta z)(t) + q(t+\mu-1)z(t+\mu-1) = f(t), \quad t \in \mathbb{N}_0.$$

Thus, $z$ satisfies the self-adjoint equation (6.11) and is therefore a solution to that equation. Since $z \in Z$, we conclude that the self-adjoint equation (6.11) has a solution $z$ satisfying $\lim_{t\to\infty} z(t) = 0$. This completes the proof. □

We conclude with an example.

*Example 6.30.* Let $f(t) := \frac{1}{(t+1)^2}$ for $t \in \mathbb{N}_0$, $q(t) \equiv 1$ for $t \in \mathbb{N}_{\mu-1}$, and $p(t) := (t+1)^2(t+1)^{\underline{\mu}}$ for $t \in \mathbb{N}_{\mu-1}$ in Theorem 6.29. Observe that each of the following holds:

(1) $p(t) > 0$ and $q(t) \geq 0$, for all $t \in \mathbb{N}_{\mu-1}$

(2) $\displaystyle\sum_{\tau=\mu}^{\infty} \frac{1}{p(\tau)} = \sum_{\tau=\mu}^{\infty} \frac{1}{(\tau+1)^2(\tau+1)^{\underline{\mu}}} < \infty$

(3) $\displaystyle\sum_{\tau=0}^{\infty} f(\tau) = \sum_{\tau=0}^{\infty} \frac{1}{(\tau+1)^2} < \infty$

(4) $\displaystyle\sum_{\tau=\mu}^{\infty} \frac{1}{p(\tau)} \left( \sum_{s=0}^{\tau-\mu} h_{\mu-1}(\tau, \sigma(s)) q(s+\mu-1) \right)$

$$= \sum_{\tau=\mu}^{\infty} \frac{(\tau)^{\underline{\mu}}}{\Gamma(\mu+1)(\tau+1)^2(\tau+1)^{\underline{\mu}}} < \infty.$$

Thus, Theorem 6.29 guarantees the existence of a solution to

$$\Delta^{\mu}_{\mu-1}(p\Delta x)(t) + q(t+\mu-1)z(t+\mu-1) = f(t), \quad t \in \mathbb{N}_0$$

such that $\lim_{t\to\infty} x(t) = 0$.

# Chapter 7
# Nonlocal BVPs and the Discrete Fractional Calculus

## 7.1 Introduction

In this chapter we discuss the concept of a nonlocal boundary value problem in the context of the discrete fractional calculus. More generally, we discuss how the nonlocal structure of the discrete fractional difference and sum operators affect their interpretation and analysis. In particular, we recall from earlier chapters that the fractional difference and sum contain *de facto* nonlocalities. For example, in the case of the discrete fractional forward difference, we have that $\Delta_a^\nu y(t)$, for $N-1 < \nu \leq N$ with $N \in \mathbb{N}$, depends not only on the value $y(t+\nu-N)$ but also on the entire collection of values $\{y(a), y(a+1), \ldots, y(t+\nu-N)\}$, for each $t \in \mathbb{Z}_{a+N-\nu}$. This means that the discrete fractional operator in some sense possesses a memory-like property, wherein the operator at a point is influenced by a linear combination of values of $y$ back to the initial time point $t = a$ itself.

The nonlocal nature described in the preceding paragraph seriously complicates the study of many potentially fundamental properties of fractional sums and differences. For example, there is, at present, no satisfactory understanding of the geometrical properties of the fractional difference. Contrast this with the integer-order setting, i.e., $\nu \in \mathbb{N}$, in which there is a complete understanding of the various geometrical implications of the sign of the fractional difference. Thus, while it is trivial to prove that $\Delta y(t) > 0$ for $t \in \mathbb{Z}$ implies that $y$ is strictly increasing on $\mathbb{Z}$, it is very *nontrivial* to decide how monotonicity is connected to the positivity or negativity of the fractional difference. Similarly, while it is equally trivial to prove that $\Delta^2 y(t) > 0$ for $t \in \mathbb{Z}$ implies that $\Delta y$ is strictly increasing on $\mathbb{Z}$ and thus that $y$ satisfies a convexity-type property, the analogue of this sort of result in the discrete fractional setting is much more difficult to obtain, and we only explore these properties to a limited extent in Sects. 7.2 and 7.3 in the sequel. And, of course, issues of monotonicity and convexity are hardly the only properties affected


C. Goodrich, A.C. Peterson, *Discrete Fractional Calculus*,
DOI 10.1007/978-3-319-25562-0_7

by the nonlocal structure of the fractional difference. The analysis of boundary value problems, for example, is also widely affected and complicated by this inherent nonlocality.

All in all, in this section we provide a collection of applications involving the nonlocal structure of the fractional difference. We also consider the problem of analyzing nonlocal boundary value problems within the context of the discrete fractional calculus. While the latter is not necessarily explicitly affected by the nonlocal nature of the fractional sum and difference operators, it does demonstrate in what way the existence of explicit nonlocalities in a boundary value problem can complicate the analysis of the problem, much as the implicit nonlocalities in the discrete fractional operates complicate the analysis of the geometrical properties of these operators.

## 7.2 A Monotonicity Result for Discrete Fractional Differences

The first result we present demonstrates that the discrete fractional difference satisfies a particular monotonicity condition—note that the results of this section can mostly be found in Dahal and Goodrich [67]. Roughly stated, see Theorem 7.1 for a precise statement, the main result of this section can be summarized as follows: Given $\nu \in (1,2)$ and a map $y : \mathbb{N}_0 \to \mathbb{R}$ satisfying

- $y(t) \geq 0$, for each $t \in \mathbb{N}_0$;
- $\Delta y(0) \geq 0$; and
- $\Delta^\nu y(t) \geq 0$, for each $t \in \mathbb{N}_{2-\nu}$;

then it holds that $y$ is increasing on its domain.

Recalling that

$$\Delta_a^\nu y(t) := \Delta^N \Delta^{\nu-N} y(t) = \Delta^N \underbrace{\left[\frac{1}{\Gamma(N-\nu)} \sum_{s=a}^{t+\nu-N} (t-s-1)^{\underline{N-\nu-1}} y(s)\right]}_{:=\Delta^{\nu-N} y(t)},$$

this result does not seem to be immediately apparent from the definition of the fractional difference. Hence, it is not obvious that the fractional order difference behaves in this way, and it highlights one of the consequences of the nonlocal structure of the fractional difference operator. In addition, that this monotonicity result holds implies some other nontrivial consequences, and we shall detail a few of these toward the conclusion of this section.

We now state and prove the monotonicity result. Observe that the proof of this result is based upon the principle of strong induction. Moreover, the reader should observe the way in which the nonlocal structure of $\Delta_0^\nu$ is explicitly utilized in the proof.

**Theorem 7.1.** *Let $y : \mathbb{N}_0 \to \mathbb{R}$ be a nonnegative function satisfying $y(0) = 0$. Fix $\nu \in (1, 2)$ and suppose that $\Delta_0^{\nu} y(t) \geq 0$ for each $t \in \mathbb{N}_{2-\nu}$. Then $y$ is increasing on $\mathbb{N}_0$.*

*Proof.* We prove this result by the principle of strong induction. To this end, observe that the base case holds somewhat trivially since we calculate

$$y(1) - y(0) = y(1) \geq 0,$$

due to the fact that $y(0) = 0$, by assumption, and the fact that $y(1) \geq 0$, also by assumption.

Now, to complete the induction step fix $k \in \mathbb{N}$ and suppose that

$$\Delta y(j-1) = y(j) - y(j-1) \geq 0,$$

for each $1 \leq j \leq k-1$. Recall that $\Delta_0^{\nu} y(t) \geq 0$ for each $t \in \mathbb{N}_{2-\nu}$, which by means of Lemma 2.33 implies that

$$\begin{aligned}
-\Delta_0^{\nu} y(2-\nu) &= \nu y(1) - y(2) \leq 0 \\
-\Delta_0^{\nu} y(3-\nu) &= \frac{1}{2}\nu(1-\nu)y(1) + \nu y(2) - y(3) \leq 0 \\
-\Delta_0^{\nu} y(4-\nu) &= \frac{1}{6}\nu(1-\nu)(2-\nu)y(1) + \frac{1}{2}\nu(1-\nu)y(2) + \nu y(3) - y(4) \\
&\leq 0 \\
&\vdots \\
-\Delta_0^{\nu} y(k-\nu) &= \frac{1}{(k-1)!}\nu(1-\nu)\cdots(k-2-\nu)y(1) + \cdots \\
&\quad + \nu y(k-1) - y(k) \leq 0,
\end{aligned} \tag{7.1}$$

for fixed $k \in \mathbb{N}$; note that in (7.1) we have used the assumption that $y(0) = 0$ to simplify suitably $\Delta^{\nu} y(j-\nu)$ for each $j$. In particular, (7.1) implies that

$$y(k) \geq \frac{1}{(k-1)!}\nu(1-\nu)\cdots(k-2-\nu)y(1) + \cdots + \nu y(k-1) \tag{7.2}$$

for fixed $k \in \mathbb{N}$. Inequality (7.2) shall be used repeatedly in the sequel.

We claim that for the value of $k$ fixed at the beginning of the preceding paragraph

$$y(k) - y(k-1) \geq 0, \tag{7.3}$$

which will complete the induction step. To prove (7.3) we first calculate, by using estimate (7.2),

$$\begin{aligned}
&y(k)-y(k-1)\\
&\geq \nu y(k-1)+\frac{1}{2}\nu(1-\nu)y(k-2)+\frac{1}{6}\nu(1-\nu)(2-\nu)y(k-3)\\
&\quad+\cdots+\frac{1}{(k-1)!}\nu(1-\nu)(2-\nu)\cdots(k-2-\nu)y(1)-y(k-1)\\
&=(\nu-1)y(k-1)+\frac{1}{2}\nu(1-\nu)y(k-2)+\frac{1}{6}\nu(1-\nu)(2-\nu)y(k-3)\\
&\quad+\cdots+\frac{1}{(k-1)!}\nu(1-\nu)(2-\nu)\cdots(k-2-\nu)y(1)\\
&=(\nu-1)y(k-1)\\
&\quad+\left(\frac{1}{2}\nu(1-\nu)y(k-2)-\frac{1}{2}\nu(1-\nu)y(k-1)\right)+\frac{1}{2}\nu(1-\nu)y(k-1)\\
&\quad\quad+\left(\frac{1}{6}\nu(1-\nu)(2-\nu)y(k-3)-\frac{1}{6}\nu(1-\nu)(2-\nu)y(k-1)\right)\\
&\quad\quad\quad+\frac{1}{6}\nu(1-\nu)(2-\nu)y(k-1)\\
&\quad\vdots\\
&\quad+\left(\frac{1}{(k-1)!}\nu(1-\nu)(2-\nu)\cdots(k-2-\nu)y(1)\right.\\
&\quad\quad\left.-\frac{1}{(k-1)!}\nu(1-\nu)(2-\nu)\cdots(k-2-\nu)y(k-1)\right)\\
&\quad\quad\quad+\frac{1}{(k-1)!}\nu(1-\nu)(2-\nu)\cdots(k-2-\nu)y(k-1). \qquad (7.4)
\end{aligned}$$

On the other hand, invoking the induction hypothesis implies that

$$\underbrace{\frac{1}{2}\nu(1-\nu)}_{<0}\underbrace{(-y(k-1)+y(k-2))}_{\leq 0}\geq 0$$

$$\underbrace{\frac{1}{6}\nu(1-\nu)(2-\nu)}_{<0}\underbrace{(-y(k-1)+y(k-3))}_{\leq 0}\geq 0$$

$$\vdots$$

$$\underbrace{\frac{1}{(k-1)!}\nu(1-\nu)(2-\nu)\cdots(k-2-\nu)}_{<0}\underbrace{(-y(k-1)+y(1))}_{\leq 0}\geq 0. \qquad (7.5)$$

Observe that in (7.5) we are using the fact that since $y(k-1) \geq y(k-2)$ it follows that $y(k-1) - y(k-3) \geq y(k-2) - y(k-3) \geq 0$, and so forth. In any case, putting the $k-2$ estimates in (7.5) into inequality (7.4) yields

$$\begin{aligned} &y(k) - y(k-1) \\ &\geq \bigg[ (\nu - 1) + \frac{1}{2}\nu(1-\nu) + \frac{1}{6}\nu(1-\nu)(2-\nu) + \dots \\ &\qquad + \frac{1}{(k-1)!}\nu(1-\nu)(2-\nu)\cdots(k-2-\nu) \bigg] y(k-1). \end{aligned}$$

Since $y(k-1) \geq 0$ by assumption, to complete the proof it suffices to show that

$$\begin{aligned} &(\nu - 1) + \frac{1}{2}\nu(1-\nu) + \frac{1}{6}\nu(1-\nu)(2-\nu) \\ &\qquad + \dots + \frac{1}{(k-1)!}\nu(1-\nu)(2-\nu)\cdots(k-2-\nu) \geq 0, \end{aligned}$$

for each $1 < \nu < 2$. To complete this final step define the $(k-1)$-th degree polynomial function $P_{k-1} : \mathbb{R} \to \mathbb{R}$ by

$$\begin{aligned} P_{k-1}(\nu) := (\nu - 1) + \frac{1}{2}\nu(1-\nu) + \frac{1}{6}\nu(1-\nu)(2-\nu) + \dots \\ + \frac{1}{(k-1)!}\nu(1-\nu)(2-\nu)\cdots(k-2-\nu). \end{aligned}$$

Then, for example,

$$\frac{2P_{k-1}(\nu)}{1-\nu} = (\nu - 2) + \frac{1}{3}\nu(2-\nu) + \dots + \frac{2}{(k-1)!}\nu(2-\nu)\cdots(k-2-\nu).$$

And, moreover,

$$\begin{aligned} \frac{6P_{k-1}(\nu)}{(1-\nu)(2-\nu)} = (\nu - 3) + \frac{1}{4}\nu(3-\nu) + \cdots \\ + \frac{6}{(k-1)!}\nu(3-\nu)\cdots(k-2-\nu). \end{aligned}$$

Continuing in this fashion we eventually arrive at

$$\frac{(k-1)!P_{k-1}(\nu)}{(1-\nu)(2-\nu)\cdots(k-2-\nu)} = -(k-1-\nu),$$

whence

$$P_{k-1}(\nu) = \frac{-1}{(k-1)!}(1-\nu)(2-\nu)\cdots(k-2-\nu)(k-1-\nu). \tag{7.6}$$

But then (7.6) implies that

$$\begin{aligned}
P_{k-1}(\nu) &= -\frac{1}{(k-1)!}(1-\nu)(2-\nu)\cdots(k-2-\nu)(k-1-\nu)\\
&= \frac{1}{(k-1)!}(-1)^2(\nu-1)(2-\nu)\cdots(k-2-\nu)(k-1-\nu)\\
&= \frac{1}{(k-1)!}(-1)^3(\nu-1)(\nu-2)(3-\nu)\cdots(k-2-\nu)(k-1-\nu)\\
&\ \ \vdots\\
&= \frac{1}{(k-1)!}(-1)^k(\nu-1)(\nu-2)\cdots(\nu-k+2)(\nu-k+1).
\end{aligned} \tag{7.7}$$

The factorization of $P_{k-1}$ given by (7.7) implies that $P_{k-1}$ has $k-1$ distinct zeros and these zeros are, in particular, $\nu = 1, 2, \ldots, k-1$. In particular, observe that when $k$ is even, it follows that $P_{k-1}(\nu) > 0$, for each $\nu \in (1, 2)$, since $(-1)^k > 0$ and the other factors will constitute a product of $k-2$ negative numbers and exactly one positive number. On the other hand, when $k$ is odd, it follows that $P_{k-1}(\nu) > 0$, for each $\nu \in (1, 2)$, since $(-1)^k < 0$ and the other factors will once again constitute a product of $k-2$ negative numbers and exactly one positive number.

We conclude that for each $k \in \mathbb{N}$ it follows that $P_{k-1}(\nu) > 0$ whenever $\nu \in (1, 2)$. And this implies that (7.3) holds. Since this completes the induction step, we obtain that $y$ is increasing for $k \in \mathbb{N}$ and $\nu \in (1, 2)$, and this completes the proof. □

Having proved the case where $y(0) = 0$, it is easy to generalize this to the case where $y(0) \geq 0$. We state this generalization as Corollary 7.2. It turns out that this generalization will be useful in the next section when we consider concavity and convexity properties of the fractional difference operator.

**Corollary 7.2.** *Let $y : \mathbb{N}_0 \to \mathbb{R}$ be a nonnegative function. Fix $\nu \in (1, 2)$ and suppose that $\Delta_0^{\nu} y(t) \geq 0$, for each $t \in \mathbb{N}_{2-\nu}$. If $\Delta y(0) \geq 0$, then $y$ is increasing on $\mathbb{N}_0$.*

*Proof.* Define the function $\tilde{y} : \mathbb{Z}_{-1} \to \mathbb{R}$ by $\tilde{y}(t) := y(t)$, if $t \neq -1$, and $\tilde{y}(-1) := 0$. Then we may apply Theorem 7.1 to $\tilde{y}$ on $\mathbb{Z}_{-1}$ and obtain that $\tilde{y}$ is increasing on $\mathbb{Z}_{-1}$; observe that it can be shown that $\Delta_{-1}^{\nu}\tilde{y} \equiv \Delta_0^{\nu} y$ so that $\Delta_{-1}^{\nu}\tilde{y}(t) \geq 0$ holds. Thus, $y$ is increasing on $\mathbb{N}_0$, as desired. □

*Remark 7.3.* It is important to point out that the original version of the paper by Dahal and Goodrich [67], in which the monotonicity results for discrete fractional operators first appeared, contained a minor error in one result. In particular, [67, Corollary 2.3] is missing the hypothesis that $\Delta y(0) \geq 0$ holds.

As it may be instructive to see why this error occurs, let us briefly explain the problem. So, to see why we cannot deduce the monotonicity of $y$ from the hypothesis $\Delta_0^{\nu} y(t) \geq 0$, $t \in \mathbb{Z}_{2-\nu}$, alone, we recall that the proof of the corollary in the original paper employed the map $\tilde{y} : \mathbb{Z}_{-1} \to \mathbb{R}$ defined by

$$\tilde{y}(t) := \begin{cases} y(t), & t \in \mathbb{N}_0 \\ 0, & t \in \{-1\} \end{cases}. \tag{7.8}$$

Note that in (7.8) the map $y$ is the same map as in the statement of Corollary 7.2 above. The goal of the proof in [67, Corollary 2.3] was to show that $\Delta_{-1}^{\nu} \tilde{y}(t) \geq 0$, for each $t \in \mathbb{Z}_{1-\nu}$, so that we could use [67, Theorem 2.2] to conclude that $\tilde{y}$ and, hence, $y$ was increasing.

To see why this does not work quite as intended, observe that

$$\begin{aligned}
\Delta_{-1}^{\nu} \tilde{y}(1-\nu) &= \frac{1}{\Gamma(-\nu)} \sum_{s=-1}^{1} ((1-\nu)-s-1)^{\underline{-\nu-1}} \tilde{y}(s) \\
&= \frac{1}{\Gamma(-\nu)} \Big[ (-\nu+1)^{\underline{-\nu-1}} \tilde{y}(-1) + (-\nu)^{\underline{-\nu-1}} \tilde{y}(0) \\
&\qquad + (-\nu-1)^{\underline{-\nu-1}} \tilde{y}(1) \Big] \\
&= -\frac{1}{2}\nu(-\nu+1)\tilde{y}(-1) - \nu\tilde{y}(0) + \tilde{y}(1) \\
&= -\nu y(0) + y(1).
\end{aligned} \tag{7.9}$$

Thus, (7.9) shows us that without additional information about $y(0)$ and $y(1)$, we cannot deduce that $\Delta_{-1}^{\nu} \tilde{y}(1-\nu) \geq 0$. Moreover, while it is true that $\Delta_{-1}^{\nu} \tilde{y}(k-\nu) \equiv \Delta_0^{\nu} y(k-\nu)$, for each $k \in \mathbb{N}_2$, this does not force $\Delta_{-1}^{\nu} \tilde{y}(1-\nu)$ to be nonnegative, as shown by (7.9) above. This is the basis of the error.

Finally, suppose that $\Delta y(0) \geq 0$, which is the necessary additional hypothesis as noted above. By a calculation similar to (7.9) we find that

$$\Delta_0^{\nu} y(2-\nu) = -\frac{1}{2}\nu(-\nu+1)y(0) - \nu y(1) + y(2) \geq 0. \tag{7.10}$$

Thus, combining (7.10) with the fact that $\Delta y(0) \geq 0$ we estimate

$$\Delta y(1) \geq (\nu-1)y(1) + \frac{1}{2}\nu(1-\nu)y(0) \geq (\nu-1)\left[1-\frac{1}{2}\nu\right]y(0). \tag{7.11}$$

Since the map $\nu \mapsto (\nu-1)\left[1-\frac{1}{2}\nu\right]$ is nonnegative for $\nu \in (1,2)$, we obtain from (7.11) that $\Delta y(1) \geq 0$. Finally, proceeding inductively from inequality (7.11), we obtain the monotonicity of $\tilde{y}$ on $\mathbb{Z}_{-1}$ and thus of $y$ on $\mathbb{N}_0$.

Recall that in the statement of both Theorem 7.1 and Corollary 7.2 we have that $y$ is nonnegative. It is easy to obtain a result similar to Theorem 7.1 in case $y$ is instead nonpositive. In addition, as with Theorem 7.1, we may obtain a corollary dual to Corollary 7.2, but we omit its statement.

**Theorem 7.4.** *Let $y : \mathbb{N}_0 \to \mathbb{R}$ be a nonpositive function satisfying $y(0) = 0$. Fix $\nu \in (1,2)$ and suppose that $\Delta_0^\nu y(t) \leq 0$, for each $t \in \mathbb{N}_{2-\nu}$. If it also holds that $\Delta y(0) \leq 0$, then $y$ is decreasing on $\mathbb{N}_0$.*

*Proof.* Let $y$ be as in the statement of this theorem. Put $z := -y$. Then $z(0) = 0$, $z$ is nonnegative on its domain, and (by the linearity of the fractional difference operator) $\Delta_0^\nu z(t) \geq 0$ for each $t \in \mathbb{N}_{2-\nu}$. Consequently, each of the hypotheses of Theorem 7.1 is satisfied. Therefore, we conclude that $z$ is increasing, whence $-z = y$ is decreasing at each $t \in \mathbb{N}_0$. And this completes the proof. □

We mention next a couple of representative consequences of Theorems 7.1 and 7.4. We begin by providing a result regarding a discrete fractional IVP, which is Corollary 7.5, and then a result about a discrete fractional BVP with (possibly) inhomogeneous boundary conditions, which is Corollary 7.6.

**Corollary 7.5.** *Let $h : [1, +\infty)_{\mathbb{N}} \times \mathbb{R} \to \mathbb{R}$ be a nonnegative, continuous function, and let $A, B \in \mathbb{R}$ be nonnegative constants. Then the unique solution of the discrete fractional IVP*

$$\begin{aligned}\Delta_0^\nu y(t) &= h(t+\nu-1, y(t+\nu-1)),\ t \in [2-\nu, +\infty)_{\mathbb{Z}_{2-\nu}}\\ y(0) &= A,\ \Delta y(0) = B\end{aligned}$$

*is increasing (and nonnegative).*

*Proof.* Simply note that the proof of Theorem 7.1 reveals that one may replace the hypothesis that $y$ is nonnegative on its domain with the hypothesis that $y(1) \geq y(0) \geq 0$. Since $A, B \geq 0$, the result follows. □

**Corollary 7.6.** *Let $h : \mathbb{Z}_{\nu-2} \times \mathbb{R} \to \mathbb{R}$ be a nonpositive function, and let $A, B \in \mathbb{R}$ be nonpositive constants. Then the unique solution of the discrete fractional IVP*

$$\begin{aligned}\Delta_{\nu-2}^\nu y(t) &= h(t+\nu-1, y(t+\nu-1))\\ y(\nu-2) &= A\\ \Delta y(\nu-2) &= B\end{aligned}$$

*is decreasing.*

Our final consequence of Theorem 7.1 deserves special mention. To contextualize the result, let us consider the problem

$$\begin{aligned}\left(\Delta_{\alpha-1}^\alpha u\right)(t) &= \lambda u(t+\alpha-1) + f(t+\alpha-1, u(t+\alpha-1)),\ t \in [0, T-1]_{\mathbb{N}_0}\\ u(\alpha-1) &= u(\alpha-1+T),\end{aligned} \tag{7.12}$$

which was studied by Ferreira and Goodrich [83]. Supposing that $f$ satisfied superlinear growth at 0 and $+\infty$ (uniformly for $t$), they proved that problem (7.12) has at least one positive solution for a range of values of the parameter $\lambda$, even if $f$ is nonnegative. Assuming $\lambda > 0$, such an occurrence is clearly impossible in case $\alpha = 1$, for then (7.12) implies that $u$ is strictly increasing, contradicting the periodic boundary conditions. So, this result demonstrates that the fractional difference can behave in unexpected ways, due precisely to its nonlocal structure.

With this somewhat aberrant result in mind, we might wonder if it is possible for the problem

$$\begin{aligned} -\Delta^{\nu}_{\nu-2}y(t) &= f(t+\nu-1, y(t+\nu-1)), t \in [0, b+1]_{\mathbb{N}_0} \\ y(\nu-2) &= 0 \\ y(\nu+b+1) &= 0 \end{aligned} \tag{7.13}$$

to have at least one positive solution if $f$ is nonpositive. Now, when $\nu = 2$, the nonpositivity of $f$ implies that $\Delta^2 y(t) \geq 0$, for each admissible $t$, from which it follows at once that if $y$ is a solution of problem (7.13), then $y(t) \leq 0$ for each $t$. This is a simple consequence of the geometrical implications of $\Delta^2 y(t) \geq 0$ together with the Dirichlet boundary conditions. However, as the discussion regarding problem (7.12) demonstrates, in the fractional setting one cannot be so sure. In fact, it would not be entirely unreasonable to suspect that perhaps that nonlocal structure of $\Delta^{\nu}_{\nu-2}$ somehow allows for a positive solution to exist in spite of the nonpositivity of $f$. Corollary 7.7 demonstrates conclusively that this particular geometric aberration is forbidden.

**Corollary 7.7.** *Let* $f : [\nu-1, \nu+b]_{\mathbb{Z}_{\nu-1}} \times \mathbb{R} \to \mathbb{R}$ *be continuous and nonpositive and* $b \in \mathbb{N}$ *a given constant. Then the discrete fractional boundary value problem* (7.13) *has no positive solution.*

We would also like to mention that it is necessary to impose *some* additional restriction beyond the positivity of the fractional difference if we hope to deduce the monotonicity of $y$. For example, in Corollary 7.2 we impose the condition $\Delta y(0) \geq 0$, which, as was explained earlier, was inadvertently omitted from the statement of the corresponding result in [67], though **all** of the other results in that paper are correct as stated. In any case, to demonstrate that the positivity of the fractional difference is *not sufficient*, we provide the following example.

*Example 7.8.* Let $f(t) = 2^{-t}$, $t \in \mathbb{N}_0$, and assume that $\frac{2+\sqrt{2}}{2} < \nu < 2$. We will show that $\Delta^{\nu}_0 f(t) \geq 0$, $t \in \mathbb{N}_{2-\nu}$, $f(t) \geq 0$ on $\mathbb{N}_0$, but $f(t)$ is not increasing on $\mathbb{N}_1$. Clearly $f(t) \geq 0$ on $\mathbb{N}_0$.

For $t = 2-\nu+k$, $k \geq 0$, we have

$$\begin{aligned} \Delta^{\nu}_0 f(t) &= \int_0^{3+k} h_{-\nu-1}(2-\nu+k, \tau+1) f(\tau)\, \Delta\tau \\ &= \sum_{i=0}^{k+2} h_{-\nu-1}(2-\nu+k, i+1) 2^{-i}. \end{aligned}$$

For $0 \leq i \leq k+2$, $1 < \nu < 2$, we have

$$\begin{aligned} h_{-\nu-1}(2-\nu+k, i+1) &= \frac{(1-\nu+k-i)^{\underline{-\nu-1}}}{\Gamma(-\nu)} \\ &= \frac{\Gamma(2-\nu+k-i)}{\Gamma(3+k-i)\Gamma(-\nu)} \\ &= \frac{(-\nu+1+k-i)\cdots(-\nu+1)(-\nu)}{(2+k-i)!}. \end{aligned} \tag{7.14}$$

It follows from (7.14) that if $k-i \geq 1$, then $h_{-\nu-1}(2-\nu+k, i+1) > 0$. When $i = k, k+1, k+2$, we have

$$h_{-\nu-1}(2-\nu+k, k+1) = \frac{\Gamma(2-\nu)}{2!\Gamma(-\nu)} = \frac{(-\nu+1)(-\nu)}{2}, \tag{7.15}$$

$$h_{-\nu-1}(2-\nu+k, k+2) = \frac{\Gamma(1-\nu)}{\Gamma(-\nu)} = -\nu, \tag{7.16}$$

$$h_{-\nu-1}(2-\nu+k, k+3) = \frac{\Gamma(-\nu)}{\Gamma(-\nu)} = 1, \tag{7.17}$$

respectively. So from (7.15), (7.16), (7.17), and the fact that $\frac{2+\sqrt{2}}{2} < \nu < 2$, we get that

$$\begin{aligned} \Delta_0^\nu f(t) &\geq \sum_{i=k}^{k+2} h_{-\nu-1}(2-\nu+k, i+1)2^{-i} \\ &= \frac{(-\nu+1)(-\nu)}{2} \cdot \frac{1}{2^k} - \frac{\nu}{2^{k+1}} + \frac{1}{2^{k+2}} \\ &= \frac{2\nu^2 - 4\nu + 1}{2^{k+2}} > 0. \end{aligned}$$

But since $f(t)$ is obviously decreasing, it follows that the some additional condition is necessary above and beyond the positivity of the fractional difference.

It should be noted that the above example is just a special case of a more general result, which we illustrate with the following example.

*Example 7.9.* Let $f(t) = \alpha^{-t}$, $t \in \mathbb{N}_0$, and assume that $\alpha > 1$. If we proceed as in the example above, we obtain the estimate

$$\begin{aligned}\Delta_0^\nu f(t) &\geq \sum_{i=k}^{k+2} h_{-\nu-1}(2-\nu+k, i+1)\alpha^{-i} \\ &= \frac{(-\nu+1)(-\nu)}{2} \cdot \frac{1}{\alpha^k} - \frac{\nu}{\alpha^{k+1}} + \frac{1}{\alpha^{k+2}} \\ &= \frac{\alpha^2\nu^2 - \nu\alpha^2 - 2\nu\alpha + 2}{2\alpha^{k+2}}.\end{aligned}$$

Let us define

$$g(\nu) := \nu^2 - \frac{\nu(2+\alpha)}{\alpha} + \frac{2}{\alpha^2}.$$

We see, therefore, that $\Delta_0^\nu f(t) > 0$ if $g(\nu) > 0$. If we solve the quadratic inequality $g(\nu) > 0$, we find that for all $\nu_0 < \nu < 2$, we have $\Delta_0^\nu f(t) > 0$, where $\nu_0 := \frac{\alpha+2+\sqrt{\alpha^2+4\alpha-4}}{2\alpha}$. Since $\nu_0 < 2$ holds because $\alpha > 1$, it follows that we obtain a family of functions $f(t) := \alpha^{-t}$ which are decreasing but for which $\Delta_0^\nu f(t) > 0$. Finally, the following table illustrates how the interval $(\nu_0, 2)$ varies with differing choices for $\alpha \in (1, \infty)$.

| $\alpha$ | 1.2 | 2 | 3 | 8 |
|---|---|---|---|---|
| $\nu_0$ | 1.950 | 1.707 | 1.510 | 1.224 |

We conclude by presenting a monotonicity theorem of a different color. The interesting point regarding this result is that we do ***not*** necessarily suppose that $\Delta f(a) \geq 0$. Rather, we replace this condition with a weaker condition on the "initial growth" of the map $t \mapsto y(t)$. Thus, this result improves the monotonicity result that was proved earlier. This new result was proved by Baoguo, Erbe, Goodrich, and Peterson, and it relies in a special way on a useful difference inequality discovered by Baoguo, Erbe, and Peterson. Thus, we first state the aforementioned technical lemma and then state and prove the existence theorem; we mention that for the interested reader, the proof of Lemma 7.10 may be found in [47]. (For readers who have read Chap. 3, it is seen that Lemma 7.10 is very much related, in the nabla setting, to Theorem 3.115.)

**Lemma 7.10.** *Assume that $\Delta_a^\nu f(t) \geq 0$, for each $t \in \mathbb{N}_{a+2-\nu}$, with $1 < \nu < 2$. Then*

$$\Delta f(a+k+1) \geq -h_{-\nu}(a+k+2-\nu, a)f(a) - \sum_{\tau=a}^{a+k} h_{-\nu}(a+k+2-\nu, \sigma(\tau))\Delta f(\tau)$$

*for each $k \in \mathbb{N}_0$, where*

$$h_{-\nu}(t, \sigma(\tau)) = \frac{(t-\tau)^{\underline{-\nu}}}{(t-\nu-\tau)!(a+k+2-\tau)!} < 0,$$

*for* $t \in \mathbb{N}_{a+2-\nu}$, $a \leq \sigma(\tau) \leq t + \nu - 1$.

**Theorem 7.11.** *Assume that* $f : \mathbb{N}_a \to \mathbb{R}$ *and that* $\Delta_a^\nu f(t) \geq 0$, *for each* $t \in \mathbb{N}_{a+2-\nu}$, *with* $1 < \nu < 2$. *If*

$$f(a+1) \geq \frac{\nu}{k+2} f(a)$$

*for each* $k \in \mathbb{N}_0$, *then* $\Delta f(t) \geq 0$, *for* $t \in \mathbb{N}_{a+1}$.

*Proof.* We prove that $\Delta f(a + k + 1) \geq 0$, for each $k \geq 0$, by the principle of strong induction. From Lemma 7.10, in case $k = 0$, together with the hypothesis $f(a+1) \geq \frac{\nu}{2} f(a)$, we estimate

$$\begin{aligned}
\Delta f(a+1) &\geq -h_{-\nu}(a+2-\nu, a) f(a) - h_{-\nu}(a+2-\nu, a+1)\Delta f(a) \\
&= -\left[\frac{\Gamma(3-\nu)}{\Gamma(3)\Gamma(-\nu+1)} f(a) + \frac{\Gamma(2-\nu)}{\Gamma(2)\Gamma(-\nu+1)} \Delta f(a)\right] \\
&= -\frac{\Gamma(2-\nu)}{\Gamma(-\nu+1)} \left[\frac{1}{2}(2-\nu) f(a) + \Delta f(a)\right] \\
&\geq -\frac{\Gamma(2-\nu)}{\Gamma(-\nu+1)} \underbrace{\left[\frac{1}{2}(2-\nu) + \frac{\nu}{2} - 1\right]}_{=0} f(a) \\
&= 0.
\end{aligned}$$

Suppose next that $k \geq 1$ and $\Delta f(a+i) \geq 0$, for $i \in \mathbb{N}_1^k$. From Lemma 7.10 together with the hypothesis $f(a+1) \geq \frac{\nu}{k+2} f(a)$ for each $k \in \mathbb{N}_0$, we estimate

$$\begin{aligned}
&\Delta f(a+k+1) \\
&\quad \geq -f(a) h_{-\nu}(a+k+2-\nu, a) - \sum_{\tau=a}^{a+k} h_{-\nu}(a+k+2-\nu, \sigma(\tau))\Delta f(\tau) \\
&\quad \geq -f(a) h_{-\nu}(a+k+2-\nu, a) - h_{-\nu}(a+k+2-\nu, a+1)\Delta f(a) \\
&\quad = -\frac{\Gamma(k+3-\nu)}{\Gamma(k+3)\Gamma(-\nu+1)} f(a) - \frac{\Gamma(k+2-\nu)}{\Gamma(k+2)\Gamma(-\nu+1)} \Delta f(a) \\
&\quad = -\underbrace{\frac{\Gamma(k+2-\nu)}{\Gamma(k+2)\Gamma(-\nu+1)}}_{>0} \left[\frac{k+2-\nu}{k+2} f(a) + \Delta f(a)\right]
\end{aligned}$$

$$\geq -\frac{\Gamma(k+2-\nu)}{\Gamma(k+2)\Gamma(-\nu+1)}\underbrace{\left[\frac{k+2-\nu}{k+2}+\frac{\nu}{k+2}-1\right]}_{=0}f(a)$$

$$= 0.$$

As this inequality implies that $f$ is monotone increasing, the proof is thus complete. □

*Remark 7.12.* We would like to point out that very recently (i.e., as of late 2015) there has been quite a bit of progress made in extending the results of this section, which were deduced by Dahal and Goodrich in late 2013. Correspondingly, there has been much activity very recently (again, as of late 2015) extending the concavity/convexity results of the next section. In addition to the preceding theorem, additional generalizations have been produced. We direct the interested reader to the forthcoming papers by Baoguo, Erbe, Goodrich, and Peterson [53, 54].

## 7.3 Concavity and Convexity Results for Fractional Difference Operators

In the previous section we demonstrated that the discrete fractional difference operator satisfied a particular monotonicity condition, and we saw how this was a direct consequence of the implicit nonlocality in the construction of the fractional difference. In this section we present some convexity and concavity results for discrete fractional difference operators. We note that the results of this section can mostly be found in a paper by Goodrich [114].

We begin by stating the main result of this section and then proceed to state and discuss several consequences of this result. Essentially, the result states that if

- $y(0) = \Delta y(0) = \cdots = \Delta^{N-3}y(0) = 0$;
- $\Delta^{N-2}y(0) \geq 0$;
- $\Delta^{N-1}y(0) \geq 0$; and
- $\Delta_0^{\mu}y(t) \geq 0$, for each $t \in \mathbb{N}_{N-\mu}$;

then $\Delta^{N-1}y(t) \geq 0$, for each $t \in \mathbb{N}_0$. In particular, if we fix $N = 3$, then we obtain a suitable convexity result. In this special case we obtain that if $y(0) = 0$, $\Delta y(0)$ is nonnegative, $\Delta^2 y(0)$ is also nonnegative, and $\Delta_0^{\mu}y(t)$ is nonnegative for each admissible $t$, then the map $t \mapsto y(t)$ is concave on its domain. So, this result essentially demonstrates that if $y$ has a bit of initial convexity, so to speak, then this is propagated provided that the sufficient auxiliary conditions are in force, as described precisely above. In some sense, this is not quite what one would expect, since it would be preferable not to have to require the condition $\Delta^2 y(0) \geq 0$. At the end of this section we shall suggest how this may be improved.

With these considerations in mind, we now state and prove the convexity result. Observe that the proof of this result and its corollaries are strongly based on an application of the monotonicity result. Thus, this collection of results gives us yet another nontrivial application of the monotonicity result.

**Theorem 7.13.** *Fix $\mu \in (N-1, N)$, for $N \in \mathbb{N}_3$ given, and let $y : \mathbb{N}_0 \to \mathbb{R}$ be a given function satisfying $\Delta^j y(0) = 0$ for each $j \in \{0, 1, 2, \ldots, N-3\}$, $\Delta^{N-2}y(0) \geq 0$, and $\Delta_0^{\mu} y(t) \geq 0$ for each $t \in \mathbb{N}_{N-\mu}$. If it also holds that $\Delta^{N-1}y(0) \geq 0$, then $\Delta^{N-1}y(t) \geq 0$, for each $t \in \mathbb{N}_0$.*

*Proof.* Define the function $w : \mathbb{N}_0 \to \mathbb{R}$ by

$$w(t) := \Delta^{N-2}y(t).$$

We show that $w$ satisfies the monotonicity theorem—namely, Corollary 7.2. To this end, put $\nu := \mu - N + 2 \in (1, 2)$. On the one hand, by Theorem 2.51 we obtain

$$\begin{aligned}
\Delta_0^{\nu} w(t) &= \Delta_0^{\nu} \Delta_0^{N-2} y(t) = \Delta_{2-\nu}^{N-2} \Delta_0^{\nu} y(t) \\
&\quad \underbrace{- \sum_{j=0}^{N-3} \frac{\Delta_0^j y(0)}{\Gamma(-\nu - N + j + 3)} t^{\underline{-\nu - N + 2 + j}}}_{=0} \\
&= \Delta_{2-\nu}^{N-2} \Delta_0^{\nu} y(t) \\
&= \Delta_0^{\nu + N - 2} y(t) \\
&= \Delta_0^{\mu} y(t) \\
&\geq 0,
\end{aligned}$$

for each $t \in \mathbb{N}_{2-\nu} = \mathbb{N}_{N-\mu}$. We also observe that

$$w(0) = \Delta^{N-2}y(0) \geq 0.$$

By Corollary 7.2 it follows that $t \mapsto w(t)$ is increasing at $t = 0$. That is to say, it holds that $\Delta^{N-1}y(0) \geq 0$. We also note that $\Delta^{N-1}y(0) = \Delta w(0) \geq 0$. Finally, by repeatedly applying Corollary 7.2 we obtain that $\Delta^{N-1}y(t) \geq 0$ for each $t \in \mathbb{N}_0$. And this completes the proof. □

*Remark 7.14.* Observe that in the proof of Theorem 7.13 we repeatedly apply Corollary 7.2 at the end of the argument. In fact, it is worth noting that one can strengthen Corollary 7.2 in precisely this way—namely, it is sufficient that $y(0) \geq 0$. In particular, one need not know *a priori* that $y$ is nonnegative, merely that $y$ is "initially" nonnegative. The nonnegativity is, in fact, then propagated. A careful proof of this assertion is left to the reader.

We now demonstrate that the hypotheses of Theorem 7.13 can be altered somewhat.

**Corollary 7.15.** *Fix $\mu \in (N-1, N)$, for $N \in \mathbb{N}_3$ given, and assume that $\Delta_0^{\mu} y(t) \geq 0$ for each $t \in \mathbb{N}_{N-\mu}$. In case $N$ is odd, assume that*

$$\begin{cases} \Delta^j y(0) < 0, & j = 0, 2, \ldots, N-3 \\ \Delta^j y(0) > 0, & j = 1, 3, \ldots, N-4 \end{cases},$$

*whereas in case $N$ is even, assume that*

$$\begin{cases} \Delta^j y(0) > 0, & j = 0, 2, \ldots, N-4 \\ \Delta^j y(0) < 0, & j = 1, 3, \ldots, N-3 \end{cases}.$$

*If in addition it holds both that $\Delta^{N-2} y(0) \geq 0$ and that $\Delta^{N-1} y(0) \geq 0$, then $\Delta^{N-1} y(t) \geq 0$, for each $t \in \mathbb{N}_0$.*

*Proof.* Observe that by the calculation in the proof of Theorem 7.13, it follows that $\Delta_0^{\nu} w(t) \geq 0$ if and only if

$$\sum_{j=0}^{N-3} \frac{\Delta^j y(0)}{\Gamma(-\nu - N + j + 3)} t^{\underline{-\nu-N+2+j}} \leq 0; \tag{7.18}$$

recall here that the inequality $\Delta_{2-\nu}^{N-2} \Delta_0^{\nu} y(t) = \Delta_0^{\mu} y(t) \geq 0$ is still assumed. It then follows from (7.18) that

$$\frac{t(t-1)\cdots(-\mu+1)}{(t+\mu)!} y(0) + \cdots + \frac{t(t-1)\cdots(-\mu+N-2)}{(t+3+\mu-N)!} \Delta^{N-3} y(0) \leq 0$$

must hold for each $t \in \mathbb{N}_{N-\mu}$.

For notational convenience we next define the map $C_j : \mathbb{N}_3 \times \mathbb{N}_{N-\mu} \to \mathbb{R}$, for $j \in \{0, 1, \ldots, N-3\}$, by

$$\begin{aligned} C_j(N, t) &:= \frac{1}{\Gamma(-\nu - N + j + 3)} t^{\underline{-\nu-N+2+j}} \\ &= \frac{\Gamma(t+1)}{\Gamma(-\nu - N + j + 3)\Gamma(t + \nu + N - 1 - j)}. \end{aligned}$$

Observe that $C_j(N, t)$ is nothing more than the coefficient of $\Delta^j y(0)$ in (7.18). Recalling that $t \in \mathbb{N}_{N-\mu}$, we may simplify the ratio of gamma functions appearing in the definition of $C_j$. In particular, it is then apparent that if $j$ is even, then $C_j(N, t) < 0$ if $N$ is even, whereas $C_j(N, t) > 0$ if $N$ is odd; moreover, this holds for each $t \in \mathbb{N}_{N-\mu}$ as a simple calculation reveals. The sign relationship is reversed if $j$ is odd, and, once again, the relationship holds for each admissible $t$. We then see that (7.18) holds provided that $\Delta^j y(0)$ satisfies the sign condition, for each admissible $j$, presented in the statement of the theorem. And this completes the proof. □

We next present three examples to illustrate the application of Corollary 7.15 in the cases where $N = 3, 4$, or 5.

*Example 7.16.* Suppose that $N = 3$. Then Corollary 7.15 demonstrates that $\Delta^2 y(t) \geq 0$, for example, provided that

$$\begin{aligned} y(0) &< 0 \\ \Delta y(0) &\geq 0 \\ \Delta^2 y(0) &\geq 0 \\ \Delta_0^\mu y(t) &\geq 0 \text{ for some } \mu \in (2,3). \end{aligned}$$

*Example 7.17.* Suppose that $N = 4$. Then Corollary 7.15 implies that $\Delta^3 y(t) \geq 0$ if it holds that

$$\begin{aligned} y(0) &> 0 \\ \Delta y(0) &\leq 0 \\ \Delta^2 y(0) &\geq 0 \\ \Delta^3 y(0) &\geq 0 \\ \Delta_0^\mu y(t) &\geq 0 \text{ for some } \mu \in (3,4). \end{aligned}$$

*Example 7.18.* Suppose that $N = 5$. Then Corollary 7.15 implies that $\Delta^4 y(t) \geq 0$ if it holds that

$$\begin{aligned} y(0) &< 0 \\ \Delta y(0) &\geq 0 \\ \Delta^2 y(0) &\leq 0 \\ \Delta^3 y(0) &\geq 0 \\ \Delta^4 y(0) &\geq 0 \\ \Delta_0^\mu y(t) &\geq 0 \text{ for some } \mu \in (4,5). \end{aligned}$$

The next corollary is immediate and provides a geometrical interpretation of Theorem 7.13 in case $\mu \in (2,3)$ and thus $N = 3$; in particular, it provides for a convexity-type result.

**Corollary 7.19.** *If* $2 < \mu < 3$ *and* $y : \mathbb{N}_0 \to \mathbb{R}$ *satisfies* $y(0) = 0$, $\Delta y(0) \geq 0$, $\Delta^2 y(0) \geq 0$, *and* $\Delta_0^{\mu} y(t) \geq 0$, *for each* $t \in \mathbb{Z}_{3-\mu}$, *then* $\Delta^2 y(t) \geq 0$, *for each* $t \in \mathbb{N}_0$.

We can also obtain an alternative version of Theorem 7.13.

**Corollary 7.20.** *Fix* $\mu \in (N-1, N)$, *for* $N \in \mathbb{N}$ *given, and let* $y : \mathbb{N}_0 \to \mathbb{R}$ *be a given function satisfying* $\Delta^j y(0) = 0$ *for each* $j \in \{0, 1, 2, \ldots, N-3\}$, $\Delta^{N-2} y(0) \leq 0$, *and* $\Delta_0^{\mu} y(t) \leq 0$ *for each* $t \in \mathbb{Z}_{N-\mu}$. *If it also holds that* $\Delta^{N-1} y(0) \leq 0$, *then* $\Delta^{N-1} y(t) \leq 0$, *for each* $t \in \mathbb{N}_0$.

*Proof.* Put $z \equiv -y$ and apply Theorem 7.13 to the function $z$. □

Finally, as a specific application of this result and to demonstrate a nontrivial consequence of Theorem 7.13 we consider the theorem in case $N = 3$. To this end, consider the following FBVP, and observe that this is a special case of the so-called $(N-1, 1)$ problem for the case $N = 3$; see [124] for additional results on a class of discrete fractional $(N-1, 1)$ problems.

$$\begin{aligned} \Delta_{\mu-3}^{\mu} y(t) &= f(t+\mu-1, y(t+\mu-1)),\ t \in \mathbb{N}_0^{b+1} =: \{0, 1, \ldots, b+1\} \\ y(\mu-3) &= 0 = \Delta y(\mu-3) \\ y(\mu+b+1) &= 0 \end{aligned} \tag{7.19}$$

**Corollary 7.21.** *If the continuous function* $f : \mathbb{N}_{\mu-1}^{\mu+b} \times \mathbb{R} \to \mathbb{R}$ *is nonnegative and* $2 < \mu < 3$, *then problem* (7.19) *has no nontrivial positive solution.*

*Proof.* We begin by noting that from the boundary conditions we clearly have both that $y(\mu-3) = 0$ and that $\Delta y(\mu-3) \geq 0$. Furthermore, we compute

$$\Delta^2 y(\mu-3) = y(\mu-1) - 2y(\mu-2) + y(\mu-3) = y(\mu-1).$$

Therefore, supposing for contradiction that $y$ is a fictitious positive solution of problem (7.19), the above calculation demonstrates that $\Delta^2 y(\mu-3) \geq 0$. Note, in addition, that by the form of the difference equation in (7.19) together with the assumption on the function $f$ we may also conclude that $\Delta_{\mu-3}^{\mu} y(t) \geq 0$ for each $t \in \mathbb{N}_0$. Thus, we may invoke Theorem 7.13 to deduce that $\Delta^2 y(t) \geq 0$ for each $t \in \mathbb{Z}_{\mu-3}$.

Now, by the contradiction assumption we have that $y$ is nontrivial, and so, it follows that for some time $t_0 \in \mathbb{Z}_{\mu-2}^{\mu+b}$ it holds that $y(t_0) > 0$. But then $\Delta y(t_0) > 0$. Since Theorem 7.13 has shown that $\Delta^2 y(t) \geq 0$ for all $t$, it follows that $\Delta y(t) \geq 0$ for each $t \in \mathbb{Z}_{t_0}^{\mu+b}$. Thus, $y(\mu+b+1) > 0$, which violates the boundary condition at the right endpoint, and so, a contradiction is obtained. Consequently, (7.19) cannot have a nontrivial positive solution, as claimed. □

*Remark 7.22.* For obvious geometrical reasons, problem (7.19) cannot have a positive solution in case $\mu = 3$. However, lacking this simple geometric intuition when $\mu \notin \mathbb{N}$, it does not appear to be plainly obvious that problem (7.19) maintains this similar solution structure. Accordingly, Corollary 7.21 demonstrates that no such aberrant or otherwise pathological behavior occurs in the fractional case.

In fact, this is of some interest since the nonlocal structure of the fractional difference can be responsible for aberrant behavior. As mentioned in the previous section, it has been shown by Ferreira and Goodrich [83, Theorem 3.13] that this nonlocality can contribute to certain boundary value problems possessing positive solutions even in the case where their integer-order counterpart does not possess nontrivial, positive solutions. Thus, Corollary 7.21 demonstrates that no such aberration occurs with respect to the fractional $(2, 1)$ problem studied above.

The reader should observe that it is certainly possible to write numerous analogues of Corollary 7.21 by repeatedly applying Theorem 7.13 for different choice of $N$. It is instructive to do this in a few cases to obtain a better sense of the implications of the result. However, we leave this as an optional exercise.

*Remark 7.23.* Due to the minor error in [67, Corollary 2.3], there are subsequently some minor errors in [114]. Fortunately, the changes required to that paper are very minor. In particular, the following minor changes must be made. It should be noted that other than including a single additional hypothesis, no changes are required to the proofs in [114]; the proofs are otherwise correct.

- In [114, Theorem 2.6, Corollary 2.8] the hypothesis $\Delta^{N-1}y(0) \geq 0$ must be added, whereas in [114, Corollary 2.11] the hypothesis $\Delta^{N-1}y(0) \leq 0$ must be added.
- In [114, Example 2.9] the hypothesis $\Delta^2 y(0) \geq 0$ must be added in the first part of the example, whereas in the second part of the example the hypothesis $\Delta^3 y(0) \geq 0$ must be added.
- In [114, Corollary 2.10] the hypothesis $\Delta^2 y(0) \geq 0$ must be added.

We would like to conclude this section, much as we did in Sect. 7.2, by pointing out that due to a flurry of recent work in the area, the basic convexity and concavity results presented earlier in this section have been able to be substantively extended in a variety of directions. As one such representative result, we present the following theorem, which was recently proved by Baoguo, Erbe, Goodrich, and Peterson; it will appear in a forthcoming paper [53], and we direct the reader to the paper for further details on this and other related results. In particular, we omit the proof of the result, but instead focus on its relationship to the results presented earlier in this section, and the way in which it improves them.

**Theorem 7.24.** *Fix $\nu \in (2, 3)$ and suppose that $\Delta_a^\nu f(t) \geq 0$ for each $t \in \mathbb{N}_{3+a-\nu}$. If for each $k \in \mathbb{N}_{-1}$ it holds that*

$$\frac{1}{-\nu+1} f(a+2) + \frac{\nu+2+k}{(\nu-1)(3+k)} f(a+1) - \frac{\nu}{(3+k)(4+k)} f(a) \leq 0, \tag{7.20}$$

*then $\Delta^2 f(t) \geq 0$ for each $t \in \mathbb{N}_{a+1}$.*

*Proof.* Omitted—see [53]. □

*Remark 7.25.* Observe that inequality (7.20) does *not* necessarily imply that $\Delta^2 f(a) \geq 0$. For example, if we put $f(a) = 0$, $f(a+1) = 1$, and $f(a+2) = 1.9$ and we also fix $\nu = \frac{5}{2} \in (2, 3)$, then we calculate

$$\frac{1}{-\nu+1} f(a+2) + \frac{\nu+1}{2(\nu-1)} f(a+1) - \frac{\nu}{6} f(a)$$
$$= -\frac{2}{3} \cdot 1.9 + \frac{7}{6} \cdot 1 - \frac{5}{12} \cdot 0 = -\frac{1}{10} < 0,$$

which shows that inequality (7.20) is satisfied in case $k = -1$; in fact, it can be shown that (7.20) is satisfied for each $k \in \mathbb{N}_{-1}$. Yet we calculate $\Delta^2 f(a) = -\frac{1}{10} < 0$.

Similarly, if we put $g(a) = 1$, $g(a+1) = 2.8$, and $g(a+2) = 4.5$ as well as again taking $\nu = \frac{5}{2}$, then we see that $\Delta^2 g(a) = -\frac{1}{10} < 0$. Yet at the same time we calculate

$$\frac{1}{-\nu+1} g(a+2) + \frac{\nu+1}{2(\nu-1)} g(a+1) - \frac{\nu}{6} g(a)$$
$$= -\frac{2}{3} \cdot 4.5 + \frac{7}{6} \cdot 2.8 - \frac{5}{12} \cdot 1 = -\frac{3}{20} < 0,$$

which shows that inequality (7.20) is satisfied in case $k = -1$, and, as can be easily shown, it holds for $k \in \mathbb{N}_{-1}$.

All in all, then, we see that condition (7.20) may be satisfied even if the map $t \mapsto f(t)$ is not convex "at" $t = a$. In particular, this means that Theorem 7.24 does *not* require that the map $t \mapsto f(t)$ be "initially convex."

## 7.4 Analysis of a Three-Point Boundary Value Problem

In the preceding two sections we demonstrated that under certain conditions the discrete fractional difference operator satisfies both monotonicity and convexity properties. We thus focused on the nonlocal structure implicit to the fractional operators. As mentioned in the introduction to this chapter, one can also study explicitly nonlocal boundary value problems. In this and the succeeding sections of this chapter, we examine a few specific examples of these so-called nonlocal boundary value problems.

We begin by examining a three-point problem in this section. This is a special case of the so-called $m$-point problem, wherein our boundary value problem has a boundary condition of the form, say,

$$y(0) = \sum_{j=1}^{m} a_j y(\xi_j),$$

where, for each $j$, the number $\xi_j$ is both nonzero and an element of the domain of $y$. In particular, then, the value of $y$ at $t = 0$ depends on the values of $y$ at time points other than $t = 0$. This then gives rise to an explicit nonlocal boundary condition. Now, in the integer-order case determining the Green's function and its properties for such a problem can be tedious, but is usually not too overly taxing. However, in the fractional case, as this section demonstrates, determining explicitly the Green's function and its properties for even the three-point problem is extremely technical. In particular, the problem we study in this section is

$$\begin{aligned} -\Delta^{\nu} y(t) &= f(t+\nu-1, y(t+\nu-1)) \\ y(\nu-2) &= 0 \\ \alpha y(\nu+K) &= y(\nu+b). \end{aligned}$$

Finally, as we shall note later in this section, if we remove the nonlocal boundary condition element by simply putting $\alpha = 0$, then we recover the Green's function for the conjugate problem as is easily checked by the reader. Moreover, most of the results in this section can be found in the paper by Goodrich [104].

So, with this context in mind, we first deduce the Green's function for the operator $-\Delta^{\nu}$ together with the boundary conditions $y(\nu-2) = 0$ and $\alpha y(\nu+K) = y(\nu+b)$, where $0 \leq \alpha \leq 1$ and $K \in [-1, b-1]_{\mathbb{Z}}$. We first present a preliminary lemma, which will prove to be rather useful in what follows. The lemma can be found in a paper by Goodrich [88].

**Lemma 7.26.** *Fix $k \in \mathbb{N}$ and let $\{m_j\}_{j=1}^{k}, \{n_j\}_{j=1}^{k} \subseteq (0, +\infty)$ such that*

$$\max_{1\leq j\leq k} m_j \leq \min_{1\leq j\leq k} n_j$$

*and that for at least one $j_0$, $1 \leq j_0 \leq k$, we have that $m_{j_0} < n_{j_0}$. Then for fixed $\alpha_0 \in (0,1)$, it follows that*

$$\left(\frac{n_1}{n_1+\alpha_0}\cdot\dots\cdot\frac{n_k}{n_k+\alpha_0}\right)\left(\frac{m_1+\alpha_0}{m_1}\cdot\dots\cdot\frac{m_k+\alpha_0}{m_k}\right) > 1.$$

*Proof.* Fix an index $j_0$, where $j_0$ is one of the indices, of which there exists at least one, for which $n_{j_0} > m_{j_0}$. Notice that as $n_{j_0} > m_{j_0}$ and $\alpha_0 > 0$, it follows that $n_{j_0}\alpha_0 > m_{j_0}\alpha_0$, whence $m_{j_0}n_{j_0} + n_{j_0}\alpha_0 > m_{j_0}n_{j_0} + m_{j_0}\alpha_0$, so that

$$\frac{m_{j_0}+\alpha_0}{m_{j_0}} > \frac{n_{j_0}+\alpha_0}{n_{j_0}},$$

whence

$$\frac{n_{j_0}}{n_{j_0}+\alpha_0}\cdot\frac{m_{j_0}+\alpha_0}{m_{j_0}} > 1.$$

But now the claim follows at once by repeating the above steps for each of the remaining $j_0 - 1$ terms and observing that the product of $j$ terms, each of which is at least unity and at least one of which exceeds unity, is greater than unity. □

In addition, for reference in the sequel and to simplify the rather formidable notational burden associated with this problem, let us make the following declarations; note that we provide the domains of these maps in the statement of Theorem 7.27.

$$\begin{aligned}
g_1(t,s) &:= \frac{1}{\Gamma(\nu)}\Bigg[ -(t-s-1)^{\underline{\nu-1}} \\
&\qquad + \frac{t^{\underline{\nu-1}}}{\Omega_0}\left[(b+\nu-s-1)^{\underline{\nu-1}} - \alpha(K+\nu-s-1)^{\underline{\nu-1}}\right]\Bigg] \\
g_2(t,s) &:= \frac{1}{\Gamma(\nu)}\left[\frac{t^{\underline{\nu-1}}}{\Omega_0}\left[(b+\nu-s-1)^{\underline{\nu-1}} - \alpha(K+\nu-s-1)^{\underline{\nu-1}}\right]\right] \\
g_3(t,s) &:= \frac{1}{\Gamma(\nu)}\left[-(t-s-1)^{\underline{\nu-1}} + \frac{t^{\underline{\nu-1}}}{\Omega_0}(b+\nu-s-1)^{\underline{\nu-1}}\right] \\
g_4(t,s) &:= \frac{1}{\Gamma(\nu)}\left[\frac{t^{\underline{\nu-1}}}{\Omega_0}(b+\nu-s-1)^{\underline{\nu-1}}\right] \\
\Omega_0 &:= (b+\nu)^{\underline{\nu-1}} - \alpha(K+\nu)^{\underline{\nu-1}}
\end{aligned} \tag{7.21}$$

**Theorem 7.27.** *Let $h : [\nu-1, \nu+b-1]_{\mathbb{N}_{\nu-1}} \to \mathbb{R}$ be given. The unique solution of the problem*

$$\begin{aligned}
-\Delta^\nu y(t) &= h(t+\nu-1) \\
y(\nu-2) &= 0 \\
\alpha y(\nu+K) &= y(\nu+b)
\end{aligned} \tag{7.22}$$

*is the function*

$$y(t) = \sum_{s=0}^{b} G(t,s)h(s+\nu-1),$$

*where $G(t,s)$ is the Green's function for the operator $-\Delta^\nu$ together with the boundary conditions in (7.22), and where*

$$G(t,s) := \begin{cases} g_1(t,s), & 0 \le s \le \min\{t-\nu, K\} \\ g_2(t,s), & 0 \le t-\nu < s \le K \le b \\ g_3(t,s), & 0 < K < s \le t-\nu \le b \\ g_4(t,s), & \max\{t-\nu, K\} < s \le b \end{cases},$$

*with $g_i(t,s)$, $1 \le i \le 4$, are as defined in (7.21) above.*

*Proof.* Omitted—see [104]. □

*Remark 7.28.* It is easy to observe that in case $\alpha = 0$, not only does problem (7.22) reduce to the usual conjugate FBVP that was considered in [31], but, moreover, the Green's function given by Theorem 7.27 reduces to the Green's function derived in [31].

We now wish to prove that the Green's function $(t, s) \mapsto G(t, s)$ in Theorem 7.27 satisfies certain properties that will prove to be of use in the sequel and are also of independent interest. We first prove an easy preliminary lemma.

**Lemma 7.29.** *Let $\Omega_0$ be as defined in* (7.21). *Then for each $K \in [-1, b-1]_{\mathbb{Z}}$, $\nu \in (1, 2]$, and $b \in \mathbb{N}$, we find that $\Omega_0 > 0$.*

*Proof.* Recall from (7.21) that $\Omega_0 = (b+\nu)^{\underline{\nu-1}} - \alpha(K+\nu)^{\underline{\nu-1}}$. It is evident that this function is decreasing in $\alpha$ for each fixed $K$, $\nu$, and $b$, and so, it suffices to show that $\Omega_0 > 0$ when $\alpha = 1$. To this end, note that $t^{\underline{\mu}}$ is increasing in $t$, whenever $0 < \mu < 1$. Since $b + \nu > K + \nu$, it immediately follows that

$$\Omega_0\big|_{\alpha=1} = (b+\nu)^{\underline{\nu-1}} - (K+\nu)^{\underline{\nu-1}} > 0,$$

which proves the claim. We observe that this same result holds even in the case where $\nu = 2$. □

**Theorem 7.30.** *Let $G$ be the Green's function given in the statement of Theorem 7.27. Then for each $(t, s) \in [\nu-2, \nu+b]_{\mathbb{N}_{\nu-2}} \times [0, b]_{\mathbb{N}_0}$, we find that $G(t, s) \geq 0$.*

*Proof.* As was mentioned at the beginning of this section, the proof of this result may be found in its entirety in [104]. However, we include the proof here for its instructive value. In particular, it shall give the reader a sense of the delicacy that is involved in arguing the properties of Green's functions associated with fractional difference operators—delicacy that is obviated if we pass to the integer-order case. Moreover, this will also give the reader a general sense for certain of the techniques that may be utilized in these sorts of arguments.

With this in mind, our program to complete the proof is to show that for each $i$, $1 \leq i \leq 4$, it holds that $g_i(t, s) > 0$ for each admissible pair $(t, s)$. To complete this program, we begin by showing both that $g_2(t, s) > 0$ and that $g_4(t, s) > 0$, as these are the easier cases. In the case of $g_2(t, s)$, observe that it suffices to show that the inequality

$$(b+\nu-s-1)^{\underline{\nu-1}} - \alpha(K+\nu-s-1)^{\underline{\nu-1}} > 0 \tag{7.23}$$

holds. Showing that (7.23) is true is equivalent to showing that

$$\frac{(b+\nu-s-1)^{\underline{\nu-1}}}{\alpha(K+\nu-s-1)^{\underline{\nu-1}}} > 1. \tag{7.24}$$

But to see that (7.24) is true for each admissible pair $(t, s)$ and each $\alpha \in (0, 1]$, notice that $t^{\underline{\mu}}$ is increasing in $t$ provided that $\mu \in (0, 1)$. Consequently, we obtain that

$$\frac{(b+\nu-s-1)^{\underline{\nu-1}}}{\alpha(K+\nu-s-1)^{\underline{\nu-1}}} \geq \frac{(b+\nu-s-1)^{\underline{\nu-1}}}{(K+\nu-s-1)^{\underline{\nu-1}}} > 1,$$

which proves (7.24) and thus (7.23). On the other hand, we note that by the form of $g_4$ given in (7.21), we obtain at once that $g_4(t, s) > 0$ since $\Omega_0 > 0$ by Lemma 7.29 and $b+\nu-s-1 > 0$ in this case. Thus, we conclude that both $g_2$ and $g_4$ are positive on each of their respective domains.

We next consider the function $g_3$. Recall from (7.21) that

$$g_3(t, s) = \frac{1}{\Gamma(\nu)}\left[-(t-s-1)^{\underline{\nu-1}} + \frac{t^{\underline{\nu-1}}}{\Omega_0}(b+\nu-s-1)^{\underline{\nu-1}}\right].$$

Evidently, to prove that $g_3(t, s) > 0$, we may instead just prove that $\Gamma(\nu)g_3(t, s) > 0$. Now, it is clear that $g_3$ is increasing in $\alpha$, for as $\alpha$ increases, $\Omega_0$ clearly decreases. In particular, then, we deduce that

$$\Gamma(\nu)g_3(t, s) \geq -(t-s-1)^{\underline{\nu-1}} + \frac{t^{\underline{\nu-1}}(b+\nu-s-1)^{\underline{\nu-1}}}{(b+\nu)^{\underline{\nu-1}}}. \tag{7.25}$$

Note that (7.25) implies that $g_3(t, s) > 0$ if and only if

$$\frac{t^{\underline{\nu-1}}(b+\nu-s-1)^{\underline{\nu-1}}}{(t-s-1)^{\underline{\nu-1}}(b+\nu)^{\underline{\nu-1}}} > 1. \tag{7.26}$$

To prove that (7.26) holds, recall that on the domain of $g_3$ it holds that $t \geq s+\nu > K+\nu$. Consequently, given a fixed but arbitrary $s_0 > K$, we have that $t = s_0+\nu+j$, for some $0 \leq j \leq b-s_0$ with $j \in \mathbb{N}_0$. But then for this number $s_0$, we may recast the left-hand side of (7.26) as

$$\begin{aligned}
&\frac{t^{\underline{\nu-1}}(b+\nu-s-1)^{\underline{\nu-1}}}{(t-s-1)^{\underline{\nu-1}}(b+\nu)^{\underline{\nu-1}}} \\
&= \frac{\Gamma(t+1)\Gamma(b+\nu-s_0)\,\Gamma(t-s_0-\nu+1)\,\Gamma(b+2)}{\Gamma(t-\nu+2)\Gamma(b-s_0+1)\,\Gamma(t-s_0)\,\Gamma(b+\nu+1)} \\
&= \frac{\Gamma(s_0+\nu+j+1)\,\Gamma(b+\nu-s_0)\,\Gamma(j+1)\Gamma(b+2)}{\Gamma(s_0+j+2)\,\Gamma(b-s_0+1)\,\Gamma(\nu+j)\Gamma(b+\nu+1)} \\
&= \frac{j!(b+1)!\,[(\nu+j+s_0)\cdots(\nu+j)]}{(s_0+j+1)!\,(b-s_0)!\,[(b+\nu)\cdots(b+\nu-s_0)]} \\
&= \frac{(b+1)\cdots(b-s_0+1)}{(b+\nu)\cdots(b+\nu-s_0)}\cdot\frac{(\nu+j+s_0)\cdots(\nu+j)}{(s_0+j+1)\cdots(j+1)}.
\end{aligned} \tag{7.27}$$

Now, observe that each of the fractions on the right-hand side of (7.27) has exactly $s_0 + 1$ factors in each of its numerator and denominator. In addition, by putting $\alpha_0 := \nu - 1 > 0$, we observe that this expression satisfies the hypotheses of Lemma 7.26. (Of course, some repetition of factors may occur between the two fractions on the right-hand side of (7.27), but these may always be canceled to obtain the form required by Lemma 7.26. Thus, we may safely ignore the existence of any possible repetition.) Consequently, we deduce from this lemma that

$$\frac{t^{\underline{\nu-1}}(b+\nu-s-1)^{\underline{\nu-1}}}{(t-s-1)^{\underline{\nu-1}}(b+\nu)^{\underline{\nu-1}}}$$
$$= \frac{(b+1)\cdots(b-s_0+1)}{(b+\nu)\cdots(b+\nu-s)} \cdot \frac{(\nu+j+s_0)\cdots(\nu+j)}{(s_0+j+1)\cdots(j+1)} > 1,$$

whence (7.26) holds. But as (7.26) holds for each admissible pair $(t, s)$, it follows at once that (7.25) holds, too, so that $g_3(t, s) > 0$, as claimed.

Finally, we show that $g_1(t, s) > 0$ on its domain, which we recall is $0 \le s \le \min\{t-\nu, K\}$. Recall from (3.1) that

$$\Gamma(\nu)g_1(t,s) = -(t-s-1)^{\underline{\nu-1}}$$
$$+ \frac{t^{\underline{\nu-1}}}{\Omega_0}\left[(b+\nu-s-1)^{\underline{\nu-1}} - \alpha(K+\nu-s-1)^{\underline{\nu-1}}\right],$$

where we again use the fact that $g_1(t, s)$ is positive if and only if $\Gamma(\nu)g_1(t, s)$ is positive. Let us pause momentarily to notice that

$$(b+\nu-s-1)^{\underline{\nu-1}} - \alpha(K+\nu-s-1)^{\underline{\nu-1}} > 0, \tag{7.28}$$

which is evidently an important condition. Note that (7.28) just follows from (7.23) above.

Notice that $g_1(t, s) > 0$ only if

$$\frac{t^{\underline{\nu-1}}}{\Omega_0}\left[(b+\nu-s-1)^{\underline{\nu-1}} - \alpha(K+\nu-s-1)^{\underline{\nu-1}}\right] > (t-s-1)^{\underline{\nu-1}}. \tag{7.29}$$

We begin by proving that the function $F : [0, 1] \to \mathbb{R}$ defined by

$$F(\alpha) := \frac{(b+\nu-s-1)^{\underline{\nu-1}} - \alpha(K+\nu-s-1)^{\underline{\nu-1}}}{(b+\nu)^{\underline{\nu-1}} - \alpha(K+\nu)^{\underline{\nu-1}}} \tag{7.30}$$

is increasing in $\alpha$ for $0 \le \alpha \le 1$. Note that a straightforward calculation shows that $F(\alpha)$ is increasing in $\alpha$ if and only if

$$\frac{(b+\nu-s-1)^{\underline{\nu-1}}(K+\nu)^{\underline{\nu-1}}}{(K+\nu-s-1)^{\underline{\nu-1}}(b+\nu)^{\underline{\nu-1}}} > 1. \tag{7.31}$$

To see that (7.31) holds, let $s_0$ be arbitrary but fixed such that each of $s_0 \in [0, b]_{\mathbb{N}_0}$ and $0 \leq s_0 \leq \min\{t - \nu, K\}$ holds. So, it follows that the left-hand side of (7.31) above satisfies

$$\begin{aligned}&\frac{(b + \nu - s_0 - 1)^{\underline{\nu-1}}(K + \nu)^{\underline{\nu-1}}}{(K + \nu - s_0 - 1)^{\underline{\nu-1}}(b + \nu)^{\underline{\nu-1}}}\\ &\qquad = \frac{(b + 1)\cdots(b - s_0 + 1)}{(b + \nu)\cdots(b + \nu - s_0)} \cdot \frac{(K + \nu)\cdots(K + \nu - s_0)}{(K + 1)\cdots(K - s_0 + 1)}.\end{aligned} \tag{7.32}$$

But it is easy to check that by putting $\alpha_0 := \nu - 1 > 0$, we may apply Lemma 7.26 to the right-hand side of (7.32) to conclude that (7.31) holds. Thus, the map $\alpha \mapsto F(\alpha)$ is increasing in $\alpha$, as desired. In particular, this implies that to prove that (7.29) is true, it suffices to check its truth in case $\alpha = 0$. In this case, we find that proving (7.29) reduces to proving that

$$\frac{t^{\underline{\nu-1}}(b + \nu - s - 1)^{\underline{\nu-1}}}{(b + \nu)^{\underline{\nu-1}}(t - s - 1)^{\underline{\nu-1}}} > 1 \tag{7.33}$$

holds. Observe that the same proof that was used to show that (7.26) held can be used to show that (7.33) holds, too. Thus, as (7.29) holds in case $\alpha = 0$, the result of (7.30)–(7.33) implies that (7.29) holds for each admissible $\alpha$. Consequently, we conclude that $g_1(t, s) > 0$, from which it follows that $g_i(t, s) > 0$ for each $i$, $1 \leq i \leq 4$. Hence, it follows that $G(t, s) \geq 0$ for each admissible pair $(t, s)$. And this concludes the proof. □

**Theorem 7.31.** *Let $G$ be the Green's function given in the statement of Theorem 7.27. In addition, suppose that for given $K \in [-1, b - 1]_{\mathbb{Z}}$ and $1 < \nu \leq 2$, we have that $\alpha$ satisfies the inequality*

$$\begin{aligned}0 &\leq \alpha\\ &\leq \min_{(t,s)\in[s+\nu,\nu+b]_{\mathbb{N}_{\nu-1}}\times[0,b]_{\mathbb{N}_0}} \left\{\frac{(b + \nu)^{\underline{\nu-1}}}{(K + \nu)^{\underline{\nu-1}}} - \frac{t^{\underline{\nu-2}}(b + \nu - s - 1)^{\underline{\nu-1}}}{(K + \nu)^{\underline{\nu-1}}(t - s - 1)^{\underline{\nu-2}}}\right\}.\end{aligned} \tag{7.34}$$

*Then for each $s \in [0, b]_{\mathbb{N}_0}$ it holds that*

$$\max_{t\in[\nu-1,\nu+b]_{\mathbb{N}_{\nu-1}}} G(t, s) = G(s + \nu - 1, s). \tag{7.35}$$

*Proof.* Our strategy is to show that $\Delta_t g_i(t, s) > 0$ for each $i = 2, 4$, and that $\Delta_t g_i(t, s) < 0$ for $i = 1, 3$. From this the claim will follow at once. To this end, we first show that the former claim holds, as this is the easier of the two cases. Note, for example, that when $i = 2$, we find by direct computation that

$$\Gamma(\nu)\Delta_t g_2(t, s) = \frac{(\nu - 1)t^{\underline{\nu-2}}}{\Omega_0}\left[(b + \nu - s - 1)^{\underline{\nu-1}} - \alpha(K + \nu - s - 1)^{\underline{\nu-1}}\right]. \tag{7.36}$$

So, it is clear from (7.36) that $\Delta_t g_2(t,s) > 0$ if and only if

$$(b+\nu-s-1)^{\underline{\nu-1}} > \alpha(K+\nu-s-1)^{\underline{\nu-1}}.$$

But as this immediately follows from (7.23)–(7.24), we deduce that

$$\Delta_t g_2(t,s) > 0,$$

as desired. On the other hand, the estimate $\Delta_t g_4(t,s) > 0$ evidently holds considering that $\Delta_t g_4(t,s) = \frac{(\nu-1)t^{\underline{\nu-2}}}{\Omega_0}(b+\nu-s-1)^{\underline{\nu-1}}$. And this concludes the analysis of $\Delta_t g_i(t,s)$ in case $i$ is even.

We next attend to $g_3(t,s)$. In particular, we claim that $\Delta_t g_3(t,s) < 0$ for each admissible pair $(t,s)$. To see that this claim holds, note that

$$\begin{aligned}\Gamma(\nu)\Delta_t g_3(t,s) &= -(\nu-1)(t-s-1)^{\underline{\nu-2}} \\ &\quad + \frac{(\nu-1)t^{\underline{\nu-2}}}{\Omega_0}(b+\nu-s-1)^{\underline{\nu-1}},\end{aligned}$$

where we have used the fact that $\Delta_t(t-s-1)^{\underline{\nu-1}} = (\nu-1)(t-s-1)^{\underline{\nu-2}}$, which may be easily verified from the definition. So, if $\Delta_t g_3$ is to be a nonpositive function, then it must hold that

$$\frac{t^{\underline{\nu-2}}(b+\nu-s-1)^{\underline{\nu-1}}}{\Omega_0} < (t-s-1)^{\underline{\nu-2}}. \tag{7.37}$$

Notice that (7.37) is true if and only if $(b+\nu)^{\underline{\nu-1}} - \alpha(K+\nu)^{\underline{\nu-1}} > \frac{t^{\underline{\nu-2}}(b+\nu-s-1)^{\underline{\nu-1}}}{(t-s-1)^{\underline{\nu-2}}}$ is true. But this latter inequality is true only if

$$-\alpha > \frac{t^{\underline{\nu-2}}(b+\nu-s-1)^{\underline{\nu-1}}}{(t-s-1)^{\underline{\nu-2}}(K+\nu)^{\underline{\nu-1}}} - \frac{(b+\nu)^{\underline{\nu-1}}}{(K+\nu)^{\underline{\nu-1}}} \tag{7.38}$$

is true. From (7.38) we see that by requiring $\alpha$ to satisfy, for each admissible $K$ and $\nu$, the estimate

$$\begin{aligned}0 &\le \alpha \\ &\le \min_{(t,s)\in[s+\nu,\nu+b]_{\mathbb{N}_{\nu-1}}\times[0,b]_{\mathbb{N}_0}} \left\{\frac{(b+\nu)^{\underline{\nu-1}}}{(K+\nu)^{\underline{\nu-1}}} - \frac{t^{\underline{\nu-2}}(b+\nu-s-1)^{\underline{\nu-1}}}{(K+\nu)^{\underline{\nu-1}}(t-s-1)^{\underline{\nu-2}}}\right\},\end{aligned} \tag{7.39}$$

it follows that (7.37) is true—that is, that $g_3(t,s) > 0$ for each admissible pair $(t,s)$. Note that restriction (7.39) above is precisely restriction (7.34), which was given in the statement of this theorem. Thus, with restriction (7.34) in place, we conclude that the map $(t,s) \mapsto \Delta_t g_3(t,s)$ will be nonpositive on its domain, as desired.

Finally, we claim that $\Delta_t g_1(t,s) < 0$ on its domain. Observe that by the definition of $g_1$ given in (7.21), we must argue that

$$\begin{aligned} &-(\nu-1)(t-s-1)^{\underline{\nu-2}} \\ &+\frac{(\nu-1)t^{\underline{\nu-2}}}{\Omega_0}\left[(b+\nu-s-1)^{\underline{\nu-1}}-\alpha(K+\nu-s-1)^{\underline{\nu-1}}\right]<0. \end{aligned} \tag{7.40}$$

But observe that

$$\begin{aligned} &-(\nu-1)(t-s-1)^{\underline{\nu-2}} \\ &+\frac{(\nu-1)t^{\underline{\nu-2}}}{\Omega_0}\left[(b+\nu-s-1)^{\underline{\nu-1}}-\alpha(K+\nu-s-1)^{\underline{\nu-1}}\right] \\ &\leq -(\nu-1)(t-s-1)^{\underline{\nu-2}}+\frac{(\nu-1)t^{\underline{\nu-2}}(b+\nu-s-1)^{\underline{\nu-1}}}{\Omega_0}. \end{aligned}$$

So, we deduce that if

$$-(\nu-1)(t-s-1)^{\underline{\nu-2}}+\frac{(\nu-1)t^{\underline{\nu-2}}(b+\nu-s-1)^{\underline{\nu-1}}}{\Omega_0}<0, \tag{7.41}$$

then inequality (7.40) holds. Now, note that we can solve for $\alpha$ in (7.41) to obtain an upper bound on $\alpha$. As this calculation is exactly the same as the one given earlier in the argument, we do not repeat it here. Instead we point out that the restriction (7.41) implies that

$$0 \leq \alpha \leq \frac{(b+\nu)^{\underline{\nu-1}}}{(K+\nu)^{\underline{\nu-1}}}-\frac{t^{\underline{\nu-2}}(b+\nu-s-1)^{\underline{\nu-1}}}{(K+\nu)^{\underline{\nu-1}}(t-s-1)^{\underline{\nu-2}}}.$$

Note that the right-hand side of (7.41) is precisely restriction (7.34). So, by assuming (7.34) we also get that (7.40) holds. Consequently, the preceding analysis shows that (7.40) holds, from which it follows that $\Delta_t g_1(t,s) > 0$ on its domain. Thus, we deduce that (7.35) holds, which completes the proof. □

Before presenting our final theorem in this section regarding the map $(t,s) \mapsto G(t,s)$, we make some definitions for convenience.

$$\gamma_1 := \frac{\left(\frac{b+\nu}{4}\right)^{\underline{\nu-1}}}{(b+\nu)^{\underline{\nu-1}}}$$

$$\gamma_2 := \frac{1}{\left(\frac{3(b+\nu)}{4}\right)^{\underline{\nu-1}}}\left[\left(\frac{3(b+\nu)}{4}\right)^{\underline{\nu-1}} - \frac{\left(\frac{3(b+\nu)}{4}-1\right)^{\underline{\nu-1}}\left[(b+\nu)^{\underline{\nu-1}} - \alpha(K+\nu)^{\underline{\nu-1}}\right]}{(b+\nu-1)^{\underline{\nu-1}}}\right]$$

$$\gamma_3 := \frac{1}{\left(\frac{3(b+\nu)}{4}\right)^{\underline{\nu-1}}}\left[\left(\frac{3(b+\nu)}{4}\right)^{\underline{\nu-1}} - \frac{\left(\frac{3(b+\nu)}{4}-1\right)^{\underline{\nu-1}}(b+\nu)^{\underline{\nu-1}}}{(b+\nu-1)^{\underline{\nu-1}}}\right]$$

We will make use of these constants in the sequel.

**Theorem 7.32.** *Let $G$ be the Green's function given in the statement of Theorem 7.27. Let $\gamma_i$, $1 \leq i \leq 3$, be defined as above. Then it follows that for each $s \in [0,b]_{\mathbb{N}_0}$*

$$\min_{t\in\left[\frac{b+\nu}{4},\frac{3(b+\nu)}{4}\right]} G(t,s) \geq \gamma \max_{t\in[\nu-2,\nu+b]_{\mathbb{N}_{\nu-2}}} G(t,s) = \gamma G(s+\nu-1,s), \tag{7.42}$$

*where*

$$\gamma := \min\{\gamma_1,\gamma_3\}, \tag{7.43}$$

*and $\gamma$ satisfies the inequality $0<\gamma<1$.*

*Proof.* To simplify the notation used in this proof, let us put, for each $1 \leq i \leq 4$,

$$\tilde{g}_i(t,s) := \begin{cases} \frac{g_i(t,s)}{g_2(s+\nu-1,s)}, & i=1,2 \\ \frac{g_i(t,s)}{g_4(s+\nu-1,s)}, & i=3,4 \end{cases}.$$

Observe that for $s \geq t-\nu+1$ and $\frac{b+\nu}{4} \leq t \leq \frac{3(b+\nu)}{4}$, it holds that

$$\tilde{g}_2(t,s) = \tilde{g}_4(t,s) = \frac{t^{\underline{\nu-1}}}{(s+\nu-1)^{\underline{\nu-1}}} \geq \frac{\left(\frac{b+\nu}{4}\right)^{\underline{\nu-1}}}{(b+\nu)^{\underline{\nu-1}}}, \tag{7.44}$$

whence from (7.44) it is clear that in the case where both $s \geq t-\nu+1$ and $t \in \left[\frac{b+\nu}{4},\frac{3(b+\nu)}{4}\right]$, it follows that

$$\min_{t\in\left[\frac{b+\nu}{4},\frac{3(b+\nu)}{4}\right]} G(t,s) \geq \gamma_1 G(s+\nu-1,s).$$

On the other hand, suppose that $s < t - \nu + 1$ and $t \in \left[\frac{b+\nu}{4}, \frac{3(b+\nu)}{4}\right]$. Then we consider two cases depending upon whether or not the pair $(t, s)$ lives in the domain of $\tilde{g}_1$ or $\tilde{g}_3$. In the case where $(t, s)$ lives in the domain of $\tilde{g}_3$, we note that by definition

$$\begin{aligned}
&\tilde{g}_3(t,s)\\
&= \frac{-(t-s-1)^{\underline{\nu-1}}\Omega_0}{(s+\nu-1)^{\underline{\nu-1}}(b+\nu-s-1)^{\underline{\nu-1}}} + \frac{t^{\underline{\nu-1}}}{(s+\nu-1)^{\underline{\nu-1}}}\\
&= \frac{1}{(s+\nu-1)^{\underline{\nu-1}}}\left[t^{\underline{\nu-1}} - \frac{(t-s-1)^{\underline{\nu-1}}\left[(b+\nu)^{\underline{\nu-1}} - \alpha(K+\nu)^{\underline{\nu-1}}\right]}{(b+\nu-s-1)^{\underline{\nu-1}}}\right]\\
&\geq \frac{1}{\left(\frac{3(b+\nu)}{4}\right)^{\underline{\nu-1}}}\\
&\quad\times\left[\left(\frac{3(b+\nu)}{4}\right)^{\underline{\nu-1}} - \frac{\left(\frac{3(b+\nu)}{4}-1\right)^{\underline{\nu-1}}\left[(b+\nu)^{\underline{\nu-1}} - \alpha(K+\nu)^{\underline{\nu-1}}\right]}{(b+\nu-1)^{\underline{\nu-1}}}\right].
\end{aligned} \tag{7.45}$$

So, it is clear from (7.45) that in case $s < t - \nu + 1$ and $t \in \left[\frac{b+\nu}{4}, \frac{3(b+\nu)}{4}\right]$, we get that $\min_{t\in\left[\frac{b+\nu}{4}, \frac{3(b+\nu)}{4}\right]} G(t,s) \geq \gamma_2 G(s+\nu-1, s)$.

Finally, suppose that $s < t - \nu + 1$, $t \in \left[\frac{b+\nu}{4}, \frac{3(b+\nu)}{4}\right]$, and that the pair $(t, s)$ lives in the domain of $\tilde{g}_1$. By using a similar calculation as in (7.45) together with the definition of $\tilde{g}_1$, we obtain the lower bound

$$\begin{aligned}
&\tilde{g}_1(t,s)\\
&= \frac{-(t-s-1)^{\underline{\nu-1}}\Omega_0}{(s+\nu-1)^{\underline{\nu-1}}\left[(b+\nu-s-1)^{\underline{\nu-1}} - \alpha(K+\nu-s-1)^{\underline{\nu-1}}\right]}\\
&\qquad + \frac{t^{\underline{\nu-1}}}{(s+\nu-1)^{\underline{\nu-1}}}\\
&\geq \frac{1}{\left(\frac{3(b+\nu)}{4}\right)^{\underline{\nu-1}}}\\
&\quad\times\left[\left(\frac{3(b+\nu)}{4}\right)^{\underline{\nu-1}} - \frac{\left(\frac{3(b+\nu)}{4}-s-1\right)^{\underline{\nu-1}}\left[(b+\nu)^{\underline{\nu-1}} - \alpha(K+\nu)^{\underline{\nu-1}}\right]}{(b+\nu-s-1)^{\underline{\nu-1}} - \alpha(K+\nu-s-1)^{\underline{\nu-1}}}\right].
\end{aligned} \tag{7.46}$$

We now need to focus on the quotient $\frac{(b+\nu)^{\underline{\nu-1}} - \alpha(K+\nu)^{\underline{\nu-1}}}{(b+\nu-s-1)^{\underline{\nu-1}} - \alpha(K+\nu-s-1)^{\underline{\nu-1}}}$ appearing on the right-hand side of (7.46). We claim that this is a decreasing function of $\alpha$.

To prove this claim, let us put

$$g(\alpha) := \frac{(b+\nu)^{\underline{\nu-1}} - \alpha(K+\nu)^{\underline{\nu-1}}}{(b+\nu-s-1)^{\underline{\nu-1}} - \alpha(K+\nu-s-1)^{\underline{\nu-1}}}, \tag{7.47}$$

where for each fixed but arbitrary $b$, $s$, $\nu$, and $K$, we have that $g : [0,1] \to [0,+\infty)$. Now, let the map $\alpha \mapsto F(\alpha)$ be defined as in (7.30) above. Note from (7.47) that

$$g(\alpha) = \frac{1}{F(\alpha)}.$$

Recall that in case $0 \le \alpha \le 1$ we have already argued that $F$ is increasing in $\alpha$. So, straightforward computations demonstrate that $g$ is decreasing in $\alpha$, for $0 \le \alpha \le 1$, as desired.

Since $g$ is decreasing in $\alpha$, we conclude that

$$\begin{aligned}
&\tilde{g}_1(t,s)\\
&\ge \frac{1}{\left(\frac{3(b+\nu)}{4}\right)^{\underline{\nu-1}}}\\
&\quad \times \left[\left(\frac{3(b+\nu)}{4}\right)^{\underline{\nu-1}} - \frac{\left(\frac{3(b+\nu)}{4}-s-1\right)^{\underline{\nu-1}}\left[(b+\nu)^{\underline{\nu-1}} - \alpha(K+\nu)^{\underline{\nu-1}}\right]}{(b+\nu-s-1)^{\underline{\nu-1}} - \alpha(K+\nu-s-1)^{\underline{\nu-1}}}\right]\\
&\ge \frac{1}{\left(\frac{3(b+\nu)}{4}\right)^{\underline{\nu-1}}}\left[\left(\frac{3(b+\nu)}{4}\right)^{\underline{\nu-1}} - \frac{\left(\frac{3(b+\nu)}{4}-s-1\right)^{\underline{\nu-1}}(b+\nu)^{\underline{\nu-1}}}{(b+\nu-s-1)^{\underline{\nu-1}}}\right]\\
&\ge \frac{1}{\left(\frac{3(b+\nu)}{4}\right)^{\underline{\nu-1}}}\left[\left(\frac{3(b+\nu)}{4}\right)^{\underline{\nu-1}} - \frac{\left(\frac{3(b+\nu)}{4}-1\right)^{\underline{\nu-1}}(b+\nu)^{\underline{\nu-1}}}{(b+\nu-1)^{\underline{\nu-1}}}\right].
\end{aligned}$$

Thus, we observe that in this case it holds that $\min_{t\in\left[\frac{b+\nu}{4},\frac{3(b+\nu)}{4}\right]} G(t,s) \ge \gamma_3 G(s+\nu-1,s)$.

Finally, note that since $\gamma_2 \ge \gamma_3$, it must hold that $\min\{\gamma_1,\gamma_2,\gamma_3\} = \min\{\gamma_1,\gamma_3\}$. Thus, we can put $\gamma := \min\{\gamma_1,\gamma_3\}$ as in (7.43). The previous part of the proof then shows that for each $s \in [0,b]_{\mathbb{N}_0}$ it holds that

$$\min_{t\in\left[\frac{b+\nu}{4},\frac{3(b+\nu)}{4}\right]} G(t,s) \ge \gamma \max_{t\in[\nu-2,\nu+b]_{\mathbb{N}_{\nu-2}}} G(t,s) = \gamma G(s+\nu-1,s), \tag{7.48}$$

and as (7.48) is (7.42), the first part of the proof is complete.

To complete the proof, it remains to show that $\gamma$, as defined in (7.43), satisfies $0 < \gamma < 1$. We first observe that $\gamma_1 < 1$. This follows from the fact that $t^{\underline{\nu-1}}$ is an increasing function in $t$ whenever $\nu \in (1,2]$. To see that this latter claim is true, simply observe that

$$\Delta\left[t^{\underline{\nu-1}}\right] = (\nu-1)\cdot\frac{\Gamma(t+1)}{\Gamma(t-\nu+3)} > 0.$$

Thus, as $\frac{b+\nu}{4} > b+\nu$ and $\left(\frac{b+\nu}{4}\right)^{\underline{\nu-1}}, (b+\nu)^{\underline{\nu-1}} \neq 0$, the claim follows. In particular, this demonstrates that

$$\gamma = \min\{\gamma_1, \gamma_3\} \leq \gamma_1 < 1. \tag{7.49}$$

On the other hand, observe that $\gamma_1 > 0$. So, it only remains to show that $\gamma_3 > 0$. Note that $\gamma_3$ is strictly positive if and only if

$$\frac{\left(\frac{3(b+\nu)}{4}\right)^{\underline{\nu-1}}(b+\nu-1)^{\underline{\nu-1}}}{\left(\frac{3(b+\nu)}{4}-1\right)^{\underline{\nu-1}}(b+\nu)^{\underline{\nu-1}}} > 1. \tag{7.50}$$

But (7.50) is true if and only if

$$\frac{(b+1)\left(\frac{3(b+\nu)}{4}\right)}{(b+\nu)\left(\frac{3(b+\nu)}{4}-\nu+1\right)} > 1 \tag{7.51}$$

holds for each admissible $b$ and $\nu$—that is, each $b \in [2,+\infty)_{\mathbb{N}}$ and $\nu \in (1,2]$.

We claim that (7.51) is true for each $b \in [2,+\infty)$ and each $\nu \in (1,2]$. To see this, for each fixed and admissible $b$, put

$$H_b(\nu) := \frac{(b+1)\left(\frac{3(b+\nu)}{4}\right)}{(b+\nu)\left(\frac{3(b+\nu)}{4}-\nu+1\right)}, \tag{7.52}$$

which is the left-hand side of inequality (7.51), and note that each of

$$H_b(1) = 1 \tag{7.53}$$

and

$$H_b(2) = \frac{(b+1)\left(\frac{3}{4}b+\frac{3}{2}\right)}{(b+2)\left(\frac{3}{4}b+\frac{1}{2}\right)} = \frac{3b+3}{3b+2} \tag{7.54}$$

holds. Now, $H_b(2) > 1$ is evidently true for each admissible $b$. Moreover, a straightforward computation shows that

$$H_b'(\nu) = \frac{3(b+1)}{(3b-\nu+4)^2}. \tag{7.55}$$

But then (7.55) demonstrates that for each $b$, it holds that the map $\nu \mapsto H_b(\nu)$ is strictly increasing in $\nu$. Therefore, as $H_b(1) = 1$ and $H_b(2) > 1$, we obtain at once that

$$H_b(\nu) > 1 \tag{7.56}$$

for each $\nu \in (1,2]$ and $b \in [2,+\infty)_{\mathbb{N}}$. But then from (7.56) we deduce that (7.51) holds, as desired.

In summary, (7.50)–(7.56) demonstrate that $\gamma_3 > 0$. But we then find that

$$\gamma = \min\{\gamma_1, \gamma_3\} > 0. \tag{7.57}$$

Putting (7.49) and (7.57) together implies that $\gamma \in (0,1)$, as claimed. And this completes the proof. □

*Remark 7.33.* Note that in case $\alpha = 0$, the result of Theorem 7.32 reduces to the results obtained in [31], as the reader may easily check.

*Remark 7.34.* For a brief investigation of the properties of the set of admissible values of $\alpha$ generated by condition (7.34) above, one may consult [104].

*Remark 7.35.* Once we have the preceding properties of the Green's function $G$ in hand, it then is standard to provide some basic existence result for the FBVP

$$\begin{aligned} -\Delta^\nu y(t) &= f(t+\nu-1, y(t+\nu-1)) \\ y(\nu-2) &= 0 \\ \alpha y(\nu+K) &= y(\nu+b), \end{aligned}$$

where $f : [0,b]_{\mathbb{N}_0} \times \mathbb{R} \to [0,+\infty)$ is a continuous map. However, since we complete this sort of analysis in the somewhat more general case of (potentially) nonlinear boundary conditions in the next section, we will not present existence theorems for the three-point problem studied in this section. We instead direct the interested reader to [104, §5] where results of this sort may be found for the three-point problem studied in this section.

## 7.5 A Nonlocal BVP with Nonlinear Boundary Conditions

In the previous section we saw how a three-point problem can be analyzed. In particular, notice that the boundary condition in that setting is linear in the sense that if we define the boundary operator $B$ defined by

$$By := \alpha y(t+\nu) + y(t+\nu+b), \quad y \in \mathbb{R}^m,$$

then $B$ is linear map from $\mathbb{R}^m$ into $\mathbb{R}$, for some $m > 1$ with $m \in \mathbb{N}$. However, there is no requirement that the boundary conditions for a given BVP be linear. In fact, if the boundary conditions are *nonlinear*, then the mathematical analysis of the problem can be very interesting and potentially challenging. For one thing, one cannot generally approach the problem in the same way—namely by determining an appropriate Green's function. Rather, an alternative but viable approach in this setting is to instead construct a new operator by taking the operator associated with the linear boundary condition problem and then suitably perturbing it. This approach will be seen in this section. In particular, we wish to consider a modification of the BVP considered in the previous section; namely, we consider in this section the problem

$$\begin{aligned}-\Delta^\nu y(t) &= f(t+\nu-1, y(t+\nu-1))\\ y(\nu-2) &= g(y)\\ y(\nu+b) &= 0\end{aligned}$$

in the case where the map $y \mapsto g(y)$ is potentially nonlinear. The results of this section may be found largely in Goodrich [92].

We begin by providing a lemma, which essentially recasts the above BVP as an appropriate summation operator. Studying the existence of solutions to the BVP will then be reduced to demonstrating the existence of nontrivial fixed points of the associated summation operator.

**Theorem 7.36.** *Let $h : [\nu-1, \ldots, \nu+b-1]_{\mathbb{N}_{\nu-1}} \to \mathbb{R}$ and $g : \mathbb{R}^{b+3} \to \mathbb{R}$ be given. A function $y$ is a solution of the discrete FBVP*

$$\begin{aligned}-\Delta^\nu y(t) &= h(t+\nu-1)\\ y(\nu-2) &= g(y)\\ y(\nu+b) &= 0\end{aligned} \tag{7.58}$$

*where $t \in [0,b]_{\mathbb{N}_0}$, if and only if $y(t)$, for each $t \in [\nu-2, \nu+b]_{\mathbb{N}_{\nu-2}}$, has the form*

$$\begin{aligned}y(t) = &-\frac{1}{\Gamma(\nu)}\sum_{s=0}^{t-\nu}(t-s-1)^{\underline{\nu-1}}h(s+\nu-1)\\ &+ t^{\underline{\nu-1}}\left[\frac{1}{(\nu+b)^{\underline{\nu-1}}\Gamma(\nu)}\sum_{s=0}^{b}(\nu+b-s-1)^{\underline{\nu-1}}h(s+\nu-1)\right.\\ &\left.-\frac{g(y)}{(b+2)\Gamma(\nu-1)}\right] + \frac{t^{\underline{\nu-2}}}{\Gamma(\nu-1)}g(y).\end{aligned} \tag{7.59}$$

*Proof.* Using the results from earlier in this text, we find that a general solution for (7.58) is the function

$$y(t) = -\Delta^{-\nu} h(t+\nu-1) + c_1 t^{\underline{\nu-1}} + c_2 t^{\underline{\nu-2}}, \tag{7.60}$$

where $t \in [\nu-2, \nu+b]_{\mathbb{N}_{\nu-2}}$. On the one hand, applying the boundary condition at $t = \nu - 2$ in (7.58) implies at once that

$$c_2 = \frac{1}{\Gamma(\nu-1)} g(y). \tag{7.61}$$

Applying the boundary condition at $t = \nu + b$ in (7.58) yields

$$\begin{aligned} 0 &= y(\nu+b) \\ &= \left[-\Delta^{-\nu} h(t)\right]_{t=\nu+b} + c_1(\nu+b)^{\underline{\nu-1}} + \frac{(\nu+b)^{\underline{\nu-2}}}{\Gamma(\nu-1)} g(y) \\ &= -\frac{1}{\Gamma(\nu)} \sum_{s=0}^{b} (\nu+b-s-1)^{\underline{\nu-1}} h(s+\nu-1) + c_1(\nu+b)^{\underline{\nu-1}} \\ &\quad + \frac{(\nu+b)^{\underline{\nu-2}}}{\Gamma(\nu-1)} g(y), \end{aligned} \tag{7.62}$$

whence (7.62) implies that

$$\begin{aligned} c_1 &= \frac{1}{(\nu+b)^{\underline{\nu-1}}\Gamma(\nu)} \sum_{s=0}^{b} (\nu+b-s-1)^{\underline{\nu-1}} h(s+\nu-1) \\ &\quad - \frac{(\nu+b)^{\underline{\nu-2}}}{(\nu+b)^{\underline{\nu-1}}\Gamma(\nu-1)} g(y) \qquad (7.63) \\ &= \frac{1}{(\nu+b)^{\underline{\nu-1}}\Gamma(\nu)} \sum_{s=0}^{b} (\nu+b-s-1)^{\underline{\nu-1}} h(s+\nu-1) \\ &\quad - \frac{1}{(b+2)\Gamma(\nu-1)} g(y). \end{aligned}$$

Consequently, using (7.60)–(7.63), we deduce that for each $t \in [\nu-2, \nu+b]_{\mathbb{N}_{\nu-2}}$ it holds that $y$ has the form given in (7.59) above. And this shows that if (7.58) has a solution, then it can be represented by (7.59) and that every function of the form (7.59) is a solution of (7.58). And this completes the proof of the theorem. □

We now recall an additional lemma that will prove to be useful later in this section.

**Lemma 7.37.** *For $t$ and $s$ for which both $(t-s-1)^{\underline{\nu}}$ and $(t-s-2)^{\underline{\nu}}$ are defined, we find that*

$$\Delta_s\left[(t-s-1)^{\underline{\nu}}\right] = -\nu(t-s-2)^{\underline{\nu-1}}.$$

*Proof.* Omitted—see [89, Lemma 2.4]. □

Finally, for $\nu \in (1,2]$ given, we provide the following lemma, which will also be of importance later in this section.

**Lemma 7.38.** *The map*

$$t \mapsto \frac{1}{\Gamma(\nu-1)}\left[t^{\underline{\nu-2}} - \frac{1}{b+2}t^{\underline{\nu-1}}\right]$$

*is strictly decreasing in $t$, for $t \in [\nu-2, \nu+b]_{\mathbb{N}_{\nu-2}}$. In addition, it holds both that*

$$\min_{t\in[\nu-2,\nu+b]_{\mathbb{N}_{\nu-2}}}\left[\frac{1}{\Gamma(\nu-1)}\left[t^{\underline{\nu-2}} - \frac{1}{b+2}t^{\underline{\nu-1}}\right]\right] = 0$$

*and that*

$$\max_{t\in[\nu-2,\nu+b]_{\mathbb{N}_{\nu-2}}}\left[\frac{1}{\Gamma(\nu-1)}\left[t^{\underline{\nu-2}} - \frac{1}{b+2}t^{\underline{\nu-1}}\right]\right] = 1.$$

*Proof.* Note that

$$\Delta_t\left[t^{\underline{\nu-2}} - \frac{1}{b+2}t^{\underline{\nu-1}}\right] = (\nu-2)t^{\underline{\nu-3}} - \frac{\nu-1}{b+2}t^{\underline{\nu-2}} < 0, \tag{7.64}$$

where the inequality in (7.64) follows from the observation that $(\nu-2)(b+2) - (t-\nu+3)(\nu-1) < 0$. It follows that the map

$$t \mapsto \frac{1}{\Gamma(\nu-1)}\left[t^{\underline{\nu-2}} - \frac{1}{b+2}t^{\underline{\nu-1}}\right]$$

is strictly decreasing in $t$ as well. Furthermore, notice both that

$$\frac{1}{\Gamma(\nu-1)}\left[t^{\underline{\nu-2}} - \frac{1}{b+2}t^{\underline{\nu-1}}\right]_{t=\nu-2} = \frac{1}{\Gamma(\nu-1)}\left[\Gamma(\nu-1) - \frac{0}{b+2}\right] = 1 \tag{7.65}$$

and that

$$\frac{1}{\Gamma(\nu-1)}\left[t^{\underline{\nu-2}}-\frac{1}{b+2}t^{\underline{\nu-1}}\right]_{t=\nu+b}=0.$$

In particular, as a consequence of (7.64)–(7.65) we see that the second claim in the statement of the theorem follows. And this completes the proof of the lemma. □

We now wish to show that under certain conditions, problem (7.58) has at least one solution. We observe that problem (7.58) may be recast as an equivalent summation equation. In particular, $y$ is a solution of (7.58) if and only if $y$ is a fixed point of the operator $T : \mathbb{R}^{b+3} \to \mathbb{R}^{b+3}$, where

$$\begin{aligned}(Ty)(t) := &-\frac{1}{\Gamma(\nu)}\sum_{s=0}^{t-\nu}(t-s-1)^{\underline{\nu-1}}f(s+\nu-1,y(s+\nu-1))\\ &+\frac{t^{\underline{\nu-1}}}{(\nu+b)^{\underline{\nu-1}}\Gamma(\nu)}\sum_{s=0}^{b}(\nu+b-s-1)^{\underline{\nu-1}}f(s+\nu-1,y(s+\nu-1))\\ &-\frac{t^{\underline{\nu-1}}g(y)}{(b+2)\Gamma(\nu-1)}+\frac{t^{\underline{\nu-2}}}{\Gamma(\nu-1)}g(y),\end{aligned} \tag{7.66}$$

for $t \in [\nu-2, \nu+b]_{\mathbb{N}_{\nu-2}}$; this observation follows from Theorem 7.36. We now use this fact to state and prove our first existence theorem.

**Theorem 7.39.** *Suppose that the maps $(t, y) \mapsto f(t, y)$ and $y \mapsto g(y)$ are Lipschitz in $y$. That is, there exist $\alpha$, $\beta > 0$ such that $|f(t, y_1) - f(t, y_2)| \le \alpha |y_1 - y_2|$ whenever $y_1, y_2 \in \mathbb{R}$, and $|g(y_1) - g(y_2)| \le \beta \|y_1 - y_2\|$ whenever $y_1, y_2 \in C([\nu-2, \nu+b]_{\mathbb{N}_{\nu-2}}, \mathbb{R})$. Then it follows that problem (7.58) has a unique solution provided that the condition*

$$2\alpha\prod_{j=1}^{b}\left(\frac{\nu+j}{j}\right)+\beta<1 \tag{7.67}$$

*holds.*

*Proof.* We will show that under the hypotheses in the statement of this theorem $T$ is a contraction mapping. To this end, we notice that for each admissible $y_1$ and $y_2$ it holds that

$$\begin{aligned}
&\|Ty_1 - Ty_2\| \\
&\leq \alpha\|y_1 - y_2\| \max_{t\in[\nu-2,\nu+b]_{\mathbb{N}_{\nu-2}}} \left[\frac{1}{\Gamma(\nu)}\sum_{s=0}^{t-\nu}(t-s-1)^{\underline{\nu-1}}\right] \\
&\quad + \alpha\|y_1 - y_2\| \max_{t\in[\nu-2,\nu+b]_{\mathbb{N}_{\nu-2}}} \left[\frac{t^{\underline{\nu-1}}}{(\nu+b)^{\underline{\nu-1}}\Gamma(\nu)}\sum_{s=0}^{b}(\nu+b-s-1)^{\underline{\nu-1}}\right] \\
&\quad + \beta\|y_1 - y_2\| \max_{t\in[\nu-2,\nu+b]_{\mathbb{N}_{\nu-2}}} \left|-\frac{t^{\underline{\nu-1}}}{(b+2)\Gamma(\nu-1)} + \frac{t^{\underline{\nu-2}}}{\Gamma(\nu-1)}\right|.
\end{aligned} \tag{7.68}$$

We now analyze each of the three terms on the right-hand side of (7.68).

We first notice, by an application of Lemma 7.37, that

$$\begin{aligned}
&\alpha\|y_1 - y_2\|\left[\frac{1}{\Gamma(\nu)}\sum_{s=0}^{t-\nu}(t-s-1)^{\underline{\nu-1}}\right] = \frac{\alpha\|y_1 - y_2\|}{\Gamma(\nu)}\left[-\frac{1}{\nu}(t-s)^{\underline{\nu}}\right]_{s=0}^{t-\nu+1} \\
&= \alpha\|y_1 - y_2\|\left[\frac{\Gamma(t+1)}{\Gamma(t-\nu+1)\Gamma(\nu+1)}\right] \leq \alpha\|y_1 - y_2\|\left[\frac{\Gamma(\nu+b+1)}{\Gamma(b+1)\Gamma(\nu+1)}\right] \\
&= \alpha\prod_{j=1}^{b}\left(\frac{\nu+j}{j}\right)\|y_1 - y_2\|.
\end{aligned} \tag{7.69}$$

So, this estimates the first term on the right-hand side of (7.68). Then another application of Lemma 7.37 reveals that

$$\begin{aligned}
&\alpha\|y_1 - y_2\|\left[\frac{t^{\underline{\nu-1}}}{(\nu+b)^{\underline{\nu-1}}\Gamma(\nu)}\sum_{s=0}^{b}(\nu+b-s-1)^{\underline{\nu-1}}\right] \\
&\leq \frac{\alpha\|y_1 - y_2\|}{\Gamma(\nu)}\sum_{s=0}^{b}(\nu+b-s-1)^{\underline{\nu-1}} = \frac{\alpha\|y_1 - y_2\|}{\Gamma(\nu)}\left[-\frac{1}{\nu}(\nu+b-s)^{\underline{\nu}}\right]_{s=0}^{b+1} \\
&= \alpha\|y_1 - y_2\|\prod_{j=1}^{b}\left(\frac{\nu+j}{j}\right),
\end{aligned} \tag{7.70}$$

which provides an upper bound for the second term appearing on the right-hand side of (7.66). Finally, we may estimate the third term on the right-hand side of (7.68) by employing Lemma 7.38 and observing that

$$\beta\|y_1 - y_2\|\left|-\frac{t^{\underline{\nu-1}}}{(b+2)\Gamma(\nu-1)} + \frac{t^{\underline{\nu-2}}}{\Gamma(\nu-1)}\right| \leq \beta\|y_1 - y_2\|. \tag{7.71}$$

Putting (7.69)–(7.71) into the right-hand side of (7.68), we conclude at once that

$$\|Ty_1 - Ty_2\| \le \left\{ 2\alpha \prod_{j=1}^{b} \left( \frac{\nu + j}{j} \right) + \beta \right\} \|y_1 - y_2\|.$$

So, by requiring condition (7.67) to hold, we find that (7.58) has a unique solution. And this completes the proof. □

By weakening the conditions imposed on the functions $f$ and $g$, we can still deduce the existence of at least one solution to (7.58). We shall appeal to the Brouwer theorem to accomplish this.

**Theorem 7.40.** *Suppose that there exists a constant $M > 0$ such that $f(t, y)$ satisfies the inequality*

$$\max_{(t,y)\in[\nu-1,\nu+b-1]_{\mathbb{N}_{\nu-1}}\times[-M,M]} |f(t,y)| \le \frac{M}{\frac{2\Gamma(\nu+b+1)}{\Gamma(\nu+1)\Gamma(b+1)} + 1} \tag{7.72}$$

*and $g(y)$ satisfies the inequality*

$$\max_{0\le\|y\|\le M} |g(y)| \le \frac{M}{\frac{2\Gamma(\nu+b+1)}{\Gamma(\nu+1)\Gamma(b+1)} + 1}. \tag{7.73}$$

*Then (7.58) has at least one solution, say $y_0$, satisfying $|y_0(t)| \le M$, for all $t \in [\nu - 2, \nu + b]_{\mathbb{N}_{\nu-2}}$.*

*Proof.* Consider the Banach space $\mathcal{B} := \{y \in \mathbb{R}^{b+3} : \|y\| \le M\}$. Let $T$ be the operator defined in (7.66). It is clear that $T$ is a continuous operator. Therefore, the main objective in establishing this result is to show that $T : \mathcal{B} \to \mathcal{B}$—that is, whenever $\|y\| \le M$, it follows that $\|Ty\| \le M$. Once this is established, the Brouwer theorem will be invoked to deduce the conclusion.

To this end, assume that inequalities (7.72)–(7.73) hold for given $f$ and $g$. For notational convenience in the sequel, let us put

$$\Omega_0 := \frac{M}{\frac{2\Gamma(\nu+b+1)}{\Gamma(\nu+1)\Gamma(b+1)} + 1}, \tag{7.74}$$

which is a positive constant. Using the notation introduced previously in (7.74), observe that

$$\begin{aligned}
&\|Ty\| \\
&\le \max_{t\in[\nu-2,\nu+b]_{\mathbb{N}_{\nu-2}}} \frac{1}{\Gamma(\nu)} \sum_{s=0}^{t-\nu} (t-s-1)^{\underline{\nu-1}} |f(s+\nu-1), y(s+\nu-1))| \\
&\quad + \max_{t\in[\nu-2,\nu+b]_{\mathbb{N}_{\nu-2}}} \left\{ \frac{t^{\underline{\nu-1}}}{(\nu+b)^{\underline{\nu-1}}\Gamma(\nu)} \right.
\end{aligned}$$

$$\times \sum_{s=0}^{b} (\nu + b - s - 1)^{\underline{\nu-1}} |f(s + \nu - 1, y(s + \nu - 1))| \Bigg\}$$

$$+ \max_{t \in [\nu-2, \nu+b]_{\mathbb{N}_{\nu-2}}} \left| -\frac{t^{\underline{\nu-1}}}{(b+2)\Gamma(\nu-1)} + \frac{t^{\underline{\nu-2}}}{\Gamma(\nu)} \right| |g(y)|$$

$$\leq \Omega_0 \max_{t \in [\nu-2, \nu+b]_{\mathbb{N}_{\nu-2}}} \left[ \frac{1}{\Gamma(\nu)} \sum_{s=0}^{t-\nu} (t - s - 1)^{\underline{\nu-1}} \right.$$

$$\left. + \sum_{s=0}^{b} \frac{t^{\underline{\nu-1}} (\nu + b - s - 1)^{\underline{\nu-1}}}{(\nu + b)^{\underline{\nu-1}} \Gamma(\nu)} \right]$$

$$+ \Omega_0 \max_{t \in [\nu-2, \nu+b]_{\mathbb{N}_{\nu-2}}} \left| -\frac{t^{\underline{\nu-1}}}{(b+2)\Gamma(\nu-1)} + \frac{t^{\underline{\nu-2}}}{\Gamma(\nu)} \right|. \tag{7.75}$$

Now, much as in the proof of Theorem 7.39 we can simplify the expression on the right-hand side of inequality (7.75). In particular, we observe that

$$\frac{1}{\Gamma(\nu)} \sum_{s=0}^{t-\nu} (t - s - 1)^{\underline{\nu-1}} + \frac{t^{\underline{\nu-1}}}{(\nu + b)^{\underline{\nu-1}} \Gamma(\nu)} \sum_{s=0}^{b} (\nu + b - s - 1)^{\underline{\nu-1}}$$

$$\leq \frac{1}{\Gamma(\nu)} \sum_{s=0}^{t-\nu} (t - s - 1)^{\underline{\nu-1}} + \frac{1}{\Gamma(\nu)} \sum_{s=0}^{b} (\nu + b - s - 1)^{\underline{\nu-1}}$$

$$\leq \frac{1}{\Gamma(\nu)} \sum_{s=0}^{b} (\nu + b - s - 1)^{\underline{\nu-1}} + \frac{1}{\Gamma(\nu)} \sum_{s=0}^{b} (\nu + b - s - 1)^{\underline{\nu-1}}$$

$$= \frac{2}{\Gamma(\nu)} \sum_{s=0}^{b} (\nu + b - s - 1)^{\underline{\nu-1}}, \tag{7.76}$$

where to obtain inequality (7.76) we have used the fact that the map $t \mapsto t^{\underline{\nu-1}}$ is increasing in $t$ since $\nu > 1$. Furthermore, it holds that

$$\sum_{s=0}^{b} (\nu + b - s - 1)^{\underline{\nu-1}} = \left[ -\frac{1}{\nu} (\nu + b - s)^{\underline{\nu}} \right]_{s=0}^{b+1} = \frac{\Gamma(\nu + b + 1)}{\nu \Gamma(b + 1)}. \tag{7.77}$$

In addition we may estimate the second term on the right-hand side of inequality (7.75) by using Lemma 7.38, which implies that

$$\max_{t \in [\nu-2, \nu+b]_{\mathbb{N}_{\nu-2}}} \left| -\frac{t^{\underline{\nu-1}}}{(b+2)\Gamma(\nu-1)} + \frac{t^{\underline{\nu-2}}}{\Gamma(\nu)} \right| = 1. \tag{7.78}$$

If we now put (7.75)–(7.78) together, then we find that

$$\begin{aligned}\|Ty\| &\le \Omega_0\left[\frac{2\Gamma(\nu+b+1)}{\Gamma(\nu+1)\Gamma(b+1)}\right]+\Omega_0\\ &= \Omega_0\left[\frac{2\Gamma(\nu+b+1)}{\Gamma(\nu+1)\Gamma(b+1)}+1\right].\end{aligned}\tag{7.79}$$

Finally, by the definition of $\Omega_0$ given earlier in (7.74), we deduce that (7.79) implies that

$$\|Ty\| \le \Omega_0\left[\frac{2\Gamma(\nu+b+1)}{\Gamma(\nu)\Gamma(b+1)}+1\right]=M.\tag{7.80}$$

Thus, from (7.80) we conclude that $T : \mathcal{B} \to \mathcal{B}$, as desired. Consequently, it follows at once by the Brouwer theorem that there exists a fixed point of the map $T$, say $y_0 \in \mathcal{B}$. But this function $y_0$ is a solution of (7.58). Moreover, $y_0$ satisfies the bound $|y_0(t)| \le M$, for each $t \in [\nu-2, \nu+b]_{\mathbb{N}_{\nu-2}}$. Thus, the proof is complete. □

We next we wish to deduce the existence of at least one positive solution to problem (7.58). To this end, we first need recall some facts about the Green's function for the problem

$$\begin{aligned}-\Delta^\nu y(t) &= f(t+\nu-1, y(t+\nu-1))\\ y(\nu-2) &= 0\\ y(\nu+b) &= 0.\end{aligned}$$

In particular, we recall the following result.

**Lemma 7.41.** *Let* $1 < \nu \le 2$. *The unique solution of the FBVP*

$$\begin{aligned}-\Delta^\nu y(t) &= h(s+\nu-1)\\ y(\nu-2) &= 0\\ y(\nu+b) &= 0\end{aligned}$$

*is given by the map* $y : [\nu-2, \nu+b]_{\mathbb{Z}_{\nu-2}} \to \mathbb{R}$ *defined by*

$$y(t) = \sum_{s=0}^{b+1} G(t,s)h(s+\nu-1),$$

*where the Green's function* $G : [\nu-2, \nu+b]_{\mathbb{Z}_{\nu-2}} \times [0,b]_{\mathbb{N}_0} \to \mathbb{R}$ *is defined by*

$$G(t,s) := \frac{1}{\Gamma(\nu)}\begin{cases}\frac{t^{\underline{\nu-1}}(\nu+b-s-1)^{\underline{\nu-1}}}{(\nu+b)^{\underline{\nu-1}}} - (t-s-1)^{\underline{\nu-1}}, & (t,s)\in T_2\\ \frac{t^{\underline{\nu-1}}(\nu+b-s-1)^{\underline{\nu-1}}}{(\nu+b)^{\underline{\nu-1}}}, & (t,s)\in T_2\end{cases},$$

*where*

$$T_1 := \{(t,s) \in [\nu-2, \nu+b]_{\mathbb{Z}_{\nu-2}} \times [0,b]_{\mathbb{N}_0} \; : \; 0 \le s < t-\nu+1 \le b+1\}$$

*and*

$$T_2 := \{(t,s) \in [\nu-2, \nu+b]_{\mathbb{Z}_{\nu-2}} \times [0,b]_{\mathbb{N}_0} \; : \; 0 \le t-\nu+1 \le s \le b+1\}.$$

**Lemma 7.42.** *The Green's function G defined in Lemma 7.41 satisfies the following conditions:*

(i) $G(t,s) > 0$ *for* $t \in [\nu-1, \nu+b]_{\mathbb{N}_{\nu-1}}$ *for* $s \in [0,b]_{\mathbb{N}}$;
(ii) $\max_{t\in[\nu-1,\nu+b]_{\mathbb{N}_{\nu-1}}} G(t,s) = G(s+\nu-1,s)$ *for* $s \in [0,b]_{\mathbb{N}}$; *and*
(iii) *There exists a number* $\gamma \in (0,1)$ *such that*

$$\min_{\frac{b+\nu}{4} \le t \le \frac{3(b+\nu)}{4}} G(t,s) \ge \gamma \max_{t\in[\nu-1,\nu+b]_{\mathbb{N}_{\nu-1}}} G(t,s) = \gamma G(s+\nu-1,s),$$

*for* $s \in [0,b]_{\mathbb{N}_0}$.

*Remark 7.43.* The proof of both Lemmas 7.41 and 7.42 are simple modifications of the proofs of [31, Theorem 3.1] and [31, Theorem 3.2], respectively. Hence, we omit the proofs.

Before defining the cone that we shall use to prove our existence theorems, we need a preliminary lemma.

**Lemma 7.44.** *If the map* $y \mapsto g(y)$ *is nonnegative, then there exists a constant* $\tilde{\gamma} \in (0,1)$ *with the property that*

$$\begin{aligned}
&\min_{t\in\left[\frac{b+\nu}{4},\frac{3(b+\nu)}{4}\right]} \sum_{s=0}^{b} G(t,s)f(s+\nu-1, y(s+\nu-1)) \\
&\quad + \min_{t\in\left[\frac{b+\nu}{4},\frac{3(b+\nu)}{4}\right]} \left[-\frac{t^{\underline{\nu-1}}}{(b+2)\Gamma(\nu-1)} + \frac{t^{\underline{\nu-2}}}{\Gamma(\nu-1)}\right] g(y) \\
&\ge \tilde{\gamma} \max_{t\in[\nu-2,\nu+b]_{\mathbb{N}_{\nu-2}}} \sum_{s=0}^{b} G(t,s)f(s+\nu-1, y(s+\nu-1)) \\
&\quad + \tilde{\gamma} \max_{t\in[\nu-2,\nu+b]_{\mathbb{N}_{\nu-2}}} \left[-\frac{t^{\underline{\nu-1}}}{(b+2)\Gamma(\nu-1)} + \frac{t^{\underline{\nu-2}}}{\Gamma(\nu-1)}\right] g(y). \qquad (7.81)
\end{aligned}$$

*Proof.* To see that this is true, observe first that by Lemma 7.42 we find $\gamma \in (0,1)$ such that

$$\min_{t\in\left[\frac{b+\nu}{4},\frac{3(b+\nu)}{4}\right]} \sum_{s=0}^{b} G(t,s)f(s+\nu-1,y(s+\nu-1))$$

$$\geq \gamma \max_{t\in[\nu-2,\nu+b]_{\mathbb{N}_{\nu-2}}} \sum_{s=0}^{b} G(t,s)f(s+\nu-1,y(s+\nu-1)). \tag{7.82}$$

Now, recall from Lemma 7.38 that the map

$$t \mapsto \frac{1}{\Gamma(\nu-1)}\left[t^{\underline{\nu-2}} - \frac{1}{b+2}t^{\underline{\nu-1}}\right]$$

is strictly decreasing in $t$ and, furthermore, is strictly positive for $t < b+\nu$. In particular, from this observation we deduce the existence of a number $M > 0$ such that

$$\min_{t\in\left[\frac{b+\nu}{4},\frac{3(b+\nu)}{4}\right]} \left[-\frac{t^{\underline{\nu-1}}}{(b+2)\Gamma(\nu-1)} + \frac{t^{\underline{\nu-2}}}{\Gamma(\nu-1)}\right] = M. \tag{7.83}$$

Note that we assume here that there exists a point $\tilde{t} \in \mathbb{N}_{\nu-2}$ such that $\frac{b+\nu}{4} \leq \tilde{t} \leq \frac{3(b+\nu)}{4}$. Additionally, we recall from Lemma 7.38 that

$$\max_{t\in[\nu-2,\nu+b]_{\mathbb{N}_{\nu-2}}} \frac{1}{\Gamma(\nu-1)}\left[t^{\underline{\nu-2}} - \frac{1}{b+2}t^{\underline{\nu-1}}\right] = 1. \tag{7.84}$$

In particular, then, (7.83)–(7.84) imply that by putting

$$\gamma_0 := M, \tag{7.85}$$

where $\gamma_0$ is clearly strictly positive, it follows from (7.83)–(7.85) that

$$\min_{t\in\left[\frac{b+\nu}{4},\frac{3(b+\nu)}{4}\right]} \left[-\frac{t^{\underline{\nu-1}}}{(b+2)\Gamma(\nu-1)} + \frac{t^{\underline{\nu-2}}}{\Gamma(\nu-1)}\right] g(y)$$

$$= \gamma_0 \cdot \max_{t\in[\nu-2,\nu+b]_{\mathbb{N}_{\nu-2}}} \left[-\frac{t^{\underline{\nu-1}}}{(b+2)\Gamma(\nu-1)} + \frac{t^{\underline{\nu-2}}}{\Gamma(\nu-1)}\right] g(y).$$

Finally, define $\tilde{\gamma}$ by

$$\tilde{\gamma} := \min\{\gamma, \gamma_0\}. \tag{7.86}$$

Evidently, definition (7.86) implies that $\tilde{\gamma} \in (0,1)$. Moreover, inequality (7.82) implies that

$$\min_{t\in\left[\frac{b+\nu}{4},\frac{3(b+\nu)}{4}\right]} \sum_{s=0}^{b} G(t,s)f(s+\nu-1,y(s+\nu-1))$$

$$+ \min_{t\in\left[\frac{b+\nu}{4},\frac{3(b+\nu)}{4}\right]} \left[-\frac{t^{\underline{\nu-1}}}{(b+2)\Gamma(\nu-1)}+\frac{t^{\underline{\nu-2}}}{\Gamma(\nu-1)}\right]g(y)$$

$$\geq \tilde{\gamma} \max_{t\in[\nu-1,\nu+b]_{\mathbb{N}_{\nu-1}}} \sum_{s=0}^{b} G(t,s)f(s+\nu-1,y(s+\nu-1))$$

$$+ \tilde{\gamma} \max_{t\in[\nu-1,\nu+b]_{\mathbb{N}_{\nu-1}}} \left[-\frac{t^{\underline{\nu-1}}}{(b+2)\Gamma(\nu-1)}+\frac{t^{\underline{\nu-2}}}{\Gamma(\nu-1)}\right]g(y), \tag{7.87}$$

which since (7.87) is (7.81) completes the proof of the lemma. □

Now, let us put

$$\eta := \frac{1}{\sum_{s=0}^{b} G(s+\nu-1,s)}$$

and

$$\lambda := \frac{1}{\sum_{s=\lceil\frac{\nu+b}{4}-\nu+1\rceil}^{\lfloor\frac{3(\nu+b)}{4}-\nu+1\rfloor} \tilde{\gamma} G\left(\left\lfloor\frac{b+1}{2}\right\rfloor+\nu,s\right)}. \tag{7.88}$$

In addition, define the set $\mathcal{K} \subseteq \mathcal{C}\left([\nu-2,\nu+b]_{\mathbb{N}_{\nu-2}},\mathbb{R}\right)$ by

$$\mathcal{K} := \left\{ y : [\nu-2,\nu+b]_{\mathbb{N}_{\nu-2}} \to \mathbb{R} : y(t) \geq 0, \right.$$
$$\left. \min_{t\in\left[\frac{b+\nu}{4},\frac{3(b+\nu)}{4}\right]} y(t) \geq \tilde{\gamma}\|y(t)\| \right\}, \tag{7.89}$$

which is a cone in the Banach space $\mathcal{C}\left([\nu-2,\nu+b]_{\mathbb{N}_{\nu-2}},\mathbb{R}\right)$, where the number $\tilde{\gamma}$ in (7.88)–(7.89) is the same number as given in Lemma 7.44 above. Moreover, we will also need in the sequel the constant

$$\eta^* := \frac{1}{2}.$$

Finally, we introduce some conditions that will be helpful in the sequel; these conditions place some control on the growth of the nonlinearity $f$ as well as the functional $g$ appearing in (7.58).

**F1:** There exists a number $r > 0$ such that $f(t, y) \leq \frac{1}{2}\eta r$ whenever $0 \leq y \leq r$.
**F2:** There exists a number $r > 0$ such that $f(t, y) \geq \lambda r$ whenever $\tilde{\gamma} r \leq y \leq r$, where $\tilde{\gamma}$ is the number provided in Lemma 7.44.
**G1:** There exists a number $r > 0$ such that $g(y) \leq \eta^* r$ whenever $0 \leq \|y\| \leq r$.

*Remark 7.45.* The operator $T$ defined in (7.66) may be written in the form

$$
\begin{aligned}
(Ty)(t) = & \sum_{s=0}^{b} G(t,s) f(s+\nu-1, y(s+\nu-1)) \\
& + \left[ -\frac{t^{\underline{\nu-1}}}{(b+2)\Gamma(\nu-1)} + \frac{t^{\underline{\nu-2}}}{\Gamma(\nu-1)} \right] g(y),
\end{aligned} \tag{7.90}
$$

where $G$ is the Green's function from Lemma 7.41. This observation is important since it allows us to use the known properties of the map $(t, s) \mapsto G(t, s)$ to obtain useful estimates in the existence argument.

With these declarations in hand, we proceed with proving an existence theorem. We begin with a preliminary lemma, however, to establish separately that $T$ in fact maps $\mathcal{K}$ into itself.

**Lemma 7.46.** *Let $T$ be defined as in* (7.90) *and $\mathcal{K}$ as in* (7.89). *Assume in addition that both $f$ and $g$ are nonnegative. Then $T(\mathcal{K}) \subseteq \mathcal{K}$.*

*Proof.* Let $T$ be the operator defined in (7.90). Observe that

$$
\begin{aligned}
& \min_{t \in \left[\frac{b+\nu}{4}, \frac{3(b+\nu)}{4}\right]} (Ty)(t) \\
& \geq \min_{t \in \left[\frac{b+\nu}{4}, \frac{3(b+\nu)}{4}\right]} \sum_{s=0}^{b} G(t,s) f(s+\nu-1, y(s+\nu-1)) \\
& \quad + \min_{t \in \left[\frac{b+\nu}{4}, \frac{3(b+\nu)}{4}\right]} \left[ -\frac{t^{\underline{\nu-1}}}{(b+2)\Gamma(\nu-1)} + \frac{t^{\underline{\nu-2}}}{\Gamma(\nu-1)} \right] g(y) \\
& \geq \tilde{\gamma} \max_{t \in [\nu-1, \nu+b]_{\mathbb{N}_{\nu-1}}} \sum_{s=0}^{b} G(t,s) f(s+\nu-1, y(s+\nu-1)) \\
& \quad + \tilde{\gamma} \max_{t \in [\nu-1, \nu+b]_{\mathbb{N}_{\nu-1}}} \left[ -\frac{t^{\underline{\nu-1}}}{(b+2)\Gamma(\nu-1)} + \frac{t^{\underline{\nu-2}}}{\Gamma(\nu-1)} \right] g(y) \\
& \geq \tilde{\gamma} \|Ty\|,
\end{aligned} \tag{7.91}
$$

where $\tilde{\gamma}$ is as defined in (7.86). Since it is obvious that $(Ty)(t) \geq 0$ for all $t$ whenever $y \in \mathcal{K}$, it follows that (7.91) establishes that $T(\mathcal{K}) \subseteq \mathcal{K}$, as desired. □

**Theorem 7.47.** *Suppose that there exists two distinct numbers $r_1$ and $r_2$, with $r_1$, $r_2 > 0$, such that conditions (F1) and (G1) hold at $r_1$ and condition (F2) holds at $r_2$. Finally, assume that each of f and g is nonnegative. Then problem* (7.58) *has a positive solution, whose norm lies between $r_1$ and $r_2$.*

*Proof.* Let $T$ be the operator defined in (7.90). It is clear that $T$ is completely continuous, and Lemma 7.46 establishes that $T(\mathcal{K}) \subseteq \mathcal{K}$. Without loss of generality, suppose that $0 < r_1 < r_2$. Define the set $\Omega_1$ by

$$\Omega_1 := \left\{y \in \mathcal{C}\left([\nu-2,\nu+b]_{\mathbb{N}_{\nu-2}}, \mathbb{R}\right) \,:\, \|y\| < r_1\right\}.$$

Then we have that for $y \in \partial\Omega_1 \cap \mathcal{K}$

$$\begin{aligned} \|Ty\| &\le \max_{t\in[\nu-2,\nu+b]_{\mathbb{N}_{\nu-2}}} \sum_{s=0}^{b} G(t,s)f(s+\nu-1, y(s+\nu-1)) \\ &\quad + \max_{t\in[\nu-2,\nu+b]_{\mathbb{N}_{\nu-2}}} \left\{\left[-\frac{t^{\underline{\nu-1}}}{(b+2)\Gamma(\nu-1)} + \frac{t^{\underline{\nu-2}}}{\Gamma(\nu-1)}\right] g(y)\right\} \\ &\le \frac{r_1\eta}{2}\sum_{s=0}^{b} G(s+\nu-1,s) \\ &\quad + g(y) \max_{t\in[\nu-2,\nu+b]_{\mathbb{N}_{\nu-2}}} \frac{1}{\Gamma(\nu-1)}\left[t^{\underline{\nu-2}} - \frac{t^{\underline{\nu-1}}}{b+2}\right] \\ &\le \frac{r_1}{2} + \frac{r_1}{2} \\ &= \|y\|. \end{aligned} \tag{7.92}$$

So, from (7.92) we conclude that $\|Ty\| \le \|y\|$ for $y \in \mathcal{K} \cap \partial\Omega_1$.

Conversely, define the set $\Omega_2$ by

$$\Omega_2 := \left\{y \in \mathcal{C}\left([\nu-2,\nu+b]_{\mathbb{N}_{\nu-2}}, \mathbb{R}\right) \,:\, \|y\| < r_2\right\}.$$

Then using Lemma 7.42, for $y \in \partial\Omega_2 \cap \mathcal{K}$ we estimate

$$\begin{aligned} &(Ty)\left(\left\lfloor\frac{b+1}{2}\right\rfloor + \nu\right) \\ &\ge \sum_{s=0}^{b} G\left(\left\lfloor\frac{b+1}{2}\right\rfloor + \nu, s\right) f(s+\nu-1, y(s+\nu-1)) \\ &\ge \lambda r_2 \sum_{s=\lceil\frac{\nu+b}{4}-\nu+1\rceil}^{\lfloor\frac{3(\nu+b)}{4}-\nu+1\rfloor} \tilde{\gamma} G\left(\left\lfloor\frac{b+1}{2}\right\rfloor + \nu, s\right) \ge r_2 = \|y\|. \end{aligned} \tag{7.93}$$

Consequently, from (7.93) we conclude that $\|Ty\| \geq \|y\|$ whenever $y \in \partial\Omega_2 \cap \mathcal{K}$. But then by an application of the well-known Krasnosel'skiĭ fixed point theorem we conclude that $T$ has a fixed point, say, $y_0 \in \mathcal{K}$. This map $t \mapsto y_0(t)$ is a positive solution to problem (7.58) since $y_0 \in \mathcal{K}$ satisfies $r_1 < \|y_0\| < r_2$. Thus, the proof is complete. □

We now provide a second result that yields the existence of at least one positive solution. In what follows, we shall assume that $f$ has the special form $f(t, y) \equiv F_1(t)F_2(y)$. Moreover, to facilitate this result, we introduce the following additional conditions on $F_2$ and $g$.

**F3:** The function $F_2$ satisfies $\lim_{y\to 0^+} \frac{F_2(y)}{y} = 0$.

**F4:** The function $F_2$ satisfies $\lim_{y\to\infty} \frac{F_2(y)}{y} = +\infty$.

**G2:** The function $g$ satisfies $\lim_{\|y\|\to 0^+} \frac{g(y)}{\|y\|} = 0$.

*Remark 7.48.* Observe that there are many nontrivial functionals $y \mapsto g(y)$ satisfying condition (G2). For example, the functional defined by $g(y) := [y(\nu+1)]^3$ clearly satisfies (G2).

**Theorem 7.49.** *Suppose that conditions (F3)–(F4) and (G2) hold. Moreover, assume that each of $F_1$, $F_2$, and $g$ is nonnegative. Then problem* (7.58) *has at least one positive solution.*

*Proof.* Because of condition (F3), there exists a number $\alpha_1 > 0$ sufficiently small such that

$$F_2(y) \leq \eta_1 y, \tag{7.94}$$

for each $y \in (0, \alpha_1]$, and where we choose $\eta_1$ sufficiently small so that

$$\eta_1 \sum_{s=0}^{b} G(s + \nu - 1, s)F_1(s) \leq \frac{1}{2} \tag{7.95}$$

holds. Similarly, condition (G2) implies that there exists a number $\alpha_2 > 0$ such that

$$g(y) \leq \eta_2 \|y\| \tag{7.96}$$

whenever $\|y\| \in (0, \alpha_2]$, and where $\eta_2$ is chosen so that

$$\eta_2 \max_{t\in[\nu-2,\nu+b]_{\mathbb{N}_{\nu-2}}} \left\{ -\frac{t^{\underline{\nu-1}}}{(b+2)\Gamma(\nu-1)} + \frac{t^{\underline{\nu-2}}}{\Gamma(\nu-1)} \right\} \leq \eta_2 \leq \frac{1}{2}. \tag{7.97}$$

Now, put $\alpha^* := \min\{\alpha_1, \alpha_2\}$ and define the set $\Omega_1$ by

$$\Omega_1 := \{y \in \mathcal{K} : \|y\| < \alpha^*\}.$$

Then it follows that for all $y \in \mathcal{K} \cap \partial\Omega_1$ inequalities (7.94)–(7.97) imply that

$$\begin{aligned}\|Ty\| &\le \max_{t\in[\nu-2,\nu+b]_{\mathbb{N}_{\nu-2}}} \sum_{s=0}^{b} G(t,s)F_1(s+\nu-1)F_2(y(s+\nu-1)) \\ &\quad + \max_{t\in[\nu-2,\nu+b]_{\mathbb{N}_{\nu-2}}} \left\{\left[\frac{t^{\underline{\nu-1}}}{(b+2)\Gamma(\nu-1)} + \frac{t^{\underline{\nu-2}}}{\Gamma(\nu-1)}\right] g(y)\right\} \\ &\le \eta_1\|y\| \sum_{s=0}^{b} G(s+\nu-1,s)F_1(s+\nu-1) + \eta_2\|y\| \\ &\le \left[\frac{1}{2}+\frac{1}{2}\right]\|y\| \\ &= \|y\|, \end{aligned} \tag{7.98}$$

whence (7.98) implies that $\|Ty\| \le \|y\|$.

On the other hand, condition (F4) implies the existence of a number $\alpha_3 > 0$ such that

$$F_2(y) \ge \eta_3 y \tag{7.99}$$

whenever $y \ge \alpha_3$. Furthermore, we can choose $\eta_3$ sufficiently large such that

$$\eta_3 \sum_{s=\lceil \frac{\nu+b}{4}-\nu+1\rceil}^{\lfloor \frac{3(\nu+b)}{4}-\nu+1\rfloor} \tilde{\gamma} G\left(\left\lfloor \frac{b+1}{2}\right\rfloor + \nu, s\right) F_1(s+\nu-1) \ge 1. \tag{7.100}$$

Put

$$\alpha^{**} := \max\left\{2\alpha^*, \frac{\alpha_3}{\tilde{\gamma}}\right\} \tag{7.101}$$

and observe that for $\|y\| = \alpha^{**}$ we estimate

$$\min_{\frac{b+\nu}{4}\le t\le \frac{3(b+\nu)}{4}} y(t) \ge \tilde{\gamma}\|y\| \ge \alpha_3. \tag{7.102}$$

Now, define the set $\Omega_2$ by

$$\Omega_2 := \{y \in \mathcal{K} : \|y\| < \alpha^{**}\}.$$

Recall from the proof of Lemma 7.38 that

$$-\frac{t^{\underline{\nu-1}}}{(b+2)\Gamma(\nu-1)} + \frac{t^{\underline{\nu-2}}}{\Gamma(\nu-1)} \ge 0, \tag{7.103}$$

for each $t \in [\nu-2, \nu+b]_{\mathbb{N}_{\nu-2}}$. And from this it follows that

$$\left[-\frac{t^{\underline{\nu-1}}}{(b+2)\Gamma(\nu-1)} + \frac{t^{\underline{\nu-2}}}{\Gamma(\nu-1)}\right] g(y) \geq 0, \tag{7.104}$$

for each $t \in [\nu-2, \nu+b]_{\mathbb{N}_{\nu-2}}$. Thus, putting (7.99)–(7.104) together, we find that for $y \in \partial\Omega_2 \cap \mathcal{K}$,

$$\begin{aligned}
&(Ty)\left(\left\lfloor\frac{b+1}{2}\right\rfloor + \nu\right) \\
&= \sum_{s=0}^{b} G\left(\left\lfloor\frac{b+1}{2}\right\rfloor + \nu, s\right) F_1(s+\nu-1)F_2(y(s+\nu-1)) \\
&\qquad + \left[-\frac{t^{\underline{\nu-1}}}{(b+2)\Gamma(\nu-1)} + \frac{t^{\underline{\nu-2}}}{\Gamma(\nu-1)}\right]_{t=\left\lfloor\frac{b+1}{2}\right\rfloor+\nu} g(y) \\
&\geq \sum_{s=\lceil\frac{\nu+b}{4}-\nu+1\rceil}^{\lfloor\frac{3(\nu+b)}{4}-\nu+1\rfloor} G\left(\left\lfloor\frac{b+1}{2}\right\rfloor + \nu, s\right) F_1(s+\nu-1)F_2(y(s+\nu-1)) \\
&\geq \eta_3 \sum_{s=\lceil\frac{\nu+b}{4}-\nu+1\rceil}^{\lfloor\frac{3(\nu+b)}{4}-\nu+1\rfloor} G\left(\left\lfloor\frac{b+1}{2}\right\rfloor + \nu, s\right) F_1(s+\nu-1)y(s+\nu-1) \\
&\geq \eta_3\|y\| \sum_{s=\lceil\frac{\nu+b}{4}-\nu+1\rceil}^{\lfloor\frac{3(\nu+b)}{4}-\nu+1\rfloor} \tilde{\gamma} G\left(\left\lfloor\frac{b+1}{2}\right\rfloor + \nu, s\right) F_1(s+\nu-1) \geq \|y\|.
\end{aligned} \tag{7.105}$$

So, from (7.105) we conclude that $\|Ty\| \geq \|y\|$ whenever $y \in \mathcal{K} \cap \partial\Omega_2$. Consequently, we deduce that $T$ has a fixed point in the set $\left(\mathcal{K} \cap \overline{\Omega_2}\right) \setminus \Omega_1$. Since this fixed point is a positive solution to (7.58), the claim follows. □

*Remark 7.50.* Observe that in case $\nu = 2$, both Theorems 7.47 and 7.49 provide results for the existence of a positive solution to the integer-order nonlocal BVP given by (7.58).

We conclude this section by providing two examples of certain of the theorems presented in this section. We begin with an example illustrating Theorem 7.39 followed by an example illustrating Theorem 7.40.

*Example 7.51.* Suppose that $\nu = \frac{11}{10}$ and $b = 10$. In addition, let us suppose that $f(t, y) := \frac{\sin y}{30+t^2}$ and that $g(y) := \frac{1}{50}\left[y(\nu+1) + y(\nu+2)\right]$. We consider the FBVP

$$
\begin{aligned}
-\Delta^{\frac{11}{10}}y(t) &= \frac{\sin\left(y\left(t+\frac{1}{10}\right)\right)}{30+\left(t+\frac{1}{10}\right)^2} \\
y(\nu-2) &= \frac{1}{50}\left[y(\nu+1)+y(\nu+2)\right] \\
y(\nu+b) &= 0. \qquad (7.106)
\end{aligned}
$$

Now, in this case inequality (7.67) is

$$
2\alpha\prod_{j=1}^{b}\left(\frac{\nu+j}{j}\right)+\beta \le 26.851\alpha+\beta < 1. \qquad (7.107)
$$

But it is not difficult to prove that each of $f$ and $g$ is Lipschitz with Lipschitz constants $\alpha = \frac{1}{30}$ and $\beta = \frac{1}{25}$, respectively. So, for these choices of $\alpha$ and $\beta$, inequality (7.107) is satisfied. Therefore, we deduce from Theorem 7.39 that problem (7.106) has a unique solution.

*Example 7.52.* Suppose that $\nu = \frac{3}{2}$, $b = 10$, and $M = 1000$. Also suppose that $f(t,y) := \frac{1}{10}te^{-\frac{1}{100}t|y|}$ and that $g(y) := \sum_{i=1}^{n} c_i y(t_i)$, where $\{t_i\}_{i=1}^{n} \subseteq [\nu-2,\nu+b]_{\mathbb{N}_{\nu-2}}$ is a strictly increasing sequence satisfying $\nu-2 \le t_1 < t_2 < \cdots < t_n \le \nu+b$ with $t_i \in \mathbb{N}_{\nu-1}$ for each $i$. (Clearly, we must take $n \le b+3$ here.) Thus, in this case problem (7.58) becomes

$$
\begin{aligned}
-\Delta^{\frac{3}{2}}y(t) &= \frac{1}{10}\left(t+\frac{1}{2}\right)e^{-\frac{1}{100}\left(t+\frac{1}{2}\right)\left|y\left(t+\frac{1}{2}\right)\right|} \\
y(\nu-2) &= \sum_{i=1}^{n} c_i y(t_i) \\
y(\nu+b) &= 0. \qquad (7.108)
\end{aligned}
$$

Furthermore, note that in this setting the Banach space $\mathcal{B}$ assumes the form $\mathcal{B} := \left\{y \in \mathbb{R}^{13} : \|y\| \le 1000\right\}$.

We claim that (7.108) has at least one solution. So, to check that the hypotheses of Theorem 7.40 hold, we note that

$$
\frac{M}{\frac{2\Gamma(\nu+b+1)}{\Gamma(\nu+1)\Gamma(b+1)}+1} = \frac{1000}{\frac{2\Gamma\left(\frac{3}{2}+10+1\right)}{\Gamma\left(\frac{5}{2}\right)\Gamma(11)}+1} \approx 11.614.
$$

It is evident that $|f(t,y)| \le \frac{23}{20} < 11.614$ whenever $y \in [-1000, 1000]$. On the other hand, if we require, say, the condition

$$
\sum_{i=1}^{n}|c_i| \le \frac{1}{100}, \qquad (7.109)
$$

then (7.109) implies that for each $y \in \mathcal{C}([\nu-2, \nu+b], \mathbb{R})$ satisfying the condition $\|y\| \leq 1000$ it holds that

$$|g(y)| \leq \sum_{i=1}^{n} |c_i|\, |y(t_i)| \leq 1000 \sum_{i=1}^{n} |c_i| \leq 10 < 11.350$$

so that $g$ satisfies condition (7.73). Thus, given restriction (7.109), we conclude from Theorem 7.40 that (7.108) has at least one solution. In particular, by the conclusion of Theorem 7.40 we deduce that this solution, say $y_0$, satisfies

$$|y_0(t)| \leq 1000, \text{ for } t \in \left[-\frac{1}{2}, \frac{23}{2}\right]_{\mathbb{Z}_{-\frac{1}{2}}}.$$

## 7.6 Discrete Sequential Fractional Boundary Value Problems

In this section we emphasize a different property of the discrete fractional difference and see how it can give rise to a suitably nonlocal problem. In particular, we consider the concept of a so-called sequential fractional boundary value problem. Recall that for fractional differences it does *not* necessarily hold that $\Delta_{a+M-\mu}^{\nu}\Delta_a^{\mu}f(t) = \Delta_a^{\nu+\mu}f(t)$, as was discussed in Chap. 2. Consequently, we may consider a so-called discrete sequential FBVP. In this case, we consider the discrete fractional boundary value problem

$$\begin{aligned} -\Delta^{\mu_1}\Delta^{\mu_2}\Delta^{\mu_3}y(t) &= f\left(t+\mu_1+\mu_2+\mu_3-1, y\left(t+\mu_1+\mu_2+\mu_3-1\right)\right) \\ y(0) &= 0 \\ y(b+2) &= 0, \end{aligned} \tag{7.110}$$

for

$$t \in [2-\mu_1-\mu_2-\mu_3, b+2-\mu_1-\mu_2-\mu_3]_{\mathbb{Z}_{2-\mu_1-\mu_2-\mu_3}},$$

and where throughout we make the assumptions that $\mu_i \in (0,1)$, for each $i = 1, 2, 3$, and that each of $1 < \mu_2 + \mu_3 < 2$ and $1 < \mu_1 + \mu_2 + \mu_3 < 2$ holds. The potential interest in problem (7.110) is that the sequence of fractional difference $\Delta^{\mu_1}\Delta^{\mu_2}\Delta^{\mu_3}y(t)$ is not necessarily equivalent to the non-sequential difference $\Delta^{\mu_1+\mu_2+\mu_3}y(t)$. Consequently, we have in the fractional setting a situation that cannot occur in the integer-order setting since $\Delta^{k_1}\Delta^{k_2}y(t) = \Delta^{k_1+k_2}y(t)$, for each $k_1$, $k_2 \in \mathbb{N}$. Moreover, this dissimilarity is a direct consequence of the implicit nonlocal structure of the fractional difference. We note that the results of this section can be found in Goodrich [99].

We begin by proving a simple proposition. This realization will be important later in this section.

**Proposition 7.53.** *Let* $y : \mathbb{N}_0 \to \mathbb{R}$ *with* $\mu \in (0, 1]$. *Then we find that*

$$\Delta^{\mu-1}y(1-\mu) = y(0).$$

*Proof.* To see that this is true, observe that $\mu - 1 \leq 0$ since $\mu \in (0, 1]$. By definition, then, it follows that

$$\begin{aligned}
\Delta^{\mu-1}y(1-\mu) &= \left[\frac{1}{\Gamma(1-\mu)}\sum_{s=0}^{t+\mu-1}(t-s-1)^{\underline{-\mu}}y(s)\right]_{t=1-\mu} \\
&= \frac{1}{\Gamma(1-\mu)}\sum_{s=0}^{0}(-\mu-s)^{\underline{-\mu}}y(s) \\
&= \frac{1}{\Gamma(1-\mu)}\cdot\Gamma(1-\mu)y(0) \\
&= y(0),
\end{aligned}$$

as claimed. □

We now provide an analysis of problem (7.110). We begin by repeatedly applying the composition rules for fractional differences to derive a representation of a solution to (7.110) as the fixed point of an appropriate operator. In the sequel, the Banach space $\mathcal{B}$ is the set of (continuous) real-valued maps from $[0, b+2]_{\mathbb{N}_0}$ when equipped with the usual maximum norm, $\|\cdot\|$. Moreover, henceforth we also put

$$\tilde{\mu} := \mu_1 + \mu_2 + \mu_3,$$

for notational convenience. Recall that in what follows we assume both that $\mu_1 + \mu_2 \in (1, 2)$ and that $\tilde{\mu} \in (1, 2)$. Finally, we give the following notation, which will also be useful in the sequel.

$$\begin{aligned}
T_1 :=\Big\{(t, s) \in [0, b+2]_{\mathbb{N}_0} \times [2-\tilde{\mu}, b+2-\tilde{\mu}]_{\mathbb{N}_{2-\tilde{\mu}}} :& \\
0 \leq s < t - \tilde{\mu} + 1 \leq b+2&\Big\}
\end{aligned}$$

$$\begin{aligned}
T_2 :=\Big\{(t, s) \in [0, b+2]_{\mathbb{N}_0} \times [2-\tilde{\mu}, b+2-\tilde{\mu}]_{\mathbb{N}_{2-\tilde{\mu}}} :& \\
0 \leq t - \tilde{\mu} + 1 \leq s \leq b+2&\Big\}
\end{aligned}$$

**Theorem 7.54.** *Let the operator* $T : \mathcal{B} \to \mathcal{B}$ *be defined by*

$$(Ty)(t) := \alpha(t)y(1) + \sum_{s=-\tilde{\mu}+2}^{b+2-\tilde{\mu}} G(t,s)f\left(s+\tilde{\mu}-1, y\left(s+\tilde{\mu}-1\right)\right), \tag{7.111}$$

*where* $\alpha : [0, b+2]_{\mathbb{N}_0} \to \mathbb{R}$ *is defined by*

$$\alpha(t) := \frac{(t-2+\mu_2+\mu_3)^{\underline{\mu_2+\mu_3-1}}}{\Gamma(\mu_2+\mu_3)} - \frac{(b+\mu_2+\mu_3)^{\underline{\mu_2+\mu_3-1}}}{(b+\tilde{\mu})^{\underline{\tilde{\mu}-1}}\,\Gamma(\mu_2+\mu_3)}(t+\tilde{\mu}-2)^{\underline{\tilde{\mu}-1}} \tag{7.112}$$

*and* $G : [0, b+2]_{\mathbb{N}_0} \times [-\tilde{\mu}+2, -\tilde{\mu}+b+2]_{\mathbb{N}_{2-\tilde{\mu}}} \to \mathbb{R}$ *is the Green's function for the non-sequential conjugate problem given by*

$$G(t,s) := \begin{cases} \frac{(t+\tilde{\mu}-2)^{\underline{\tilde{\mu}-1}}(b+1-s)^{\underline{\tilde{\mu}-1}}}{(b+\tilde{\mu})^{\underline{\tilde{\mu}-1}}} - (t-s-1)^{\underline{\tilde{\mu}-1}}, & (t,s) \in T_1 \\ \frac{(t+\tilde{\mu}-2)^{\underline{\tilde{\mu}-1}}(b+1-s)^{\underline{\tilde{\mu}-1}}}{(b+\tilde{\mu})^{\underline{\tilde{\mu}-1}}}, & (t,s) \in T_2 \end{cases}. \tag{7.113}$$

*Then whenever* $y \in \mathcal{B}$ *is a fixed point of* $T$, *it follows that* $y$ *is a solution of problem* (7.110).

*Proof.* To begin the proof notice that by the operational properties deduced in Chap. 2 we may write

$$\begin{aligned}
&\Delta^{\mu_1}\Delta^{\mu_2}\Delta^{\mu_3}y(t)\\
&= \Delta^{\mu_1}\left[\Delta^{\mu_2+\mu_3}y(t) - \frac{y(0)}{\Gamma(-\mu_2)}(t-1+\mu_3)^{\underline{-\mu_2-1}}\right]\\
&= \Delta^{\mu_1}\left[\Delta^{\mu_2+\mu_3}y(t)\right] - \frac{y(0)}{\Gamma(-\mu_2)}\Delta^{\mu_1}\left[(t-1+\mu_3)^{\underline{-\mu_2-1}}\right]\\
&= \Delta^{\tilde{\mu}}y(t) - \frac{y(0)}{\Gamma(-\mu_2)}\cdot\frac{\Gamma(-\mu_2)}{\Gamma(-\mu_2-\mu_1)}(t-1+\mu_3)^{\underline{-\mu_2-\mu_1-1}}]\\
&\quad - \sum_{j=0}^{1}\left[\frac{\Delta^{j-2+\mu_2+\mu_3}y(2-\mu_2-\mu_3)}{\Gamma(-\mu_1-2+j+1)}(t-2+\mu_2+\mu_3)^{\underline{-\mu_1-2+j}}\right]\\
&= \Delta^{\tilde{\mu}}y(t) - \frac{\Delta^{\mu_2+\mu_3-2}y(2-\mu_2-\mu_3)}{\Gamma(-\mu_1-1)}(t-2+\mu_2+\mu_3)^{\underline{-\mu_1-2}}\\
&\quad - \frac{\Delta^{\mu_2+\mu_3-1}y(2-\mu_2-\mu_3)}{\Gamma(-\mu_1)}(t-2+\mu_2+\mu_3)^{\underline{-\mu_1-1}}\\
&\quad - \frac{y(0)}{\Gamma(-\mu_2-\mu_1)}(t-1+\mu_3)^{\underline{-\mu_2-\mu_1-1}}.
\end{aligned} \tag{7.114}$$

Now, the same argument as in Proposition 7.53 shows that

$$\Delta^{\mu_2+\mu_3-2} y\,(2-\mu_2-\mu_3) = y(0). \tag{7.115}$$

On the other hand, note that by the definition of the fractional sum, keeping in mind that $\mu_2+\mu_3-2<0$, we obtain that

$$\begin{aligned}
\Delta^{\mu_2+\mu_3-1} y(t) &= \Delta\Delta^{\mu_2+\mu_3-2} y(t)\\
&= \Delta_t\left[\frac{1}{\Gamma(2-\mu_2-\mu_3)}\sum_{s=0}^{t-2+\mu_2+\mu_3}(t-s-1)^{\underline{1-\mu_2-\mu_3}} y(s)\right]\\
&= \frac{1}{\Gamma(2-\mu_2-\mu_3)}\sum_{s=0}^{t-1+\mu_2+\mu_3}(t-s)^{\underline{1-\mu_2-\mu_3}} y(s)\\
&\quad - \frac{1}{\Gamma(2-\mu_2-\mu_3)}\sum_{s=0}^{t-2+\mu_2+\mu_3}(t-s-1)^{\underline{1-\mu_2-\mu_3}} y(s).
\end{aligned} \tag{7.116}$$

So, from (7.116), we obtain

$$\begin{aligned}
&\Delta^{\mu_2+\mu_3-1} y\,(2-\mu_2-\mu_3)\\
&= \frac{1}{\Gamma(2-\mu_2-\mu_3)}\sum_{s=0}^{1}(2-\mu_2-\mu_3-s)^{\underline{1-\mu_2-\mu_3}}\, y(s)\\
&\quad - \frac{1}{\Gamma(2-\mu_2-\mu_3)}\sum_{s=0}^{0}(1-\mu_2-\mu_3-s)^{\underline{1-\mu_2-\mu_3}}\, y(s)\\
&= \frac{1}{\Gamma(2-\mu_2-\mu_3)} y(0)\left[(2-\mu_2-\mu_3)^{\underline{1-\mu_2-\mu_3}} - (1-\mu_2-\mu_3)^{\underline{1-\mu_2-\mu_3}}\right]\\
&\quad + \frac{1}{\Gamma(2-\mu_2-\mu_3)}(1-\mu_2-\mu_3)^{\underline{1-\mu_2-\mu_3}}\, y(1).
\end{aligned} \tag{7.117}$$

Putting (7.115) and (7.117) into (7.114), we deduce that

$$\begin{aligned}
&\Delta^{\mu_1}\Delta^{\mu_2}\Delta^{\mu_3} y(t)\\
&= \Delta^{\tilde{\mu}} y(t) - \frac{[y(1)+(1-\mu_2-\mu_3)\,y(0)]}{\Gamma(-\mu_1)}(t-2+\mu_2+\mu_3)^{\underline{-\mu_1-1}}\\
&\quad - \frac{y(0)}{\Gamma(-\mu_1-1)}(t-2+\mu_2+\mu_3)^{\underline{-\mu_1-2}}\\
&\quad - \frac{y(0)}{\Gamma(-\mu_2-\mu_1)}(t-1+\mu_3)^{\underline{-\mu_2-\mu_1-1}},
\end{aligned} \tag{7.118}$$

where we have made some routine simplifications. Now, by the boundary conditions in (7.110) we find that (7.118) reduces to

$$\Delta^{\mu_1}\Delta^{\mu_2}\Delta^{\mu_3}y(t) = \Delta^{\tilde{\mu}}y(t) - \frac{(t-2+\mu_2+\mu_3)^{\underline{-\mu_1-1}}}{\Gamma(-\mu_1)}y(1). \tag{7.119}$$

Inverting the problem (7.110), we find by means of (7.119) that

$$\begin{aligned} y(t) = -\Delta^{-\tilde{\mu}}&\left[-\frac{(t-2+\mu_2+\mu_3)^{\underline{-\mu_1-1}}}{\Gamma(-\mu_1)}y(1)\right] \\ &- \Delta^{-\tilde{\mu}}f(t+\tilde{\mu}-1, y(t+\tilde{\mu}-1)) \\ &\quad + c_1(t+\tilde{\mu}-2)^{\underline{\tilde{\mu}-1}} + c_2(t+\tilde{\mu}-2)^{\underline{\tilde{\mu}-2}} \end{aligned} \tag{7.120}$$

holds.

Now, continuing from (7.120), it is clear that the boundary condition $y(0) = 0$ implies that $c_2 = 0$. On the other hand, the boundary condition $y(b+2) = 0$, implies that

$$\begin{aligned} 0 = c_1(b+\tilde{\mu})^{\underline{\tilde{\mu}-1}} &+ \frac{y(1)}{\Gamma(\mu_2+\mu_3)}(b+\mu_2+\mu_3)^{\underline{\mu_2+\mu_3-1}} \\ &- \frac{1}{\Gamma(\tilde{\mu})}\sum_{s=-\tilde{\mu}+2}^{b+2-\tilde{\mu}}(b+1-s)^{\underline{\tilde{\mu}-1}}f(s+\tilde{\mu}-1, y(s+\tilde{\mu}-1)) \end{aligned} \tag{7.121}$$

From (7.121), we deduce that

$$\begin{aligned} c_1 = &-\frac{(b+\mu_2+\mu_3)^{\underline{\mu_2+\mu_3-1}}}{(b+\tilde{\mu})^{\underline{\tilde{\mu}-1}}\Gamma(\mu_2+\mu_3)}y(1) \\ &+ \frac{1}{\Gamma(\tilde{\mu})}\sum_{s=-\tilde{\mu}+2}^{b+2-\tilde{\mu}}\frac{(b+1-s)^{\underline{\tilde{\mu}-1}}}{(b+\tilde{\mu})^{\underline{\tilde{\mu}-1}}}f(s+\tilde{\mu}-1, y(s+\tilde{\mu}-1)). \end{aligned}$$

At last, substituting the values of $c_1$ and $c_2$ into (7.120), we conclude that

$$y(t) = \alpha(t)y(1) + \sum_{s=-\tilde{\mu}+2}^{b+2-\tilde{\mu}} G(t,s)f(s+\tilde{\mu}-1, y(s+\tilde{\mu}-1)), \tag{7.122}$$

where $\alpha$ is as defined in (7.112) above and the map $(t,s) \mapsto G(t,s)$ is as defined in (7.113) above. Now, if $(Ty)(t)$ is defined by the right-hand side of (7.122), i.e., we define $T : \mathcal{B} \to \mathcal{B}$ as in the statement of this theorem, then it is clear that $T$ satisfies the boundary value problem (7.110). And this completes the proof. □

We next state an easy proposition regarding the Green's function, $(t,s) \mapsto G(t,s)$, appearing in the operator $T$, as defined above.

**Proposition 7.55.** *The Green's function $(t,s) \mapsto G(t,s)$ given in Theorem 7.54 satisfies:*

(i) $G(t,s) \geq 0$ *for each* $(t,s) \in [0,b+2]_{\mathbb{N}_0} \times [2-\tilde{\mu}, b+2-\tilde{\mu}]_{\mathbb{N}_{2-\tilde{\mu}}}$;
(ii) $\max_{t\in[0,b+2]_{\mathbb{N}_0}} G(t,s) = G(s+\tilde{\mu}-1,s)$ *for each* $s \in [2-\tilde{\mu}, b+2-\tilde{\mu}]_{\mathbb{N}_{2-\tilde{\mu}}}$; *and*
(iii) *there exists a number* $\gamma \in (0,1)$ *such that*

$$\min_{[\frac{b}{4},\frac{3b}{4}]_{\mathbb{N}_0}} G(t,s) \geq \gamma \max_{t\in[0,b+2]_{\mathbb{N}_0}} G(t,s) = \gamma G(s+\tilde{\mu}-1,s),$$

*for* $s \in [2-\tilde{\mu}, b+2-\tilde{\mu}]_{\mathbb{N}_{2-\tilde{\mu}}}$.

*Proof.* Omitted—see [99] for details. □

We next require a preliminary lemma regarding the behavior of $\alpha$ appearing in (7.112) above.

**Lemma 7.56.** *Let $\alpha$ be defined as in* (7.112). *Then* $\alpha(0) = \alpha(b+2) = 0$. *Moreover,* $\|\alpha\| \in (0,1)$.

*Proof.* That $\alpha(0) = \alpha(b+2) = 0$ is obvious. On the other hand, to show that $0 < \|\alpha\| < 1$, we argue as follows.

We show first that $\alpha(t) > 0$, for all $t \in [1,b+1]_{\mathbb{N}}$. To this end, let us first note that

$$\begin{aligned}
\alpha(t) \\
&= \frac{(t-2+\mu_2+\mu_3)^{\underline{\mu_2+\mu_3-1}}}{\Gamma(\mu_2+\mu_3)} - \frac{(b+\mu_2+\mu_3)^{\underline{\mu_2+\mu_3-1}}}{(b+\tilde{\mu})^{\underline{\tilde{\mu}-1}}\Gamma(\mu_2+\mu_3)}(t+\tilde{\mu}-2)^{\underline{\tilde{\mu}-1}} \\
&= \frac{\Gamma(t+\mu_2+\mu_3-1)}{\Gamma(t)\Gamma(\mu_2+\mu_3)} - \frac{\Gamma(b+\mu_2+\mu_3+1)\Gamma(t+\tilde{\mu}-1)}{\Gamma(b+\tilde{\mu}+1)\Gamma(\mu_2+\mu_3)\Gamma(t)} \\
&= \frac{\Gamma(t+\mu_2+\mu_3-1)\Gamma(b+\tilde{\mu}+1) - \Gamma(t+\tilde{\mu}-1)\Gamma(b+\mu_2+\mu_3+1)}{\Gamma(t)\Gamma(\mu_2+\mu_3)\Gamma(b+\tilde{\mu}+1)}.
\end{aligned} \tag{7.123}$$

Therefore, $\alpha(t) > 0$, for each $t \in [1,b+1]_{\mathbb{N}}$, if and only if

$$\Gamma(t+\mu_2+\mu_3-1)\Gamma(b+\tilde{\mu}+1) > \Gamma(t+\tilde{\mu}-1)\Gamma(b+\mu_2+\mu_3+1) \tag{7.124}$$

for each $t \in [1,b+1]_{\mathbb{N}}$. Now, (7.124) is equivalent to

$$\frac{\Gamma(t+\mu_2+\mu_3-1)\Gamma(b+\tilde{\mu}+1)}{\Gamma(t+\tilde{\mu}-1)\Gamma(b+\mu_2+\mu_3+1)} > 1.$$

But since

$$\frac{\Gamma(t+\mu_2+\mu_3-1)\,\Gamma(b+\tilde{\mu}+1)}{\Gamma(t+\tilde{\mu}-1)\,\Gamma(b+\mu_2+\mu_3+1)}=\frac{(b+\tilde{\mu})\cdots(t+\tilde{\mu}-1)}{(b+\mu_2+\mu_3)\cdots(t+\mu_2+\mu_3-1)} \tag{7.125}$$

and the right-hand side of (7.125) is clearly greater than unity, it follows that (7.124) holds, and so, we conclude from (7.123)–(7.125) that $\alpha(t) > 0$, for $t \in [1, b+1]_{\mathbb{N}}$, as claimed.

On the other hand, to argue that $\alpha(t) < 1$, for $t \in [0, b+2]_{\mathbb{N}_0}$, we begin by recasting $\alpha(t)$ in a different form. In particular, define $\mu_0 \in (1,2)$ by

$$\mu_0 := \mu_2 + \mu_3. \tag{7.126}$$

Then it follows that

$$\tilde{\mu} = \mu_0 + \mu_1. \tag{7.127}$$

Therefore, putting (7.126)–(7.127) into the definition of $\alpha$ provided in (7.112) we conclude that

$$\alpha(t) = \frac{(t-2+\mu_0)^{\underline{\mu_0-1}}}{\Gamma(\mu_0)} - \frac{(b+\mu_0)^{\underline{\mu_0-1}}(t+\mu_0+\mu_1-2)^{\underline{\mu_0+\mu_1-1}}}{(b+\mu_0+\mu_1)^{\underline{\mu_0+\mu_1-1}}\,\Gamma(\mu_0)}. \tag{7.128}$$

Now, consider the map

$$t \mapsto \frac{(t+\mu_0+\mu_1-2)^{\underline{\mu_0+\mu_1-1}}}{(b+\mu_0+\mu_1)^{\underline{\mu_0+\mu_1-1}}} \tag{7.129}$$

appearing in the second addend on the right-hand side of (7.128). Since

$$\begin{aligned}\frac{(t+\mu_0+\mu_1-2)^{\underline{\mu_0+\mu_1-1}}}{(b+\mu_0+\mu_1)^{\underline{\mu_0+\mu_1-1}}}\\ =\frac{(b+1)\cdots(t+1)(t)}{(b+\mu_0+\mu_1)\cdots(t+\mu_0+\mu_1)\,(t+\mu_0+\mu_1-1)},\end{aligned} \tag{7.130}$$

we see from (7.130) that for each fixed but arbitrary $b$, $t$, and $\mu_0$, the map defined in (7.129) decreases as $\mu_1$ increases. Consequently, for fixed but arbitrary $b$, $t$, and $\mu_0$ we conclude that

$$\begin{aligned}\alpha(t) &< \frac{(t-2+\mu_0)^{\underline{\mu_0-1}}}{\Gamma(\mu_0)} - \left[\frac{(b+\mu_0)^{\underline{\mu_0-1}}(t+\mu_0+\mu_1-2)^{\underline{\mu_0+\mu_1-1}}}{(b+\mu_0+\mu_1)^{\underline{\mu_0+\mu_1-1}}\,\Gamma(\mu_0)}\right]_{\mu_1=1}\\ &= \frac{(t-2+\mu_0)^{\underline{\mu_0-1}}}{\Gamma(\mu_0)} - \frac{(b+\mu_0)^{\underline{\mu_0-1}}(t+\mu_0-1)^{\underline{\mu_0}}}{(b+\mu_0+1)^{\underline{\mu_0}}\,\Gamma(\mu_0)}\end{aligned}$$

$$= \frac{(t-2+\mu_0)^{\underline{\mu_0-1}}}{\Gamma(\mu_0)} - \frac{\Gamma(b+\mu_0+1)\,\Gamma(t+\mu_0)\,\Gamma(b+2)}{\Gamma(b+2)\Gamma(t)\Gamma(\mu_0)\,\Gamma(b+\mu_0+2)}$$

$$= \frac{(t-2+\mu_0)^{\underline{\mu_0-1}}}{\Gamma(\mu_0)} - \frac{\Gamma(t+\mu_0)}{(b+\mu_0+1)\,\Gamma(t)\Gamma(\mu_0)}. \tag{7.131}$$

Now, from (7.131), we see that $\alpha(t) < 1$ if and only if

$$\frac{\Gamma(t+\mu_0-1)}{\Gamma(\mu_0)\,\Gamma(t)} - \frac{\Gamma(t+\mu_0)}{(b+\mu_0+1)\,\Gamma(t)\Gamma(\mu_0)} \leq 1, \tag{7.132}$$

which is itself equivalent to

$$\frac{(b+\mu_0+1)\,\Gamma(t+\mu_0-1)\,\Gamma(t)\Gamma(\mu_0)}{\Gamma(\mu_0)\,\Gamma(t)\,[(b+\mu_0+1)\,\Gamma(\mu_0)\,\Gamma(t)+\Gamma(t+\mu_0)]} \leq 1. \tag{7.133}$$

Inequality (7.133) is equivalent to

$$\frac{(b+\mu_0+1)\,\Gamma(t+\mu_0-1)}{(b+\mu_0+1)\,\Gamma(\mu_0)\,\Gamma(t)+\Gamma(t+\mu_0)} \leq 1. \tag{7.134}$$

We claim that (7.134) holds for each triple $(b,t,\mu_0) \in \mathbb{N} \times [1,b+1]_{\mathbb{N}_0} \times (1,2)$.

To prove this latter claim, we rewrite left-hand side of inequality (7.134) in the following way:

$$\frac{(b+\mu_0+1)\,\Gamma(t+\mu_0-1)}{(b+\mu_0+1)\,\Gamma(\mu_0)\,\Gamma(t)+\Gamma(t+\mu_0)} = \frac{\Gamma(t+\mu_0-1)}{\Gamma(\mu_0)\,\Gamma(t)+\frac{\Gamma(t+\mu_0)}{b+\mu_0+1}}$$

$$= \frac{1}{\frac{\Gamma(\mu_0)\Gamma(t)}{\Gamma(t+\mu_0-1)} + \frac{t+\mu_0-1}{b+\mu_0+1}}.$$

Then inequality (7.134) is equivalent to

$$\frac{\Gamma(\mu_0)\,\Gamma(t)}{\Gamma(t+\mu_0-1)} + \frac{t+\mu_0-1}{b+\mu_0+1} \geq 1. \tag{7.135}$$

Now, each of the addends on the left-hand side of (7.135) is nonnegative. In addition, we observe that

$$\frac{\Gamma(\mu_0)\,\Gamma(t)}{\Gamma(t+\mu_0-1)} \geq 1, \tag{7.136}$$

for each admissible $t$ and $\mu_0$ since $t > t+\mu_0-1$, noting that in the case where $\mu_0 = 1$ we get equality in (7.136). But then (7.136) implies (7.135), which in turn implies that (7.132) holds.

In summary, for each admissible triple $(b, t, \mu_0)$, we conclude that $\alpha(t) < 1$. Moreover, based on the discussion regarding $\mu_1$ given in (7.129)–(7.130), we have actually shown something stronger—namely, that for each fixed but arbitrary $b$, $t$, and $\mu_0$, it holds that

$$\sup_{\mu_1 \in (0,1)} \alpha(t; b, \mu_0) < 1. \tag{7.137}$$

Thus, (7.137) implies that $\alpha(t) < 1$, for each fixed but arbitrary 4-tuple $(b, t, \mu_0, \mu_1) \in \mathbb{N} \times [1, b+2]_{\mathbb{N}} \times (1, 2) \times (0, 1)$. Since we earlier showed that $\alpha(t) > 0$ whenever $t \neq 0, b+2$, we conclude that

$$\|\alpha\| < 1,$$

as desired. And this completes the proof. □

*Remark 7.57.* As we mentioned in the introduction to this section, note that Theorem 7.54 shows that problem (7.110) is not necessarily the same as the conjugate problem studied in [31]. In fact, there is a *de facto* nonlocal nature to problem (7.110) as evidenced by the explicit appearance of $y(1)$ in the operator $T$, as defined above. As remarked above, this is an interesting complication that cannot occur in the integer-order setting.

As an application of the preceding analysis, we now provide a typical existence theorem for problem (7.110). The basic argument is similar to those presented elsewhere in this book—e.g., Sect. 7.7. However, the appearance of the term $y(1)$ in the operator $T$ does add some interest.

So, let us next provide some standard assumptions on the nonlinearity. For simplicity's sake, we assume that $f(t, y) := a(t)g(y)$; here, it is assumed that $a$ is continuous and not zero identically on $[0, b+2]_{\mathbb{N}_0}$. We also assume (H1) and (H2) below. These assumptions are standard superlinear growth assumptions on $g$ at both 0 and $+\infty$.

**H1:** We find that $\lim_{y \to 0^+} \frac{g(y)}{y} = 0$.

**H2:** We find that $\lim_{y \to \infty} \frac{g(y)}{y} = +\infty$.

We also need to define a suitable cone in which to look for fixed points of $T$. In particular, we consider the cone $\mathcal{K} \subseteq \mathcal{B}$, defined by

$$\mathcal{K} := \left\{ y \in \mathcal{B} : y \geq 0, \min_{t \in \left[\frac{b}{4}, \frac{3b}{4}\right]_{\mathbb{N}}} y(t) \geq \gamma^* \|y\| \right\}, \tag{7.138}$$

where $\gamma^* \in (0, 1)$ is a constant to be determined later. Note that in (7.138) the constant $\gamma^*$ is ***not*** the same as the constant $\gamma$ appearing in part 3 of Proposition 7.55. However, it does satisfy $0 < \gamma^* < 1$, as will be demonstrated in the proof of

Lemma 7.58 below. We first show that the cone $\mathcal{K}$ is invariant under the operator $T$. We then argue that conditions (H1)–(H2) imply, as is well known in the integer-order case (e.g., [77]), that problem (7.110) has at least one positive solution.

**Lemma 7.58.** *Let $T$ be the operator defined in* (7.111) *and $\mathcal{K}$ the cone defined in* (7.138). *Then $T(\mathcal{K}) \subseteq \mathcal{K}$.*

*Proof.* Evidently when $y \in \mathcal{K}$, it follows that $(Ty)(t) \geq 0$, for each $t$. On the other hand, we observe that

$$\begin{aligned}
&\min_{t\in\left[\frac{b}{4},\frac{3b}{4}\right]_{\mathbb{N}}} (Ty)(t) \\
&\geq \gamma_0 y(1)\|\alpha\| + \gamma \sum_{s=-\tilde{\mu}+2}^{b+2-\tilde{\mu}} G(s+\tilde{\mu}-1,s) f(s+\tilde{\mu}-1, y(s+\tilde{\mu}-1)) \\
&\geq \gamma^* \left[ y(1)\|\alpha\| + \sum_{s=-\tilde{\mu}+2}^{b+2-\tilde{\mu}} G(s+\tilde{\mu}-1,s) f(s+\tilde{\mu}-1, y(s+\tilde{\mu}-1)) \right] \\
&\geq \gamma^* \|Ty\|,
\end{aligned} \tag{7.139}$$

where the number $\gamma$ appearing in (7.139) is the same number $\gamma$ as in part 3 of Proposition 7.55. Furthermore, the number $\gamma_0 > 0$ appearing in (7.139) is defined by

$$\gamma_0 := \frac{\min_{t\in\left[\frac{b}{4},\frac{3b}{4}\right]_{\mathbb{N}}} \alpha(t)}{\|\alpha\|}.$$

We may then define $\gamma^*$ by

$$\gamma^* := \min\{\gamma_0, \gamma\},$$

where $0 < \gamma^* < 1$. Thus, whenever $y \in \mathcal{K}$, it follows that $Ty \in \mathcal{K}$, as desired. And this completes the proof. □

**Theorem 7.59.** *Assume that $f$ satisfies conditions (H1)–(H2). Then problem* (7.110) *has at least one positive solution.*

*Proof.* First of all, note that $T$ is trivially completely continuous in this setting. Second of all, recall from Lemma 7.56 that $\alpha(t) < 1$, for all $t \in [0, b+2]_{\mathbb{N}_0}$. Therefore, we may select $\varepsilon > 0$ so that $\alpha(t) < \varepsilon < 1$ holds for all admissible $t$. Given this $\varepsilon$, we may, by way of condition (H1), select $\eta_1 > 0$ sufficiently small so that both

$$g(y) \leq \eta_1 y \tag{7.140}$$

and

$$\eta_1 \sum_{s=-\tilde{\mu}+2}^{b+2-\tilde{\mu}} G\left(s+\tilde{\mu}-1, s\right) a(s) \leq 1-\varepsilon \tag{7.141}$$

hold for all $0 < y < r_1$, where $r_1 := r_1(\eta_1)$. Next put

$$\Omega_1 := \{y \in \mathcal{B} : \|y\| < r_1\}.$$

Let $y \in \partial\Omega_1 \cap \mathcal{K}$ be arbitrary but fixed. Then upon combining (7.140)–(7.141) we estimate

$$\begin{aligned}
\|Ty\| &\leq y(1) \max_{t\in[0,b+2]_{\mathbb{N}_0}} \alpha(t) + \max_{t\in[0,b+2]_{\mathbb{N}_0}} \sum_{s=-\tilde{\mu}+2}^{b+2-\tilde{\mu}} G(t,s)a(s)g\left(y\left(s+\tilde{\mu}-1\right)\right) \\
&< \varepsilon y(1) + \sum_{s=-\tilde{\mu}+2}^{b+2-\tilde{\mu}} G\left(s+\tilde{\mu}-1, s\right) a(s)\eta_1 y(s) \\
&\leq \varepsilon\|y\| + \|y\| \cdot \eta_1 \sum_{s=-\tilde{\mu}+2}^{b+2-\tilde{\mu}} G\left(s+\tilde{\mu}-1, s\right) a(s) \\
&\leq \|y\|,
\end{aligned} \tag{7.142}$$

whence (7.142) implies that $T$ is a cone contraction on $\partial\Omega_1 \cap \mathcal{K}$.

On the other hand, from condition (H2) we may select a number $\eta_2 > 0$ such that both

$$\eta_2 \sum_{s=-\tilde{\mu}+2}^{b+2-\tilde{\mu}} \gamma^* G\left(s+\tilde{\mu}-1, s\right) a(s) > 1$$

and

$$g(y) > \eta_2 y$$

hold whenever $y > r_2 > 0$, for some sufficiently large number $r_2 := r_2(\eta_2) > 0$. Define the number $r_2^* > 0$ by

$$r_2^* := \left\{2r_1, \frac{r_2}{\gamma^*}\right\}$$

and put

$$\Omega_2 := \{y \in \mathcal{B} : \|y\| < r_2^*\}.$$

Recall that for $y \in \mathcal{K}$, we must have $y(1) \geq 0$, and that from Lemma 7.56 we know also that $\alpha(t) \geq 0$, for all $t \in [0, b+2]_{\mathbb{N}_0}$. Then it is not difficult to show (see, for example, a similar argument in [94]) that

$$\|Ty\| \geq \|y\|,$$

whenever $y \in \partial\Omega_2 \cap \mathcal{K}$, so that $T$ is a cone expansion on $\partial\Omega_2 \cap \mathcal{K}$.

In summary, by once again appealing to Krasnosel'skiĭ's fixed point theorem we obtain the existence of a function $y_0 \in \mathcal{K} \cap \left(\overline{\Omega_2} \setminus \Omega_1\right)$ such that $Ty_0 = y_0$, where $y_0$ is a positive solution to problem (7.110). And this completes the proof. □

We now briefly comment on a couple of extensions of the preceding results. In particular, let us consider the following sequential fractional difference

$$\Delta^{\mu_n} \cdots \Delta^{\mu_1} y(t),$$

where $\mu_j \in (0,1)$ for each $j = 1, \ldots, n$, under a couple of different additional assumptions on the collection $\{\mu_j\}_{j=1}^{n}$. For notational simplicity in the sequel, we define

$$\tilde{\mu}_j^+ := \sum_{k=1}^{j} \mu_k$$

and

$$\tilde{\mu}_j^- := \sum_{k=n-j}^{n-1} \mu_k.$$

We continue to use the symbol $\tilde{\mu}$ to denote the sum $\sum_{j=1}^{n} \mu_j$.

**Proposition 7.60.** *Assume that* $0 < \sum_{j=1}^{n-1} \mu_j < 1$ *and* $1 < \sum_{j=1}^{n} \mu_j < 2$. *Then it follows that*

$$\begin{aligned}
&\Delta^{\mu_n} \cdots \Delta^{\mu_1} y(t) \\
&= \Delta^{\tilde{\mu}_n^+} y(t) \\
&\quad - \left[\frac{\left(t-1+\tilde{\mu}_{n-1}^+\right)^{\underline{-\mu_n-1}}}{\Gamma(-\mu_n)} - \sum_{j=1}^{n-2} \frac{\left(t-1+\tilde{\mu}_j^+\right)^{\underline{-\tilde{\mu}_{n-j+1}^- - \mu_n - 1}}}{\Gamma\left(-\tilde{\mu}_{n-j+1}^- - \mu_n\right)}\right] y(0).
\end{aligned}$$

*Proof.* We note first that

$$
\begin{aligned}
&\Delta^{\mu_n} \cdots \Delta^{\mu_3}\left[\Delta^{\mu_2}\Delta^{\mu_1} y(t)\right] \\
&\quad = \Delta^{\mu_n} \cdots \Delta^{\mu_3}\left[\Delta^{\tilde{\mu}_2^+} y(t) - \frac{\Delta^{\mu_1-1} y\left(1-\mu_1\right)}{\Gamma\left(-\mu_2\right)}\left(t-1+\mu_1\right)^{\underline{-\mu_2-1}}\right] \\
&\quad = \Delta^{\mu_n} \cdots \Delta^{\mu_4}\left[\Delta^{\tilde{\mu}_3^+} y(t)\right. \\
&\qquad - \frac{\Delta^{\mu_1+\mu_2-1} y\left(1-\mu_1-\mu_2\right)}{\Gamma\left(-\mu_3\right)}\left(t-1+\mu_1+\mu_2\right)^{\underline{-\mu_3-1}} \\
&\qquad \left. - \frac{\Delta^{\mu_1-1} y\left(1-\mu_1\right)}{\Gamma\left(-\mu_2-\mu_3\right)}\left(t-1+\mu_1\right)^{\underline{-\mu_2-\mu_3-1}}\right].
\end{aligned}
$$

Now, inductively repeating this process results in the following equality:

$$
\begin{aligned}
&\Delta^{\mu_{n-1}} \cdots \Delta^{\mu_1} y(t) \\
&\quad = \Delta^{\tilde{\mu}_{n-1}^+} y(t) - \sum_{j=1}^{n-2}\left[\frac{\Delta^{\tilde{\mu}_j^+-1} y\left(1-\tilde{\mu}_j^+\right)}{\Gamma\left(-\tilde{\mu}_{n-j-1}^-\right)}\left(t-1+\tilde{\mu}_j^+\right)^{\underline{-\tilde{\mu}_{n-j-1}^- -1}}\right].
\end{aligned}
$$

So, it follows that

$$
\begin{aligned}
&\Delta^{\mu_n} \cdots \Delta^{\mu_1} y(t) \\
&\quad = \Delta^{\mu_n}\left\{\Delta^{\tilde{\mu}_{n-1}^+} y(t) - \sum_{j=1}^{n-2}\left[\frac{\Delta^{\tilde{\mu}_j^+-1} y\left(1-\tilde{\mu}_j^+\right)}{\Gamma\left(-\tilde{\mu}_{n-j-1}^-\right)}\left(t-1+\tilde{\mu}_j^+\right)^{\underline{-\tilde{\mu}_{n-j-1}^- -1}}\right]\right\} \\
&\quad = \Delta^{\tilde{\mu}_n^+} y(t) - \frac{\Delta^{-1+\tilde{\mu}_{n-1}^+} y\left(1-\tilde{\mu}_{n-1}^+\right)}{\Gamma\left(-\mu_n\right)}\left(t-1+\tilde{\mu}_{n-1}^+\right)^{\underline{-\mu_n-1}} \\
&\qquad + \sum_{j=1}^{n-2}\left[\frac{\Delta^{\tilde{\mu}_j^+-1} y\left(1-\tilde{\mu}_j^+\right)}{\Gamma\left(-\tilde{\mu}_{n-j-1}^-\right)} \cdot \frac{\Gamma\left(-\tilde{\mu}_{n-j-1}^-\right)}{\Gamma\left(-\tilde{\mu}_{n-j-1}^- - \mu_n\right)}\left(t-1+\tilde{\mu}_j^+\right)^{\underline{-\tilde{\mu}_{n-j-1}^- -\mu_n-1}}\right] \\
&\quad = \Delta^{\tilde{\mu}_n^+} y(t) \\
&\qquad - \left[\frac{\left(t-1+\tilde{\mu}_{n-1}^+\right)^{\underline{-\mu_n-1}}}{\Gamma\left(-\mu_n\right)} - \sum_{j=1}^{n-2}\frac{\left(t-1+\tilde{\mu}_j^+\right)^{\underline{-\tilde{\mu}_{n-j+1}^- -\mu_n-1}}}{\Gamma\left(-\tilde{\mu}_{n-j+1}^- - \mu_n\right)}\right] y(0),
\end{aligned}
$$

as claimed, which completes the proof. □

Our next proposition provides for a more direct generalization of problem (7.110) considered earlier.

**Proposition 7.61.** *Suppose that* $0 < \sum_{j=1}^{n-2} \mu_j < 1$, $1 < \sum_{j=1}^{n-1} \mu_j < 2$, *and* $1 < \sum_{j=1}^{n} \mu_j < 2$. *Then we find that*

$$\begin{aligned}
&\Delta^{\mu_n} \cdots \Delta^{\mu_1} y(t) \\
&= \Delta^{\tilde{\mu}} y(t) - \frac{\left(t-2+\tilde{\mu}_{n-1}^{+}\right)^{\underline{-\mu_n-1}}}{\Gamma\left(-\mu_n\right)} y(1) \\
&\quad - \sum_{j=1}^{n-2}\left[\frac{1}{\Gamma\left(-\tilde{\mu}_{n-j-1}^{-}-\mu_n\right)}\left(t-1+\tilde{\mu}_j^{+}\right)^{\underline{-\tilde{\mu}_{n-j-1}^{-}-\mu_n-1}}\right] y(0) \\
&\quad - \left[\frac{\left(t-2+\tilde{\mu}_{n-1}^{+}\right)^{\underline{-\mu_n-1}}}{\Gamma\left(-\mu_n\right)}\left(1-\tilde{\mu}_{n-1}^{+}\right) - \frac{\left(t-2+\tilde{\mu}_{n-1}^{+}\right)^{\underline{-\mu_n-2}}}{\Gamma\left(-\mu_n-1\right)}\right] y(0).
\end{aligned}$$

*Proof.* We first write

$$\begin{aligned}
&\Delta^{\mu_n} \cdots \Delta^{\mu_1} y(t) \\
&= \Delta^{\mu_n}\left\{\Delta^{\tilde{\mu}_{n-1}^{+}} y(t) - \sum_{j=1}^{n-2}\left[\frac{\Delta^{\tilde{\mu}_j^{+}-1} y\left(1-\tilde{\mu}_j^{+}\right)}{\Gamma\left(-\tilde{\mu}_{n-j-1}^{-}\right)}\left(t-1+\tilde{\mu}_j^{+}\right)^{\underline{-\tilde{\mu}_{n-j-1}^{-}-1}}\right]\right\} \\
&= \Delta^{\mu_n} \Delta^{\tilde{\mu}_{n-1}^{+}} y(t) \\
&\quad - \sum_{j=1}^{n-2}\left[\frac{\Delta^{\tilde{\mu}_j^{+}-1} y\left(1-\tilde{\mu}_j^{+}\right)}{\Gamma\left(-\tilde{\mu}_{n-j-1}^{-}\right)} \Delta^{\mu_n}\left[\left(t-1+\tilde{\mu}_j^{+}\right)^{\underline{-\tilde{\mu}_{n-j-1}^{-}-1}}\right]\right] \\
&= \Delta^{\tilde{\mu}} y(t) - \sum_{k=0}^{1} \frac{\Delta^{j-2+\tilde{\mu}_{n-1}^{+}} y\left(2-\tilde{\mu}_{n-1}^{+}\right)}{\Gamma\left(-\mu_n-1+j\right)}\left(t-2+\tilde{\mu}_{n-1}^{+}\right)^{\underline{-\mu_n-2+j}} \\
&\quad - \sum_{j=1}^{n-2}\left[\frac{\Delta^{\tilde{\mu}_j^{+}-1} y\left(1-\tilde{\mu}_j^{+}\right)}{\Gamma\left(-\tilde{\mu}_{n-j-1}^{-}\right)}\right. \\
&\quad \left.\times \frac{\Gamma\left(-\tilde{\mu}_{n-j-1}^{-}\right)}{\Gamma\left(-\tilde{\mu}_{n-j-1}^{-}-\mu_n\right)}\left(t-1+\tilde{\mu}_j^{+}\right)^{\underline{-\tilde{\mu}_{n-j-1}^{-}-\mu_n-1}}\right].
\end{aligned}$$

Now notice both that

$$\frac{\Delta^{-2+\tilde{\mu}_{n-1}^{+}} y\left(2-\tilde{\mu}_{n-1}^{+}\right)}{\Gamma\left(-\mu_n-1\right)}\left(t-2+\tilde{\mu}_{n-1}^{+}\right)^{\underline{-\mu_n-2}} = \frac{\left(t-2+\tilde{\mu}_{n-1}^{+}\right)^{\underline{-\mu_n-2}}}{\Gamma\left(-\mu_n-1\right)} y(0)$$

and that

$$\frac{\Delta^{-1+\tilde{\mu}_{n-1}^{+}}y\left(2-\tilde{\mu}_{n-1}^{+}\right)}{\Gamma\left(-\mu_n\right)}\left(t-2+\tilde{\mu}_{n-1}^{+}\right)^{\underline{-\mu_n-1}}$$
$$=\frac{\left(t-2+\tilde{\mu}_{n-1}^{+}\right)^{\underline{-\mu_n-1}}}{\Gamma\left(-\mu_n\right)}\left[\left(1-\tilde{\mu}_{n-1}^{+}\right)y(0)+y(1)\right].$$

So, we conclude that

$$\begin{aligned}
&\Delta^{\mu_n}\cdots\Delta^{\mu_1}y(t)\\
&=\Delta^{\tilde{\mu}}y(t)-\frac{\left(t-2+\tilde{\mu}_{n-1}^{+}\right)^{\underline{-\mu_n-1}}}{\Gamma\left(-\mu_n\right)}y(1)\\
&\quad-\sum_{j=1}^{n-2}\left[\frac{1}{\Gamma\left(-\tilde{\mu}_{n-j-1}^{-}-\mu_n\right)}\left(t-1+\tilde{\mu}_j^{+}\right)^{\underline{-\tilde{\mu}_{n-j-1}^{-}-\mu_n-1}}\right]y(0)\\
&\quad-\left[\frac{\left(t-2+\tilde{\mu}_{n-1}^{+}\right)^{\underline{-\mu_n-1}}}{\Gamma\left(-\mu_n\right)}\left(1-\tilde{\mu}_{n-1}^{+}\right)-\frac{\left(t-2+\tilde{\mu}_{n-1}^{+}\right)^{\underline{-\mu_n-2}}}{\Gamma\left(-\mu_n-1\right)}\right]y(0).
\end{aligned}$$

And this completes the proof. □

Propositions 7.60 and 7.61 again demonstrate that the sequential problems are (potentially) different than the non-sequential problems and, in particular, isolate these differences. Furthermore, with Propositions 7.60 and 7.61 in hand, we can write down a number of existence results for sequential discrete FBVPs. But we omit their statements here.

## 7.7 Systems of FBVPs with Nonlinear, Nonlocal Boundary Conditions

In this section we shall demonstrate how we can apply our analysis of nonlocal discrete fractional boundary value problems to systems of such problems. Essentially, other than modifying the Banach space and associated cone in which we work, the analysis is very similar. In particular, we are interested in the system

$$\begin{aligned}
-\Delta^{\nu_1}y_1(t)&=\lambda_1a_1\left(t+\nu_1-1\right)f_1\left(y_1\left(t+\nu_1-1\right),y_2\left(t+\nu_2-1\right)\right)\\
-\Delta^{\nu_2}y_2(t)&=\lambda_2a_2\left(t+\nu_2-1\right)f_2\left(y_1\left(t+\nu_1-1\right),y_2\left(t+\nu_2-1\right)\right),
\end{aligned}\tag{7.143}$$

for $t\in[0,b]_{\mathbb{N}_0}$, subject to the boundary conditions

$$y_1(\nu_1-2)=\psi_1(y_1),\, y_2(\nu_2-2)=\psi_2(y_2)$$
$$y_1(\nu_1+b)=\phi_1(y_1),\, y_2(\nu_2+b)=\phi_2(y_2), \tag{7.144}$$

where $\lambda_i>0$, $a_i:\mathbb{R}\to[0,+\infty)$, $\nu_i\in(1,2]$ for each $1\le i\le 2$, and for each $i$ we have that $\psi_i$, $\phi_i:\mathbb{R}^{b+3}\to\mathbb{R}$ are given functionals. We shall also assume that $f_i:[0,+\infty)\times[0,+\infty)\to[0,+\infty)$ is continuous for each admissible $i$. One point of interest to which we wish to draw the reader's attention is the fact that because it may well occur that $\nu_1\neq\nu_2$, it follows, due to the inherent domain shifting of the operator $\Delta_0^\nu$, that the two functions $y_1$ and $y_2$ appearing in (7.143) may be defined on different domains. Evidently, this cannot occur in the integer-order problem—i.e., when $\nu_1,\nu_2\in\mathbb{N}$. Problem (7.143)–(7.144) was originally studied by Goodrich [94], and the results of this section may be found in that paper.

We now wish to fix our framework for the study of problem (7.143)–(7.144). First of all, we let $\mathcal{B}_i$ represent the Banach space of all maps from $[\nu_i-2,\ldots,\nu_i+b]_{\mathbb{N}_{\nu_i-2}}$ into $\mathbb{R}$ when equipped with the usual maximum norm, $\|\cdot\|$. We shall then put

$$\mathcal{X}:=\mathcal{B}_1\times\mathcal{B}_2.$$

By equipping $\mathcal{X}$ with the norm

$$\|(y_1,y_2)\|:=\|y_1\|+\|y_2\|,$$

it follows that $(\mathcal{X},\|\cdot\|)$ is a Banach space, too—see, for example, [74].

Next we wish to develop a representation for a solution of (7.143)–(7.144) as the fixed point of an appropriate operator on $\mathcal{X}$. To accomplish this we present some adaptations of results from [31] that will be of use here. Because the proofs of these lemmas are straightforward, we omit them.

**Lemma 7.62 ([31]).** *Let $1<\nu\le 2$ and $h:[\nu-1,\nu+b-1]_{\mathbb{N}_{\nu-1}}\to\mathbb{R}$ be given. The unique solution of the FBVP $-\Delta^\nu y(t)=h(t+\nu-1)$, $y(\nu-2)=0=y(\nu+b)$ is given by $y(t)=\sum_{s=0}^{b}G(t,s)h(s+\nu-1)$, where $G:[\nu-2,\nu+b]_{\mathbb{N}_{\nu-2}}\times[0,b]_{\mathbb{N}_0}\to\mathbb{R}$ is defined by*

$$G(t,s):=\begin{cases}\frac{t^{\underline{\nu-1}}(\nu+b-s-1)^{\underline{\nu-1}}}{\Gamma(\nu)(\nu+b)^{\underline{\nu-1}}}-(t-s-1)^{\underline{\nu-1}}, & 0\le s<t-\nu+1\le b\\ \frac{t^{\underline{\nu-1}}(\nu+b-s-1)^{\underline{\nu-1}}}{\Gamma(\nu)(\nu+b)^{\underline{\nu-1}}}, & 0\le t-\nu+1\le s\le b\end{cases}.$$

**Lemma 7.63 ([31]).** *The Green's function $G$ given in Lemma 7.62 satisfies:*

(i) *$G(t,s)\ge 0$ for each $(t,s)\in[\nu-2,\nu+b]_{\mathbb{N}_{\nu-2}}\times[0,b]_{\mathbb{N}_0}$;*
(ii) *$\max_{t\in[\nu-2,\nu+b]_{\mathbb{N}_{\nu-2}}}G(t,s)=G(s+\nu-1,s)$ for each $s\in[0,b]_{\mathbb{N}_0}$; and*
(iii) *there exists a number $\gamma\in(0,1)$ such that*

$$\min_{\frac{b+\nu}{4}\le t\le\frac{3(b+\nu)}{4}}G(t,s)\ge\gamma\max_{t\in[\nu-2,\nu+b]_{\mathbb{N}_{\nu-2}}}G(t,s)=\gamma G(s+\nu-1,s),$$

*for $s\in[0,b]_{\mathbb{N}_0}$.*

Now consider the operator $S : \mathcal{X} \to \mathcal{X}$ defined by

$$S(y_1, y_2)(t_1, t_2) := (S_1(y_1, y_2)(t_1), S_2(y_1, y_2)(t_2)), \tag{7.145}$$

where we define $S_1 : \mathcal{X} \to \mathcal{B}_1$ by

$$\begin{aligned} S_1(y_1, y_2)(t_1) \\ &:= \alpha_1(t_1)\psi_1(y_1) + \beta_1(t_1)\phi_1(y_1) \\ &\quad + \lambda_1 \sum_{s=0}^{b} G_1(t_1, s)\, a_1(s + \nu_1 - 1) f_1(y_1(s + \nu_1 - 1), y_2(s + \nu_2 - 1)) \end{aligned}$$

and $S_2 : \mathcal{X} \to \mathcal{B}_2$ by

$$\begin{aligned} S_2(y_1, y_2)(t_2) \\ &:= \alpha_2(t_2)\psi_2(y_2) + \beta_2(t_2)\phi_2(y_2) \\ &\quad + \lambda_2 \sum_{s=0}^{b} G_2(t_2, s)\, a_2(s + \nu_2 - 1) f_2(y_1(s + \nu_1 - 1), y_2(s + \nu_2 - 1)); \end{aligned}$$

note that, for $j = 1, 2$, we define the maps $\alpha_j, \beta_j : [\nu_j - 2, \nu_j + b]_{\mathbb{Z}_{\nu_j - 2}} \to \mathbb{R}$ by

$$\alpha_j(t) := \frac{1}{\Gamma(\nu_j - 1)}\left[t^{\underline{\nu_j - 2}} - \frac{1}{b+2} t^{\underline{\nu_j - 1}}\right]$$

$$\beta_j(t) := \frac{t^{\underline{\nu_j - 1}}}{(\nu + b)^{\underline{\nu_j - 1}}},$$

which occur in the definitions of $S_1$ and $S_2$ above. Moreover, the map $(t, s) \mapsto G_j(t, s)$ is precisely the map $(t, s) \mapsto G(t, s)$ as given in Lemma 7.62 with $\nu$ replaced by $\nu_j$. We claim that whenever $(y_1, y_2) \in \mathcal{X}$ is a fixed point of the operator $S$, it follows that the pair of functions $y_1$ and $y_2$ is a solution to problem (7.143)–(7.144).

**Theorem 7.64.** *Let $f_j : \mathbb{R}^2 \to [0, +\infty)$ and*

$$\psi_j, \phi_j \in C\left([\nu_j - 2, \nu_j + b]_{\mathbb{N}_{\nu_j - 2}}, \mathbb{R}\right)$$

*be given, for $j = 1, 2$. If $(y_1, y_2) \in \mathcal{X}$ is a fixed point of $S$, then the pair of functions $y_1$ and $y_2$ is a solution to problem* (7.143)–(7.144).

*Proof.* Omitted—see [94]. □

The following lemma and its associated corollary are of particular importance in the sequel. Because the proofs of each of these are straightforward, we omit them.

**Lemma 7.65.** *For each $j = 1, 2$, the function $t_j \mapsto \alpha_j(t_j)$ is decreasing in $t_j$, for $t_j \in [\nu_j - 2, \nu_j + b]_{\mathbb{N}_{\nu_j-2}}$. Also, it holds both that*

$$\min_{t_j \in [\nu_j-2, \nu_j+b]_{\mathbb{N}_{\nu_j-2}}} \alpha_j(t_j) = 0$$

*and*

$$\max_{t_j \in [\nu_j-2, \nu_j+b]_{\mathbb{N}_{\nu_j-2}}} \alpha_j(t_j) = 1.$$

*On the other hand, for each $j = 1, 2$, the function $t_j \mapsto \beta_j(t_j)$ is strictly increasing in $t_j$, for $t_j \in [\nu_j - 2, \nu_j + b]_{\mathbb{N}_{\nu_j-2}}$. In addition, it holds that*

$$\min_{t_j \in [\nu_j-2, \nu_j+b]_{\mathbb{N}_{\nu_j-2}}} \beta_j(t_j) = 0$$

*and that*

$$\max_{t_j \in [\nu_j-2, \nu_j+b]_{\mathbb{N}_{\nu_j-2}}} \beta_j(t_j) = 1.$$

**Corollary 7.66.** *Let $j = 1, 2$ be given. Put $I_j := \left[\frac{b+\nu_j}{4}, \frac{3(b+\nu_j)}{4}\right]$. Then there exist constants $M_{\alpha_j}, M_{\beta_j} \in (0, 1)$ such that*

$$\min_{t_j \in I_j} \alpha_j(t_j) = M_{\alpha_j} \|\alpha_j\|$$

*and*

$$\min_{t_j \in I_j} \beta_j(t_j) = M_{\beta_j} \|\beta_j\|.$$

Let us conclude this section with a remark.

*Remark 7.67.* Observe that unlike in the case of the integer-order problem (i.e., when $\nu_1 = \nu_2 = 2$), in the fractional-order problem we encounter a significant problem with respect to the domains of the various operators insofar as it may occur that $\mathbb{Z}_{\nu_1-2} \neq \mathbb{Z}_{\nu_2-2}$. As has been noted with different problems in previous sections, this complication arises in the discrete fractional calculus due to the domain shifting of the fractional forward difference and sum operators.

We now present the first of two theorems for the existence of at least one positive solution to problem (7.143)–(7.144). Note that for this first existence result we shall not assume that either $\psi_i(y_i)$ or $\phi_i(y_i)$, with $i = 1, 2$, is nonnegative for all $y_i \geq 0$. Rather, we shall make some other assumptions about these functionals.

So, let us now present the conditions that we shall assume henceforth. We note that conditions (F1) and (F2) are essentially the same conditions given by Henderson et al. [121]. Moreover, condition (L1) is essentially the same condition (up to a constant multiple) as given in [121, Theorem 3.1].

**F1:** There exist numbers $f_1^*$ and $f_2^*$, with $f_1^*, f_2^* \in (0, +\infty)$, such that

$$\lim_{y_1+y_2\to 0^+} \frac{f_1(y_1, y_2)}{y_1 + y_2} = f_1^* \text{ and } \lim_{y_1+y_2\to 0^+} \frac{f_2(y_1, y_2)}{y_1 + y_2} = f_2^*.$$

**F2:** There exist numbers $f_1^{**}$ and $f_2^{**}$, with $f_1^{**}, f_2^{**} \in (0, +\infty)$, such that

$$\lim_{y_1+y_2\to +\infty} \frac{f_1(y_1, y_2)}{y_1 + y_2} = f_1^{**} \text{ and } \lim_{y_1+y_2\to +\infty} \frac{f_2(y_1, y_2)}{y_1 + y_2} = f_2^{**}.$$

**G1:** For each $j = 1, 2$, the functionals $\psi_j$ and $\phi_j$ are linear. In particular, we assume both that

$$\psi_j(y_j) = \sum_{i=\nu_j-2}^{\nu_j+b} c^j_{i-\nu_j+2} y_j(i)$$

and that

$$\phi_j(y_j) = \sum_{k=\nu_j-2}^{\nu_j+b} d^j_{k-\nu_j+2} y_j(k),$$

for constants $c^j_{i-\nu_j+2}, d^j_{k-\nu_j+2} \in \mathbb{R}$.

**G2:** For each $j = 1, 2$, we have both that

$$\sum_{i=\nu_j-2}^{\nu_j+b} c^j_{i-\nu_j+2} G_j(i, s) \geq 0$$

and that

$$\sum_{k=\nu_j-2}^{\nu_j+b} d^j_{k-\nu_j+2} G_j(k, s) \geq 0,$$

for each $s \in [0, b]_{\mathbb{N}_0}$, and in addition that

$$\sum_{i=\nu_j-2}^{\nu_j+b} c^j_{i-\nu_j+2} + \sum_{k=\nu_j-2}^{\nu_j+b} d^j_{k-\nu_j+2} \leq \frac{1}{4}.$$

**G3:** We have that each of $\psi_i(\alpha_i)$, $\psi_i(\beta_i)$, $\phi_i(\alpha_i)$, and $\phi_i(\beta_i)$ is nonnegative for each admissible $i$—that is, $i = 1, 2$.

**L1:** The constants $\lambda_1$ and $\lambda_2$ satisfy

$$\Lambda_1 < \lambda_i < \Lambda_2,$$

for each $i$, where

$$\Lambda_1 := \max\left\{\frac{1}{2}\left[\sum_{s=0}^{b}\gamma G_1\left(\left\lfloor\frac{b+1}{2}\right\rfloor + \nu_1, s\right)a_1(s+\nu_1-1)f_1^{**}\right]^{-1},\right.$$
$$\left.\frac{1}{2}\left[\sum_{s=0}^{b}\gamma G_2\left(\left\lfloor\frac{b+1}{2}\right\rfloor + \nu_2, s\right)a_2(s+\nu_2-1)f_2^{**}\right]^{-1}\right\}$$

and

$$\Lambda_2 := \min\left\{\frac{1}{4}\left[\sum_{s=0}^{b}G_1(s+\nu_1-1, s)\,a_1(s+\nu_1-1)f_1^{*}\right]^{-1},\right.$$
$$\left.\frac{1}{4}\left[\sum_{s=0}^{b}G_2(s+\nu_2-1, s)\,a_2(s+\nu_2-1)f_2^{*}\right]^{-1}\right\},$$

where $\gamma \in (0, 1)$ is a constant defined by

$$\gamma := \min\left\{M_{\alpha_1}, M_{\alpha_2}, M_{\beta_1}, M_{\beta_2}, \gamma_1, \gamma_2\right\},$$

where $M_{\alpha_1}$, $M_{\alpha_2}$, $M_{\beta_1}$, and $M_{\beta_2}$ each comes from Corollary 7.66 and $\gamma_1$ and $\gamma_2$ are associated by Lemma 7.63 with $G_1$ and $G_2$, respectively. Recall that these are defined on possibly different time scales.

In what follows we shall also make use of the cone

$$\mathcal{K} := \Bigg\{(y_1, y_2) \in \mathcal{X} : y_1, y_2 \geq 0,$$
$$\min_{(t_1,t_2)\in\left[\frac{b+\nu_1}{4}, \frac{3(b+\nu_1)}{4}\right]\times\left[\frac{b+\nu_2}{4}, \frac{3(b+\nu_2)}{4}\right]}\left[y_1(t_1) + y_2(t_2)\right] \geq \gamma\|(y_1, y_2)\|,$$
$$\psi_j(y_j) \geq 0,\ \phi_j(y_j) \geq 0, \text{ for each } j = 1, 2\Bigg\}, \tag{7.146}$$

where $\gamma$ is defined exactly as in the statement of condition (L1) above. This cone is essentially a modification of the type of cone introduced by Infante and Webb [159]. Clearly, we have that $\mathcal{K} \subseteq \mathcal{X}$. In order to show that $S$ has a fixed point in $\mathcal{K}$, we must first demonstrate that $\mathcal{K}$ is invariant under $S$—that is, $S(\mathcal{K}) \subseteq \mathcal{K}$. This we now show.

**Lemma 7.68.** *Let $S : \mathcal{X} \to \mathcal{X}$ be the operator defined as in* (7.145). *Then $S : \mathcal{K} \to \mathcal{K}$.*

*Proof.* Suppose that $(y_1, y_2) \in \mathcal{K}$. We show first that

$$\min_{(t_1,t_2)\in\left[\frac{b+\nu_1}{4},\frac{3(b+\nu_1)}{4}\right]\times\left[\frac{b+\nu_2}{4},\frac{3(b+\nu_2)}{4}\right]} \left[S_1(y_1,y_2)(t_1)+S_2(y_1,y_2)(t_2)\right]$$
$$\geq \gamma \|S(y_1,y_2)\|,$$

whenever $(y_1, y_2) \in \mathcal{K}$.

So note that

$$\begin{aligned}
&\min_{t_1\in\left[\frac{b+\nu_1}{4},\frac{3(b+\nu_1)}{4}\right]} S_1(y_1,y_2)(t_1)\\
&\geq M_{\alpha_1}\|\alpha_1\|\phi_1(y_1)+M_{\beta_1}\|\beta_1\|\psi_1(y_1)\\
&\quad+\lambda_1\sum_{s=0}^{b}G_1(t_1,s)\,a_1(s+\nu_1-1)f_1(y_1(s+\nu_1-1),y_2(s+\nu_1-1))\\
&\geq \tilde{\gamma}_1 \max_{t_1\in[\nu_1-2,\nu_1+b]}\Bigg[\alpha_1(t_1)\phi_1(y_1)+\beta_1(t_1)\psi_1(y_1)\\
&\quad+\lambda_1\sum_{s=0}^{b}G_1(t_1,s)\,a_1(s+\nu_1-1)f_1(y_1(s+\nu_1-1),y_2(s+\nu_1-1))\Bigg]\\
&= \tilde{\gamma}_1\|S_1(y_1,y_2)\|,
\end{aligned} \tag{7.147}$$

where $\widetilde{\gamma_1} := \min\left\{M_{\alpha_1}, M_{\beta_1}, \gamma_1\right\}$, whence

$$\min_{t_1\in\left[\frac{b+\nu_1}{4},\frac{3(b+\nu_1)}{4}\right]} S_1(y_1,y_2)(t_1) \geq \tilde{\gamma}_1\|S_1(y_1,y_2)\|, \tag{7.148}$$

as desired. In an entirely similar manner to (7.147), we deduce that

$$\min_{t_2\in\left[\frac{b+\nu_2}{4},\frac{3(b+\nu_2)}{4}\right]} S_2(y_1,y_2)(t_2) \geq \tilde{\gamma}_2\|S_2(y_1,y_2)\|, \tag{7.149}$$

where $\tilde{\gamma}_2 := \min\left\{M_{\alpha_2}, M_{\beta_2}, \gamma_2\right\}$.

Now, put $\gamma := \min\{\tilde{\gamma}_1, \tilde{\gamma}_2\}$. Consequently, from (7.148)–(7.149) it follows that

$$\begin{aligned}
&\min_{(t_1,t_2)\in\left[\frac{b+\nu_1}{4},\frac{3(b+\nu_1)}{4}\right]\times\left[\frac{b+\nu_2}{4},\frac{3(b+\nu_2)}{4}\right]} \left[S_1(y_1,y_2)(t_1) + S_2(y_1,y_2)(t_2)\right] \\
&\quad \geq \min_{(t_1,t_2)\in\left[\frac{b+\nu_1}{4},\frac{3(b+\nu_1)}{4}\right]\times\left[\frac{b+\nu_2}{4},\frac{3(b+\nu_2)}{4}\right]} S_1(y_1,y_2)(t_1) \\
&\qquad + \min_{(t_1,t_2)\in\left[\frac{b+\nu_1}{4},\frac{3(b+\nu_1)}{4}\right]\times\left[\frac{b+\nu_2}{4},\frac{3(b+\nu_2)}{4}\right]} S_2(y_1,y_2)(t_2) \\
&\quad \geq \left(\tilde{\gamma}_1\|S_1(y_1,y_2)\| + \tilde{\gamma}_2\|S_2(y_1,y_2)\|\right) \\
&\quad \geq \left(\gamma\|S_1(y_1,y_2)\| + \gamma\|S_2(y_1,y_2)\|\right) \\
&\quad = \gamma\|\left(S_1(y_1,y_2), S_2(y_1,y_2)\right)\| \\
&\quad = \gamma\|S(y_1,y_2)\|.
\end{aligned} \tag{7.150}$$

So, from (7.150) we conclude that whenever $(y_1, y_2) \in \mathcal{X}$, we find that

$$\begin{aligned}
&\min_{(t_1,t_2)\in\left[\frac{b+\nu_1}{4},\frac{3(b+\nu_1)}{4}\right]\times\left[\frac{b+\nu_2}{4},\frac{3(b+\nu_2)}{4}\right]} \left[S_1(y_1,y_2)(t_1) + S_2(y_1,y_2)(t_2)\right] \\
&\quad \geq \gamma\|S(y_1,y_2)\|,
\end{aligned}$$

as desired.

We next show that for each $j = 1,2$ we have $\psi_j\left(S_j(y_1,y_2)\right) \geq 0$ whenever $(y_1,y_2) \in \mathcal{K}$. Indeed, first note that

$$\begin{aligned}
&\psi_j\left(S_j(y_1,y_2)\right) \\
&\quad = \lambda_j \sum_{s=0}^{b} \sum_{i=\nu_j-2}^{\nu_j+b} \Big\{ c^j_{i-\nu_j+2} G_j(i,s) a_j\left(s+\nu_j-1\right) \\
&\quad \times f_j\left(y_1(s+\nu_1-1), y_2(s+\nu_2-1)\right) + \psi_j(\alpha_j)\,\psi_j(y_j) + \psi_j(\beta_j)\,\phi_j(y_j) \Big\}.
\end{aligned} \tag{7.151}$$

But by assumptions (G2) and (G3) together with the nonnegativity of $f_j(y_1,y_2)$ and the fact that $(y_1,y_2) \in \mathcal{K}$, we find from (7.151) that

$$\psi_j\left(S_j(y_1,y_2)\right) \geq 0,$$

for each $j = 1,2$. An entirely dual argument, which we omit, shows that

$$\phi_j\left(S_j(y_1,y_2)\right) \geq 0,$$

too, whenever $(y_1,y_2) \in \mathcal{K}$ and $j = 1,2$.

Finally, it is clear from the definitions of both $S_1$ and $S_2$ that

$$S_1(y_1, y_2)(t_1) \geq 0 \text{ and } S_2(y_1, y_2)(t_2) \geq 0,$$

for each $t_1$ and $t_2$, whenever $(y_1, y_2) \in \mathcal{K}$. Therefore, we conclude that whenever $(y_1, y_2) \in \mathcal{K}$, it follows that $S(y_1, y_2) \in \mathcal{K}$. Thus, $S : \mathcal{K} \to \mathcal{K}$, as desired. And this completes the proof. □

We now prove the first of our two main existence theorems, which we label Theorem 7.69.

**Theorem 7.69.** *Suppose that conditions (F1)–(F2), (G1)–(G3), and (L1) hold. Then problem* (7.143)–(7.144) *has at least one positive solution.*

*Proof.* We have already shown in Lemma 7.68 that $S : \mathcal{K} \to \mathcal{K}$. Furthermore, it is evident that $S$ is completely continuous.

We begin by observing that by condition (L1) there exists a number $\varepsilon > 0$ such that each of

$$\max\left\{\frac{1}{2}\left[\sum_{s=0}^{b}\gamma G_1\left(\left\lfloor\frac{b+1}{2}\right\rfloor + \nu_1, s\right)a_1(s+\nu_1-1)\left(f_1^{**}-\varepsilon\right)\right]^{-1},\right.$$
$$\left.\frac{1}{2}\left[\sum_{s=0}^{b}\gamma G_2\left(\left\lfloor\frac{b+1}{2}\right\rfloor + \nu_2, s\right)a_2(s+\nu_2-1)\left(f_2^{**}-\varepsilon\right)\right]^{-1}\right\} \leq \lambda_1, \lambda_2 \tag{7.152}$$

and

$$\lambda_1, \lambda_2 \leq \min\left\{\frac{1}{4}\left[\sum_{s=0}^{b} G_1(s+\nu_1-1, s)\,a_1(s+\nu_1-1)\left(f_1^{*}+\varepsilon\right)\right]^{-1},\right.$$
$$\left.\frac{1}{4}\left[\sum_{s=0}^{b} G_2(s+\nu_2-1, s)\,a_2(s+\nu_2-1)\left(f_2^{*}+\varepsilon\right)\right]^{-1}\right\}. \tag{7.153}$$

holds. Now, given this number $\varepsilon$, by condition (F1) it follows that there exists some number $r_1^* > 0$ such that

$$f_1(y_1, y_2) \leq \left(f_1^* + \varepsilon\right)(y_1 + y_2), \tag{7.154}$$

whenever $\|(y_1, y_2)\| < r_1$. Similarly, by condition (F2) and for the same number $\varepsilon$, there exists a number $r_1^{**} > 0$ such that

$$f_2(y_1, y_2) \leq \left(f_2^* + \varepsilon\right)(y_1 + y_2), \tag{7.155}$$

whenever $\| (y_1, y_2) \| < r_2$. Then by putting $r_1 := \min \{r_1^*, r_1^{**}\}$, we find that each of (7.154) and (7.155) is true whenever $\| (y_1, y_2) \| < r_1$. This suggests defining the set $\Omega_1 \subseteq \mathcal{X}$ by

$$\Omega_1 := \{(y_1, y_2) \in \mathcal{X} : \| (y_1, y_2) \| < r_1\}, \tag{7.156}$$

which we shall use momentarily.

Now, let $\Omega_1$ be as in (7.156) above. Then for $(y_1, y_2) \in \mathcal{K} \cap \partial\Omega_1$ we find that

$$\begin{aligned}
&\|S_1 (y_1, y_2) \| \\
&= \max_{t_1 \in [\nu_1 - 2, \nu_1 + b]_{\mathbb{N}_{\nu_1 - 2}}} \Bigg| \alpha_1 (t_1) \phi_1 (y_1) + \beta_1 (t_1) \psi_1 (y_1) \\
&\quad + \lambda_1 \sum_{s=0}^{b} G_1 (t_1, s) a_1 (s + \nu_1 - 1) f_1 (y_1 (s + \nu_1 - 1), y_2 (s + \nu_2 - 1)) \Bigg| \\
&\leq r_1 \left[ \sum_{i=\nu_1 - 2}^{\nu_1 + b} c^1_{i - \nu_1 + 2} + \sum_{k=\nu_1 - 2}^{\nu_1 + b} d^1_{k - \nu_1 + 2} \right] \\
&\quad + \lambda_1 \sum_{s=0}^{b} G_1 (s + \nu_1 - 1, s) a_1 (s + \nu_1 - 1) (f_1^* + \varepsilon) \| (y_1, y_2) \| \\
&\leq \| (y_1, y_2) \| \left[ \frac{1}{4} + \lambda_1 \sum_{s=0}^{b} G_1 (s + \nu_1 - 1, s) a_1 (s + \nu_1 - 1) (f_1^* + \varepsilon) \right],
\end{aligned} \tag{7.157}$$

where we use the fact that $S_1 (y_1, y_2)$ is nonnegative whenever $(y_1, y_2) \in \mathcal{K}$. However, by the choice of $\lambda_1$ as given in (7.152)–(7.153), we deduce from (7.157) that

$$\|S_1 (y_1, y_2) \| \leq \frac{1}{2} \| (y_1, y_2) \|. \tag{7.158}$$

We note that by an entirely dual argument we may estimate

$$\|S_2 (y_1, y_2) \| \leq \frac{1}{2} \| (y_1, y_2) \|. \tag{7.159}$$

Thus, by combining estimates (7.152)–(7.159) we deduce that for $(y_1, y_2) \in \mathcal{K} \cap \partial\Omega_1$ we have

$$\|S (y_1, y_2) \| \leq \frac{1}{2} \| (y_1, y_2) \| + \frac{1}{2} \| (y_1, y_2) \| = \| (y_1, y_2) \|.$$

Now, let $\varepsilon > 0$ be the same number selected at the beginning of this proof. Then by means of condition (F2) we can find a number $\tilde{r}_2 > 0$ such that

$$f_1(y_1, y_2) \geq \left(f_1^{**} - \varepsilon\right)(y_1 + y_2) \tag{7.160}$$

and

$$f_2(y_1, y_2) \geq \left(f_2^{**} - \varepsilon\right)(y_1 + y_2), \tag{7.161}$$

whenever $y_1 + y_2 \geq \tilde{r}_2$. Put

$$r_2 := \max\left\{2r_1, \frac{\tilde{r}_2}{\gamma}\right\}, \tag{7.162}$$

where, as before, we take

$$\gamma := \min\{\tilde{\gamma}_1, \tilde{\gamma}_2\}.$$

Moreover, define the set $\Omega_2 \subseteq \mathcal{X}$ by

$$\Omega_2 := \{(y_1, y_2) \in \mathcal{X} : \|(y_1, y_2)\| < r_2\}. \tag{7.163}$$

Note that if $(y_1, y_2) \in \mathcal{K} \cap \partial\Omega_2$, then it follows that

$$\begin{aligned} y_1(t_1) + y_2(t_2) &\geq \min_{(t_1,t_2)\in\left[\frac{b+\nu_1}{4}, \frac{3(b+\nu_1)}{4}\right]\times\left[\frac{b+\nu_2}{4}, \frac{3(b+\nu_2)}{4}\right]} [y_1(t_1) + y_2(t_2)] \\ &\geq \gamma \|(y_1, y_2)\| \\ &\geq \tilde{r}_2. \end{aligned} \tag{7.164}$$

Now, define the numbers $0 < \sigma_1 < \sigma_2$ by

$$\sigma_1 := \max\left\{\left\lceil \frac{\nu_1 + b}{4} - \nu_1 + 1\right\rceil, \left\lceil \frac{\nu_2 + b}{4} - \nu_2 + 1\right\rceil\right\}$$

and

$$\sigma_2 := \min\left\{\left\lfloor \frac{3(\nu_1 + b)}{4} - \nu_1 + 1\right\rfloor, \left\lfloor \frac{3(\nu_2 + b)}{4} - \nu_2 + 1\right\rfloor\right\};$$

we assume in the sequel that $b$ is sufficiently large so that $[\sigma_1, \sigma_2] \cap \mathbb{N}_0 \neq \varnothing$. Then for each $(y_1, y_2) \in \mathcal{K} \cap \partial\Omega_2$ we estimate

$$
\begin{aligned}
& S_1(y_1, y_2)\left(\left\lfloor\frac{b+1}{2}\right\rfloor + \nu_1\right) \\
&= \sum_{i=\nu_1-2}^{\nu_1+b} c^1_{i-\nu_1+2} y_1(k) + \sum_{k=\nu_1-2}^{\nu_1+b} d^1_{k-\nu_1+2} y_1(k) \\
&\quad + \lambda_1 \sum_{s=0}^{b}\Bigg[G_1\left(\left\lfloor\frac{b+1}{2}\right\rfloor + \nu_1, s\right) \\
&\quad \times a_1(s+\nu_1-1) f_1(y_1(s+\nu_1-1), y_2(s+\nu_2-1))\Bigg] \\
&\ge \lambda_1 \sum_{s=\sigma_1}^{\sigma_2} G_1\left(\left\lfloor\frac{b+1}{2}\right\rfloor + \nu_1, s\right) a_1(s+\nu_1-1)\left(f_1^{**}-\epsilon\right)\gamma\left[\|y_1\| + \|y_2\|\right] \\
&\ge \frac{1}{2}\|(y_1, y_2)\|,
\end{aligned}
\tag{7.165}
$$

where to arrive at the first inequality in (7.165) we have used the positivity assumption imposed on each of $\psi_1$ and $\phi_1$ whenever $(y_1, y_2) \in \mathcal{K}$. Thus, we conclude from (7.165) that

$$
\|S_1(y_1, y_2)\| \ge \frac{1}{2}\|(y_1, y_2)\|. \tag{7.166}
$$

In a completely similar way, it can be shown that

$$
\|S_2(y_1, y_2)\| \ge \frac{1}{2}\|(y_1, y_2)\|. \tag{7.167}
$$

Consequently, (7.160)–(7.167) imply that

$$
\|S(y_1, y_2)\| \ge \|(y_1, y_2)\|, \tag{7.168}
$$

whenever $(y_1, y_2) \in \mathcal{K} \cap \partial\Omega_2$.

Finally, notice that (7.160) implies that the operator $S$ is a cone compression on $\mathcal{K} \cap \partial\Omega_1$, whereas (7.168) implies that $S$ is a cone expansion on $\mathcal{K} \cap \partial\Omega_2$. Consequently we conclude that $S$ has a fixed point, say $\left(y_1^*, y_2^*\right) \in \mathcal{K}$. As $\left(y_1^*, y_2^*\right)$ is a positive solution of (7.143)–(7.144), the theorem is proved. □

*Remark 7.70.* Note that in the preceding arguments it is important that each of $\gamma_1$ and $\gamma_2$ (and thus $\gamma$) is a constant. That $\gamma$ is constant here is a reflection of the fact that the Green's function $G$ satisfies a sort of Harnack-like inequality. Interestingly, however, in the continuous fractional setting, this may (see [90]) or may not (see [46]) be true. This is one of the differences one may observe between the discrete and continuous fractional calculus.

*Remark 7.71.* It is clear that Theorem 7.69 could be readily extended to the case of $n$ equations and $2n$ boundary conditions.

We now wish to present an alternative method for deducing the existence of at least one positive solution to problem (7.143)–(7.144). In particular, instead of using the cone given in (7.146), we shall now revert to a more traditional cone, whose use can be found in innumerable papers. An advantage of this approach is that it shall allow us to weaken hypothesis (G1). However, we achieve this increased generality at the expense of having to assume *a priori* the positivity of each of these functionals for all $y \geq 0$. In particular, for the second existence result we make the following hypotheses.

**G4:** For $i = 1, 2$ we have that

$$\lim_{\|y_i\| \to 0^+} \frac{\psi_i(y_i)}{\|y_i\|} = 0.$$

**G5:** For each $i = 1, 2$ we have that

$$\lim_{\|y_i\| \to 0^+} \frac{\phi_i(y_i)}{\|y_i\|} = 0.$$

**G6:** For each $i = 1, 2$ we have that $\psi_i(y_i)$ and $\phi_i(y_i)$ are nonnegative for all $y_i \geq 0$.

**L2:** The constants $\lambda_1$ and $\lambda_2$ satisfy

$$\Lambda_1 < \lambda_i < \Lambda_2,$$

for each $i = 1, 2$, where

$$\Lambda_1 := \max\left\{ \frac{1}{2}\left[\sum_{s=0}^{b} \gamma G_1\left(\left\lfloor \frac{b+1}{2} \right\rfloor + \nu_1, s\right) a_1(s+\nu_1-1) f_1^{**}\right]^{-1}, \right.$$
$$\left. \frac{1}{2}\left[\sum_{s=0}^{b} \gamma G_2\left(\left\lfloor \frac{b+1}{2} \right\rfloor + \nu_2, s\right) a_2(s+\nu_2-1) f_2^{**}\right]^{-1} \right\}$$

and

$$\Lambda_2 := \min\left\{ \frac{1}{3}\left[\sum_{s=0}^{b} G_1(s+\nu_1-1, s)\, a_1(s+\nu_1-1) f_1^{*}\right]^{-1}, \right.$$
$$\left. \frac{1}{3}\left[\sum_{s=0}^{b} G_2(s+\nu_2-1, s)\, a_2(s+\nu_2-1) f_2^{*}\right]^{-1} \right\},$$

where $f_i^*$ and $f_i^{**}$ retain their earlier meaning from conditions (F1)–(F2), for each $i = 1, 2$. Moreover, $\gamma$ is defined just as it was earlier in this section.

*Remark 7.72.* Observe that there do exist nontrivial functionals satisfying conditions (G4) and (G5). For example, consider the functional given by

$$\phi(y) := [y(t_0)]^6,$$

where $t_0$ is some number in the domain of $y$. Then it is clear that

$$0 \leq \lim_{\|y\| \to 0+} \frac{[y(t_0)]^6}{\|y\|} \leq \lim_{\|y\| \to 0+} \frac{[y(t_0)]^6}{y(t_0)} = \lim_{\|y\| \to 0+} [y(t_0)]^5 = 0,$$

from which it follows that $\phi$ satisfies conditions (G4)–(G5); this specifically relies upon the fact that $\phi(y)$ is nonnegative for all $y \geq 0$.

We now present our second existence theorem of this section.

**Theorem 7.73.** *Suppose that conditions (F1)–(F2), (G4)–(G6), and (L2) hold. Then problem* (7.143)–(7.144) *has at least one positive solution.*

*Proof.* Begin by noting that by condition (L2) that there is $\varepsilon > 0$ such that

$$\max\left\{\frac{1}{2}\left[\sum_{s=0}^{b} G_1\left(\left\lfloor\frac{b+1}{2}\right\rfloor + \nu_1, s\right) a_1(s+\nu_1-1)\left(f_1^{**}-\varepsilon\right)\right]^{-1}, \frac{1}{2}\left[\sum_{s=0}^{b} G_2\left(\left\lfloor\frac{b+1}{2}\right\rfloor + \nu_2, s\right) a_2(s+\nu_2-1)\left(f_2^{**}-\varepsilon\right)\right]^{-1}\right\} \leq \lambda_1, \lambda_2 \tag{7.169}$$

and

$$\lambda_1, \lambda_2 \leq \min\left\{\frac{1}{3}\left[\sum_{s=0}^{b} G_1(s+\nu_1-1, s)\, a_1(s+\nu_1-1)\left(f_1^{*}+\varepsilon\right)\right]^{-1}, \frac{1}{3}\left[\sum_{s=0}^{b} G_2(s+\nu_2-1, s)\, a_2(s+\nu_2-1)\left(f_2^{*}+\varepsilon\right)\right]^{-1}\right\}. \tag{7.170}$$

Now, for the number $\varepsilon$ determined by (7.169)–(7.170), it follows from conditions (G4)–(G5) there exists a number $\eta_1 > 0$ such that

$$\phi_1(y_1) \leq \varepsilon \|y_1\|, \tag{7.171}$$

whenever $\|y_1\| \leq \eta_1$, and there exists a number $\eta_2 > 0$ such that

$$\psi_1(y_2) \leq \varepsilon \|y_1\|, \tag{7.172}$$

whenever $\|y_1\| \leq \eta_2$. Put

$$\eta^* := \min\{\eta_1, \eta_2\}.$$

We conclude that whenever $\|(y_1, y_2)\| < \eta^*$, each of (7.171) and (7.172) holds.

Now, for the same $\varepsilon > 0$ given in the first paragraph of this proof, we find that there exists a number $\eta_3$ such that

$$f_1(y_1, y_2) \leq \left(f_1^* + \epsilon\right)(y_1 + y_2), \tag{7.173}$$

whenever $\|(y_1, y_2)\| < \eta_3$. Thus, by putting

$$\eta^{**} := \min\{\eta^*, \eta_3\},$$

we get that (7.171), (7.172), and (7.173) are collectively true.

So, define the set $\Omega_1 \subseteq \mathcal{X}$ by

$$\Omega_1 := \{(y_1, y_2) \in \mathcal{X} \ : \ \|(y_1, y_2)\| < \eta^{**}\}.$$

Then whenever $(y_1, y_2) \in \mathcal{K} \cap \partial\Omega_1$ we have, for any $t_1 \in [\nu_1 - 2, \nu_1 + b]_{\mathbb{N}_{\nu_1 - 2}}$,

$$\begin{aligned}
&S_1(y_1, y_2)(t_1) \\
&\leq \varepsilon\|y_1\| + \varepsilon\|y_1\| \\
&\qquad + \lambda_1 \sum_{s=0}^{b} G_1(t_1, s)\, a_1(s + \nu_1 - 1) f_1(y_1(s + \nu_1 - 1), y_2(s + \nu_2 - 1)) \\
&\leq 2\varepsilon\|y_1\| + \lambda_1 \sum_{s=0}^{b} G_1(s + \nu_1 - 1, s)\, a_1(s + \nu_1 - 1)\left(f_1^* + \varepsilon\right)\|(y_1, y_2)\| \\
&\leq \left[2\varepsilon + \frac{1}{3}\right]\|(y_1, y_2)\|,
\end{aligned} \tag{7.174}$$

where we have used condition (L2) together with (7.170). An entirely dual argument reveals that

$$S_2(y_1, y_2)(t) \leq \left[2\varepsilon + \frac{1}{3}\right]\|(y_1, y_2)\|. \tag{7.175}$$

Therefore, putting (7.174)–(7.175) together we conclude that

$$\|S(y_1, y_2)\| \le \|(y_1, y_2)\|,$$

whenever $(y_1, y_2) \in \mathcal{K} \cap \partial\Omega_1$ and $\varepsilon$ is chosen sufficiently small, which may be assumed without loss of generality.

To complete the proof, we may give an argument essentially identical to the second half of the proof of Theorem 7.69. We omit this, and so, the proof is complete. □

*Remark 7.74.* As with Theorem 7.69, it is clear how the results of this section can be extended to the case in which (7.143) is replaced with $n$ equations and boundary conditions (7.144) are extended to $2n$ boundary conditions in the obvious way. As with the corresponding generalization of Theorem 7.69, however, we omit the details of this extension.

We conclude by providing an explicit numerical example in order to illustrate the application of Theorem 7.69. This is the same example as the one presented in [94].

*Example 7.75.* Consider the boundary value problem

$$\begin{aligned} -\Delta^{1.3} y_1(t) &= \lambda_1 a_1\left(t+\frac{3}{10}\right) f_1\left(y_1\left(t+\frac{3}{10}\right), y_2\left(t+\frac{7}{10}\right)\right) \\ -\Delta^{1.7} y_2(t) &= \lambda_2 a_2\left(t+\frac{7}{10}\right) f_2\left(y_1\left(t+\frac{3}{10}\right), y_2\left(t+\frac{7}{10}\right)\right), \end{aligned} \tag{7.176}$$

subject to the boundary conditions

$$\begin{aligned} y_1\left(\frac{-7}{10}\right) &= \frac{1}{12} y_1\left(\frac{13}{10}\right) - \frac{1}{25} y_1\left(\frac{53}{10}\right) \\ y_1\left(\frac{213}{10}\right) &= \frac{1}{30} y_1\left(\frac{83}{10}\right) - \frac{1}{100} y_1\left(\frac{73}{10}\right) \\ y_2\left(-\frac{3}{10}\right) &= \frac{1}{40} y_2\left(\frac{17}{10}\right) - \frac{1}{150} y_2\left(\frac{77}{10}\right) \\ y_2\left(\frac{217}{10}\right) &= \frac{1}{17} y_2\left(\frac{47}{20}\right) - \frac{1}{30} y_2\left(\frac{107}{20}\right), \end{aligned} \tag{7.177}$$

where we take

$$a_1(t) := e^{t-4},$$

$$a_2(t) := e^{t-4},$$

$$f_1(y_1,y_2) := 5000e^{-y_2}(y_1+y_2) + (y_1+y_2), \text{ and}$$
$$f_2(y_1,y_2) := 7500e^{-y_1}(y_1+y_2) + (y_1+y_2),$$

with $f_1, f_2 : [0,+\infty) \times [0,+\infty) \to [0,+\infty)$. It is clear from the statement of problem (7.176)–(7.177) that we have made the following declarations.

$$\begin{aligned}
\psi_1(y_1) &:= \frac{1}{12}y_1\left(\frac{13}{10}\right) - \frac{1}{25}y_1\left(\frac{53}{10}\right)\\
\phi_1(y_1) &:= \frac{1}{30}y_1\left(\frac{83}{10}\right) - \frac{1}{100}y_1\left(\frac{73}{10}\right)\\
\psi_2(y_2) &:= \frac{1}{40}y_2\left(\frac{17}{10}\right) - \frac{1}{150}y_2\left(\frac{77}{10}\right)\\
\phi_2(y_2) &:= \frac{1}{17}y_2\left(\frac{47}{20}\right) - \frac{1}{30}y_2\left(\frac{107}{20}\right)
\end{aligned} \tag{7.178}$$

Note, in addition, that $y_1$ is defined on the set

$$\left\{-\frac{7}{10}, \frac{3}{10}, \ldots, \frac{213}{10}\right\} \subseteq \mathbb{Z}_{-\frac{7}{10}}^{\frac{213}{10}},$$

whereas $y_2$ is defined on the set

$$\left\{-\frac{3}{10}, \frac{7}{10}, \ldots, \frac{217}{10}\right\} \subseteq \mathbb{Z}_{-\frac{3}{10}}^{\frac{217}{10}},$$

and we note that $\mathbb{Z}_{-\frac{7}{10}}^{\frac{213}{10}} \cap \mathbb{Z}_{-\frac{3}{10}}^{\frac{217}{10}} = \varnothing$, as, toward the beginning of this section, we indicated could occur in the study of problem (7.143)–(7.144). In particular, we have chosen $\nu_1 = \frac{13}{10}$, $\nu_2 = \frac{17}{10}$, and $b = 20$. We shall select $\lambda_1$ and $\lambda_2$ below.

We next check that each of conditions (F1)–(F2), (G1)–(G3), and (L1) holds. It is easy to check that (F1)–(F2) hold. On the other hand, since (7.178) reveals that each of the functionals is linear in $y_1$ and $y_2$, we conclude at once that (G1) holds. On the other hand, to see that conditions (G2)–(G3) hold, observe both that

$$\frac{1}{12} + \frac{1}{25} + \frac{1}{30} + \frac{1}{100} = \frac{1}{6} \leq \frac{1}{4}$$

and that

$$\frac{1}{40} + \frac{1}{150} + \frac{1}{17} + \frac{1}{30} = \frac{421}{3400} \leq \frac{1}{4}.$$

Furthermore, additional calculations show both that

$$\sum_{i=\nu_j-2}^{\nu_j+b} c^j_{i-\nu_j+2} G_j(i,s) \geq 0$$

and that

$$\sum_{k=\nu_j-2}^{\nu_j+b} d^j_{k-\nu_j+2} G_j(k,s) \geq 0,$$

for each $j = 1, 2$. So, we conclude that condition (G2) holds. Finally, one can compute the following estimates.

$$\begin{aligned}
\psi_1(\alpha_1) &\approx 0.012 \\
\psi_1(\beta_1) &\approx 0.012 \\
\psi_2(\alpha_2) &\approx 0.012 \\
\psi_2(\beta_2) &\approx 0.0012 \\
\phi_1(\alpha_1) &\approx 0.00091 \\
\phi_1(\beta_1) &\approx 0.018 \\
\phi_2(\alpha_2) &\approx 0.015 \\
\phi_2(\beta_2) &\approx 0.000099
\end{aligned}$$

Consequently, condition (G3) is satisfied.

Finally, we check condition (L1) to determine the admissible range of the parameters, $\lambda_i$ for $i = 1, 2$. To this end, recall from [31, Theorem 3.2] that the constant $\gamma$ in Lemma 7.63 is

$$\gamma := \min\left\{\frac{1}{\left(\frac{3(b+\nu)}{4}\right)^{\underline{\nu-1}}} \times \left[\left(\frac{3(b+\nu)}{4}\right)^{\underline{\nu-1}} - \frac{\left(\frac{3(b+\nu)}{4}-2\right)^{\underline{\nu-1}}(\nu+b+1)^{\underline{\nu-1}}}{(\nu+b-1)^{\underline{\nu-1}}}\right], \frac{\left(\frac{b+\nu}{4}\right)^{\underline{\nu-1}}}{(b+\nu)^{\underline{\nu-1}}}\right\}. \tag{7.179}$$

Thus, using the definition of $\gamma$ provided by (7.179), we estimate that

$$\begin{aligned}
\Lambda_1 &\approx \max\left\{f_1^{**}\cdot 3.288\times 10^{-7}, f_2^{**}\cdot 1.0322\times 10^{-7}\right\} \\
&= \max\left\{3.288\times 10^{-7}, 1.0322\times 10^{-7}\right\} \\
&= 3.337\times 10^{-7},
\end{aligned}$$

whereas

$$\begin{aligned}\Lambda_2 &\approx \min\left\{f_1^* \cdot 1.871 \times 10^{-9}, f_2^* \cdot 1.363 \times 10^{-9}\right\}\\ &= \min\left\{5001 \cdot 1.871 \times 10^{-9}, 7501 \cdot 1.363 \times 10^{-9}\right\}\\ &= \min\left\{9.357 \times 10^{-6}, 1.022 \times 10^{-5}\right\}\\ &= 9.357 \times 10^{-6}.\end{aligned}$$

So, suppose that

$$\lambda_1, \lambda_2 \in \left[3.337 \times 10^{-7}, 9.357 \times 10^{-6}\right].$$

Then we conclude from Theorem 7.69 that problem (7.176)–(7.177) has at least one positive solution. And this completes the example.

*Remark 7.76.* A similar example could be provided for Theorem 7.73.

*Remark 7.77.* We note that a class of functions satisfying conditions (F1)–(F2) are given by the function $f : \overline{\mathbb{R}_+^n} \to [0, +\infty)$ defined by

$$f(\mathbf{x}) := C_1 e^{-g(\mathbf{x})} \nabla \cdot \mathbf{H}(\mathbf{x}),$$

where $g : \overline{\mathbb{R}_+^n} \to [0, +\infty)$, $C_1 > 0$ is a constant, and $\mathbf{H} : \overline{\mathbb{R}_+^n} \to \overline{\mathbb{R}_+^n}$ is the vector field defined by

$$\mathbf{H}(\mathbf{x}) := \sum_{i=1}^{n} \frac{1}{2} x_i^2 \mathbf{e}_i,$$

where $\mathbf{e}_i$ is the $i$-th vector in the standard ordered basis for $\mathbb{R}^n$; note that by the notation $\overline{\mathbb{R}_+^n}$ we mean the closure of the open positive cone in $\mathbb{R}^n$—i.e., we put

$$\overline{\mathbb{R}_+^n} := \{\mathbf{x} \in \mathbb{R}^n : x_i \geq 0 \text{ for each } 1 \leq i \leq n\} \subseteq \mathbb{R}^n.$$

More trivially, we remark that the collection of functions defined $L(y_1, y_2) = ay_1 + ay_2^3$, for $a > 0$, satisfies (F1)–(F2).

## 7.8 Concluding Remarks

In this chapter we have demonstrated several ways in which nonlocal elements may occur in the discrete fractional calculus. Such elements may arise explicitly, as is the case in the nonlocal BVP setting. On the other hand, the fractional sum and difference themselves contain nonlocal elements, and this considerably complicates the analysis and interpretation of fractional operators.

In closing we wish to draw attention to the fact that due to this implicit nonlocal structure, there are many open questions regarding the interpretation, particularly geometric, of the discrete fractional difference. For instance, as alluded to earlier, only recently has there been any development in our understanding of how the sign of, say $\Delta_0^\nu y(t)$ is related, for various ranges of $\nu$, to the behavior of $y$ itself. Yet in spite of these recent developments, it seems that there is likely much to be discovered in this arc of research.

Moreover, above and beyond pure geometrical implications, the nonlocal structure embedded within $\Delta_0^\nu y$ affects negatively the analysis of boundary and initial value problems insofar as the attendant analysis is much more complicated and there still remain some very basic open questions. For example, as we have seen in this chapter, even the elementary problem of analyzing a particular Green's function associated with a given boundary value operator is very nontrivial, often requiring arguments that while elementary are nonetheless technical. Furthermore, many fundamental areas of study in the integer-order difference calculus presently do not possess satisfactory analogues in the fractional-order setting. Among these is oscillation theory, which has no satisfactory fractional-order analogue. On the one hand, this is rather remarkable in recognition of the centrality of such results in the integer-order theory. On the other hand, however, given the tremendous complexity that the nonlocal structure of $\Delta_0^\nu y$ creates, perhaps it is unsurprising that such gaps exist. As with some of the other questions surrounding the discrete fractional calculus, it is unclear at present whether this gap can ultimately be filled in an at once satisfactory and elegant manner.

All in all, this section has shown a few of the ways in which nonlocalities may arise in the setting of boundary value problems. Moreover, we have seen how the implicit nonlocal structure of the discrete fractional sum and difference complicate in surprising ways their analysis. Finally, we hope that the reader has gained a sense of some of the open and unanswered questions in the discrete fractional calculus, questions whose solutions appear to be at once greatly complicated and substantively enriched by the nonlocal structure of fractional operators. As a concluding point, we wish to note that the interested reader may consult any of the following references for additional information on not only local and nonlocal boundary value problems, but also on other related topics in the discrete calculus that we have touched upon in this and other chapters [1, 2, 5, 6, 8–12, 14–30, 44, 45, 48, 51, 55–61, 68–73, 75, 79–82, 84–86, 97, 98, 100–103, 105–113, 115–118, 120, 122, 126–130, 132, 133, 136, 138, 140–144, 148–151, 154–158, 160–166, 168–172]:

## 7.9 Exercises

**7.1.** Prove the result mentioned in Remark 7.14.

# Solutions to Selected Problems

## Chapter 1

**1.6:** $\Gamma(\frac{5}{2}) = \frac{3}{4}\sqrt{\pi}$

**1.10:** For integers $m$ and $n$ satisfying $m > n \geq 0$, $\binom{n}{m} = 0$

**1.11:** (i) $\binom{t}{t} = 1$ for $t \neq -1, -2, -3, \cdots$

(ii) $\binom{\frac{1}{2}}{\frac{3}{2}} = \frac{4}{3\pi}$

(iii) $\binom{\frac{1}{2}}{\frac{3}{2}} = 0$ (by convention)

(iv) $\binom{\sqrt{2}+2}{\sqrt{2}} = \frac{4+3\sqrt{2}}{2}$

**1.14:** (ii) $e_p(t, 0) = \frac{1}{2}(t+2)^{\underline{2}}, \quad t \in \mathbb{N}_0$

(iv) $e_p(t, 0) = \frac{30(t+1)}{(t+5)(t+6)}, t \in \mathbb{N}_0$

**1.16:** $y(80) = \$487.54$

**1.17:** 5.732 hours

**1.31:** (i) $\frac{n(n+1)}{2}$

(ii) $\frac{n^2(n+1)^2}{4}$

**1.33:** (i) $y(t) = c_1 3^t + c_2 4^t, \quad t \in \mathbb{N}_0$

(iii) $y(t) = c_1 e_1(t, a) \cos_1(t, a) + c_2 e_1(t, a) \sin_1(t, a), \quad t \in \mathbb{N}_a$

(iv)

$$\begin{aligned} y(t) &= c_1 \cos_1(t, 0) + c_2 \sin_1(t, 0) \\ &= a_1 \left(\sqrt{2}\right)^t \cos\left(\frac{\pi t}{4}\right) + a_2 \left(\sqrt{2}\right)^t \sin\left(\frac{\pi t}{4}\right) \end{aligned}$$

(v) $y(t) = c_1(-2)^t + c_2 t(-2)^t, \quad t \in \mathbb{N}_0$

(vi) $y(t) = c_1(-3)^t + c_2 t(-3)^t$


C. Goodrich, A.C. Peterson, *Discrete Fractional Calculus*,
DOI 10.1007/978-3-319-25562-0

**1.34:** (ii) $y(t) = c_1(2\sqrt{2})^t \cos(\frac{\pi}{4}t) + c_2(2\sqrt{2})^t \sin(\frac{\pi}{4}t), \quad t \in \mathbb{Z}$
(iii) $y(t) = c_1 2^t + c_2 t 2^t + c_3(-3)^t, \quad t \in \mathbb{Z}$
**1.37:** $D(t) = -\frac{1}{2} + \frac{3}{2}(3)^t$
**1.38:** $D(t) = 2^t \cos(\frac{2\pi}{3}t) - \frac{1}{\sqrt{3}} 2^t \sin(\frac{2\pi}{3}t), \quad t \in \mathbb{N}_1$
**1.39:** $D(t) = 2^t \cos(\frac{\pi}{3}t) + \frac{1}{\sqrt{3}} 2^t \sin(\frac{\pi}{3}t), \quad t \in \mathbb{N}_1$
**1.40:** (i) $y(t) = c_1 2^t + c_2 3^t + c_3 4^t, \quad t \in \mathbb{N}_0$
(ii) $y(t) = c_1 3^t + c_2 t 3^t + c_3(-2)^t, \quad t \in \mathbb{N}_0$
**1.42:** $y(n) = \frac{2}{3} 2^n + \frac{1}{3}(-1)^n, \quad n \geq 2$
**1.43:** (iii) $u(t) = A2^t \cos\left(\frac{2\pi}{3}t\right) + B2^t \sin\left(\frac{2\pi}{3}t\right), \quad t \in \mathbb{N}_0$
**1.44:** (i) $u(t) = A + B(-9)^t, \quad t \in \mathbb{N}_0$
(ii) $u(t) = A3^t \cos\left(\frac{\pi}{2}t\right) + B3^t \sin\left(\frac{\pi}{2}t\right), \quad t \in \mathbb{N}_0$
(iii) $u(t) = A3^t + B(-2)^t + Ct(-2)^t, \quad t \in \mathbb{N}_0$
**1.45:** (iv) $u(t) = A3^t + Bt(3)^t + C(-3)^t, \quad t \in \mathbb{N}_0$
**1.46:** (i) $y(t) = A + B4^t + \frac{1}{12} t 4^t$
(ii) $y(t) = A3^t + B(-2)^t + \frac{1}{14} 5^t$
**1.47:** (ii) $y(t) = A4^t + B(-1)^t + \frac{1}{20} t(4)^t$
(iii) $y(t) = A(2)^t + B4^t - 3^t$
**1.48:** (i) $\frac{1}{2} 3^t(t^2 - 3t + 3) + C$
(iii) $\binom{t}{2} \cdot \binom{t}{3} - t\binom{t+1}{4} + \binom{t+2}{5} + C$
**1.49:** $\frac{1}{4}(n+1)5^{n+1} - \frac{1}{16}5^{n+1} + \frac{5}{16}$
**1.50:** (i) $\binom{t}{6} \cdot \binom{t}{2} - t\binom{t+1}{7} + \binom{t+2}{8} + C$
(ii) $-\frac{t}{2(t+1)(t+2)} - \frac{1}{2(t+2)} + C$
**1.53:** $y(n) = 2^n - 1$
**1.54:** $y(t) = 50{,}000[(1.04)^t - 1]$
**1.55:** $y(t) = 63{,}000[(1.05)^t - 1], y(27.11) \approx 64.5$
**1.58:** (ii) $y(t) = \frac{1}{4} 4^t \binom{t}{6} + A4^t$
(iv) $y(t) = At + \frac{1}{2}t^3 - \frac{3}{2}t^2$
**1.63:** (i) $y(t) = At! \sum_{s=0}^{t-1} \frac{2^s}{(s+1)!} + Bt!, \quad t \in \mathbb{N}_0$
(ii) $u(t) = \alpha(t-1)! + \beta(t-1)! \sum_{s=1}^{t-1} \frac{2^s}{s!}, \quad t \in \mathbb{N}_1$
**1.64:** (iii) $y(t) = c_1 6^{t-a} + c_2 6^{t-a} \sum_{\tau=a}^{t-1} \frac{1}{\tau+5}, \quad t \in \mathbb{N}_a$
**1.65:** (iii) $y(t) = c_1 4^t + c_2 4^t \sum_{\tau=1}^{t-1} \frac{\tau}{4^\tau}, \quad t \in \mathbb{N}_1$
**1.66:** (i) $y(t) = A(t-2)^{\underline{-1}} + B(t-3)^{\underline{-2}}, \quad t \in \mathbb{N}_2$
(ii) $y(t) = A(t-2)^{\underline{-4}} + B(t-3)^{\underline{-4}} \sum_{k=5}^{t-1} \frac{1}{k-4}, \quad t \in \mathbb{N}_5$
**1.69:**

$$y(t) = \frac{1}{20}\, t^{\underline{5}}, \quad t \in \mathbb{N}_0$$

**1.70:**

$$x(t) = \frac{1}{p^2} - \frac{1}{p^2} \cos_p(t, 0)$$

**1.80:** (iv)

$$u(t) = c_1 \begin{bmatrix} 1 \\ \frac{1}{4} + \frac{1}{2}t - \frac{1}{4}(-1)^t \\ -\frac{1}{4} + \frac{1}{2}t + \frac{1}{4}(-1)^t \end{bmatrix} + c_2 \begin{bmatrix} 0 \\ \frac{1}{2} + \frac{1}{2}(-1)^t \\ \frac{1}{2} - \frac{1}{2}(-1)^t \end{bmatrix}$$
$$+ c_3 \begin{bmatrix} 0 \\ \frac{1}{2} - \frac{1}{2}(-1)^t \\ \frac{1}{2} + \frac{1}{2}(-1)^t \end{bmatrix}$$

**1.81:** (ii) $u(t) = \begin{bmatrix} 2^t + t2^{t-1} \\ \frac{1}{2}3^t + \frac{3}{2} \end{bmatrix}$

**1.86:** (ii) The trivial solution is globally asymptotically stable on $\mathbb{N}_a$.

**1.90:** $\mu_{1,2} = \frac{3}{2} \pm \frac{\sqrt{5}}{2}$

**1.91:** $\mu_1 = \mu_2 = \frac{7}{4}$

## Chapter 2

**2.10:** (i) $y(t) = 4(3)^t - 2(4)^t, \quad t \in \mathbb{N}_0$
(iii) $y(t) = 3(2)^t - 3(4)^t + \frac{5}{2}t(4)^t, \quad t \in \mathbb{N}_0$

**2.11:**

$$u(t) = \cos\left(\frac{\pi}{2}t\right), \quad v(t) = \sin\left(\frac{\pi}{2}t\right), \quad t \in \mathbb{N}_0$$

**2.12:** (i) $y(t) = 2(3)^t + (3)^{t-5}u_5(t), \quad t \in \mathbb{N}_0$
(ii) $y(t) = 4(6)^t + 3\left(-\frac{1}{3} + \frac{1}{5}\right)u_{60}(t), \quad t \in \mathbb{N}_0$

**2.13:**

(ii) $y(t) = 3$
(iii) $y(t) = -1 + 2^t$

**2.14:** (i) $y(t) = 4^t, \quad t \in \mathbb{N}_0$
(ii) $y(t) = \frac{1}{2}(3)^t + \frac{1}{2}(5)^t, t \in \mathbb{N}_0$

**2.17:** (ii) $\Delta_a^{\frac{2}{3}} 1 - \frac{1}{\Gamma(\frac{1}{3})}(t-a)^{\underline{-\frac{2}{3}}}$

**2.21:** (i) $\frac{3}{2}(t-1)^{\underline{-\frac{1}{2}}}$
(iii) $\frac{15\sqrt{\pi}}{8}(t^2 - 7t + 12)$.

**2.24:** (i) $x(t) = \frac{\Gamma(3)}{\Gamma(5.7)}t^{\underline{4.7}} \approx 0.0276t^{\underline{4.7}}, \quad t \in \mathbb{N}_{-0.3}$.

**2.26:**

$$[h_1(\cdot, a) * e_p(\cdot, a)](t) = \frac{1}{p^2}e_p(t, a) - \frac{1}{p}h_1(t, a) - \frac{1}{p^2}, \quad t \in \mathbb{N}_a$$

**2.27:**

$$[e_p(\cdot,a) * e_q(\cdot,a)](t) = \frac{1}{p-q}e_p(t,a) + \frac{1}{q-p}e_q(t,a), \quad t \in \mathbb{N}_a$$

## Chapter 3

**3.16:** (i) $u(t) = c_1 + c_2\left(\frac{1}{4}\right)^t$
**3.17:** (ii) $x(t) = c_1 2^{a-t} + c_2(-4)^{a-t}t$
**3.18:** (i) $x(t) = c_1 \cos\left[\frac{\pi}{2}(t-a)\right] + c_2 \sin\left[\frac{\pi}{2}(t-a)\right], \quad t \in \mathbb{N}_a$
**3.24:** (i) $y(t) = -\frac{1}{4} + \frac{1}{2}t + \frac{1}{4}3^{-t}, \quad t \in \mathbb{N}_0$
(iii) $y(t) = \frac{1}{6}(t-2)^{\overline{3}}, \quad t \in \mathbb{N}_2$
**3.29:** (i)

$$\begin{aligned}\nabla_a^{-2}\mathrm{Cosh}_3(t,a) &= \frac{1}{9}\mathrm{Cosh}_3(t,a) - \frac{1}{9}\\ &= \frac{1}{18}(-2)^{a-t} + \frac{1}{18}4^{a-t} - \frac{1}{9}, \quad t \in \mathbb{N}_a\end{aligned}$$

**3.35:** (i) $x(t) = \pi + \frac{6}{\sqrt{\pi}}t^{0.5}$

## Chapter 4

**4.2:** $D_q f(t) = (1+q)t - 2s$
**4.7:**

(ii) $y(t) = \frac{1}{21}(t-a)(t-2a)(t-4a)$

**4.8:** (i) $y(t) = e_3(t,a) - 2e_2(t,a)$
(iv) $y(t) = e_3(t,a)\cos_{\frac{4}{1+3\mu}}(t,a) - \frac{1}{4}e_3(t,a)\sin_{\frac{4}{1+3\mu}}(t,a)$

## Chapter 5

**5.10:** $y(t) = \frac{1}{26}(t-2)(t-8)(t-26)$

# Bibliography


1. Agrawal, O.P.: Formulation of Euler-Lagrange equations for fractional variational problems. J. Math. Anal. Appl. **272**, 368–379 (2002)
2. Agarwal, R.P., Lakshmikantham, V., Nieto, J.J.: On the concept of solution for fractional differential equations with uncertainty. Nonlinear Anal. **72**, 2859–2862 (2009)
3. Ahrendt, K., Castle, L., Holm, M., Yochman, K.: Laplace transforms for the nabla-difference operator and a fractional variation of parameters formula. Commun. Appl. Anal. **16**, 317–347 (2012)
4. Ahrendt, K., DeWolf, L., Mazurowski, L., Mitchell, K., Rolling, T., Veconi, D.: Boundary value problems for a self-adjoint Caputo nabla fractional difference equation, submitted
5. Almeida, R., Torres, D.F.M.: Calculus of variations with fractional derivatives and fractional integrals. Appl. Math. Lett. **22**, 1816–1820 (2009)
6. Al-Salam, W.A.: Some fractional $q$-integrals and $q$-derivatives. Proc. Edinburgh Math. Soc. (2) **15**, 135–140 (1966)
7. Anastassiou, G.A.: Nabla discrete fractional calculus and nabla inequalities. Math. Comput. Model. **51**, 562–571 (2010)
8. Anastassiou, G.A.: Principles of delta fractional calculus on time scales and inequalities. Math. Comput. Model. **52**, 556–566 (2010)
9. Anastassiou, G.A.: Foundations of nabla fractional calculus on time scales and inequalities. Comput. Math. Appl. **59**, 3750–3762 (2010)
10. Anastassiou, G.A.: Right nabla discrete fractional calculus. Int. J. Differ. Equ. **6**, 91–104 (2011)
11. Anastassiou, G.A.: $q$-fractional inequalities. Cubo **13**, 61–71 (2011)
12. Anastassiou, G.A.: Elements of right delta fractional calculus on time scales. J. Concr. Appl. Math. **10**, 159–167 (2012)
13. Anastassiou, G.A.: Nabla fractional calculus on time scales and inequalities. J. Concr. Appl. Math. **11**, 96–111 (2013)
14. Anderson, D.R.: Taylor polynomials for nabla dynamic equations on time scales. Panamer. Math. J. **12**, 17–27 (2002)
15. Anderson, D.R.: Solutions to second-order three-point problems on time scales. J. Differ. Equ. Appl. **8**, 673–688 (2002)
16. Anderson, D.R.: Existence of solutions for first-order multi-point problems with changing-sign nonlinearity. J. Differ. Equ. Appl. **14**, 657–666 (2008)
17. Anderson, D.R.: Taylor's formula for conformable fractional derivatives, preprint
18. Anderson, D.R., Avery, R.I.: An even-order three-point boundary value problem on time scales. J. Math. Anal. Appl. **291**, 514–525 (2004)




C. Goodrich, A.C. Peterson, *Discrete Fractional Calculus*,
DOI 10.1007/978-3-319-25562-0

19. Anderson, D.R., Avery, R.I.: Fractional-order boundary value problem with Sturm-Liouville boundary conditions. Electron. J. Differ. Equ. **2015**, 10 pp. (2015)
20. Anderson, D.R., Hoffacker, J.: Even order self adjoint time scale problems. Electron. J. Differ. Equ. **24**, 9 pp. (2005)
21. Anderson, D.R., Hoffacker, J.: A stacked delta-nabla self-adjoint problem of even order. Math. Comput. Model. **38**, 481–494 (2003)
22. Anderson, D.R., Hoffacker, J.: Existence of solutions for a cantilever beam problem. J. Math. Anal. Appl. **323**, 958–973 (2006)
23. Anderson, D.R., Tisdell, C.C.: Third-order nonlocal problems with sign-changing nonlinearity on time scales. Electron. J. Differ. Equ. **2007**, 1–12 (2007)
24. Anderson, D.R., Zhai, C.: Positive solutions to semi-positone second-order three-point problems on time scales. Appl. Math. Comput. **215**, 3713–3720 (2010)
25. Arara, A., Benchohra, M., Hamidi, N., Nieto, J.J.: Fractional order differential equations on an unbounded domain. Nonlinear Anal. **72**, 580–586 (2010)
26. Agarwal, R.P.: ertain fractional $q$-integral and $q$-derivative. Proc. Camb. Phil. Soc. **66**, 365–370 (1969)
27. Atici, F.M., Guseinov, G.: On Green's functions and positive solutions for boundary value problems on time scales. J. Comput. Appl. Math. **141**, 75–99 (2002)
28. Atici, F.M., Eloe, P.W.: Fractional $q$-calculus on a time scale. J. Nonlinear Math. Phys. **14**, 333–344 (2007)
29. Atici, F.M., Eloe, P.W.: Discrete fractional calculus with the nabla operator. Electron. J. Qual. Theory Differ. Equ. **3**, 1–12 (2009)
30. Atici, F.M., Şengül, S.: Modeling with fractional difference equations. J. Math. Anal. Appl. **369**, 1–9 (2010)
31. Atici, F.M., Eloe, P.W.: Two-point boundary value problems for finite fractional difference equations. J. Differ. Equ. Appl. **17**, 445–456 (2011)
32. Atici, F.M., Eloe, P.W.: A transform method in discrete fractional calculus. Int. J. Differ. Equ. **2**, 165–176 (2007)
33. Atici, F.M., Eloe, P.W.: Linear systems of fractional nabla difference equations. Rocky Mountain J. Math. **41**, 353–370 (2011)
34. Atici, F.M., Eloe, P.W.: Initial value problems in discrete fractional calculus. Proc. Am. Math. Soc. **137**, 981–989 (2009)
35. Atici, F.M., Eloe, P.W.: Discrete fractional calculus with the nabla operator. Electron. J. Qual. Theory Differ. Equ., Spec. Ed. I **3**, 1–12 (2009)
36. Atici, F.M., Eloe, P.W.: Linear forward fractional difference equations. Commun. Appl. Anal. **19**, 31–42 (2015)
37. Auch, T.: Developement and application of difference and fractional calculus on discrete time scales. PhD Disertation, University of Nebraska-Lincoln (2013)
38. Auch, T., Lai, J., Obudzinski, E., Wright, C.: Discrete $q$-calculus and the $q$-Laplace transform. Panamer. Math. J. **24**, 1–19 (2014)
39. Auch, T.: Discrete calculus on a scaled number line. Panamer. Math. J., to appear
40. Awasthi, P.: Boundary value problems for discrete fractional equations. PhD Dissertation, University of Nebraska-Lincoln (2013)
41. Awasthi, P.: Existence and uniqueness of solutions of a conjugate fractional boundary value problem. Commun. Appl. Anal. **16**, 529–540 (2012)
42. Awasthi, P.: Boundary value problems for a discrete self-adjoint equation. Panamer. Math. J. **23**, 13–34 (2013)
43. Awasthi, P., Erbe, L., Peterson, A.: Existence and uniqueness results for positive solutions of a nonlinear fractional difference equation. Commun. Appl. Anal. **19**, 61–78 (2015)
44. Babakhani, A., Daftardar-Gejji, V.: Existence of positive solutions of nonlinear fractional differential equations. J. Math. Anal. Appl. **278**, 434–442 (2003)
45. Bai, Z.: On positive solutions of nonlocal fractional boundary value problem. Nonlinear Anal. **72**, 916–924 (2010)

46. Bai, Z., Lü, H.: Positive solutions for boundary value problem of nonlinear fractional differential equation. J. Math. Anal. Appl. **311**, 495–505 (2005)
47. Baoguo, J., Erbe, L., Peterson, A.: Two monotonicity results for nabla and delta fractional differences. Arch. Math. (Basel) **104**, 589–597 (2015)
48. Baoguo, J., Erbe, L., Peterson, A.: Convexity for nabla and delta fractional differences. J. Differ. Equ. Appl. **21**, 360–373 (2015)
49. Baoguo, J., Erbe, L., Peterson, A.: Some relations between the Caputo fractional difference operators and integer order differences. Electron. J. Differ. Equ. **2015**, 1–7 (2015)
50. Baoguo, J., Erbe, L., Peterson, A.: Asymptotic behavior of solutions of fractional nabla $q$-difference equations. Submitted 2015
51. Baoguo, J., Erbe, L., Peterson, A.: Comparison theorems and asymptotic behavior of solutions of discrete fractional equations. Electron. J. Qual. Theory Differ. Equ. to appear
52. Baoguo, J., Erbe, L., Peterson, A.: Comparison theorems and asymptotic behavior of solutions of Caputo fractional equations. Submitted 2015
53. Baoguo, J., Erbe, L., Goodrich, C.S., Peterson, A.: The relation between nabla fractional differences and nabla integer differences, Filomat. to appear
54. Baoguo, J., Erbe, L., Goodrich, C.S., Peterson, A.: Monotonicity results for delta fractional differences revisited, Math. Slovaca. to appear
55. Bastos, N.R.O., Mozyrska, D., Torres, D.F.M.: Fractional derivatives and integrals on time scales via the inverse generalized Laplace transform. Int. J. Math. Comput. **11**, 1–9 (2011)
56. Bastos, N.R.O., Ferreira, R.A.C., Torres, D.F.M.: Necessary optimality conditions for fractional difference problems of the calculus of variations. Discrete Contin. Dyn. Syst. **29**, 417–437 (2011)
57. Bastos, N.R.O., Ferreira, R.A.C., Torres, D.F.M.: Discrete-time fractional variational problems. Signal Process. **91**, 513–524 (2011)
58. Bastos, N.R.O., Torres, D.F.M.: Combined delta-nabla sum operator in discrete fractional calculus. Commun. Frac. Calc. **1**, 41–47 (2010)
59. Benchohra, M., Henderson, J., Ntouyas, S.K., Ouahab, A.: Existence results for fractional order functional differential equations with infinite delay. J. Math. Anal. Appl. **338**, 1340–1350 (2008)
60. Benchohra, M., Hamani, S., Ntouyas, S.K.: Boundary value problems for differential equations with fractional order and nonlocal conditions. Nonlinear Anal. **71**, 2391–2396 (2009)
61. Bohner, M., Chieochan, R.: Floquet theory for $q$-difference equations. Sarajevo J. Math. **8**, 1–12 (2012)
62. Bohner, M., Peterson, A.: Dynamic Equations on Time Scales: An Introduction with Application. Birkhäuser, Boston, MA (2001)
63. Bohner, M., Peterson, A.: Advances in Dynamic Equations on Time Scales. Birkhäuser, Boston, MA (2003)
64. Brackins, A.: Boundary value problems for nabla fractional difference equations. PhD Dissertation, University of Nebraska-Lincoln (2014)
65. Čermák, J., Nechvátal, L.: On $(q, h)$-analogue of fractional calculus. J. Nonlinear Math. Phys. **17**, 51–68 (2010)
66. Čermák, J., Kisela, T., Nechvátal, L.: Discrete Mittag-Leffler functions in linear fractional difference equations. Abstr. Appl. Anal. **2011**(Article ID 565067), 21 p. (2011)
67. Dahal, R., Goodrich, C.S.: A monotonicity result for discrete fractional difference operators. Arch. Math. (Basel) **102**, 293–299 (2014)
68. Dahal, R., Goodrich, C.S.: Erratum to: A monotonicity result for discrete fractional difference operators. Arch. Math. (Basel) **104**, 599–600 (2015)
69. Dahal, R., Duncan, D., Goodrich, C.S.: Systems of semipositone discrete fractional boundary value problems. J. Differ. Equ. Appl. **20**, 473–491 (2014)
70. Devi, J.V.: Generalized monotone method for periodic boundary value problems of Caputo fractional differential equation. Commun. Appl. Anal. **12**, 399–406 (2008)
71. Diaz, J.B., Olser, T.J.: Differences of fractional order. Math. Comput. **28**, 185–202 (1974)

72. Diethelm, K., Ford, N.: Analysis of fractional differential equations. J. Math. Anal. Appl. **265**, 229–248 (2002)
73. Došlý, O.: Reciprocity principle for even-order dynamic equations with mixed derivatives. Dyn. Syst. Appl. **16**, 697–708 (2007)
74. Dunninger, D., Wang, H.: Existence and multiplicity of positive solutions for elliptic systems. Nonlinear Anal. **29**, 1051–1060 (1997)
75. Eloe, P., Neugebauer, J.: Conjugate points for conjugate differential equations. Fract. Calc. Appl. Anal. **17**, 855–871 (2014)
76. Erbe, L., Mert, R., Peterson, A., Zafer, A.: Oscillation of even order nonlinear delay dynamic equations on time scales. Czech. Math. J. **63**, 265–279 (2013)
77. Erbe, L.H., Wang, H.: On the existence of positive solutions of ordinary differential equations. Proc. Am. Math. Soc. **120**, 743–748 (1994)
78. Estes, A.: Discrete calculus. Undergraduate Honors Thesis, University of Nebraska-Lincoln, 2013
79. Estes, A.: Discrete calculus on mixed time scales. Panamer. Math. J. **23**, 23–46 (2013)
80. Ferreira, R.A.C.: Positive solutions for a class of boundary value problems with fractional $q$-differences. Comput. Math. Appl. **61**, 367–373 (2011)
81. Ferreira, R.A.C.: A discrete fractional Gronwall inequality. Proc. Am. Math. Soc. **140**, 1605–1612 (2012)
82. Ferreira, R.A.C.: Existence and uniqueness of solution to some discrete fractional boundary value problems of order less than one. J. Differ. Equ. Appl. **19**, 712–718 (2013)
83. Ferreira, R.A.C., Goodrich, C.S.: Positive solution for a discrete fractional periodic boundary value problem. Dyn. Contin. Discrete Impuls. Syst. Ser. A Math. Anal. **19**, 545–557 (2012)
84. Ferreira, R.A.C., Goodrich, C.S.: On positive solutions to fractional difference inclusions. Analysis (Berlin) **35**, 73–83 (2015)
85. Ferreira, R.A.C., Torres, D.F.M.: Fractional $h$-difference equations arising from the calculus of variations. Appl. Anal. Discrete Math. **5**, 110–121 (2011)
86. Feza Güvenilir, A., Kaymakçalan, B., Peterson, A.C., Taş, K.: Nabla discrete fractional Grüss type inequality. J. Inequal. Appl. **2014**(86), (2014). doi:10.1186/1029-242X-2014-86
87. Gautschi, W.: Zur numerik rekurrenten relationen. Computing **9**, 107–126 (1972)
88. Goodrich, C.S.: Solutions to a discrete right-focal boundary value problem. Int. J. Differ. Equ. **5**, 195–216 (2010)
89. Goodrich, C.S.: Continuity of solutions to discrete fractional initial value problems. Comput. Math. Appl. **59**, 3489–3499 (2010)
90. Goodrich, C.S.: Existence of a positive solution to a class of fractional differential equations. Appl. Math. Lett. **23**, 1050–1055 (2010)
91. Goodrich, C.S.: Some new existence results for fractional difference equations. Int. J. Dyn. Syst. Differ. Equ. **3**, 145–162 (2011)
92. Goodrich, C.S.: Existence and uniqueness of solutions to a fractional difference equation with nonlocal conditions. Comput. Math. Appl. **61**, 191–202 (2011)
93. Goodrich, C.S.: A comparison result for the fractional difference operator. Int. J. Differ. Equ. **6**, 17–37 (2011)
94. Goodrich, C.S.: Existence of a positive solution to a system of discrete fractional boundary value problems. Appl. Math. Comput. **217**, 4740–4753 (2011)
95. Goodrich, C.S.: Existence of a positive solution to a first-order $p$-Laplacian BVP on a time scale. Nonlinear Anal. **74**, 1926–1936 (2011)
96. Goodrich, C.S.: On positive solutions to nonlocal fractional and integer-order difference equations. Appl. Anal. Discrete Math. **5**, 122–132 (2011)
97. Goodrich, C.S.: A comparison result for the fractional difference operator. Int. J. Differ. Equ. **6**, 17–37 (2011)
98. Goodrich, C.S.: Existence of a positive solution to systems of differential equations of fractional order. Comput. Math. Appl. **62**, 1251–1268 (2011)
99. Goodrich, C.S.: On discrete sequential fractional boundary value problems. J. Math. Anal. Appl. **385**, 111–124 (2012)

100. Goodrich, C.S.: The existence of a positive solution to a second-order $p$-Laplacian BVP on a time scale. Appl. Math. Lett. **25**, 157–162 (2012)
101. Goodrich, C.S.: Positive solutions to boundary value problems with nonlinear boundary conditions. Nonlinear Anal. **75**, 417–432 (2012)
102. Goodrich, C.S.: On a fractional boundary value problem with fractional boundary conditions. Appl. Math. Lett. **25**, 1101–1105 (2012)
103. Goodrich, C.S.: Nonlocal systems of BVPs with asymptotically superlinear boundary conditions. Comment. Math. Univ. Carolin. **53**, 79–97 (2012)
104. Goodrich, C.S.: On a discrete fractional three-point boundary value problem. J. Differ. Equ. Appl. **18**, 397–415 (2012)
105. Goodrich, C.S.: On nonlocal BVPs with boundary conditions with asymptotically sublinear or superlinear growth. Math. Nachr. **285**, 1404–1421 (2012)
106. Goodrich, C.S.: Nonlocal systems of BVPs with asymptotically sublinear boundary conditions. Appl. Anal. Discrete Math. **6**, 174–193 (2012)
107. Goodrich, C.S.: On discrete fractional boundary value problems with nonlocal, nonlinear boundary conditions. Commun. Appl. Anal. **16**, 433–446 (2012)
108. Goodrich, C.S.: On a first-order semipositone discrete fractional boundary value problem. Arch. Math. (Basel) **99**, 509–518 (2012)
109. Goodrich, C.S.: On nonlinear boundary conditions satisfying certain asymptotic behavior. Nonlinear Anal. **76**, 58–67 (2013)
110. Goodrich, C.S.: On a nonlocal BVP with nonlinear boundary conditions. Res. Math. **63**, 1351–1364 (2013)
111. Goodrich, C.S.: Positive solutions to differential inclusions with nonlocal, nonlinear boundary conditions. Appl. Math. Comput. **219**, 11071–11081 (2013)
112. Goodrich, C.S.: On semipositone discrete fractional boundary value problems with nonlocal boundary conditions. J. Differ. Equ. Appl. **19**, 1758–1780 (2013)
113. Goodrich, C.S.: Existence of a positive solution to a nonlocal semipositone boundary value problem on a time scale. Comment. Math. Univ. Carolin. **54**, 509–525 (2013)
114. Goodrich, C.S.: A convexity result for fractional differences. Appl. Math. Lett. **35**, 58–62 (2014)
115. Goodrich, C.S.: Systems of discrete fractional boundary value problems with nonlinearities satisfying no growth conditions. J. Differ. Equ. Appl. **21**, 437–453 (2015)
116. Graef, J., Webb, J.R.L.: Third order boundary value problems with nonlocal boundary conditions. Nonlinear Anal. **71**, 1542–1551 (2009)
117. Gray, H.L., Zhang, N.: On a new definition of the fractional difference. Math. Comp. **50**, 513–529 (1988)
118. Guseinov, G.: Self-adjoint boundary value problems on time scales and symmetric Green's functions. Turkish J. Math. **29**, 365–380 (2005)
119. Hein, J., McCarthy, S., Gaswick, N., McKain, B., Speer, K.: Laplace transforms for the nabla-difference operator. Panamer. Math. J. **21**, 79–96 (2011)
120. Henderson, J., Ntouyas, S.K., Purnaras, I.K.: Positive solutions for systems of generalized three-point nonlinear boundary value problems. Commen. Math. Univ. Carolin. **49**, 79–91 (2008)
121. Henderson, J., Ntouyas, S.K., Purnaras, I.K.: Positive solutions for systems of nonlinear discrete boundary value problem. J. Differ. Equ. Appl. **15**, 895–912 (2009)
122. Higgins, R., Peterson, A.: Cauchy functions and Taylor's theorem for time scales. In: Proceedings of the 6-th ICDEA in Augsburg, Germany, 2001, new Progress in Difference Equations, Aulbach, B., Elaydi, S., Ladas, G. (eds.): J. Differ. Equ. Appl. 299–308 (2003)
123. Holm, M.: Sum and difference compositions in discrete fractional calculus. Cubo **13**, 153–184 (2011)
124. Holm, M.: Solutions to a discrete, nonlinear, $(N-1, 1)$ fractional boundary value problem. Int. J. Dyn. Syst. Differ. Equ. **3**, 267–287 (2011)
125. Holm, M.: The theory of discrete fractional calculus: development and application. PhD Dissertation, University of Nebraska-Lincoln (2011)

126. Infante, G.: Nonlocal boundary value problems with two nonlinear boundary conditions. Commun. Appl. Anal. **12**, 279–288 (2008)
127. Infante, G., Pietramala, P.: Existence and multiplicity of non-negative solutions for systems of perturbed Hammerstein integral equations. Nonlinear Anal. **71**, 1301–1310 (2009)
128. Infante, G., Pietramala, P.: Eigenvalues and non-negative solutions of a system with nonlocal BCs. Nonlinear Stud. **16**, 187–196 (2009)
129. Infante, G., Pietramala, P.: A third order boundary value problem subject to nonlinear boundary conditions. Math. Bohem. **135**, 113–121 (2010)
130. Jagan Mohan, J., Deekshitulu, G.V.S.R.: Fractional order difference equations. Int. J. Differ. Equ. **2012**(Art. ID 780619), 11 pp. (2012)
131. Kac, V., Cheung, P.: Quantum calculus. Universitext, Springer, New York (2002)
132. Kang, P., Wei, Z.: Three positive solutions of singular nonlocal boundary value problems for systems of nonlinear second-order ordinary differential equations. Nonlinear Anal. **70**, 444–451 (2009)
133. Kaufmann, E.R., Raffoul, Y.N.: Positive solutions for a nonlinear functional dynamic equation on a time scale. Nonlinear Anal. **62**, 1267–1276 (2005)
134. Kelley, W.G., Peterson, A.C.: Difference Equations: An Introduction with Applications, First Edition. Academic Press, New York (1991)
135. Kelley, W.G., Peterson, A.C.: Difference Equations: An Introduction with Applications, Second Edition. Academic Press, New York (2001)
136. Kelley, W.G., Peterson, A.C.: The Theory of Differential Equations: Classical and Qualitative, First Edition. Pearson Prentice Hall, Upper Saddle River (2004)
137. Kelley, W.G., Peterson, A.C.: The Theory of Differential Equations: Classical and Qualitative, Second Edition. Universitext, Springer, New York (2010)
138. Kirane, M., Malik, S.A.: The profile of blowing-up solutions to a nonlinear system of fractional differential equations. Nonlinear Anal. **73**, 3723–3736 (2010)
139. Knuth, D.: Two notes on notation. Am. Math. Monthly **99**, 403–422 (1992)
140. Kong, L., Kong, Q.: Positive solutions of higher-order boundary-value problems. Proc. Edinb. Math. Soc. (2) **48**, 445–464 (2005)
141. Kuttner, B.: On differences of fractional order. Proc. Lond. Math. Soc. (3) **7**, 453–466 (1957)
142. Lakshmikantham, V., Vatsala, A.S.: Basic theory of fractional differential equations. Nonlinear Anal. **69**, 2677–2682 (2008)
143. Li, X., Han, Z., Sun, S.: Existence of positive solutions of nonlinear fractional $q$-difference equation with parameter, preprint
144. Malinowska, A., Torres, D.F.M.: Generalized natural boundary conditions for fractional variational problems in terms of the Caputo derivative. Comput. Math. Appl. **59**, 3110–3116 (2010)
145. Mert, R.: Oscillation of higher-order neutral dynamic equations on time scales. Adv. Differ. Equ. (68), 11 pp. (2012)
146. Miller, K.S., Ross, B.: Fractional Difference Calculus. In: Proceedings of the International Symposium on Univalent Functions, Fractional Calculus and their Applications, Nihon University, Koriyama, Japan, 1988, 139–152; Ellis Horwood Ser. Math. Appl, Horwood, Chichester (1989)
147. Miller, K.S., Ross, B.: An Introduction to the Fractional Calculus and Fractional Difference Equations. Wiley, New York (1993)
148. Mittag-Leffler, G.M.: Sur la nouvelle fonction $E_\alpha(x)$. C.R. Acad. Sci. Paris **137**, 554–558 (1903)
149. Munkhammar, J.D.: Fractional calculus and the Taylor-Riemann series. Rose-Hulman Undergraduate Math. J. **6**, (2005)
150. Nagai, A.: On a certain fractional $q$-difference and its Eigen function. J. Nonlinear Math. Phys. **10**, 133–142 (2003)
151. Nieto, J.J.: Maximum principles for fractional differential equations derived from Mittag-Leffler functions. Appl. Math. Lett. **23**, 1248–1251 (2010)

152. Oldham, K., Spanier, J.: The Fractional Calculus: Theory and Applications of Differentiation and Integration to Arbitrary Order. Dover Publications, Mineola, New York (2002)
153. Podlubny, I.: Fractional Differential Equations. Academic Press, New York (1999)
154. Schuster, H.G. (ed.): Reviews of Nonlinear Dynamics and Complexity. Wiley-VCH, Berlin (2008)
155. Su, X.: Boundary value problems for a coupled system of nonlinear fractional differential equations. Appl. Math. Lett. **22**, 64–69 (2009)
156. Wang, G., Zhou, M., Sun, L.: Fourth-order problems with fully nonlinear boundary conditions. J. Math. Anal. Appl. **325**, 130–140 (2007)
157. Wang, J., Zhou, Y.: A class of fractional evolution equations and optimal controls. Nonlinear Anal. Real World Appl. **12**, 262–272 (2011)
158. Wang, P., Wang, Y.: Existence of positive solutions for second-order m-point boundary value problems on time scales. Acta Math. Appl. Sin. Engl. Ser. **22**, 457–468 (2006)
159. Webb, J.R.L., Infante, G.: Positive solutions of nonlocal boundary value problems: a unified approach. J. Lond. Math. Soc. (2) **74**, 673–693 (2006)
160. Webb, J.R.L.: Nonlocal conjugate type boundary value problems of higher order. Nonlinear Anal. **71**, 1933–1940 (2009)
161. Webb, J.R.L.: Solutions of nonlinear equations in cones and positive linear operators. J. Lond. Math. Soc. (2) **82**, 420–436 (2010)
162. Webb, J.R.L.: Remarks on a non-local boundary value problem. Nonlinear Anal. **72**, 1075–1077 (2010)
163. Wei, Z., Li, Q., Che, J.: Initial value problems for fractional differential equations involving Riemann-Liouville sequential fractional derivative. J. Math. Anal. Appl. **367**, 260–272 (2010)
164. Xu, X., Jiang, D., Yuan, C.: Multiple positive solutions for the boundary value problem of a nonlinear fractional differential equation. Nonlinear Anal. **71**, 4676–4688 (2009)
165. Yang, Z.: Positive solutions to a system of second-order nonlocal boundary value problems. Nonlinear Anal. **62**, 1251–1265 (2005)
166. Yang, Z.: Positive solutions of a second-order integral boundary value problem. J. Math. Anal. Appl. **321**, 751–765 (2006)
167. Zeidler, E. (Wadsack, P.R. trans.): Nonlinear Functional Analysis and its Applications: Part 1: Fixed-Point Theorems. Springer, New York (1986)
168. Zhang, S.: Positive solutions to singular boundary value problem for nonlinear fractional differential equation. Comput. Math. Appl. **59**, 1300–1309 (2010)
169. Zhao, X., Ge, W.: Unbounded solutions for a fractional boundary value problem on the infinite interval. Acta Appl. Math. **109**, 495–505 (2010)
170. Zhou, Y., Jiao, F., Li, J.: Existence and uniqueness for fractional neutral differential equations with infinite delay. Nonlinear Anal. **71**, 3249–3256 (2009)
171. Zhou, Y., Jiao, F., Li, J.: Existence and uniqueness for *p*-type fractional neutral differential equations. Nonlinear Anal. **71**, 2724–2733 (2009)
172. Zhou, Y., Jiao, F.: Nonlocal Cauchy problem for fractional evolution equations. Nonlinear Anal. Real World Appl. **11**, 4465–4475 (2010)

[illegible] Applications [illegible]
[illegible] New York (2010)
[illegible] New York (2009)
[illegible]

[illegible] boundary value problems [illegible]

[illegible]

[illegible] Nonlinear Anal. [illegible]

[illegible] boundary value problems [illegible]

[illegible] Nonlinear Anal. [illegible]

[illegible] Appl. [illegible]

[illegible] boundary value problem [illegible]

[illegible]

[illegible] Zeidler, E.: Nonlinear Functional Analysis and Its Applications [illegible] Springer, New York (1986)

[illegible] differential equations [illegible]

[illegible]

[illegible]

[illegible]

# Index




C. Goodrich, A.C. Peterson, *Discrete Fractional Calculus*,
DOI 10.1007/978-3-319-25562-0

**T**



**U**



**V**

Printed in the United States
By Bookmasters